About the Author

Author of the bestselling first edition of the *Handy History Answer Book,* Rebecca Ferguson helped develop the Handy Answer Book® series and a number of books for Visible Ink Press during the last 15 years, including titles on noteworthy legal trials, Hispanic/Latino culture, and the Olympics. She recently contributed to the Britannica Student Encyclopedia, a part of Britannica Online, as well as The Lincoln Library of American History. Her love of history runs deep. "History," Rebecca says, "teaches us to appreciate the scope of human endeavor and experience."

She lives with her husband and young son near Chicago.

The Handy Answer Book® Series

The Handy Answer Book for Kids
(and Parents)

The Handy Biology Answer Book

The Handy Bug Answer Book

The Handy Dinosaur Answer Book

The Handy Geography Answer Book

The Handy Geology Answer Book

The Handy History Answer Book

The Handy Math Answer Book

The Handy Ocean Answer Book

The Handy Physics Answer Book

The Handy Politics Answer Book

The Handy Presidents Answer Book

The Handy Religion Answer Book

The Handy Science Answer Book

The Handy Space Answer Book

The Handy Sports Answer Book

The Handy Weather Answer Book

THE HANDY HISTORY ANSWER BOOK

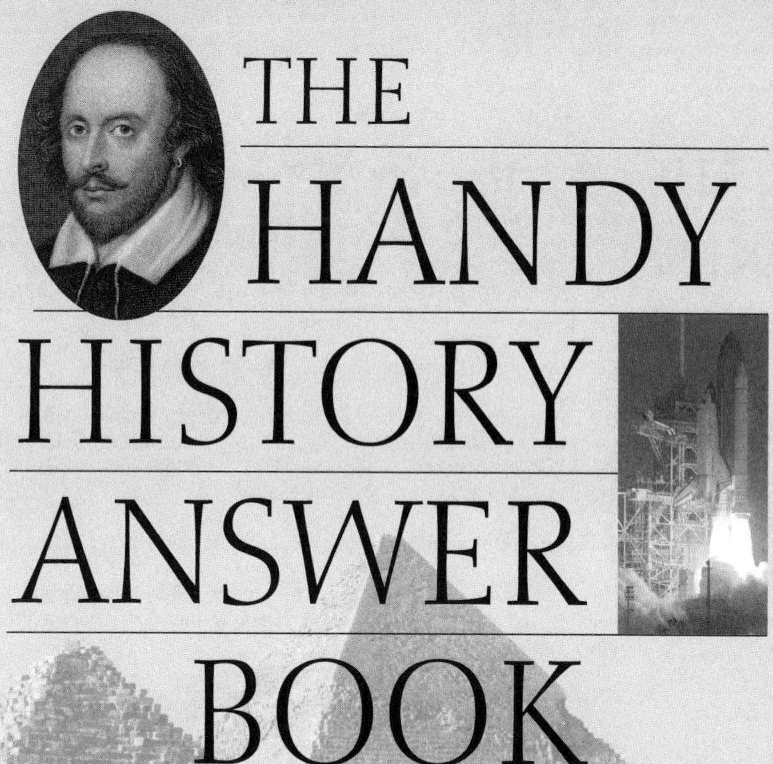

THE HANDY HISTORY ANSWER BOOK

SECOND EDITION

Rebecca Ferguson

Detroit

THE HANDY HISTORY ANSWER BOOK

Photo Credits
AP/Wide World: pages 9, 28, 33, 40, 50, 63, 65, 93, 99, 100, 102, 152, 155, 158, 165, 168, 170, 184, 187, 197, 202, 204, 208, 212, 216, 270, 272, 274, 289, 310, 312, 349, 350, 352, 367, 384, 394, 400, 406, 408, 413, 418, 423, 427, 434, 447, 487, 489, 494, 500, 502, 547, 559, 562, 570.

Getty Images: pages 18, 42, 108, 122, 176, 242, 333, 342, 441, 464, 534, 545.

The Library of Congress: pages 5, 14, 27, 29, 31, 71, 74, 83, 88, 115, 125, 135, 137, 141, 148, 151, 163, 230, 234, 238, 246, 263, 265, 268, 276, 281, 285, 292, 298, 306, 335, 345, 370, 378, 398, 420, 448, 451, 459, 471, 474, 478, 483, 507, 513, 524, 526, 543, 552, 553, 555, 557, 566, 572, 576, 580, 583, 592.

Cover images reproduced by permission of AP and Library of Congress.

Visible Ink Press®
43311 Joy Rd. #414
Canton, MI 48187-2075

Visible Ink Press is a trademark of Visible Ink Press LLC.

Most Visible Ink Press books are available at special quantity discounts when purchased in bulk by corporations, organizations, or groups. Customized printings, special imprints, messages, and excerpts can be produced to meet your needs. For more information, contact Special Markets Director, Visible Ink Press, at www.visibleink.com.

Art Director: Mary Claire Krzewinski
Typesetting: The Graphix Group
ISBN: 1-57859-170-8

Cataloging-in-Publication Data is on file with the Library of Congress.

Printed in the United States of America

10 9 8 7

Contents

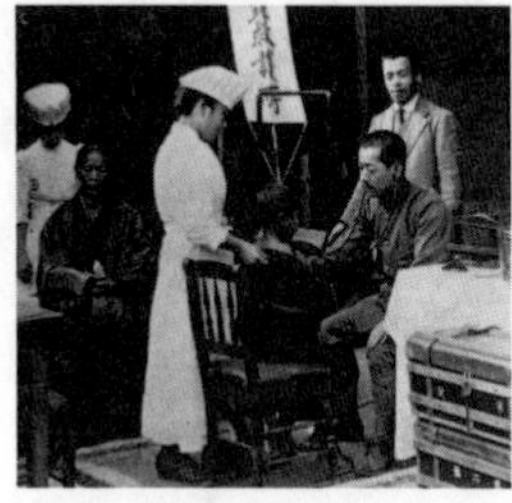

Introduction

When I was asked to research and write the first edition of this book, my first thought was that history—with decade upon decade of fugitive dates, people, and places—was particularly well-suited to the Handy Answer Book® format. My second thought was that it could not be done. History was too big a subject to squeeze itself into a finite series of questions and answers, neatly arranged by subject, trimmed and bound into a manageable size. But soon, very soon, the questions suggested themselves: What was the Pax Romana…the Trail of Tears…the Gunpowder Plot…the Oath of the Tennis Court…the Boxer Rebellion…the Battle of the Bulge…VE Day…Black Friday? Who were the Huns…the War Hawks…the Big Four…the Robber Barons? Who was Carrie Nation…Alex de Toqueville…Clara Barton…Balanchine? What were the Crusades…the Rights of Man…the Boer Wars…the carpetbaggers?

And so the first edition came into print—and stayed in print, the audience keeping it in demand. The comments I have heard and read from readers have been truly gratifying. *The Handy History Answer Book* is being used as a study aid for high school and college students; as an introduction to broad subjects, such as the world wars, for younger students; as a refresher for history buffs; and as pleasure reading for the curious-minded of all ages. Readers have called it "nifty," "handy-dandy," and "history in a nutshell." The book found its audience.

So, a second edition was called for. Once again, the questions presented themselves in a hurry. I used the opportunity of a second edition to cover the astonishing events of the early 2000s, including the devastating impact of Hurricane Katrina, which struck the Gulf Coast of the United States as this book went into final production (the preliminary toll of the storm's ravages are captured here and the situation continues to unfold as this book is published). I also took the opportunity to fill gaps that I perceived to exist in the coverage of some earlier events, update subjects that have not neatly concluded, and to change the record. On this last point, it's not true that history is unchanging. New documents are discovered, public opinion shifts, scholars develop new interpretations. Just ask Alexander Graham Bell's descendants.

The Handy History Answer Book is a resource for learning—for brushing up on the events, terms, and history-makers many of us remember but somehow can't always neatly define. This book is not intended to be a comprehensive work on history; it can't be done in a single volume. Still, the minds that conceived of it thought that reasonably intelligent people ought to have a convenient place to go to look up those devilish questions that have a way of cropping up in everyday conversation and reading.

While *The Handy History Answer Book* focuses on Western civilization, it is impossible to not take into account events in the East. Eastern influences are certainly evident in the West. And vice versa. It turns out that the idea of a global village is not so very new. This being so, the reader will find a number of questions and answers regarding Eastern events, ideas, and innovations. Further, since the readership is largely U.S.-educated, the emphasis is on those events that are most-studied in the American classroom.

Herewith: a new edition, revised and updated, to find its place on the desks, bedside tables, and bookshelves of fact-finders.

Perhaps no one has ever issued a better apology than the great Carl Sandburg, who in his prefatory remarks to *The American Songbag* wrote, "This book was begun in the depths of humility, and ended likewise with the murmur, 'God, be merciful to me, a sinner.'" I, too, sincerely apologize for any factual errors in these pages. Every effort was made to verify the accuracy of the information herein.

Peace in our time.

Acknowledgments

Thanks to Marty Connors, who approached me with the idea, and to all my friends and colleagues at Visible Ink Press, especially Roger Janecke, Terri Schell, and Christa Gainor, who have done everything in their might to support the project.

In Diane Sawinski I had not only a smart, experienced editor, but photo researcher, sounding board, and fellow mom. The good-looking cover and clean, highly readable format are the hard work of talented art designer Mary Claire Krzewinski and ace typesetter Marco Di Vita. Kudos to Larry Baker for creating such an extensive and thorough index and to Bob Huffman for his top-notch work on the book's images.

Ted and Gilbert, thanks, guys, for letting me do this work. Go team.

Chronology of Selected Events

Prehistoric Era

2 million B.C.

Paleolithic (Old Stone) Age begins; it lasts until 10,000 B.C.

Earliest forms of art and communication are used (c. 35,000 B.C.).

c. 50,000 B.C.

North America begins to be settled by primitive man; migrants (hunter-gatherers in pursuit of game) likely travel over the Bering Strait waterway via a great land or ice bridge between Asia and North America. Period of migration lasts until about 40,000 B.C. and is followed by a second period, between 26,000 and 8000 B.C.

c. 10,000 B.C.

Neolithic (New Stone) Age begins; it lasts until 3300 B.C.

c. 8000 B.C.

Ancient man migrates across the Americas, reaching as far as Tierra del Fuego, the southernmost part of South America.

c. 3300 B.C.

Bronze Age begins; it lasts until 2500 B.C.

c. 3000 B.C.

Egypt, one of the world's oldest civilizations, develops in the Nile River valley; the kingdom lasts until 332 B.C., when it is conquered by the Macedonians under Alexander the Great.

Greek civilization begins to flourish in the Mediterranean.

c. 2500 B.C.

Iron Age begins; it lasts until 2000 B.C.

The Classical Age

2000 B.C.

The Classical Age begins, dominated first by the Greeks and, later, the Romans.

c. 1400 B.C.

The Hebrew prophet Moses leads the Israelites out of captivity in Egypt.

The first five books of the Bible are written.

c. 1200 B.C.

10-year Trojan War begins when the Myceneans attack the city of Troy, considered key to the profitable Black Sea trade.

c. 1000 B.C.

Mayan culture begins to flourish in Central America.

700s B.C.

Rome is established.

500s B.C.

Roman Republic is born (509 B.C.).

Celts, an Indo-European people, spread across present-day France, Italy, Portugal, Spain, and the British Isles. By the 200s B.C., they venture as far as Greece. Ultimately they are absorbed by the Roman Empire, with the exception of Ireland and other isolated pockets.

400s B.C.

Greek civilization reaches its height.

Greek philosopher Socrates lays the foundation for Western thought.

300s B.C.

Macedonian king Alexander the Great amasses an empire stretching from Egypt to India, spreading Greek ideas and customs throughout.

200s B.C.

Punic Wars, a series of conflicts waged by the Romans to gain territory, begin in Italy.

100s B.C.

Roman armies conquer Macedonia.

Punic Wars end; the Romans are victorious.

60s B.C.

Julius Caesar is elected consul and the First Triumvirate is formed to govern Rome.

40s B.C.

Caesar is murdered by his countrymen (44 B.C.).

20s B.C.

The Roman Empire is established by the ruler Augustus (also known as Octavian), and the 200 years of the Pax Romana begin. The Roman Empire lasts nearly 500 years.

A.D. 30s

Jesus is crucified (c. 30 A.D.).

200s

Roman armies have by now conquered so many peoples that the Roman Empire stretches across Europe and includes the entire Mediterranean coast of Africa as well as parts of the Middle East.

300s

Roman emperor Constantine the Great converts to Christianity (c. 312) and makes the church legal; Christians regain freedom of worship.

Constantine the Great moves the capital of the Roman Empire to Byzantium (later called Constantinople, which is today Istanbul, Turkey), shifting the focus from West to East and paving the way for the Byzantine Empire.

Upon death of emperor Theodosius the Great (395), the Roman Empire is divided into East (centered in Byzantium) and West (centered in Rome).

The Middle Ages

400s

Under leadership of Attila, the Huns rule much of Eastern Europe; after his death (453), their dominance ends.

Rome suffers repeated attacks at the hands of various Germanic tribes and finally falls in 476, marking the end of the West Roman Empire; the East Roman Empire, which has remained Christian but has been significantly influenced by the East, survives as the Byzantine Empire.

500s

Byzantine Empire, centered in Constantinople (Istanbul, Turkey), grows in strength and influence.

600s

Islam is founded by Muhammad, who is believed to be a prophet of Allah (God).

Feudal system begins to be established as a way of organizing and protecting communities: a lord grants his subjects land, and in return they provide him with services, including military protection.

800s

On Christmas Day 800, Charlemagne is crowned ruler of the Holy Roman Empire, a loose confederation of German and Italian states; but after his death in 814, the empire lapses.

During China's Tang dynasty, the first book is published: *Diamond Sutra.*

900s

Otto I is crowned emperor of the Holy Roman Empire, a confederation of Western European states that will last until 1806.

1000s

Seafaring Norseman Leif Ericsson arrives at Newfoundland or Nova Scotia, Canada (1001), becoming the first European to set foot on North American soil.

Pope Urban II announces (1095) the first of the Christian Crusades, wars to recover the Holy Land (Palestine) from the Muslims.

Norman invasion: French duke William of Normandy (a.k.a. William the Conqueror) sails across the English Channel (1066) and invades Anglo-Saxon England.

1100s

Feudalism takes hold in France and spreads into England, Spain, and other parts of the Christian world.

Europe's first university is formed at Bologna, Italy (1158).

1200s

Genghis Khan amasses his empire in the East.

Marco Polo travels to the East (1270), where he remains for some 25 years before returning to his native Venice, bringing back fantastic accounts of his journey.

Magna Carta is signed at Runnymede, England (1215).

The Crusades, a series of military expeditions, end; the Christian goal of permanently recovering the Holy Land of Palestine is not realized, but trade routes have been established, new markets opened, and shipbuilding has been improved—paving the way for the age of exploration.

THE RENAISSANCE, THE AGE OF EXPLORATION, AND THE REFORMATION

1300s

The Renaissance takes hold in Europe; it lasts until the 1600s. Arts and letters flourish during the period.

Aztec Indians establish city of Tenochtitlán (c. 1325) on future site of Mexico City.

Hundred Years' War begins (1337); England and France fight intermittently until 1453; England loses all claims to lands on the European continent.

1400s

Byzantine Empire falls to the Ottoman Turks (1453).

Gutenberg builds his first printing press (1440s).

Age of exploration begins as Roman Catholic powers Spain and Portugal send explorers in search of new trade routes to India and the Far East.

Under the sponsorship of Spanish monarchs, Italian explorer Christopher Columbus voyages west (1492) in search of a trade route to the East and lands in the Caribbean islands (West Indies).

Italian navigator Amerigo Vespucci reaches the Western Hemisphere (1497–1503) and is later credited with being the first European explorer to realize he had arrived in the New World, which will be named for him: *America* is derived from Amerigo.

1500s

Portuguese navigator Gaspar de Corte-Real makes landfall and explores the coasts of Labrador and Newfoundland, Canada (1500); it is the first authenticated European landing on the North American mainland.

Reformation begins (1517) when theology professor Martin Luther nails his Ninety-Five Theses to the door of the Castle Church at Wittenburg in Saxony, Germany.

Spaniards, led by Hernán Cortés, arrive in central Mexico (1519); by 1521 they suppress the Aztecs and claim Mexico as a viceroyalty of Spain.

Incas of South America are conquered by the Spaniards (1530s).

Ottoman (Turkish) Empire reaches its height, spreading Islamic culture in the East and into Europe.

THE ENLIGHTENMENT AND THE SCIENTIFIC REVOLUTION

1600s

Not to be outdone by Spain's colonialism in the New World, England establishes settlements along the eastern seaboard of the North American mainland. France also settles North America, claiming regions for its crown and prompting a series of colonial wars (1689–1763) with England. England emerges the victor.

Galileo advocates (1613) the controversial Copernican system of the universe, proposing that Earth revolves around the sun.

The Scientific Revolution is under way; the era is marked by key discoveries and rapid advances in astronomy, anatomy, mathematics, and physics; science courses become part of school curricula.

Ottoman Empire begins a 300-year decline.

Peace of Westphalia (1648) ends the Thirty Years' War, helping establish Protestantism in Europe.

Oliver Cromwell is named Lord Protector of England (1653), interrupting the English monarchy.

Charles II ascends the English throne (1660), beginning the Restoration; absolutism of the monarchy is reestablished, but monarchs clash with a more powerful Parliament.

Harvard, America's first university, is chartered (1636).

Tsar Peter the Great rules Russia (1682–1725), introducing western European civilization and elevating Russia to the status of great European power.

British Parliament compels King William and Queen Mary to accept the Bill of Rights (1689), asserting the Crown no longer has absolute power and must rule through Parliament; England's constitutional monarchy is founded.

The Revolutionary Era: Wars and the Birth of Industry

1700s

Act of Union (1707) joins England and Scotland.

German physicist Daniel Fahrenheit invents mercury thermometer (1714).

Danish navigator Vitus Bering crosses the narrow strait separating Asia and North America (1728).

Trial of New York City printer John Peter Zenger (1735) lays foundation for freedom of the press.

English inventor John Harrison presents his ship's chronometer to London's Board of Longitude (1736); the device, which goes through several improvements, affords explorers and traders more accurate navigation.

First golf club is formed (1744), in Edinburgh, Scotland.

Powerful earthquake strikes Lisbon, Portugal (November 1, 1755). Felt across Europe, the quake generates debate among philosophers who try to explain why God would destroy that particular city, which was then the seat of the Holy Inquisition, on All Saints' Day.

French and Indian War (1754–63) is fought in North America; Britain emerges as the victor.

Seven Years' War (1756–63) is fought as European powers vie for supremacy.

England gains control of India (1757).

Tsarina Catherine the Great rules Russia (1762–96).

James Watt patents the first practical steam-powered engine (1769).

French military engineer Nicolas-Joseph Cugnot builds a steam-powered road vehicle (1769), beginning developments that lead to the introduction of the automobile.

Industrial Revolution begins (mid-1700s) in Great Britain with the introduction of power-driven machinery and spreads to western Europe and America.

Boston Tea Party (1773) and other acts of colonial rebellion spark the American Revolution (1775–83).

Thomas Paine writes *Common Sense* (1776), promoting the idea that democracy is the only form of government that can guarantee natural rights; the pamphlet galvanizes support for American Revolution.

Declaration of Independence is issued (1776) by representatives of the 13 American colonies.

Scottish economist Adam Smith writes *The Wealth of Nations* (1776), proposing a system of natural liberty in trade and commerce, the cornerstone of capitalism.

Vermont prohibits slavery (1777).

Articles of Confederation take effect (March 1, 1781) when the last state (Maryland) ratifies them; the document is later replaced with the Constitution.

Treaty of Paris ends the American Revolution (September 3, 1783).

Cotton-spinning machine is invented (1783) in Great Britain.

First hot-air balloon flights (1783), in Paris.

Constitutional Convention is convened at Philadelphia (1787) and the U.S. Constitution is drafted; it will be ratified by the states the following year and will go into effect in 1789.

French Revolution (1789–99) begins with the Oath of the Tennis Court and the storming of the Bastille.

Spinning mills are introduced in the United States (1790) by English-born mechanist and businessman Samuel Slater, launching the American textiles industry and creating great demand for southern-grown cotton.

Baseball is popular enough with the American public to be the subject of a Pittsfield, Massachusetts, town ordinance (1791).

Bill of Rights (1791) is added to the U.S. Constitution.

Eli Whitney invents the cotton gin (1793).

U.S. forces of General Anthony Wayne defeat the Shawnee at Fallen Timbers, Ohio (1793).

France's Reign of Terror (1793–94) is led by revolutionary leader Maximilien Robespierre.

English engineer Richard Trevithick constructs a working model of a locomotive engine (1797).

English physician Edward Jenner announces he has developed the vaccine (1798).

Coup d'État of 18th Brumaire (1799): Napoleon Bonaparte rises to power in France.

1800s

Trained scientists develop new technologies, including farm machinery and equipment for textile manufacturing and transportation, fueling the Industrial Revolution.

Manifest Destiny takes hold, resulting in U.S. expansionism through the acquisition of lands (Louisiana Territory, Florida, Texas, California, etc.) by purchase, warfare, and treaties.

U.S. labor movement has its origins (early 1800s) as workers begin organizing.

Act of Union unites Ireland with England and Scotland, forming the United Kingdom (1801).

In what came to be known as the Louisiana Purchase (1803), the United States bought from France the Louisiana Territory.

Napoleon Bonaparte declares himself emperor of France (December 2, 1804).

Louis and Clark expedition (1804–06) explores western United States.

Holy Roman Empire ends in the Confederation of the Rhine (1806), which brings most of the German states under French domination, a result of the Napoleonic Wars.

1810s

Work begins (1811) on the National Road, the first U.S. government road; the first of a long series of federal transportation projects that knit the nation together.

War of 1812 is fought between the United States and Britain.

Treaty of Ghent (December 24, 1814) officially ends the War of 1812.

Napoleon is defeated by a European coalition in the Battle of Waterloo (1815), ending his reign.

1820s

Missouri Compromise (1820): Missouri is admitted to the Union as a slave state and Maine as a free state; territories north of the 36th parallel, with the exception of Missouri, are free. The compromise intended, but failed, to settle the slavery question.

Greek War for Independence is fought (1821–29).

Lyceum movement begins (1826); the decades-long movement promotes establishment of public schools, libraries, and museums in the United States.

Erie Canal is completed (1825), spurring settlement of U.S. interior.

New York Stock Exchange opens (1825) at 11 Wall Street, New York City.

1830s

Abolition movement gains strength in the United States (1830s).

First commercially successful reaper is built (1831) by Virginia-born inventor Cyrus Hall McCormick.

Texas War of Independence (1836).

England's Queen Victoria begins her long reign (1837–1901).

Trail of Tears (1838): American Indians are forced westward by the U.S. government to make way for white settlers.

1840s

Mexican War is fought (1846–48) over U.S. annexation of Texas.

Ireland experiences the Great Famine (1845–48), prompting widespread immigration to the United States.

Gold Rush begins in California (1848).

First women's rights convention is held (1848) in Seneca Falls, New York, launching the American women's suffragist movement.

1850s

Compromise of 1850 fails to settle the slavery issue in the United States.

Bessemer process is developed (1850s); it is the first method for making steel cheaply and in large quantities.

America's first department stores are established (1850s–1880s) after the Parisian model Bon Marché (est. 1838).

U.S. owns all territory of present-day contiguous states (1853); Alaska is added in 1867 and Hawaii in 1898.

Crimean War (1853–56) is fought between Russian forces and the allied armies of Britain, France, the Ottoman Empire (present-day Turkey), and Sardinia (part of present-day Italy).

What becomes known as the Comstock Lode is discovered (1857) in Mount Davidson, Nevada; it is the richest silver mine in the U.S.

Bleeding Kansas (1858): deadly conflicts between abolitionists and pro-slavery factions.

Charles Darwin publishes *On the Origin of Species* (1858).

Mexico's War of Reform (1858–61).

U.S. oil industry begins when retired railroad conductor Edwin L. Drake drills a well (1859) near Titusville, Pennsylvania.

Great Atlantic & Pacific Tea Company is set up (1859) in New York City; A&Ps proliferate rapidly, launching chain store concept.

1860s

First practical internal-combustion engine is built (1860); the diesel engine follows two years later.

Civil War (1861–65) is fought in the United States.

Sioux uprising (August-September 1862) in southwestern Minnesota.

Red Cross is founded (1864), as part of the first Geneva Conventions (1864, 1906, 1929, 1949).

Two years after the Emancipation Proclamation, Congress passes the Thirteenth Amendment, banning slavery throughout the United States (1865).

American Civil War ends when Confederate states surrender (April 9, 1865).

President Abraham Lincoln is shot (April 14, 1865); he dies the next day.

Reconstruction begins (1865).

Europe's Austro-Hungarian monarchy is established (1867); it lasts until 1918.

Articles of impeachment are brought against President Andrew Johnson over political and ideological differences between him and Congress (February 1868).

Transcontinental railroad is completed (May 10, 1869) in the United States.

1870s

Fifteenth Amendment is passed (1870), giving all citizens equal protection under the law (which meant to extend suffrage to black men).

Panic of 1873: Monetary crisis in the U.S. begins period of economic depression, which launches the Progressive movement, seeking wide-ranging reforms.

Alexander Bell invents the telephone (1875); Italian-American inventor Antonio Meucci has already been working on transmitting voice over wire since the 1860s (it is not until 2002 that the U.S. Congress officially recognizes Meucci as the inventor of the telephone).

Custer's Last Stand: Battle of Little Bighorn (June 25, 1876).

Late 1800s

President James Garfield is shot (July 2, 1881); he dies from the wounds in September.

Haymarket Square Riot in Chicago (May 1886).

Thomas Edison invents the automatic telegraphy machine, stock-ticker machine, incandescent light bulb, the phonograph, and more (late 1800s).

Gas-powered automobile is invented (late 1800s).

Ellis Island (New York) opens as a processing center for immigrants (January 1, 1892).

Chinese-Japanese War is fought over control of Korea (1894–95).

Radio is invented (1895).

England's Queen Victoria celebrates her Diamond Jubilee (1897), a high point of the Victorian Age.

Zionism is founded (late 1890s); the movement seeks a homeland for Israel.

Spanish-American War (1898) is fought over the liberation of Cuba.

French chemists-physicists Pierre and Marie Curie discover radium (1898).

First Hague Convention is held (1899); it and a subsequent convention (1907) outline laws and customs of war.

William McKinley becomes the first U.S. president to ride in a car—a Stanley Steamer (1899).

The Twentieth Century

1900s

Boxer Rebellion in China (1900).

President William McKinley is shot (September 6, 1901); he dies September 14.

Nobel prizes are first awarded (1901).

At Kill Devil Hills, North Carolina, brothers Wilbur and Orville Wright make the world's first flight in a power-driven, heavier-than-air machine—the airplane (1903).

Russo-Japanese War is fought over interests in China and Korea (1904–05).

Russian Revolution begins with Bloody Sunday, January 22, 1905.

Irish nationalist movement Sinn Fein is organized (1905).

Ford introduces the Model T (1908).

NAACP (National Association for the Advancement of Colored People) is founded (1909).

Robert E. Peary and his expedition reach the North Pole (1909).

1910s

Norwegian explorer Roald Amundsen and his expedition reach the South Pole (1911).

Revolutionary leader Emiliano Zapata helps overthrow the Mexican government of Porfirio Diaz (1910), beginning the bloody 10-year Mexican Revolution.

Titanic sinks (1912).

American automaker Henry Ford invents the moving assembly line (1913), revolutionizing the production of consumer goods and ushering in the consumer age.

World War I, known as the Great War, is fought in Europe (1914–18).

Bolshevik Revolution (1917), a.k.a. October Revolution, ends tsarist rule in Russia and begins Communist era. It is followed by Red Terror (to c. 1920), a period of Communist coercion and civil unrest.

Germany agrees to an armistice and the Central powers surrender, drawing World War I to a close (1918); an estimated 10 million lives have been lost and 20 million have been injured.

Paris Peace Conference (1919) redraws European boundaries as part of WWI settlement. In an effort to keep Germany in check, the Treaty of Versailles metes out severe punishment to the former power.

In response to Temperance Movement (est. mid-1800s), the U.S. Congress passes the Eighteenth Amendment (1919), making prohibition federal law; but enforcement is difficult and bootlegging becomes its own industry, dominated by organized crime (to 1933, when the amendment was repealed).

May Fourth movement emerges in China (1919).

1920s

Roaring Twenties, also called the Jazz Age, is marked by extreme optimism in the United States.

Nineteenth Amendment to the U.S. Constitution is ratified (1920), granting women the right to vote.

New York City bomb explosion (September 16, 1920) rips through J. P. Morgan Bank building; anarchists are thought responsible, but no one is ever charged with the crime.

British parliament passes Government of Ireland Act, creating Northern Ireland out of the six mostly Protestant counties of Ulster; 26 southern counties refuse to accept the legislation, forming the Irish Free State (1921), later called the Republic of Ireland.

As one of the losing Central powers in World War I, the Ottoman Empire is officially dissolved (1922).

Benito Mussolini takes power in Italy (1922); institutes programs of economic and social regimentation.

Union of Soviet Socialist Republics (U.S.S.R.), also called the Soviet Union, is formed (1922).

Teapot Dome scandal is revealed (1922), alleging political favors granted by President Warren Harding's administration.

James Joyce's *Ulysses* is published (1922) in Paris; by 1928 it is listed as obscene in the United States. The ban is lifted in a court challenge and the novel goes on to masterpiece status.

Beer hall *putsch* (1923): Adolph Hitler and nine others attempt a coup in Munich. In a highly publicized trial (1924), they are found guilty; Hitler spends a five-year prison sentence penning *Mein Kampf.*

Scopes "monkey trial" (1925): scientific thought takes on religious fundamentalism in a Tennessee courtroom.

Court martial of Billy Mitchell (1925): World War I general is charged with insubordination for openly criticizing the U.S. military for a lack of air power readiness.

Harlem Renaissance begins (1925); intense period of creative activity by black artists pushes African American writing and music to the fore.

Aviator Charles Lindbergh makes the first solo nonstop transatlantic flight (1927).

Ford takes the Model T out of production (1927); the mass-market automobile, first produced in 1908, is credited with helping define America.

Post–World War I peacekeeping efforts result in the Kellogg-Briand Pact (1928): 15 nations agree to settle conflicts by diplomacy rather than military might; eventually 62 nations ratify the agreement, but it does not prevent another World War.

British women win the right to vote in all elections (1928).

Scottish bacteriologist Alexander Fleming discovers penicillin (1928).

U.S. stock market crashes on Black Tuesday (October 29, 1929): overproduction, limited foreign markets, credit overexpansion, and stock market speculation combined to create a financial crisis that will last until World War II.

1930s

Great Depression grips United States and impacts the world economy (to 1939).

Soviet leader Joseph Stalin conducts aggressive collectivization drive and "purges"; anyone opposed to his hard-line Communist regime is sent to a gulag (1930s).

Japan invades Manchuria (1931).

Amelia Earhart becomes the first woman to fly solo across the Atlantic Ocean (1932).

Prohibition is repealed in the United States (1933).

President Roosevelt begins (1933) his "fireside chats," radio addresses to reassure the American public during the Great Depression.

Adolf Hitler rises to power (1933) in Germany, promising to restore the nation to its prewar stature.

Holocaust begins (1933); Hitler leads Nazi Germany's systematic persecution of Jews as part of the "final solution." By 1945 more than 6 million Jews are killed.

Enrico Fermi announces (1934) he has discovered elements beyond uranium; it is later shown that he split the atom.

Dust Bowl devastates the Great Plains states (1934).

National Labor Relations Act (1935) strengthens unions in the United States.

Communist leader Mao Tse-tung leads Long March (1934–35) across China to Shaanxi (Shensi) Province, where his Red Army establishes a stronghold.

Italy occupies Ethiopia (1935–36) and, later, Albania (1939).

First Lady Eleanor Roosevelt begins writing (1936) her nationally syndicated newspaper column, "My Day."

Germany, Italy, and Japan form Axis powers alliance (1936).

Spanish Civil War (1936–39).

Aviator Amelia Earhart disappears (July 1937) while attempting an around-the-world flight.

Nanking Massacre (December 1937–January 1938): Japanese royal army sweeps into eastern Chinese city, killing and torturing hundreds of thousands of civilians.

Sino-Japanese War begins (1937); the conflict is later absorbed by World War II (1939–45).

Anschluss: Germany annexes Austria (March 1938).

Munich Pact allows Germany to march into Sudetenland region of Czechoslovakia and occupy it (October 1938).

Albert Einstein writes (August 1939) letter to President Roosevelt, urging him to launch a government program to study nuclear energy.

Germany claims the rest of Czechoslovakia (March 1939) and Nazi troops march into Poland (September 1, 1939), beginning World War II (to 1945).

Regularly scheduled U.S. television begins (April 1939).

1940s

Soviet Union consists of 15 republics (1940): Armenia, Azerbaijan, Belorussia, Estonia, Georgia, Kazakhstan, Kirghiz, Latvia, Lithuania, Moldavia, Russia, Tajikistan, Turkmenistan, Ukraine, and Uzbekistan.

World War II is fought in Europe, Asia, and the South Pacific.

Japanese bomb U.S. military installations at Pearl Harbor, Hawaii (December 7, 1941); U.S. enters World War II.

American general Dwight Eisenhower leads Allied forces in invasion of Normandy, France (June 6, 1944); Allied victory, which comes at a dear price with many casualties, proves to be a turning point in World War II.

Antibiotics are first produced (1944).

Yalta Conference (February 1945): allies Roosevelt, Churchill, and Stalin meet in Soviet Union.

Germany surrenders (May 7, 1945), ending the war in Europe; V-E Day (May 8, 1945) marks end of fighting in Europe.

United States drops atomic bombs on Japan—Hiroshima (August 6, 1945) and Nagasaki (August 9, 1945).

Japan surrenders (August 14, 1945), ending World War II. The enormous war involved an estimated three-fourths of the world's population, and a total of 110 million people served in the military. More than 6 million Jews were killed during the Holocaust, an estimated 25 million died in the fighting, and 30 million civilians were killed.

Japan signs terms of surrender (September 2, 1945); V-J Day marks end of war in the Pacific.

United Nations is chartered (1945).

Nuremberg trials get under way in Germany to try Nazi military leaders of war crimes and atrocities (November 25, 1945).

Former British prime minister Winston Churchill warns of an "iron curtain" of Soviet totalitarianism dividing Europe (1946).

Postwar negotiations secure India's independence from Britain (1947); sovereign state of Pakistan also established.

Modern-day nation of Israel is established by decree (May 1948).

Berlin airlift (1948–49) brings in food and supplies to West Berliners after the Soviets blocked off all roads leading to the city.

NATO is formed (1949).

Soviet Union explodes its first nuclear bomb (1949), beginning East-West arms race.

Germany is divided into two nations (1949): the Western-influenced, democratic West Germany and the Soviet-dominated East Germany, part of the Eastern bloc.

1950s

Distrust deepens in the Cold War between Communist Eastern bloc (Soviet-dominated) countries and democratic Western powers led by the United States.

Republican senator Joseph McCarthy claims (early 1950) to possess a list of known communists in the U.S. State Department; accusation launches series of congressional inquiries, "McCarthy era."

Chinese forces invade Tibet (1950).

UNIVAC, the first computer to handle numeric and alphabetical data with equal facility, is developed (1951).

Korean War is fought (1950–53) between Soviet-occupied Communist North Korea and U.S.-occupied South Korea.

New Zealand adventurer Sir Edmund Hillary and Nepalese Sherpa Tenzing Norgay lead the first successful summit of Mount Everest, the world's highest peak (1953).

Americans Julius and Ethel Rosenberg, found guilty of conspiracy, are electrocuted (1953) as tens of thousands protest in New York's Union Square.

Scientists develop a model for DNA structure (1953).

Brown v. *the Board of Education* (1954): Landmark Supreme Court case rules that school segregation in the United States is unconstitutional.

New York's Ellis Island is closed as a processing center for immigrants (1954).

France loses colonial possession Indochina (1954); international conference divides the region along the 17th parallel, creating (Communist) North Vietnam and South Vietnam.

Vietnam War (1954–75) begins when Communist-led guerrillas, the Viet Cong, try to topple South Vietnam's government.

Warsaw Pact (1955): Eastern bloc nations form an alliance.

Emmett Till is killed in Deep South (1955); grisly murder of the black teen sets off civil rights movement.

Southern black leaders organize the Montgomery Bus Boycott (1955–56), initiating civil rights protests.

Soviet Union launches *Sputnik* satellite (1957), beginning "space race" with the United States.

Nikita Khrushchev rises to power in Soviet Union (1958); initiates détente—plan of peaceful coexistence with the West.

The United States launches its first satellite, *Explorer 1* (1958).

1960s

Boynton v. *Virginia* (1960): Supreme Court rules that public facilities are for the use of all citizens, regardless of color.

Construction begins on Berlin Wall (1951).

Soviet cosmonaut Yuri Gagarin becomes the first person in space, orbiting Earth in the spaceship *Vostok I* (launched April 12, 1961).

U.S. government backs the disastrous Bay of Pigs invasion (1961) of Cuba.

Alan Shepard becomes first American in space (May 5, 1961) in a suborbital flight aboard *Freedom 7* spacecraft.

The United States puts a man into orbit (February 20, 1962): astronaut John Glenn orbits Earth three times in *Friendship 7.*

Cuban Missile Crisis (1962) heightens worries that the Cold War will turn into an all-out nuclear conflict; situation is resolved when Soviets comply with U.S. demands to remove missiles from the tiny island nation off Florida's coast.

Rachel Carson's *Silent Spring* (1962) launches the environmental movement.

Cosmonaut Valentina Tereshkova-Nikolaeva becomes the first woman in space, orbiting aboard the *Vostok 6* (launched June 16, 1963).

March on Washington (August 28, 1963): More than a quarter million people demand civil rights, hear Martin Luther King Jr.'s "I have a dream" speech.

Birmingham, Alabama's Sixteenth Street Baptist Church is bombed by white supremacists during Sunday services at the black church (September 15, 1963).

President John F. Kennedy is assassinated (November 22, 1963).

Betty Friedan's *Feminine Mystique* is published (1963), launching the feminist movement.

Palestinian Liberation Organization is formed (1964); it regards Israel as an illegal country and is determined to establish an independent Palestine.

Soviet Premier Nikita Khrushchev is ousted (1964); Leonid Brezhnev rises to power, begins rebuilding Soviet military might.

National Organization for Women (NOW) is founded (1966).

Tragedy strikes American space program: Three astronauts die in launch pad fire (January 27, 1967).

Arab-Israeli War (1967) results in Israeli takeover and occupation of Gaza.

Civil rights leader Martin Luther King Jr. is assassinated (April 4, 1968).

Apollo 11 astronaut Neil Armstrong is the first person to walk on the moon (July 20, 1969).

More than half a million American troops (1969) have been sent to South Vietnam to fight in the Vietnam War. In the United States, protesters stage demonstrations against American involvement in the conflict.

Mexico's "dirty war" targets left-wing reformists (late 1960s to 1970s).

1970s

NASA's "successful failure": ground control and flight crew work to rescue Apollo 13's astronauts after an on-board explosion (April 1970).

Khmer Rouge guerrilla force, supported by Communists from neighboring Vietnam, wages war to topple U.S.-supported government in Cambodia (1970–75).

Strategic Arms Limitation Treaty (SALT I) is signed (1972).

Five men are caught breaking into Democratic Party's national headquarters at Washington, D.C.'s Watergate complex (July 1972); the break-in becomes a full-blown political scandal for the Nixon administration.

Arab terrorist group Black September kills 11 Israeli athletes during the Summer Olympics in Munich (September 5, 1972).

Last U.S. troops leave Vietnam (March 1973).

Middle East remains volatile; Arab (PLO) forces and Israeli troops square off in 1973–74 and again in 1978.

President Nixon resigns from office (August 1974) in wake of Watergate scandal.

G-8 Summit is born with the meeting of the Group of Six, representatives from France, Germany, Italy, Japan, the United Kingdom, and the United States (March 1975); Canada and Russia later join.

South Vietnam (April 30, 1975) surrenders to North Vietnam, ending Vietnam War. North Vietnam unifies the countries as the Socialist Republic of Vietnam.

The killing fields (1975–79): an estimated 2 million people die in a Khmer Rouge-led collectivization drive in Communist Cambodia.

Toxicity at Love Canal, New York, makes headlines (1976–80s).

Two Boeing 747 airliners collide on the runway on Tenerife, in the Canary Islands, killing 583 (March 27, 1977); it is the worst airplane accident in history.

Polish Cardinal Karol Wojtyla is named Pope John Paul II on (October 16, 1978), becoming the first non-Italian head of the Roman Catholic Church in 455 years.

Camp David accords, a peace pact in the Arab-Israeli conflict, is signed in Washington, D.C. (March 26, 1979).

Near meltdown at Three Mile Island (Pennsylvania) nuclear power station (March 1979).

Soviet troops invade Afghanistan to bolster support for a pro-communist government (1979); the conflict wears on for 10 years and becomes a rallying point for Islamic extremists, who back the Afghan rebels.

Pope John Paul II visits (June 1979) his native Poland, calling for a free nation and a new kind of "solidarity"; he effectively launches a movement that leads to the downfall of communism in his homeland.

1980s

Mikhail Gorbachev becomes head of the Communist Party and leader of the Soviet Union (1985), ending rule of Stalin-trained leaders. Gorbachev institutes policies of economic development at home and of openness (*glasnost*) to the West.

U.S. space shuttle *Challenger* explodes shortly after takeoff from Cape Canaveral, Florida (January 28, 1986).

Iran-Contra affair shakes the Reagan presidency (1986–87).

Unrest continues in the Middle East: In the Israeli-occupied Gaza Strip and West Bank, Arab uprisings, called the Intifada, occur (1987–88).

Berlin Wall is dismantled (November 1989) as a wave of democratization sweeps Eastern Europe.

1990s

Soviet era draws to a close in Eastern Europe as multiparty elections are held in Romania, Czechoslovakia, Hungary, East Germany, and Bulgaria (1990).

East Germany and West Germany are reunified (1990).

Iraqi forces invade Kuwait (August 1990).

Persian Gulf War is fought (1991) after Saddam Hussein refuses to comply with international demands that he withdraw Iraqi troops from neighboring Kuwait.

Soviet Union dissolves (1991).

Apartheid is abolished in South Africa (1991).

Separatist factions fight for the independence of the tiny Russian republic of Chechnya (early 1990s into 2000s).

Representatives of 12 European nations sign Maastricht Treaty (November 1992), paving way for European Union (EU).

Bosnian War is fought (1992–95).

World Trade Center in New York City is bombed by an Islamic extremist group, killing six and injuring hundreds (February 26, 1993).

Black Hawk Down incident in Mogadishu, Somalia (October 1993).

North American Free Trade Agreement (NAFTA) goes into effect (January 1, 1994).

Zapatista uprising in Chiapas, Mexico, coincides with NAFTA's start and draws attention to plight of region's indigenous peoples (January 1994).

Genocide claims nearly 1 million lives in Rwanda (April–July 1994).

Internationally brokered peace accords (May 1994) between the PLO and Israel attempt to bring an end to conflict in the Middle East by allowing for Palestinian self-rule in the region.

Alfred P. Murrah Federal Building in Oklahoma City is bombed by members of right-wing militant group, killing 168 people and injuring hundreds more (April 19, 1995).

Pathfinder lands on Mars (July 4, 1997) and deploys robotic rover Sojourner, which collects "staggering amount of data" about the planet.

Good Friday peace accord is signed (April 1998) by Catholic and Protestant leaders in Northern Ireland who agree to form a multiparty administration; disarmament remains a sticking point to moving this agreement forward.

U.S. embassies in Kenya and Tanzania are bombed (August 1998) by al Qaeda terrorists, killing 258 people and injuring thousands.

Construction on the International Space Station begins with the launch of the space capsule *Zarya* (November 1998).

U.S. House of Representatives bring 11 counts of impeachment against President Bill Clinton (December 1998).

For the first time in more than 130 years, U.S. Senate hears charges against a president in impeachment trials (January–February 1999); Bill Clinton is acquitted of both perjury and obstruction of justice.

Ethnic unrest in Kosovo results in NATO bombings of Yugoslavia (1999) in an attempt to pressure the Serbian-dominated Yugoslav government to accept a Western-backed peace plan and end mass expulsion of ethnic Albanians from the province.

Tensions heighten (1998) between Pakistan and India over the disputed Kashmir region. Around the world, fears of a conflict between the two nations mount as both have demonstrated nuclear capabilities.

The Twenty-first Century

Anthropological finds lead researchers to conclude that human ancestors first walked the Earth nearly 6 million years ago, or 2 million years earlier than had been thought.

Second Intifada begins in the Middle East, hampering peace efforts between Israel and Palestine (2000).

Sporadic violence in Northern Ireland and a failure to meet a 2000 disarmament deadline render a permanent solution to the Irish Question elusive.

Dot-com bubble bursts (March 10, 2000), ending 1990s boom economy, the longest economic expansion in U.S. history.

Terrorists attack targets in the United States (September 11, 2001), hijacking four planes and crashing them into the twin towers of the World Trade Center and into the Pentagon; one plane goes down in rural Pennsylvania. About 3,000 people die in the attacks, which are tied to the al Qaeda network.

Anthrax is circulated in the U.S. mail (October–November 2001); the public fears another al Qaeda attack, but investigators point to a domestic perpetrator.

U.S. forces lead an invasion of Afghanistan, launching the military initiative of the War on Terror (October 2001).

Corporate scandals rock an already unstable U.S. economy (2001–04).

The euro goes into circulation as the accepted currency in 11 European countries (January 2002).

So-called Beltway Sniper (later found to be two snipers) strikes fear into the American public through a series of random killings around Washington, D.C. (October 2002).

Crisis in Darfur, Sudan, begins (February 2003).

U.S. space shuttle *Columbia* is lost upon its reentry into Earth's atmosphere (February 1, 2003).

U.S. forces lead an invasion of Iraq (March 2003); government officials call it another phase of the War on Terror.

Roadmap to Peace, a permanent two-state solution the Palestinian-Israeli conflict, is announced by the White House (March 2003).

Blackout in eastern U.S. and Canada affects 50 million people (August 14, 2003).

California experiences worst wildfires in its history (October 2003).

NASA begins collecting data from Mars Rover project (2004).

Explosions on Madrid commuter trains during morning rush hour kill 191 people and injure 1,800 (March 11, 2004); Islamic terrorists are responsible.

Tsunamis in Southeast Asia claim more than 150,000 lives and level towns and villages (December 26, 2004).

Kyoto Protocol, environmental agreement signed by 141 nations, goes into effect (February 2005).

Pope John Paul II dies (April 2, 2005); he is succeeded by German Cardinal Joseph Ratzinger, who becomes Pope Benedict XVI (April 19).

Al Qaeda in Europe claims responsibility for train bombings in London, which kill more than 50 and injure hundreds (July 7, 2005).

Hurricane Katrina devastates New Orleans and the coasts of Mississippi and Alabama (August 2005).

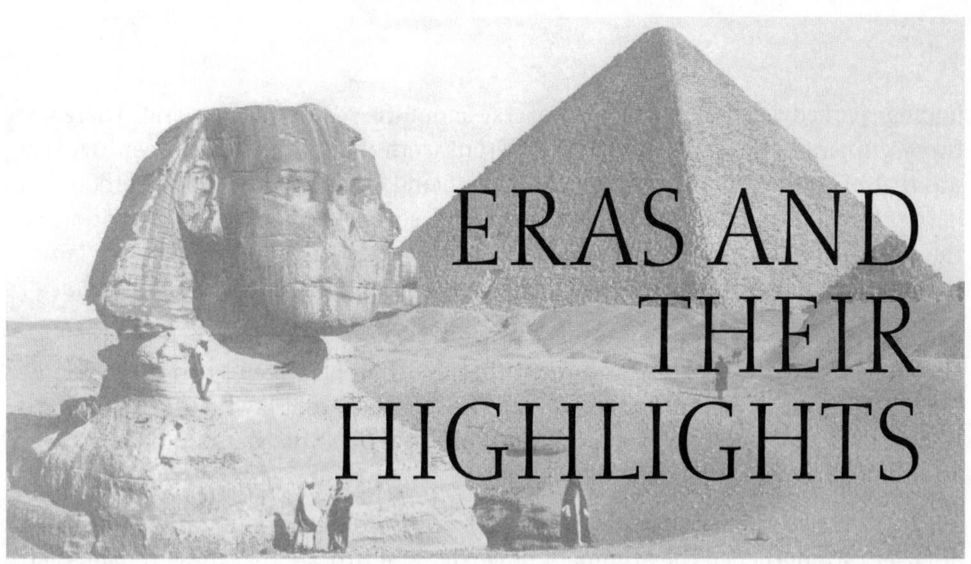

When did **history begin**?

History began when human beings were able to document events in writing; anything prior to the advent of writing is commonly referred to as prehistoric. Humans began to write approximately 3,500 years before Christ, and the Classical Age, which began with the rise of Greek civilization (c. 2000 B.C.), is considered the beginning of history.

PREHISTORIC ERA: 2,000,000 TO 2000 B.C.

What is the **Prehistoric era**?

The term *Prehistoric era* refers to the time before written history began, so it encompasses the Stone Age (Paleolithic and Neolithic ages), the Bronze Age, and the Iron Age. The Prehistoric era spans the time from about 2,000,000 B.C. to roughly 2000 B.C., when the Classical Age began with the rise of the Greek and Roman empires.

What is **Neanderthal Man**?

The term refers to a hominid who walked the Earth in the Middle Paleolithic Age (during the Old Stone Age). The term *hominid* refers to an upright-walking primate that is an extinct ancestor to man. A hominid can be an ancestor of "true" man (modern man) or a relative, such as a modern primate. (In this context, the term *man* is used to refer to both males and females of the genus *Homo*.)

Neanderthal Man was discovered in 1856 near Düsseldorf, Germany, when workers came across the skull and skeletal remains of what appeared to be a human. The finding sparked discussion and controversy about the nature of the being. There were two arguments: the skull, markedly different from that of nineteenth-century man, was that of a pathologically deformed human being (an individual who was thought to have suffered from severe bone disease or some sort of congenital malformation); or, the skull belonged to an "early" man. This latter view was supported by the famous English naturalist Thomas Henry Huxley (1825–1895) in his book *Man's Place in Nature* (1863).

Another advocate for the argument that the skull belonged to an early man was French surgeon and anthropologist Paul Broca (1824–1880), who accepted Charles Darwin's (1809–1882) theory of evolution. In this light, Broca argued that the Neanderthal skull was a key to human evolution.

In 1886 two similar skeletons and some stone tools were found in Belgium. This discovery strengthened the argument of Huxley and Broca—that these remains actually belonged to man's early ancestors. Excavations from 1890 to 1892 on the island of Java (in Southeast Asia) furthered the argument and, for the most part, settled the controversy: A number of fossil remains were found in the banks along the Solo River. Named *Pithecanthropus erectus* by their discoverer, Dutch paleontologist Marie Eugène F. T. Dubois (1858–1940), the findings were popularly known as Java Man.

Subsequent findings, including that of the so-called Peking Man in the summer of 1923 in China, produced evidence that approximately 70,000 to 11,000 years ago there were groups of the Neanderthal race in Europe, western Asia, and northern Africa. At some point they disappeared and were replaced by another type of man; the cause of this disappearance is unknown.

When did **humankind first walk on** the face of the **Earth**?

For decades after the discovery of "Lucy" (in the 1970s), it was thought that humankind first walked the Earth about 3 million years ago. But fossil finds in the mid-1990s pushed the evolutionary start-point for humans to more than 4 million years ago. Further astonishing finds in the early 2000s led researchers to conclude that human ancestors first walked the Earth nearly 6 million years ago.

In November 1974 American Donald C. Johanson (1943–) made one of paleoanthropology's most widely publicized finds when he discovered a partial skeleton at Hadar, Ethiopia. More than 3 million years old, the female skeleton was the most complete hominid fossil ever found, but the skull was not recovered. The creature stood three and a half feet tall and, although apelike, had definitely walked upright. When Johanson officially announced his find in 1979, "Lucy" (named for the Beatles song "Lucy in the Sky with Diamonds," which was popular in the camp at the time the fossil was found) became known as the mother of all humankind. (Her sex was confirmed by the pelvic bones.) Since she was an erect-walker, the finding gave certainty to theories that hominids walked erect at 3 million years B.C.

After the discovery of Lucy, older hominid fossils were unearthed by researchers in the field. In 1994 anthropologist Meave Leakey (1942–) and her associates found fossils of a 4.1-million-year-old bipedal species near Lake Turkana, Kenya. These were designated *Australopithecus anamensis.* (Technically, the first fossils of the *Australopithecus anamensis* were discovered in 1965 by a Harvard expedition in Kenya, but they were not identified as such until later finds were made.) Also in 1994, University of California at Berkeley paleoanthropologist Tim White and an international team published their 1993 discovery of the fossils of a chimplike animal in Ethiopia; they classified it as *Ardipithecus ramidus.* These fossils were dated to nearly 4.4 million years ago. Fossils discovered between 1997 and 2001 in the Middle Awash region of Ethiopia were determined to be nearly 6 million years old. In 2001 Yohannes Haile-Selassie (Cleveland Museum) and coauthors Tim White and Gen Suwa (University of Tokyo Museum) reported the monumental findings in the journal *Nature*: The hominid named *Ardipithecus kadabba* was thought to "represent the first species on the human branch of the family tree just after the evolutionary split [from chimpanzees]."

What is the **Stone Age**?

What people commonly refer to as the Stone Age is actually two ages: The Old Stone Age (about 2,000,000 B.C. to about 10,000 B.C.) and the New Stone Age (c. 10,000 to c. 3300 B.C.). It was during these periods that humans used stone tools.

During the Old Stone Age, also called the Paleolithic Age, man was evolving from his apelike ancestors to modern-looking hunter-gatherers. Early modern man's progress continued to the end of the Old Stone Age, around 10,000 B.C. Then, as the Ice Age ended and the Earth warmed, the hunter-gatherers again revolutionized their way of life. They opted for a more settled existence in which they could exercise greater control over their food supplies. With the coming of the New Stone Age, or Neolithic Age, humans turned to agriculture.

The New Stone Age brought profound changes in the development of humans. Neolithic man learned to produce food rather than collect it. People were no longer dependent on hunting, fishing, and gathering wild fruit and nuts for subsistence. They learned to cultivate crops, domesticate animals, make pottery, weave textiles from fiber and hair, and produce more sophisticated tools and weapons by hammering, grinding, and polishing granite, jasper, and other hard stone. More substantial houses and communities, even fortified villages, came into being, laying the foundation for the great civilizations that would follow.

Near the end of the New Stone Age, craftsmen in the Middle East learned to make tools and weapons from metal. The world's earliest known manmade copper objects—beads, pins, and awls—were fabricated in Turkey and Iran around 8000 B.C. Archaeological evidence points to copper mining in the Balkans around 5000 B.C. From there the technology probably spread west, reaching the Alps about a thousand years later and marking the beginning of the Copper Age (c. 4000–2200 B.C.).

Was there really such a thing as the **Copper Age**?

Some texts do not refer to a Copper Age, moving directly from the Stone Age to the Bronze Age. In fact, the Copper Age (c. 4000–2200 B.C.) overlapped with both the end of the Old Stone Age (the Neolithic Age) and the Bronze Age and is marked by man's use of copper as a material for toolmaking.

What is the **Bronze Age**?

The Bronze Age (c. 3300–2500 B.C.) is the period of human culture when man began using bronze metal to make objects—principally, tools. The Neolithic Age slowly came to an end as various cultures in Eurasia that had depended on wood, stone, and bone for tools began to develop the techniques for metallurgy. Bronze proved to be an excellent material for making tools and weapons. People in the Middle East learned to produce bronze by mixing tin and copper (hence, the transition years between the Neolithic Age and the Bronze Age are sometimes referred to as the Copper Age). Bronze had considerable hardness, strength, and density, and it proved more reliable and durable than the stone, wood, and bone tools that had been in use. The Bronze Age lasted until the beginning of the Iron Age.

When did the **Iron Age** begin?

The real advent of the Iron Age came not with the discovery of metal (in about 2500 B.C.), but with the invention of the process of casing or steeling it, probably about 1500 B.C. This happened when it was learned that by repeatedly reheating wrought iron in a charcoal fire and then hammering it, it not only became harder than bronze but kept its hardness after long use. (Wrought iron was discovered accidentally when smiths found that by hammering the small beadlike pieces of iron left as a residue after smelting copper they could form the iron particles into a mass. This kind of wrought iron, however, was good only for decorative purposes, and for more than a thousand years after 2500 B.C., iron remained a precious ornamental metal. Bronze, which was harder and capable of being sharpened to a fine cutting edge, continued to be the metal for tools and weapons.)

The next technological improvement, which again meant a further hardening of the metal, was the process of quenching it, which involved repeatedly plunging the hot iron into cold water. It was only after this series of discoveries and inventions that the significant impact of iron on culture and civilization was appreciably felt.

Because bronze was scarce, it was also costly. Consequently, it was not until iron came into use that humans extended their control over nature. For this reason, iron has been called the "democratic metal." Widespread use of iron tools meant a general increase in living standards. For example, the use of iron axes brought about the clearing of forests, and therefore new land came under cultivation. Other significant devel-

During the Old Kingdom of ancient Egypt commerce prospered and the arts flourished, as evidenced by the Great Pyramids at Giza and the Great Sphinx.

opments included the application of iron tools to sheep shearing and cloth cutting, and the invention of the lathe, the most fundamental machine tool.

The Iron Age lasted until the beginning of the Classical Age (c. 2000 B.C.).

What were the **hallmarks of ancient Egypt**?

One of the world's oldest civilizations, ancient Egypt developed about 3000 B.C., or 5,000 years ago, in the Nile River valley; it lasted until 332 B.C., when it was conquered by Alexander the Great (356–323 B.C.). In that time, Egypt was ruled by 30 dynasties. Most of those dynasties fall into three kingdoms: Old Kingdom (during the third millennium B.C.), Middle Kingdom (early second millennium B.C.), and New Kingdom (mid-second millennium B.C.); the kingdoms were followed by intermediate periods, which were times of weakened government or foreign domination.

The First Dynasty was founded by Menes in 3110 B.C. after he united rival kingdoms of Upper and Lower Egypt under his rule and established the capital at Memphis (the present-day village of Mit Rahina, 14 miles south of Cairo, in northern Egypt). During the Old Kingdom commerce prospered and the arts flourished, as evidenced by the Great Pyramids at Giza (including the Great Sphinx), which were begun during the Fourth Dynasty (c. 2500 B.C.). The Old Kingdom was followed by a 258-year (intermediate) period of weak rulers and anarchy, which was ended when Amenemhet I rose to power in 1991 B.C., reunifying Egypt and beginning the Middle Kingdom. During the Middle Kingdom, Egypt launched imperialistic campaigns, expanding its territory and conquering Palestine and Syria in the east. About 1720 B.C., Semitic nomads entered Egypt and wrested power from the pharaohs, establishing the fifteenth through the seventeenth dynasties—a peaceful and prosperous period. But the Egyptians expelled this foreign influence (c. 1570) to establish the New Kingdom: the 200 years that followed were the height of Egyptian civilization, with the cities of Thebes and Memphis regarded as the political, commercial, and cultural centers of the known world.

Ancient Egyptians invented a calendar, created a form of hieroglyphic writing, and developed papyrus (paper made from the papyrus plant). Situated along the Nile and south of the Mediterranean Sea, Egyptians also produced early seagoing vessels. But it is their buildings for which this ancient group is renowned: In addition to the Great Pyramids at Giza, the impressive relics that have been discovered include those at Abu Simbel, where King Ramses II (c. 1250–? B.C.) had two temples built out of rock during his reign (1304–1237 B.C.); numerous ruins and tombs at Abydos; a complex of temples and shrines at Karnak (part of the site of ancient Thebes); and temples and other buildings at Luxor (also part of ancient Thebes).

During the last 700 years of ancient Egypt (c. 1085–332 B.C.), the kingdom increasingly came under foreign domination, which weakened it to the point that Alexander the Great (356–323 B.C.) was able to claim it without struggle in 332 B.C.

THE CLASSICAL AGE: 2000 B.C. to A.D. 500

What is the **Classical Age**?

The Classical Age refers to the ancient Roman and Greek worlds, roughly 2000 B.C. to A.D. 500. The Classical Age followed the Prehistoric era and preceded the Middle Ages. During this period, the ancient Greeks and Romans made contributions to literature, philosophy, science, the arts, and letters that are still relevant today.

How did **Greek civilization** begin?

Ancient Greek civilization began with the Minoans. Europe's first advanced civilization, the Minoans were a prosperous and peaceful people who flourished on the Mediterranean island of Crete from about 3000 to 1450 B.C. The Minoans built structures from stone, plaster, and timbers; painted walls with brilliant frescoes; made pottery; wove and dyed cloth; cultivated the land (they are believed to be the first people to produce an agricultural surplus, which they exported); constructed stone roads and bridges; and built highly advanced drainage systems and aqueducts. (At Knossos, the royal family had a system for showers and even had toilets that could be flushed.) Minoans were a sophisticated people who loved music and dance, games and entertainment.

What is the **Mycenaean Age**?

The Minoans were either conquered by or succeeded by the Mycenaeans, who were mainland Greeks: In about 1450 B.C., Crete was struck by a 200-foot tsunami (or seismic wave), which is thought to either have completely destroyed the island or to have weakened it to a point that it could be overtaken. The Mycenaeans flourished from about 1650 B.C. to 1200 B.C., a time known as the Mycenaean Age, carrying forth the culture and skills they had learned from the Minoans (who had been their neighbors). The Mycenaeans were skilled horsemen, charioteers, and accomplished sailors who ruled the Aegean. Mycenaean culture revolved around its fortified palaces, called *acropolises* (top cities). Its cities included Argos, Corinth, Sparta, and the then-small cities of Athens and Thebes.

In about 1200 B.C. the Mycenaeans attacked the city of Troy, which was considered the key to the profitable Black Sea trade, thus launching the Trojan War. After 10 years of fighting (a period that is recounted by Greek poet Homer [c. 850–? B.C.] in the *Iliad*), the Mycenaeans were victorious. But soon their period of triumph ended as the Dorian peoples (from the northwestern part of the Greek mainland) overran most of the Peloponnesus (the southern peninsula of Greece). The Dorians, aided by the superiority of the iron sword, flooded southward, where they sacked and burned the great Mycenaean cities and conquered the wealthy sea traders, throwing Greece into the period known as the Dark Ages, or Archaic Period, which lasted from 1100 to about 800 B.C.

What were the **Dark Ages** of ancient Greece?

After the Dorians conquered the Mycenaeans in 1100 B.C., these nomadic peoples thrust Greece into a period of decline that lasted more than 300 years. The Dorians rejected the life of the great Mycenaean cities in favor of their nomadic shepherding and hunting life. A tribal people, they possessed a harsh sense of justice, and the period was marked by feuds between clans. Men typically carried weapons—now made of iron (it was the Dorians who brought the new, more durable metal from the north, ending the Bronze Age in Greece).

During this Dark Age, there is little evidence of Greek civilization; the script used by the Mycenaeans disappeared, and art, which had prospered during the Mycenaean Age, declined. Under Dorian rule numerous Mycenaean cities were abandoned, and many regions and islands seem to have been depopulated. There is no evidence of trade with other countries. Poverty had overtaken the Greeks.

As the Dorians took possession of the Greek mainland, a few Mycenaean communities survived in remote areas. Many Mycenaeans fled eastward to Athens, which became a haven for those who hoped for a return to the former civilization. Other Mycenaeans crossed the Aegean and settled on the coast of Asia Minor. Most of these refugees spoke Ionian Greek.

A lasting legacy of the Dark Ages of Greece is its mythology. As Ionian Greeks attempted to hold on to the refined civilization of the Bronze Age, they commemorated the greatness of the past in song and verse, including Greek poet Homer's (c. 850–? B.C.) *Iliad* and *Odyssey.* These epics were combined with eighth-century poet Hesiod's *Theogony,* an account of the creation of the universe and the generations of the gods, to give rise to a new Greek religion based on the god Zeus and 11 other gods who were believed to reside on Mount Olympus in northeastern Greece. The Greek gods were later adopted by the Romans and given different names.

What was the **golden age** of ancient Greece?

It is the period of classical Greek civilization that followed the so-called Dark Ages of Greece, which came to an end about 800 B.C. Over time the Dorians had become more settled, and they gradually revived trade and culture on mainland Greece. The self-governing city-state (*polis*) evolved, including the military center of Sparta and Athens, which became a center for the arts, education, and democracy. This was the beginning of the great Hellenic period of classical Greek civilization. Greek civilization reached its height in Athens during the mid-400s B.C., a period of outstanding achievement known as the Golden Age.

What are the hallmarks of **classical Greek civilization**?

The classical Greeks, who called themselves the Hellenes and their land Hellas, influenced western civilization more than any other people. Their contributions to every field of endeavor remain with us today, more than 2,000 years later.

Greek thought shaped science, medicine, philosophy, art, literature, architecture and engineering, mathematics, music, drama, language, and politics. The classical Greeks believed in individual freedom, reasoning, and truth, and that everything should be done in moderation. They also held that people should find time for both work and play and should balance the life of the mind with the exercise of the body.

The knowledge that became the Greek legacy had its beginnings in the settlements established in Asia Minor (the peninsula between the Black Sea and the Mediterranean, and which today is occupied by Turkey) after the Dorians invaded the Greek mainland. The Phoenician alphabet, an early alphabet developed by Semitic

The Parthenon, seen at the top of the Acropolis, in Athens, Greece, was constructed between 447 and 432 B.C.

peoples in the ancient maritime country of Phoenicia (present-day western Syria and Lebanon), was acquired by the Greeks, who adapted it to their language. They began using it to record Greek poet Homer's (c. 850–? B.C.) oral epics (*Iliad* and *Odyssey*) and the works of other Greek poets and historians.

Among the great Greek philosophers are Socrates, Plato, and Aristotle. Greek literature includes the epic poetry of Homer as well as the passionate love poems of Sappho. The Greeks gave humankind the tragedies of Aeschylus, Sophocles, and Euripides, which continue to be studied by students today, along with the comedies of Aristophanes and Menander. The classical Greeks loved to speak, and oratory is considered by some to be their highest form of prose. Orators known to the modern world are Antiphon, Lysias, Isocrates, and Demosthenes.

Herodotus, called the "father of history," left the modern world with an account of the Persian Wars (500–449 B.C.), a conflict between the Greek city-states and the Persian Empire. The Greeks also gave humankind the "father of medicine" in physician Hippocrates, who taught that doctors should use reason to determine the cause of illness and should study the patient's appearance, behavior, and lifestyle to diagnose and treat. (The "Hippocratic Oath," versions of which are still sworn by medical students graduating today, is attributed to Hippocrates.) Greek scientists include Thales and

Pythagoras; scientist-philosophers include Leucippus and Democritus. And of course, the Greeks gave modern culture the Olympic Games.

Did the **Roman Republic** precede the Roman Empire?

Yes, the Roman Republic, which for centuries afterward was considered the model form of a balanced government, was established in 509 B.C. The Roman Empire was not established until 27 B.C. when Augustus (also known as Octavian; 63 B.C.–A.D. 14) became its first ruler. In brief, the development of ancient Rome is as follows.

In 753 B.C. the city of Rome was established. (Legend has it, the city was founded by brothers Romulus and Remus.) Situated on wooded hills above the Tiber River, about 15 miles from the sea, Rome enjoyed the advantages of access to trade routes while having natural protection from aggressors. The city was defensible. Agriculture prospered in the area, as did other economic endeavors including manufacturing and mining.

In 509 B.C. the Republic was established by noblemen. The government was headed by two elected officials who were called consuls. Since they shared power, a certain measure of balance was ensured in that either one could veto the actions of the other. And the posts were brief: each elected official served for only one year. These heads of state were guided by the Roman Senate, which was made up of senior statesmen. There were also assemblies in which the people had a voice.

In 390 B.C. Rome was captured and sacked by the Gauls (a Celtic people from western Europe), who were able to hold it for a short time. About 300 B.C. the Romans came into contact with the Greeks, adopting not only some of their ideas, but their mythology as well. The Greek gods and goddesses were soon given Roman names.

By 275 B.C. Rome controlled most of the Italian peninsula. Their homeland stable, the Romans set their sights on overseas expansion, and between 264 and 146 B.C. fought the Punic Wars in order to gain territory. They conquered the Mediterranean islands of Sicily, Sardinia, and Corsica; part of Gaul; much of Spain; and Carthage (in northern Africa).

In the last century B.C. Rome entered a period that is considered the height of their civilization. But about the middle of that century, the Republic was torn by civil wars. After 20 years of fighting, the Roman Empire was formed in 27 B.C. when Augustus (Octavian) became the first emperor. While vestiges of the Republic were maintained, the emperor held supreme authority, nominating the consuls and appointing senators, controlling the provinces, and heading the army. The civilian assemblies were still in place, but had for the most part lost their voice in government.

The Roman Empire lasted nearly 500 years. By the third century A.D. Roman armies had conquered so many peoples that the empire stretched across Europe and included the entire Mediterranean coast of Africa as well as parts of the Middle East. During this time of power and expansion, trade thrived over a vast network of roads and sea routes, which extended to China, India, and Africa. Coins, made of gold, silver, copper, and bronze, were issued and controlled by the Roman government.

In 395, upon the death of emperor Theodosius the Great (347–395), the Roman Empire was divided into two: East and West. In 476, after suffering a series of attacks from nomadic Germanic tribes, Rome fell.

What was the **Pax Romana**?

The Pax Romana was the height of the Roman Empire, a period of peace that lasted from 27 B.C. to A.D. 180, roughly two centuries. During this time, no other country or force was strong enough to challenge the Roman Empire, so citizens turned their attention to commerce, learning, the arts, and literature, all of which flourished.

What is the legacy of **ancient Rome**?

Since the Romans borrowed and adapted the ideas of the Greeks, with whom they had come into contact about 300 B.C. and later conquered in 146 B.C., the culture of ancient Rome is sometimes called Greco-Roman. Over the course of centuries, Romans spread their ideas throughout their vast empire.

They also developed a legal code, which outlined basic principles while remaining flexible enough that lawyers and judges could interpret the laws, taking into consideration local customs and practices. The code later became the model for legal systems in Europe and in Latin America. Further, Roman armies built a network of roads, aqueducts, and tunnels, putting in place an infrastructure that outlasted the empire itself. Latin, the Roman language, remained the language of educated Europeans for more than 1,000 years, while the Latin-based (or Romance) languages of Italian, French, and Spanish took over everyday communication. The economic system put in place during the height of the Roman Empire, with a centrally controlled money supply, also had lasting effect.

Though the empire crumbled by A.D. 476, its cultural, social, and economic establishments continued to have validity well into the Middle Ages (500–1350).

What is the difference between the **Roman Empire** and the Holy Roman Empire?

Roughly four and a half centuries separated the two empires, both of which were comprised of vast regions of western and central Europe. The Roman Empire was established in 27 B.C., when Augustus (also known as Octavian; 63 B.C.–A.D. 14), the grandnephew, adopted son, and chosen heir of Julius Caesar (100–44 B.C.), became emperor. His reign lasted until A.D. 476, when Rome fell to Germanic tribes.

The Holy Roman Empire (H.R.E.) began in the mid-900s A.D., when Otto I (912–973) of Germany gained control of most of northern and central Italy. Pope John XII (c. 937–964) crowned Otto emperor in 962. In the 1200s the area of power officially became known as the Holy Roman Empire. The H.R.E. was dismantled July 12, 1806, in the Confederation of the Rhine, which brought most of the German states under French domination—the result of the Napoleonic wars. But even after Napoleon Bonaparte (1769–1821) was permanently ousted as head of France in 1815, there were no attempts to reinstate the Holy Roman Empire.

What happened to **the Celts** during the Roman Empire?

The Celts were an Indo-European people who by 500 B.C. had spread across what is now France, Italy, Portugal, Spain, and the British Isles, and by 200 B.C. they had expanded as far as present-day Bulgaria and Greece. When the Romans conquered much of Europe (about 300 B.C.), many Celts were absorbed into the Roman Empire. However, those Celts living in Ireland, Scotland, Wales, southwest England, and Brittany (in northwestern France) were able to maintain their cultures, and it is in these regions that people of Celtic origin still live today.

What is known about the **Celts prior to the Roman Empire**?

Before Europe was conquered by Rome, Celts, who were themselves divided into smaller tribes, had become rather advanced in many ways. Their society was divided among three classes: commoners, the educated, and aristocrats. They formed loose federations of tribes, raised crops and livestock, used the Greek alphabet to write their own language, and were among the first peoples in northern Europe to make iron. They also developed a form of metalwork that most people today recognize as Celtic, called *La Tène.* They never formed one united nation, however, so that when Roman armies swept across Europe, the Celtic tribes were overrun.

THE MIDDLE AGES: 500 TO 1350

What are the characteristics of **medieval times**?

Although the Middle Ages were shadowed by poverty, ignorance, economic chaos, bad government, and the plague, it was also a period of cultural and artistic achievement. For example, the university originated in medieval Europe (the first university was established in 1158 in Bologna, Italy). The period was marked by the belief, based on the Christian faith, that the universe is an ordered world, ruled by an infinite and all-

What are the Dark Ages and how did they get that name?

The Dark Ages usually refer to the historical period in Europe from about A.D. 500 to 1350, also known as the Middle Ages, or medieval period (medieval is from the Latin *medium aevum,* meaning "middle age"). The Middle Ages followed the collapse of the Roman Empire (in A.D. 476, which signaled the end of the Classical Age) and preceded the Renaissance.

The term *Dark Ages* is the legacy of seventeenth-century historians who considered the period a barbaric interruption of a tradition that began in ancient Greece and continued through the European Renaissance. Modern scholars have tried to correct this view, but the popular perception of the medieval period as "dark" still remains. (The term *Dark Ages* is also used to refer to a Greek historical period, from about 1100 B.C. to about 700 B.C., which was considered to be a period of decline.)

knowing God. This belief persisted even through the turmoil of wars and social upheavals, and it is evident in the soaring Gothic architecture (such as the Cathedral of Chartres, France), the poetry of Dante Alighieri (1265–1321), the philosophy of St. Thomas Aquinas (1225–1274), the Gregorian chant, and the music of such composers as Guillaume de Machaut (c. 1300–1377).

Who were **the Huns**?

The Huns were a nomadic central Asian people who, in the middle of the fourth century A.D., moved westward. They first defeated the Alani (a group in the Caucasus Mountain region, between the Black and the Caspian Seas), and then conquered and drove out the Goths. Unified by the ruler Attila in 434, the Huns gained control of a large part of central and eastern Europe by about 450. The Italian countryside was ravaged in the process, and many people sought refuge on the numerous islands in the Lagoon of Venice; the settlement later became the city of Venice.

With the death of Attila in 453, the subjects of the Huns revolted and defeated them. The Huns were later absorbed into the various peoples of Europe.

Who were **the barbarians**?

The term is used to refer to any of the Germanic tribes that, during the middle of the first millennium A.D. (beginning about 400), repeatedly attacked Rome, eventually conquering it and dividing the territories of the West Roman Empire into many kingdoms (while what was the East Roman Empire survived as the Byzantine Empire). The Ger-

A powerful ruler of the Huns, Attila led his people to gain control of a large part of Central and Eastern Europe.

manic tribes included the Goths, the Vandals, the Franks, and the Lombards.

Who were **the Goths**?

The Goths were a group of Germanic tribes who originated in what is now Scandinavia. As early as the third century A.D., the Goths invaded the eastern provinces of the Roman Empire. During the following century, they divided into two groups: a western group known as the Visigoths, who predominated north of the lower Danube River, in central Europe; and an eastern group known as the Ostrogoths, who were situated north of the Black Sea, between Europe and Asia. Along with other tribal Germanic peoples, the Goths brought the downfall of the Roman Empire.

What happened to **the Ostrogoths**?

The eastern division of the Goths, the Ostrogoths were overrun and absorbed by the Huns in 370. When the powerful Hun leader, Attila, died in 453, subjects of the Huns revolted and the Ostrogoths regained their freedom. Theodoric (c. 454–526) became the ruler of the Ostrogoths in 493, and it was under his leadership that the group invaded northern Italy, where they remained. In the middle of the following century, they were overthrown by the armies of the Byzantine Empire. Like other Germanic tribes, the Ostrogoths were absorbed into the various groups of Europe.

What happened to **the Visigoths**?

When both the Visigoths and Ostrogoths were attacked by the Huns in 370, the Visigoths fared better, many of them fleeing into a Roman province. In 378 the Visigoths rebelled against the Roman authorities. On horseback, they fought the battle of Adrianople (in present-day Turkey), destroying a Roman army and killing Rome's eastern emperor, Valens (c. 328–378). The Visigoths' introduction of the cavalry (troops trained to fight on horseback) as part of warfare determined European military, social, and political development for the next thousand years.

After the battle of Adrianople, the Visigoths moved into Italy, and under the leadership of their ruler, Alaric (c. 370–410), sacked Rome in 410, an event that signaled the beginning of the decline of the Roman Empire. After the success of the Visigoths, one tribe after another invaded the empire.

The Visigoths continued westward into Gaul, and there set up a monarchy that consisted of much of France and Spain and was centered in Toulouse. But in 507 they were driven out by the Franks, and the Visigoths withdrew into the Iberian Peninsula (present-day Spain and Portugal). Toledo was established as the capital of the Visigoth kingdom in 534. Roderick (or Rodrigo), the last of the Visigoth kings in Spain, was defeated and killed in 711 during a battle with the Muslims (Moors), who invaded from northern Africa. The Muslims went on to rule most of the Iberian Peninsula until the mid-1400s.

Who were **the Vandals**?

Like the Goths, the Vandals were a Germanic people who originated in an area south of the Baltic Sea in what is now Scandinavia. By A.D. 100 they had moved into the southern region of (present-day) Poland. But there they eventually found themselves threatened by the Huns, and so they began moving westward late in the fourth century. Early in the fifth century, the Vandals overran Gaul (in western Europe), Spain, and northern Africa, where they eventually settled. Between 428 and 477 the Vandals were ruled by the powerful King Genseric. Under his reign, they ravaged Rome (in 455). Their pillage was so thorough that the word *vandal* is used to describe anyone who willfully destroys property that is not theirs. In 533 and 534, like the Ostrogoths, the Vandals were defeated by armies of the Byzantine Empire.

Who were **the Franks**?

The Franks were another Germanic people who divided into two branches: the Salians, who settled near the lower Rhine River, near the North Sea; and the Ripuarians, who moved into what is now Germany, along the middle Rhine River.

In 359 the Franks entered into the Roman Empire as allies, but in 481 Clovis (c. 466–511) gained the Salerian Frank kingship. By 486 he had begun a campaign of aggression, conquering Romans, Gauls, Visigoths, and other groups. Under this cruel and cunning king, the Franks soon controlled all of Europe—from the Mediterranean to the English Channel, and from the Pyrenees Mountains to the Rhine River. Even after Clovis's death, the Franks maintained their stronghold in the region, which is how France eventually got its name.

Though Clovis was a powerful ruler, he was succeeded as king of the Franks by the even more powerful Charlemagne (also called Charles the Great; 742–814), who ruled from 771 to 814, creating a vast empire. In 800 Pope Leo III (c. 750–816) crowned him Emperor of the West, thus initiating the Holy Roman Empire. It was after Charlemagne that the empire of the Franks began to break up, becoming the kingdoms of France, Germany, and Italy.

How were **the Gauls** related to the Celts?

The ancient Gauls were a Celtic people who spoke forms of the Celtic language. They occupied the ancient country of Gaul, a region west of the Rhine River and north of the Pyrenees Mountains (an area that today consists of France, Belgium, Luxembourg, part of Germany, and part of the Netherlands). The Gauls were led by priests, who were called Druids. By 390 B.C. the Gauls had moved southward, across the Alps and into Italy. In the third century B.C., they battled the powerful Romans and were briefly successful. Ultimately, however, they were defeated, becoming subjects of Rome. Later, under Julius Caesar, the Romans conquered all of Gaul, so that by 50 B.C. the region became part of the Roman Empire. Five centuries later, Gaul was overrun by the Franks, for whom the region was named. Thus, French people today are descendants of the Gauls. Also, the Galatians (one of the Christian peoples to whom the Apostle Paul wrote while he was in jail) were descendants of the Gauls who settled in Macedonia and Asia Minor (the peninsula between the Black Sea and the Mediterranean, which today is occupied by Turkey).

Who were **the Lombards**?

The Lombards, too, were a Germanic tribe; they are believed to have originated on an island in the Baltic Sea. In the last century B.C., the Lombards moved into Germany and gradually continued southward so that by A.D. 500, they were settled in present-day Austria. From 568 to the mid-700s, they controlled much of Italy, posing a serious threat to the papal supremacy so that in 754, Pope Stephen II (d. 757) appealed to the powerful Franks for help. By this time the Franks were ruled by Pepin III (called Pepin the Short; c. 714–768), who was able to defeat the Lombards. The northern region of Italy, Lombardy, is named for them.

Who were **the Saxons**?

Saxons were a Germanic people who in the second century lived in southern Jutland, in the area of present-day Denmark and northwestern Germany. During the next two centuries the Saxons raided the coastal areas of the North Sea. By about 400, they had reached northern Gaul, the ancient country that occupied the area west and south of the Rhine River, west of the Alps, and north of the Pyrenees Mountains (roughly modern-day France, Belgium, Luxembourg, part of Germany, and part of the Netherlands). By about 450, as Roman rule was declining, the Saxons had reached England, where they merged with the Angles and began setting up Anglo-Saxon kingdoms. The Anglo-Saxons dominated England until it was conquered by Danish Vikings (under the leadership of Canute) in 1016.

Who were **the Angles**?

The Angles were yet another Germanic tribe; they originated in Schleswig in northwestern Germany. Like other Germanic peoples, they were on the move by the fifth century. Arriving in England, the Angles joined the Saxons (also a Germanic-speaking people), after which time they together became known as Anglo-Saxons, a term referring to any non-Celtic settler of Britain.

Who were **the Vikings**?

The Vikings, also called Norsemen, were fierce, seafaring warriors who originated in Scandinavia. Beginning in the late 700s they raided England, France, Germany, Ireland, Scotland, Italy, Russia, and Spain. Great shipbuilders, the Vikings also reached Greenland, Iceland, and probably even North America long before the Europeans. (Ruins of a Norse settlement were found on the northeastern coast of Newfoundland, Canada.) The Vikings were converted to Christianity during the tenth and eleventh centuries, about the same time that the kingdoms of Norway, Denmark, and Sweden were established. Under the Danish leader Canute, Vikings conquered England in 1016 and ruled it as part of Denmark until 1042.

Who were **the Danes**?

The Danes are the Scandinavian people of Denmark. Their origins date as far back as 100,000 years. Beginning at about the time of Christ (c. 30 A.D.), the Danes were organized into communities that were governed by local chieftains. During the sixth and seventh centuries they participated in the Viking (pirate) raids of England, France, the Netherlands, Belgium, and Luxembourg (these last three are sometimes referred to as the Low Countries). In about 950 Denmark was united by King Harold II (also called Harald Bluetooth; c. 910–c. 985), who 10 years later was converted to Christianity and thereafter fostered its spread. The Vikings gained control of England in 1016, and the Dane warrior Canute became king of England (and of Denmark). The Danes ruled England until 1042.

Who were **the Normans**?

The Normans, like the Danes, originated in Scandinavia and were Vikings. In the mid-800s they invaded northern France, ousting the Franks who had conquered the region 400 years earlier. There they stayed, in the region that came to be known as Normandy (or Normandie). In 1066 the Norman duke William (the Conqueror; 1027–1087) sailed across the channel and claimed the English throne, uniting Normandy with the English kingdom. This arrangement lasted until 1204 when French King Philip Augustus (1165–1223) reclaimed the territory. England took it back during the fifteenth century, but in 1450 Normandy was permanently restored to France, becoming a province. Norman descendants still live there today, and their influence is also evident across the English Channel, where the Norman nobility intermixed with the Anglo-Saxons.

William the Conqueror in battle. In 1066 the Norman duke sailed across the English Channel and claimed the throne of England.

What were **lords**?

Lords (or seigniors) were wealthy landowners during the Middle Ages (500–1350). By about the ninth century, much of western Europe was divided into huge estates, called manors. These were self-sufficient estates that were held by a lord (members of the clergy could also be lords). The lord would lease land to peasants who would farm it; in return, the peasants would pay the lord in taxes, in services, or in kind (with crops or goods). In addition to farmland, a manor would typically have meadow, woodland, and a small village. The lord presided over the entire manor and all the people living there. As the administrator of the land, he collected taxes and presided over legal matters. But the manors were not military entities; in other words, the lord did not promise protection to the peasants living on his land. As such, the manors were purely socioeconomic (as opposed to fiefs, which were social, economical, and political units).

What were **fiefs**?

A fief was an estate that was owned, governed, and protected by a lord. A fief consisted of several manors (each of which might have had its own lord) and their villages, along with all buildings on the land as well as the peasants (serfs) who worked the land, served at court, or took up arms on behalf of the lord. The lord of the fief, called a feudal lord, would secure the allegiance of the manorial lords (sometimes called seigniors), who would in turn secure the allegiance of the peasants. In short, land was exchanged for loyalty; this was feudalism, the political and economic system of the Middle Ages (500–1350). The word *feud* is of Germanic origin and means "fee"; in repayment for the land they lived on and for the protection they received from the lord, serfs were expected to pay the lord fees—in the form of money (taxes), services, or goods.

The feudal system arose in the seventh century. It was suspended during the Carolingian Empire, which began in 751 when a series of powerful kings (including Charlemagne, or Charles the Great, 742–814) united much of western Europe. But after Charlemagne died in 814, his grandsons fought each other and later divided the vast kingdom among themselves. Each of their territories later came under attack, dissolving Carolingian rule. By the ninth century feudalism had replaced the Carolingian Empire as the political and economic entity governing medieval life. Feudalism lasted into the fifteenth century.

What were **serfs**?

Serfs were the peasants who lived on either a manor or a fief, the two organizing entities of the Middle Ages (500–1350). They performed labor and were bound to the lord of the property where they lived and worked. A serf was somewhere between a free person and a slave: Though the word serf is derived from the Latin word *servus,* meaning "slave," the serfs had certain rights, which were governed not by law but by custom. One such custom was that a serf who could escape his lord for one year and a day was then considered free. A free peasant was called a *villein,* a village commoner.

What were **vassals**?

In the Middle Ages (500–1350) a vassal was anyone who was under the protection of another and therefore owed and avowed not only allegiance but a payment of some sort to their protector. Peasants (serfs and village commoners) were always vassals to a lord—whether it was the lord of the manor or the lord of the fief. But the lord of the manor was himself a vassal—to the lord of the fief. As kingdoms were created, with many fiefs within their jurisdiction, the feudal lords became the vassals of the kings.

What were **knights**?

In the Middle Ages (500–1350) knights were armed and mounted warriors who were also landholders; in other words, they were noblemen who took up arms. A knight might have been the lord of a manor who vowed to fight for the feudal lord (the lord of the fief where the manor was situated). During the Carolingian Age (during the eighth and ninth centuries), when a monarchy was established in western Europe, a feudal lord might also have been a knight in service to the king. In times of war, any man who pledged loyalty to his lord and took up arms on his behalf would become a knight, who later might receive lands from the lord or king as repayment. During the Crusades (1096–1291) knights were made of men who were not landowners—they were instead designated by primogeniture (the eldest son would bear the honor of becoming a knight). Knighthood was traditionally conferred by a blow on the shoulder with the flat side of a sword. Feudal knighthood ended by the sixteenth century. Today, the only vestige of this tradition is found in Britain, where knighthood is an honorific designation conferred by the king or queen on a noble or commoner for extraordinary achievement.

What was a **papal state**?

A papal state was a manor or fief where the lord was a member of the clergy. In 754 Carolingian king Pepin the Short (c. 714–768) granted extensive lands to the pope. These territories, which included much of what is now the Mediterranean coast of France as well as most of central Italy, were organized by the Roman Catholic Church into states. These states played an important role in medieval life. Like the fiefs, the papal states collected taxes and maintained courts of law. Also, like their secular counterparts, they were prone to war and invasion. Thus members of the Roman Catholic Church were temporal as well as spiritual leaders during the Middle Ages (500–1350). The last of the territories held by the church was in central Italy and included Rome. After 1871 these lands were claimed by Italy. The resulting land dispute (sometimes referred to as the Roman Question) was settled by the Lateran Treaty (1929), which created the sovereign state known as Vatican City.

What was the **church's role during the Middle Ages**?

Though the Roman Catholic Church became increasingly involved in secular concerns during the Middle Ages (500–1350), it played a much larger part in medieval European life. Missionaries converted many of the Germanic tribes; thus, the church was influential in civilizing these so-called barbarians. Further, churches throughout Europe housed travelers and served as hospitals for the sick. Monasteries and cathedrals became centers of learning.

What was the **Investiture Struggle**?

Also called the Investiture Controversy, it is the name for the power struggle between kings and popes during the Middle Ages (500–1350). Since the papal states played the same role in medieval society as the other states (fiefs and manors, which were held by kings), their lords, who were members of the clergy, eventually became subject to the same human weaknesses that guided the feudal lords and kings—namely, corruption and greed. Popes became powerful and worldly leaders. The struggle for supremacy peaked in 1075 when Pope Gregory VII (c. 1020–1085), who was trying to protect the church from the influence of Europe's powerful leaders, issued a decree against lay investitures, meaning that no one except the pope could name bishops or heads of monasteries. German King Henry IV (1050–1106), who was engaged in a power struggle with Saxon nobles at the time, took exception to Gregory's decree and challenged it, asserting that the kings should have the right to name the bishops. (This was an important point of disagreement, since kings wanted to be in the favor of the pope, and popes were selected from among the bishops. So, it was not purely a religious issue; political power was also at stake.) Henry was excommunicated by the pope. Though he later sought—and was granted—forgiveness by Gregory, the struggle did not end there. Henry soon regained political support, deposed Gregory (in 1084), and set up an antipope (Clement III), who, in turn, crowned him Holy Roman Emperor. The debate over whose right it was to invest clergymen with the symbol of office continued through much of the Middle Ages.

What was the **Holy Roman Empire**?

The Holy Roman Empire (sometimes abbreviated H.R.E.) was a loose federation of German and Italian states, originally formed on Christmas of A.D. 800, when Charlemagne, or Charles the Great (742–814), was crowned emperor of the Romans by Pope Leo III (c. 750–816). But after Charlemagne's death, the empire lapsed and was not fully reinstated until Otto I (912–973) was crowned emperor. Though the empire was strongly associated with the Roman Catholic Church, disputes between emperors and popes began in the mid-1100s. In 1250 Pope Innocent IV (d. 1254) was successful in gaining independence from the empire for the Italian city-states.

Later in the history of the Holy Roman Empire, the House of Habsburg rose to power. But after the 1648 signing of the Peace of Westphalia, which recognized the sovereignty of all the states of the empire, the title of Holy Roman Emperor was for the most part an honorific one. With the exception of a five-year period (1740–45), the family continued to hold power until 1806, when Emperor Francis II (1768–1835) declared the end of the Holy Roman Empire.

THE RENAISSANCE: 1350 TO c. 1600

Why is **the Renaissance** considered **a time of rebirth**?

The term *renaissance* is from the French word for "rebirth," and the period from A.D. 1350 to 1600 in Europe was marked by the resurrection of classical Greek and Roman ideals; the flourishing of art, literature, and philosophy; and the beginning of modern science.

Italians in particular believed themselves to be the true heirs to Roman achievement. For this reason, it was natural that the Renaissance began in Italy, where the ruins of ancient civilization provided a constant reminder of their classical past and where subsequent artistic movements (such as Gothic) had never taken firm hold.

How did the **Renaissance begin**?

Social and political developments in the late Middle Ages gave rise to the spirit of the Renaissance. The Crusades (1096–1291)—the military expeditions undertaken by Christian powers to win the holy land from the Muslims—brought Europeans into contact with other cultures and most importantly with Byzantine civilization. The remnant of the East Roman Empire, Byzantium had preserved the knowledge of ancient times. In addition, many texts thought to have been destroyed during the tribal ransacking of the West Roman Empire (in the fifth century A.D.) remained preserved in various translations throughout the Middle East. So it was during the Crusades that some of these were brought back to Europe, where classical scholars undertook the task of deciphering the West's cultural past.

In northern Italy, a series of city-states developed independent of the larger empires to the north and south of them. These small states—Florence, Rome, Venice, and Milan, among others—gained prosperity through trade and banking, and as a result, a wealthy class of businessmen emerged. These community leaders admired and encouraged creativity, patronizing artists who might glorify their commercial achievement with great buildings, paintings, and sculptures. The most influential patrons of the arts were the Medicis, a wealthy banking family in Florence. Members of the Medici family supported many important artists, including Botticelli and Michelangelo. Guided by the Medici patronage, Florence became the most magnificent city of the period.

One way patrons encouraged art was to sponsor competitions in order to spur artists to more significant achievement. In many cases, the losers of these contests went on to greater fame than the winners. After his defeat in the competition to create the bronze doors of the Baptistery of Florence Cathedral, architect Filippo Brunelleschi (1377–1446) made several trips to take measurements of the ruined buildings of ancient Rome. When he returned to Florence, he created the immense *il duomo* (dome) of the Santa Maria del Fiore Cathedral, a classically influenced structure that became the first great monument of the Renaissance.

How is **the attitude of the Renaissance** characterized?

The artists and thinkers of the Renaissance, like the ancient Greeks and Romans, valued earthly life, glorified man's nature, and celebrated individual achievement. These new attitudes combined to form a new spirit of optimism, the belief that man was capable of accomplishing great things.

This outlook was the result of the activities of the wealthy mercantile class in northern Italy, who, aside from supporting the arts and letters, also began collecting the classical texts that had been forgotten during the Middle Ages (500–1350). Ancient manuscripts were taken to libraries, where scholars from around Europe could study them. The rediscovery of classical texts prompted a new way of looking at the world. During the Middle Ages, scholars argued that the meaning of life on Earth lay primarily in its relation to an afterlife. Therefore, they believed that art for its own sake had no value, and they even frowned on the recognition of individual talent. (For this reason, many of the great works of the Middle Ages were created anonymously or were hidden from public view.)

In contrast, Renaissance artists and thinkers studied classical works for the purpose of imitating them. As an expression of their new optimism, Renaissance scholars embraced the study of classical subjects that addressed human, rather than scientific, concerns. These "humanities," as they came to be called, included language and literature, art, history, rhetoric, and philosophy. Above all, humanists, those who espoused the values of this type of education, believed in man's potential to become well versed in many areas. Thus, a "Renaissance man" is anyone whose talents span a variety of disciplines. During the Renaissance, people in all disciplines began using critical skills as a means of understanding everything from nature to politics.

Which artists and thinkers are considered **the greatest minds of the Renaissance**?

The great writers of the Renaissance include the Italian poet Petrarch (1304–1374), who became the first great writer of the Renaissance and was one of the first proponents of the concept that a "rebirth" was in progress; Florentine historian Niccolò Machiavelli (1469–1527), who wrote the highly influential work *The Prince* (1513); English dramatist and poet William Shakespeare (1564–1616), whose works many view as the culmination of Renaissance writing; Spain's Miguel de Cervantes (1547–1616), who penned *Don Quixote* (1605), the epic masterpiece that gave birth to the modern novel; and Frenchman François Rabelais (c. 1483–1553), who is best known for writing the five-volume novel *Gargantua and Pantagruel*.

The great artists of the Renaissance include the Italian painters/sculptors Sandro Botticelli (1445–1510), whose works include *The Birth of Venus*; Leonardo da Vinci (1452–1519), whose *Mona Lisa* and *The Last Supper* are among the most widely studied works of art; Michelangelo Buonarroti (1475–1564), whose sculpture *David* became the symbol of the new Florence; and Raphael Sanzio (1483–1520), whose *School of Athens* is considered by art historians to be the complete statement of the High Renaissance.

How did the Renaissance spread from Italy to the rest of Europe?

Eventually the ideas born in Italy during the 1300s spread northward, which is at least in part attributable to German inventor Johannes Gutenberg's printing press (c. 1440–1450). Before long, the spirit and ideas that were taking hold in Italy reached France, Germany, England, and the Netherlands, where the Renaissance continued into the 1600s.

One of the most important figures of the northern Renaissance was the Dutch humanist Desiderius Erasmus (c. 1466–1536), whose book *In Praise of Folly* (1509) is a blistering criticism of the clergy, scholars, and philosophers of his day.

Another notable figure of the northern Renaissance was Englishman Sir Thomas More (1478–1535), who was a statesman and adviser to the king. More's *Utopia,* published in 1516, criticizes the times by envisioning an ideal society in which land is communally held, men and women alike are educated, police are unnecessary, politicians are honest, and where there is religious tolerance.

The works of Flemish artist Jan van Eyck (1395–1441), including his groundbreaking portrait *Man in a Red Turban* (1433), demonstrate that the principles of the Renaissance were felt as strongly in northern Europe as they were in Italy.

When did **European colonialism** begin?

Seeking out colonies for economic benefit dates back to ancient times: Even the Romans ruled colonies in Europe, the Middle East, and Africa. But colonialism took hold during the Renaissance, between 1400 and 1600, when powerful European countries sent explorers to find new lands and forge new trade routes.

Portugal and Spain sought sea routes to India and the Far East. In the process, the Portuguese gained control of what is now Brazil; they also established trading posts in West Africa, India, and Southeast Asia. Spain gained control of most of Latin America and vast regions of what is now the United States. However, the Dutch, English, and French also had influence in these areas. And of course, the English eventually established thirteen colonies in North America. The English also became a strong influence in India and Africa, while the Dutch gained control of the Indonesian islands (which became the Dutch East Indies).

The results of colonialism were many. While trade was expanded and there was an enormous exchange of raw goods, the colonies were also rife with conflict: Indigenous peoples were killed or forcibly displaced from their lands, and foreign powers fought with each other for control of the same areas (the British and French fought four wars in North America between 1689 and 1763 alone). Further, Europeans brought to these new lands their own languages, religions, and systems of government, imposing their culture, beliefs, and ideologies on native peoples.

THE ENLIGHTENMENT AND THE SCIENTIFIC REVOLUTION: 1600s TO 1700s

What was **the Enlightenment**?

The Enlightenment, which is also referred to as the Age of Reason (or alternately as the Age of Rationalism), was a period when European philosophers emphasized the use of reason as the best method for learning the truth. Beginning in the 1600s and lasting through the 1700s, philosophers such as Jean-Jacques Rousseau (1712–1778), Voltaire (1694–1778), and John Locke (1632–1704) explored issues in education, law, and politics. They published their thoughts, issuing attacks on social injustice, religious superstition, and ignorance. Their ideas fanned the fires of the American and French revolutions in the late 1700s.

Hallmarks of the Age of Reason include the idea of the universal truth (two plus two always equals four, for example); the belief that nature is vast and complex but well ordered; the belief that humankind possesses the ability to understand the universe; the philosophy of Deism, which holds that God created the world and then left it alone; and the concept of the rational will, which posits that humans make their own choices and plans, and therefore, do not have a fate thrust upon them.

While the Age of Reason proved to be a flurry of intellectual activity that resulted in the publication of several encyclopedias of knowledge, toward the end of the eighteenth century a shift occurred. During this time Europeans began to value passion over reason, giving rise to the romantic movement and ending the Age of Reason. (This change in outlook is evident in English novelist Jane Austen's *Sense and Sensibility.*)

Nevertheless, the philosophies put forth during the Age of Reason were critical to the development of Western thought. The celebration of individual reason during this era was perhaps best expressed by René Descartes (1596–1650), who refused to believe anything unless it could be proved. His statement, "I think, therefore I am," sums up the feelings of skeptical and rational inquiry that characterized intellectual thought during the era.

Are the Enlightenment and **the scientific revolution** the same?

The two terms describe interrelated and sequential European intellectual movements that took place from the 1500s to the 1800s. Together, the movements shaped an era that would lay the foundations of modern western civilization, foundations that required the use of reason, or rational thought, to understand the universe, nature, and human relations. During this period, many of the greatest minds in Europe developed new scientific, mathematical, philosophical, and social theories.

Scientists came to believe that observation and experimentation would allow them to discover the laws of nature. Thus, the scientific method emerged, which required tools. Soon the microscope, thermometer, sextant, slide rule, and other instruments were invented. Scientists working during this time included Sir Isaac Newton (1642–1727), Joseph Priestley (1733–1804), and René Descartes (1596–1650). The era witnessed key discoveries and saw rapid advances in astronomy, anatomy, mathematics, and physics. The advances had an impact on education: universities introduced science courses to the curricula, and elementary and secondary schools followed suit. As people became trained in science, new technologies emerged: complicated farm machinery and new equipment for textile manufacturing and transportation was developed, paving the way for the Industrial Revolution.

THE INDUSTRIAL REVOLUTION: 1700s TO 1800s

When did the **Industrial Revolution** begin?

The Industrial Revolution began in Great Britain during the 1700s, and by the early 1800s it had spread to western Europe and the United States. It was brought about by the introduction of steam-power-driven machinery to manufacturing. By the close of the 1800s most finished goods, which had once been made by hand or by simple machines, were produced in quantity by technologically advanced machinery.

What were the **effects of the Industrial Revolution**?

The dawn of the Industrial Revolution spelled the end of home- or workshop-based production. Factories were built to house the new machines, causing a population shift from rural to developing urban areas by the mid-1800s, as people went where the work was. Factory owners turned to child labor, and in the United States, to the steady influx of immigrants to run the machinery in their plants. As industry grew, it required financial institutions that could provide money for expansion, thus giving birth to a new breed of wealthy business leaders— including the extraordinarily prosperous "robber barons," the industrial and financial tycoons of the late nineteenth century.

But as industry evolved, government and policy changes did not keep pace: Serious social, political, and economic problems resulted, including poor and often dan-

During the Industrial Revolution factory owners commonly hired child laborers to work long hours for little pay.

gerous working conditions, exploitation of workers (including child laborers), overcrowded housing, pollution, corruption, industry monopolies, and a widening gap between the rich and the poor. Change was slow to come, but social activism and government reforms in the late 1800s and during the Progressive Era of the early 1900s, much of which centered around trade unions, alleviated these problems. The rapid development of industry caused sweeping social changes: The Western world, which had long been agriculturally based, became an industrial society, where goods and services are the primary focus.

THE TWENTIETH CENTURY

What was the world like **between the World Wars**?

Before World War II began in 1939, World War I (1914–18) was referred to as the Great War, and understandably so given its enormous impact. The price of the conflict was paid in human casualties: more than 10 million soldiers died and another 20 mil-

Relief programs helped ease hunger during the Great Depression: Here, a long line of people wait for their ration of a sandwich and cup of coffee in Times Square, New York City.

lion were wounded. Civilian deaths were equally devastating, resulting from widespread hunger and flu epidemics (this was in the days preceding the advent of penicillin to treat complications from influenza).

During the 1920s, with the death toll of the Great War a recent memory, the civilized world enjoyed a period of relative peace. In the United States, the decade was known as the Roaring Twenties—ten years of prosperity and even frivolity, despite ratification of the Eighteenth Amendment to the U.S. Constitution (1919), which prohibited the sale and consumption of "intoxicating liquors."

At the close of the Great War, the League of Nations had been established to handle disputes among countries and avoid another major conflict. This was followed in 1928 by the Kellogg-Briand Pact (also called the Pact of Paris), in which 15 nations agreed to settle conflicts by diplomacy rather than military might; eventually 62 nations ratified the agreement.

Meanwhile, in Germany an extreme sense of nationalism was taking hold: The Treaty of Versailles, which had ended the Great War, seriously weakened the nation, allowing the rise of the Nazi Party, led by Adolf Hitler (1889–1945). Nazis were determined to see their beloved homeland rise to power once again, and they found a ready following among the German people.

In 1929 a general downturn began in the world economy, triggered by the U.S. stock market crash of late October. As the United States fell into the severe and sustained economic crisis known as the Great Depression, other industrialized nations—including Germany—also felt the impact. Unemployment jumped to record levels in many countries, and a lack of social welfare programs resulted in the destitution of numerous families. Politicians and economists alike searched for solutions to the crisis, many turning to anti–free trade or isolationist policies to "protect their own."

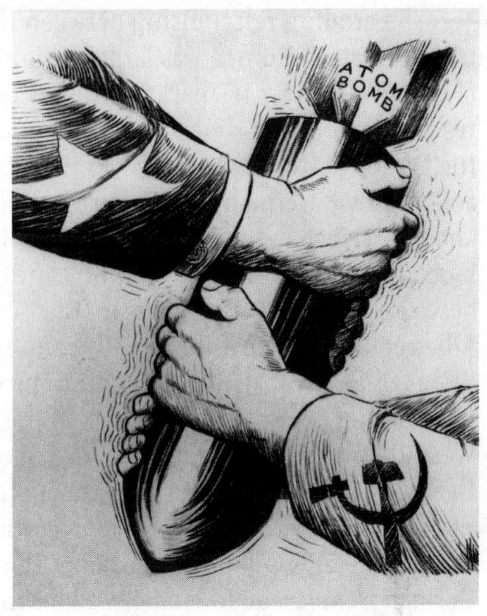

A Cold War-era cartoon, "Handle with Care," illustrating the opposing superpowers of the United States and Soviet Union, each grasping for an atomic bomb.

Meanwhile, postwar efforts to keep peace were proving ineffective: Japan invaded Manchuria (China) in 1931; Italy conquered Ethiopia in 1936; civil war raged in Spain (1936–39); India became the site of a bitter struggle between British rulers and Indian nationalists; and the Sino-Japanese War (which began in 1937 and would be absorbed by the outbreak of World War II) was fought in Asia. A dangerous alliance was forged when Germany, Italy, and Japan formed the Axis powers in 1936. Also, by 1933, Germany had become a totalitarian state known as the Third Reich. In 1938 Nazi armies took their first steps toward gaining supremacy when they marched into Austria and claimed it, setting the stage for World War II (1939–45).

When did the **Cold War** begin?

In the years following the conclusion of World War II (1939–45), the nations of Western Europe and the United States became alarmed by Soviet advances into Eastern Europe, and many Europeans and Americans voiced concerns that Communists, led by the Soviet Union, were plotting to take over the world. Political leaders in England, the United States, and elsewhere referred to this new menace in grim terms. In March 1946 former British prime minister Winston Churchill (1874–1965) warned of an "Iron Curtain" of Soviet totalitarianism that had divided the European continent, and in 1947 U.S. president Harry S. Truman (1884–1972) announced a policy of containment of Communist incursion into other countries. This policy came to be known as the Truman Doctrine, and it remained an integral part of American foreign policy for the next 40 years, ultimately leading to the nation's involvement in the Korean War (1950–53) and the Vietnam War (1954–75).

The eroding relationship between the Western powers and the Soviet-led countries of Eastern Europe was largely brought on by disagreements over Germany. At the close of World War II, marked differences of opinion on what to do with Germany had resulted in a plan for joint government of the nation by the Allies—the Soviet Union, the United States, Britain, and France. But the arrangement quickly proved unworkable. By 1948 Germany was in serious economic straits, and the United States, Britain, and France began to discuss uniting their zones. The Soviets responded by ordering a blockade of land and water traffic into Berlin, control of which had been divided between the Allies after the war (the Soviets controlled East Berlin, while the other Allies controlled West Berlin). To counter the blockade, Great Britain and the United States ordered an airlift operation to provide food and other supplies to the people of West Berlin, alleviating the effects of the 11-month Soviet blockade. In 1949 the East-West differences resulted in the formal division of Germany into two countries: West Germany, formed by the zones occupied by the United States, Great Britain, and France, was allowed to form a democratic government, and it became officially known as the Federal Republic of Germany. The same year, East Germany (also known as the German Democratic Republic) was formed out of the Soviet zones and was folded into the "Eastern bloc" countries.

By 1949, the year that the Soviet Union exploded its first nuclear bomb, the world had been roughly divided into two camps: the United States and its democratic allies, which included the nations of Western Europe and other anticommunist governments; and the Eastern bloc, the Soviet Union and its satellite countries. These camps were soon girded by formal political alliances. In 1949 a military alliance known as the North Atlantic Treaty Organization (NATO) was formed by 12 nations (the United States, Great Britain, France, Italy, Norway, Portugal, Iceland, Denmark, Canada, Belgium, Luxembourg, and the Netherlands). By 1955 three more countries—Greece, Turkey, and West Germany—joined the alliance. The Soviet Union responded by creating the Council for Mutual Economic Assistance (COMECON) in 1949 and the Warsaw Pact in 1955. COMECON was an effort to coordinate economic and industrial activities among Communist nations, while the Warsaw Pact was a military agreement between the Soviets and the Communist governments of Eastern Europe. The Cold War was on.

What were the **hallmarks of the Cold War era**?

At home, the hysteria of the Cold War era reached its height with the so-called McCarthyism of the 1950s; historian Doris Kearns Goodwin described it as "one of the most destructive chapters in American political history." In early 1950 Republican senator Joseph McCarthy (1908–1957) of Wisconsin claimed to possess a list of more than two hundred known communists in the U.S. State Department. The startling accusation launched congressional inquiries conducted by the senator's subcommittee and the House Committee to Investigate Un-American Activities. Suspicions of communist subversion ran high—even in Hollywood, where a "blacklist" named those who were believed to have been involved in the Communist Party. McCarthy never produced his laundry list of offenders in the State Department, and the sorry chapter was closed when, on live television, the senator's bitter attacks went too far: In televised hearings in 1954, the senator took on the U.S. Army, determined to ferret out what he believed was a conspiracy to cover up a known communist in the ranks. Faced with McCarthy's slanderous line of questioning, Army counsel Joseph Welch (1890–1960) delivered a reply that finally disarmed McCarthy, saying "Have you no sense of decency, sir? If there is a God in heaven, your attacks will do neither you nor your cause any good." The retort was met with applause in the courtroom, heralding the end of the Communist-in-our-midst hysteria.

In the 1950s the hysteria of the Cold War reached its height when Wisconsin senator Joseph McCarthy (foreground) made aggressive accusations of communist infiltration of the U.S. government.

But the Cold War deepened during the course of the 1950s, as distrust on both sides was increased by the shadow of possible nuclear destruction. Both the United States and the Soviet Union funneled vast resources into the development of weapons systems, as each side believed deterrence would determine the victor in the Cold War: It would be won by the nation able to create weapons so powerful that the other nation would be deterred from attacking. The military build-up became an all-out arms race.

Competition between the Eastern bloc and the West spilled over into athletics, the arts, and the sciences. In 1957 the Soviets beat the West into space with the launch of the first artificial satellite, Sputnik, which they followed in 1961 by completing the first successful manned space launch. The United States responded by stepping up its space program and vowing to put a man on the moon.

Events in the early 1960s heightened tensions between the two sides, causing many to fear the war would turn hot: When an American U-2 spy plane was shot down over the Soviet Union and its pilot was captured in 1960, the United States was forced to admit to conducting a program of aerial reconnaissance; in 1961, the U.S.-backed invasion of Cuba, known as the Bay of Pigs, failed, revealing American involvement in the plot; also that year, the Berlin Wall was built to stop the flow of emigrants out of East Germany, and becoming a visible symbol of the division between East and West; and in 1962 U.S. president John Kennedy (1917–1963) and Soviet premier Nikita Khrushchev (1894–1971) squared off in the Cuban Missile Crisis—a full-scale conflict was averted through diplomacy.

Later in the decade and into the 1970s, tensions relaxed and both sides began agreeing to limit the arms race, signing the Nuclear Non-Proliferation Treaty in 1968 and the Strategic Arms Limitation Treaty (SALT I) in 1972, and agreeing to the Helsinki Accords in 1975. But East and West remained suspicious and watchful of each other into the 1980s. Most observers agree the Cold War did not come to an end until the fall of communism in Eastern Europe (c. 1990).

What led to **the decline of communism** in Eastern Europe?

Anticommunist sentiment among Eastern Europeans was bolstered by the actions and policies of Soviet leader Mikhail Gorbachev (1931–). When Gorbachev took office in 1985, the Soviet economy was in decline. In order to reverse the trend, he advocated dramatic reforms to move the economy away from the government-controlled (communist) system and toward a decentralized system, similar to those of Western democracies. Gorbachev's efforts to modernize the Soviet Union were not limited to the economy; he further proposed a reduction in the power of the Communist Party, which had controlled the country since 1917. Gorbachev's programs for reform were termed *perestroika* (meaning "restructuring"). In the meantime, Gorbachev opened up relations with the West, which included visits with U.S. president Ronald Reagan (1911–2004), who strongly supported the Soviet leader's programs. Gorbachev referred to his policy of openness as *glasnost*. Both Russian terms quickly caught on around the world. While the economic reforms produced a slow and painful change for the Soviet people and Gorbachev had many detractors (including government officials), he also had many supporters—both inside and outside the Soviet Union.

People in other Eastern European countries watched with interest the Soviet move toward a more democratic system. Strikes in Poland had begun as early as 1980, where workers formed a free labor union called Solidarity. But the following year, the Communist leaders of the Soviet Union pressured the Polish government to put an end to the movement—which it did. After Gorbachev became head of the Soviet Union and initiated sweeping changes, the reform movements in other countries soon realized that the Soviets under Gorbachev would no longer take hard-handed tactics toward anticommunist efforts in other countries. In 1989 the Polish government ceased to prohibit Solidarity, and the Communist party there lost influence. The same was true in Hungary, East Germany, and Czechoslovakia. By the end of the decade, most of the Eastern European Communist governments were overthrown in favor of democratic-oriented governments. The transition was effected differently in each country: the "overthrow" in Czechoslovakia was so peaceful that it was called the Velvet Revolution; while in Romania, a bloody revolt ensued, and hard-line communist dictator Nicolae Ceausescu (1918–1989) was executed.

In 1990 multiparty elections were held in Romania, Czechoslovakia, Hungary, East Germany, and Bulgaria. The noncommunist party that was put in power in East

East German border guards watch as demonstrators pull down a segment of the Berlin Wall. The wall was removed in 1990, and East Germany reunited with West Germany.

Germany agreed to unification with West Germany, again creating one Germany on October 3, 1990. That same year Gorbachev received the Nobel peace prize for his contributions to world peace.

When was the **Berlin Wall dismantled**?

The barrier wall surrounding West Berlin began coming down in November 1989, as a wave of democratization swept Europe.

The concrete, electrically fortified wall was first built in 1961 as a barbed wire and cinder block structure. Communist East German leader Walter Ulbricht (1893–1973) convinced Soviet premier Nikita Khrushchev (1894–1971) that the wall was needed to prevent people from fleeing communist Eastern Europe. (Before the wall was erected, an estimated 2.5 million people had fled to the free world through West Berlin; after its completion, perhaps 5,000 managed to escape. Hundreds died trying.) When the wall was complete, it had an average height of 12 feet and ran more than 100 miles, along which there were posts where armed East German guards stood sentinel, preventing their countrymen from escaping to the West. The wall completely surrounded West Berlin and divided the German capital between East and West, communism and the free world.

The wall was a symbol of communism's oppression and of the Cold War. On June 26, 1963, President John F. Kennedy delivered his memorable "I am a Berliner" speech in its shadows, saying, "There are some who say communism is the wave of the future...Let them come to Berlin." He went on to say that the wall was "a vivid demonstration of the failure of the communist system," and that though democracy is not perfect, democratic nations had "never had to put up a wall to keep our people in." On June 12, 1987, President Ronald Reagan (1911–2004) addressed West Berliners at the wall's Brandenburg Gate; his now-famous speech was audible on the East Berlin side of the wall as well. There, Reagan issued a challenge to Soviet leader Mikhail Gorbachev (1931–), saying "if you seek peace, if you seek prosperity for the Soviet Union and Eastern Europe, if you seek liberalization...[c]ome here to this wall...Mr. Gorbachev, tear down this wall!"

East Germany's communist government was finally toppled in October 1989. On November 9 restrictions between the two Berlins were lifted, and the wall was opened. The resulting celebration brought the wall down, with gleeful Berliners chipping away at the barrier; it was gradually dismantled. By 2005 only a few sections of the wall and some watch towers still existed—the capital no longer divided, the country a unified, democratic Germany.

How did the **Soviet Union** fall apart?

Soviet leader Mikhail Gorbachev's (1931–) liberal reforms met with the opposition of conservative communist officials who were angered by the hardships produced by the transition to a free market economy and dissatisfied with the Soviet Union's loss of influence over neighboring countries, where communism had fallen by 1990. On August 19, 1991, communists attempted to overthrow Gorbachev as president of the Soviet Union. Though the effort failed in the face of widespread public opposition, it nevertheless weakened Gorbachev's leadership. Soon the 15 Soviet republics declared independence but indicated their willingness to become part of a loose confederation of former Soviet republics. Though Gorbachev tried to prevent the complete dissolution of the Soviet Union, on December 8, 1991, the republics of Russia, Ukraine, and Belorussia (Belarus) broke away completely from the Soviet Union and formed the Commonwealth of Independent States. All the remaining republics, except Georgia, followed suit. On December 25, Gorbachev resigned as president, and the Soviet Union ceased to exist.

With continuing reforms in Eastern Europe and the end of the Soviet Union, by 1992 the only communist-dominated countries that remained were, with the exception of Cuba, Asian nations: China, Laos, North Korea, and Vietnam.

EARLY TWENTY-FIRST CENTURY

How is the **current era characterized**?

Ask the question of most any observer, and the answer would include catchphrases like "global marketplace," "global village," or "globalization." Modern communications and transportation connect people as never before—businesses enjoy broader markets for their goods and services, manufacturing facilities and jobs are located far from the offices of the companies that market them, and people of many nationalities, races, and religions have more and more contact with one another every day, for business and pleasure. Some observers worry that this contact will blur rich cultural differences, diluting diversity; others say that globalization will bring tolerance and increase understanding. Whatever the case, we are living our lives on a global stage: The things we buy and use are as likely to carry "Made in China" or "Made in Mexico" labels as they are any other; people continents away talk to each other not just over the telephone, but via cell phones, instant messages, e-mail, and Internet chat rooms; we have an international forum—the Internet—for buying, selling, and publishing; and we can get almost anywhere in the world with all due haste. Everything travels faster today, including ideas.

The upsides are many. Modern communications and transportation made it possible, for example, for the world to mobilize aid to victims of the Southeast Asian tsunami of December 2004. But there are downsides as well. Critics say globalization is fueling the exploitation of workers in developing nations, contributing to a modern slave trade, rapidly depleting resources, and wiping out environmental diversity. These are some of the reasons protesters have demonstrated outside meetings of the World Trade Organization, why some people opposed the North American Free Trade Agreement (NAFTA) and oppose the pending CAFTA (Central American Free Trade Agreement), and why many people dislike that American popular culture is marketed around the world. The problem could be distilled to this: diversity versus homogeneity. And the resulting culture clashes are not restricted to the realm of scholarly thought; they are making headline news. The enemy is no longer the strong-armed, nuclear-fortified, absolutist government of the Cold War era (though, as of mid-2005, North Korea remained a serious concern); the enemy, as the U.S. State Department reminds us, is any group of individuals with extreme views.

Who is the enemy of the day? Terrorists. News stories reinforce that this catch-all term does not only include the 9/11 hijackers and their al Qaeda associates, the Madrid and London train bombers, and suicide bombers in Gaza, but antigovernment extremists who blew up a federal building in Oklahoma City, a still-unknown distributor of deadly anthrax, and animal rights activists who routinely damage property to make a point. Terrorism is a decades- (some might argue, centuries-) old problem. Remember Birmingham, Alabama's Sixteenth Street Baptist Church, bombed in 1963 by white supremacists during Sunday services. Remember the 1972 summer Olympics in Munich, when the Arab terrorist group Black September killed 11 Israeli athletes held captive in the Olympic village. Remember the October 1985 hijacking of the *Achille Lauro* cruise ship by members of the Palestinian Liberation Front, who killed a wheelchair-bound American Jew. Remember the December 1988 bombing of Pan Am Flight 103, which came down in Lockerbie, Scotland, claiming 270 lives. The list is long and growing.

The shocking events of September 11, 2001, centered the world's attention on the problem of well-financed terrorist networks that can turn the freedoms of democracy against civilians—and who are feared to have in their possession or access to powerful weapons of mass destruction (WMD). The events of that day launched what the U.S. government calls the Global War on Terrorism, a long battle that even the president acknowledged may not be winnable in a traditional sense. This is a new kind of war, fought not against a nation, but against anarchists halfway around the world, and in our midst. Globalization and terrorism are the twin concerns of the modern era.

RELIGION

How are **religions classified**?

Religion, which is a system of beliefs that usually centers on whatever is beyond the known or the natural, is commonly divided between elementary forms and higher religions.

Some elementary forms, or "traditional religions," are animism (the belief in spirits in nature or in natural objects), ancestor worship (revering the spirits of the dead), and totemism (belief in a mystical relationship between a group of people and an emblem). Animists might believe in spirits living in the sea or in the mountains. Believers in ancestor worship will both honor and fear the spirits of dead family members—for if they are neglected by the living, these spirits are believed to be able to bring harm to their descendants. Within clans, some tribal peoples adopt totems, such as a lion or a turtle. Totemists will be careful throughout their lives not to harm the animal or object that serves as their clan's emblem.

Higher religions are those that embody a concept of transcendence. Higher religions are classified as polytheistic (believing in many gods), dualistic (believing in equally powerful gods of good and evil), monotheistic (believing in one god), and pantheistic (believing in god as the forces and laws of the universe, or the worship of all gods). Religions are further classified as revealed or nonrevealed. Revealed religions are those that followers know through divine agency; both Christianity and Islam are revealed religions. Nonrevealed religions, such as Buddhism, Brahmanism, or Taoism, are known only through inquiry.

What did **ancient civilizations believe**?

In the absence of scientific knowledge, ancient civilizations—including the Greeks, the Romans, the Egyptians, the Aztecs, and the Mayas—created mythologies to

Why is mythology so widely studied?

The mythologies of various peoples are studied for their religious meanings, for their similarities to each other, and as a way of understanding the culture that originated them. Through the centuries, scholars have arrived at various conclusions about mythology. Scottish anthropologist Sir James Frazer (1854–1941) studied folklore and religion and found parallel beliefs between the systems of primitive cultures and Christianity. He published his theories in the highly influential work titled *The Golden Bough* (1890), proposing that all myths center on the cycles of nature and birth, death, and resurrection. Polish-born English anthropologist Bronislaw Malinowski (1884–1942) asserted that myths are nothing more than the validation of accepted social behavior and patterns within a culture. Swiss psychiatrist Carl Jung (1875–1961) posited that all cultures have unconsciously formed the same mythic symbols or motifs (called archetypes). To many theologians, mythologies are viewed simply as corruptions of the Bible.

Whatever their meaning or however they are interpreted, myths have figured prominently in literature—from the fifth-century B.C. works of Greek tragedian Aeschylus to the twentieth-century works of poets T. S. Eliot and Wallace Stevens.

explain origination (how they came into existence as a people); the existence of good and evil; the natural cycle of the seasons; weather; and the motions of the sun, the moon, and the stars. Such natural phenomena were explained by a body of stories that centered around gods, goddesses, and heroes.

For example, the Romans, who largely adapted Greek mythology, believed that gods and goddesses had power not only over agriculture, but also over all aspects of life. They worshiped Ceres as the goddess of the harvest, Vesta as the goddess of the hearth and home, and Jupiter (who later became their supreme god and protector) as the god of the weather. These gods are traced to the Greek beliefs in Demeter as the goddess of the harvest and of fertility, Hestia as the goddess of the hearth (she symbolized security and happiness), and Zeus—the supreme Greek god who was believed to rule from his court on Mount Olympus, and was a symbol for power, rule, and law. The ancient Greeks were also the source of what is perhaps one of the most well known myths of western civilization: Pandora's box. They believed that all that is bad or evil was once enclosed in a box, which was opened by Pandora (who, according to Greek mythology, was the first woman on Earth), releasing evil into the world. Greek mythology was preserved in the works of the poets Homer (c. 850–? B.C.) and Hesiod (c. 800 B.C.).

Other frequently studied mythologies include the Vedic (Indian), Egyptian, and Mesopotamian.

How old is **Buddhism**?

One of the great Asian religions, Buddhism dates back to ancient times: It was founded in India in the sixth and fifth centuries B.C. by Siddhartha Guatama (c. 563–c. 483 B.C.), who became known as Buddha, or Enlightened One. Born in what is today Nepal, Siddhartha's father is described in stories as a king or a warrior prince, and the family lived in luxury. When he was 29 years old, Siddhartha had a series of four visions that prompted him to leave his wife, young son, and the palace and venture out in search of spiritual enlightenment and truth. He wandered for six years, traveling to the ancient kingdom of Magadha (in present-day India). During this time he led a life of extreme austerity and even self-torture. He finally decided that his ascetic life would not lead him to truth, and he abandoned his practice of self-denial. One day, when he was 35 years old, Siddhartha went to meditate under a banyan (shade) tree, also called a bodhi tree. There he claimed to achieve enlightenment. Thereafter, Buddha traveled through the Ganges River valley, teaching meditation and adherence to moral conduct as the way to enlightenment. He established a community of monks to continue his work.

Buddhism is the world's fourth-largest religion (after Christianity, Islam, and Hinduism). Most of the estimated 360 million Buddhists in the world today live in Asia (Sri Lanka, Southeast Asia, and Japan, predominately).

What are the basic **beliefs of Buddhism**?

Buddha taught the Four Noble Truths: existence is suffering; the cause of suffering is desire; through total transcendence (called nirvana) one may suspend suffering; and to end suffering one must follow a certain path—called the Eightfold Noble Path. The path prescribes moral conduct, specifically: 1) knowing the truth; 2) resolving to resist evil; 3) using speech properly—so as not to hurt others; 4) demonstrating respect for life, morality, and property through ones actions; 5) working in a job that does not harm others; 6) making an effort to free one's mind of evil; 7) mindfulness (or controlling one's thoughts and feelings); and 8) concentration. In some places, the beliefs and practices of Buddhism are combined with those of Hinduism and the Shinto religion.

Who is the **Dalai Lama**?

The Dalai Lama is the spiritual leader of Lamaism, a Buddhist sect in Tibet. The title, first bestowed on a Lamaist leader in 1578, translates as "teacher whose wisdom is great as the ocean." The title has been bestowed 13 times since: Followers believe that the Dalai Lama is continually reborn. Monks go in search of a young boy who was born at about the same time as the Dalai Lama died. They seek other signs of continuity as well before identifying the "new" Dalai Lama, who the monks educate and train to assume the leadership role.

The Dalai Lama, pictured in December 2004, was awarded the Nobel peace prize in 1989 for his nonviolent efforts to free Tibet from China.

What is the **controversy over Tibet**?

During the seventh and eighth centuries A.D., Tibet emerged as a powerful Buddhist kingdom. But beginning in the thirteenth century, it came under the rule of the Mongols, and in the eighteenth century the Manchu dynasty took control of the area. Beginning in 1912 Tibet again knew a brief independence. But in 1950 the country was invaded by Communist China. After years of turmoil, which included an anti-Chinese rebellion in 1959, Tibet was made an autonomous region within Communist China in 1965. But the autonomy is nominal, and Buddhist culture in Tibet has been all but destroyed.

Buddhists fled Tibet, some of them taking refuge in India—which is where the current Dalai Lama (the fourteenth) leads a nonviolent struggle to free Tibet from China. In 1989 the Dalai Lama was awarded the Nobel peace prize for his efforts.

What is **Zen**?

Also called Zen Buddhism, it is a sect of Buddhism practiced predominately in Japan and China. The religion is based on meditation rather than on the strict moral doctrine of Buddhism. It was founded in China in the fifth century A.D. by Bodhidharma, an Indian Buddhist monk and missionary. He taught that sudden enlightenment can be achieved through the practice of meditation (or "wall-gazing"). The religion defines enlightenment as the direct seeing of one's original nature.

What is **karma**?

Karma is a belief shared by several Asian religions including Buddhism and Hinduism. The basic concept is that the position one holds in this life is a result of one's actions and conduct in previous lives or incarnations. Therefore, actions and thoughts in this life can influence one's future destiny. The goal of many Eastern religions is to be freed from the cycle of karma by following certain religious practices. The word originates from the Sanskrit *karma,* meaning "work" or "fate."

What is **Hinduism**?

Hinduism, an ancient religion that originated about 1500 B.C. and developed over thousands of years, encompasses the beliefs and practices of the numerous religious sects of India. Each sect has its own philosophy and form of worship. It is a polytheistic religion—believing in many gods as well as in one great god, or universal spirit, called Brahman. Hindu doctrine is centered on sacred scriptures including the Veda, which includes the Upanishads (dialogues describing sacrificial rituals and religious ceremonies), the Puranas (stories about Hindu gods and goddesses), and the Mahabharata, which contains the Bhagavad Gita (or "the Song of the Lord," in which the god Krishna discusses the meaning of life with a warrior-prince on the eve of battle). Hindus worship alone at temples, which are dedicated to divinities. Followers also practice yoga, believing that it leads to knowledge and union with God.

What is **Shinto**?

Shinto is the dominant religion of Japan. Its traditions call for the reverence of ancestors, prayer, and the observance of rituals. It is polytheistic, believing in many gods (kami), who are thought to be the forces behind nature as well as behind human conditions such as sickness, healing, and creativity. Followers of the Shinto religion believe these spirits are housed in shrines. Private shrines are erected in homes while public shrines can be highly elaborate, including multiple buildings as well as gardens. The latter are the goals of many religious pilgrimages; pilgrims pray and make offerings (of money and flowers) to the spirits.

Originating in Japan in ancient times, Shinto has an interesting modern history: In 1882 religious organizations were divided into two groups—state shrines and sectarian shrines. State Shinto was controlled by the government, which went so far as to proclaim divine origins for the Japanese emperor. After World War II (1939–45), state Shinto crumbled and Emperor Hirohito (1901–1989) was compelled to renounce his divinity. Sectarian Shinto religion still thrives in Japan today, where it has more than 3 million followers.

When was the **Bible** written?

The Old Testament, or Hebrew Bible, which is called the Book of Books because of its profound influence, was written thousands of years ago. Scholars disagree about exactly when or how the books were written, but it is generally accepted that the earliest parts of the Bible originated during the time of Moses (or Moshe), the Hebrew prophet and leader, about the fourteenth to the thirteenth century B.C. Moses is believed to be the author of the first five books of the Bible, called the Pentateuch or the Law of Moses. The books include Genesis, Exodus, Leviticus, Numbers, and Deuteronomy. These are believed to have been revealed to Moses by God.

Jews and Christians consider the Ten Commandments to be the summary of divine law, handed down by God to the prophet Moses.

Scholars believe that much of the material in the Bible was recited aloud as part of an oral tradition long before it was written down, which complicates the matter of assigning dates to the various books. The books of the prophets (Joshua, Isaiah, Samuel, Jeremiah, Ezekiel, etc.) are believed to have been written and collected during the kingdoms of Israel (c. 1020–722 B.C.) and Judah (c. 933–586 B.C.) and shortly thereafter.

Psalms, Proverbs, Song of Solomon (or Song of Songs), Ruth, Ecclesiastes, and Chronicles were written in the time of King David and King Solomon during the tenth century B.C. or shortly after 1000 B.C. The books of Job, Lamentations, Esther, Daniel, Ezra, and Nehemiah are believed to have appeared between 600 B.C. and 100 B.C.

The 27 books of the New Testament, upon which Christianity is based, were written on papyrus scrolls, none of which remain. Dating the writing of the New Testament is difficult but the historical events referred to in the books indicate they were written shortly after the time of Christ and probably before A.D. 100.

What is the difference between the **Hebrew Bible** and the **Old Testament**?

The Hebrew Bible is made up of 24 books. The Old Testament used by Christians consists of the same books as those of the Hebrew Bible, but they are arranged differently and many books are divided, resulting in more books in the Old Testament than in the Hebrew Bible.

Among Christians, the Old Testament varies between Protestantism and Roman Catholicism: Protestants include 39 books in the Old Testament while Roman Catholics add seven books, called the Apocrypha, to their version, for a total of 46 books. The books of the Apocrypha resemble those of the Old Testament, but since they were written later than most of the Old Testament (probably 300 B.C.–A.D. 70), both Protestants and Jews treat them separately. The Bibles used by all three religions—Judaism, Roman Catholicism, and Protestantism—begin with the same seven books: Genesis, Exodus, Leviticus, Numbers, Deuteronomy, Joshua, and Judges. The first five of these were written by Moses.

What are the **divisions of the Hebrew Bible**?

The Hebrew Bible is divided into three main sections called the Law (or Torah), the Prophets, and the Writings (or Hagiographa). The Hebrew Bible is accepted by Jews as sacred: The word *testament* comes from the ancient word *testamentum,* meaning "covenant with God." Much of the Hebrew Bible recounts Jewish history, demonstrating faithful observance to their agreement with God.

The Law consists of those five books written by Moses, and these recount creation, early traditions, the lives of the patriarchs of Israel, early events of the Israelites, and entrance into the Promised Land. *Torah* translates as "teaching," and Jews (as do Christians) look to these first five books of the Bible for guidance. The Prophets consists of the books of the Former Prophets (books of Joshua, Judges, 1st and 2nd Samuel, and 1st and 2nd Kings) and the Latter Prophets (Isaiah, Jeremiah, Ezekiel, and the Twelve—the teachings of 12 other prophets). The books of the Prophets chronicle historical events, but according to the Jewish tradition these books also teach that people must obey God's laws. The Writings consist of 13 books, which are believed to have been written by poets and teachers.

What are the **Dead Sea Scrolls**?

The scrolls are ancient manuscripts of great historical and religious importance. They were found in dry riverbed caves on the northwestern side of the Dead Sea (a salt lake situated between Israel, the West Bank, and Jordan). More than 800 scrolls have been found, with the most famous discoveries made in 1947. The Dead Sea Scrolls were found miles apart at a number of different sites, including Khirbat Qumran in the West Bank (formerly Israel). The texts date to different centuries but include fragments of every book of the Hebrew Bible (or Old Testament) except the Book of Esther. Some texts are almost identical to Bible texts used today, showing that much of the Old Testament is the same as it was 2,000 years ago.

Why is the **Exodus from Egypt** so well known?

The Exodus of the Jews from Egypt, an event recorded in the Bible and corroborated by historical evidence, is well known to Jews and non-Jews alike, since the event, which Jews commemorate with Passover, came to symbolize departure from oppressive conditions, for the Israelites had been enslaved in Egypt.

According to the Bible (the Old Testament books of Genesis and Exodus), the ancestors of the Israelites had settled in the Nile River Delta of Egypt 430 years before the Exodus, or about the seventeenth century B.C. A change of dynasty caused the Israelites, because of their growing numbers, to be looked upon with hostility. They were put to work as state serfs in the construction of cities.

The biblical account tells about a series of confrontations between Ramses II, the pharaoh of Egypt, and the prophet Moses, who had been commissioned by God to lead his people to the Promised Land. The pharaoh refused to allow the Israelites to leave Egypt but was later convinced, through a series of divine signs including the Ten Plagues, to permit them to go. The flight from Egypt and the Promised Land have come to symbolize freedom and the realization of hopes.

Is it true that **Moses** stuttered?

Yes, the Bible indicates that Moses had a speech impairment, which caused him to fear the mission that God gave him. In the book of Exodus (4:10), just after Moses has been called by God and told to go to the pharaoh and lead the Israelites out of Egypt, he appeals to the Lord, saying, "If you please, Lord, I have never been eloquent … I am slow of speech and tongue," and, later (4:13), "If you please, Lord, send someone else." God reassures Moses and instructs him to go meet his brother, Aaron the Levite, who is eloquent and will speak to the people for Moses.

What were the **Ten Plagues**?

The Ten Plagues were delivered by God to the people of Egypt as a way of convincing the pharaoh that he was to let the Israelites go—so Moses could lead them out of bondage and into the Promised Land. The plagues were, in the order described in Exodus: 1) water turned into blood, 2) frogs, 3) gnats, 4) flies, 5) pestilence, 6) boils, 7) hail, 8) locusts, 9) darkness, and 10) death of the firstborn. It was not until the Tenth Plague had been delivered that the pharaoh agreed to let the Israelites leave.

What are the **Ten Commandments**?

Also called the Decalogue, the Ten Commandments appear in the Bible in the book of Exodus (20:2–17) and in the book of Deuteronomy (5:6–21). They are considered the summary of divine law as handed down by God to Moses, who not only heard them but received them in the form of writing on two stone tablets as he stood atop Mount Sinai (what is known today as Gebel Musa, on Egypt's Sinai Peninsula), where he had been summoned by God.

According to the Bible, the commandments, which are paramount in the ethical systems of Judaism and Christianity, are these: 1) Do not have any other god besides the Lord God; 2) Do not have or worship idols (carved images); 3) Do not make wrong use of the name of the Lord your God (or, do not take the name of the Lord in vain); 4) Keep the Sabbath day holy; 5) Honor your mother and your father; 6) Do not commit murder; 7) Do not commit adultery; 8) Do not steal; 9) Do not give false evidence against your neighbor; and 10) Do not covet your neighbor's household or lust after your neighbor's spouse.

What happened to the tablets on which the Ten Commandments were written?

Chapter 32 of the book of Exodus explains that while Moses was up on Mount Sinai receiving the Ten Commandments from God, the Israelites who were waiting below grew impatient and soon prevailed upon Aaron to "make us a god who will be our leader; as for the man Moses who brought us out of the land of Egypt, we do not know what has happened to him." And so Aaron did what the people wanted; he proceeded to collect gold jewelry from the Israelites and then fashioned the jewelry into a golden calf. When Moses came down the mountain, he saw the people rejoicing and dancing before the calf. Enraged, Moses threw the two tablets (which were inscribed on both sides) to the ground, smashing them. Later, the Bible explains (in Exodus 34:1–4) that God instructed Moses to carve new stone tablets "like the other ones," which he did. These tablets were placed in the Ark of the Covenant, as God had told Moses.

People sometimes see the Ten Commandments as divided into two groups: the first group covers the relationship between people and God and the second covers relationships among people.

What is the **Ark of the Covenant**?

According to Judeo-Christian tradition, the Ark of the Covenant is a decorative box that holds the tablets containing the Ten Commandments, also called the Law. The Ark was constructed by skilled craftsmen in Sinai; it was made exactly to God's specifications, which are first mentioned in the Bible in Exodus 25:10-22. The Israelites carried the Ark with them into the Promised Land, where King Solomon (tenth century B.C.) built a permanent temple to house it. When the Babylonians captured Jerusalem in 586 B.C., the Ark was either destroyed or taken. If it survived, its location is a mystery that has intrigued scholars and archeologists, as well as popular culture. However, some scholars believe the Ark never really existed.

Is it true that **Moses** never entered the **Promised Land**?

Yes, according to the Bible, Moses, who was 80 years old when he led the Israelites out of slavery in Egypt, never entered the Promised Land of Canaan.

After parting the Red Sea (which had swallowed up the Egyptian soldiers who pursued the Israelites as they fled), Moses led the people into the desert, embarking on what turned out to be a 40-year journey to Canaan. During this time, which is also described as 40 years of wandering, the Israelites lived under harsh conditions. This led them to distrust a God they could not see and to challenge Moses's authority. But Moses frequently appealed to God for help, and the Bible reports that he received it. In one instance (Numbers 20:22–12), God directed Moses to get water for the parched Israelites and their livestock by ordering a rock to yield its water. But instead of speaking to the rock, Moses struck it twice with his staff. Even though water gushed out of the rock, Moses had failed to follow God's instructions. Therefore, God later said to Moses: "Because you were not faithful to me...you shall not lead this community into the land I will give them."

Eventually the Israelites reached Canaan (ancient Palestine). But God did not allow Moses to enter the Promised Land—he only viewed it from atop Mount Pisgah (in present-day Jordan) before he died.

What is **a prophet**?

Most broadly, a prophet is any person who tells what will happen. But in a religious context, prophets are people who preach what they believe has been divinely revealed to them. In this way, they are seen as instruments of God and are thought to possess profound spiritual and moral insight. Moses was a Hebrew prophet, as were Isaiah and Jeremiah, among others. Muhammad was an Islamic prophet. And Christians regard Jesus as a prophet.

The Bible (Old Testament) also warns of false prophets, in other words, prophets who do not speak the truth. For example, Deuteronomy 13:1–3 says, "Should a prophet or a peddlar of dreams appear among you and offer you a sign or portent, and call on you to go after other gods whom you have not known and to worship them, of not heed the words of that prophet or dreamer."

What does **"messiah"** mean?

Messiah is the Hebrew word for "anointed one." In the broadest sense, a messiah is the professed or accepted leader of some hope or cause. Jesus is believed by Christians to be the Messiah whose arrival was foretold by the Hebrew prophets in the Old Testament. For example, in Isaiah 9:6, it states that a child will be born among the people, "a son is given to us; he will bear the symbol of dominion...and his title will be Wonderful Counselor, Mighty Hero, Eternal Father, Prince of Peace."

Why did people first believe that **Jesus of Nazareth** was **the Messiah**?

In his lifetime, Jesus of Nazareth (c. 6 B.C.–c. A.D. 30) was believed to be the Messiah for many reasons. These reasons are explained in detail in the New Testament, particularly in the four Gospels, Matthew, Mark, Luke, and John.

First, Jesus was thought to have been miraculously conceived by Mary, a virgin. In Luke 2:8–20, an angel is described as having spoken to shepherds tending their fields, telling them that "in this night in the city of David, a Savior is born unto you…a Messiah, the Lord," and goes on to say that the sign (of the Messiah) is that they will find a baby wrapped in swaddling clothes. Also in Luke (2:36–38), a prophetess arrives and proclaims Mary's baby to be the "child to all who were looking for the liberation of the Jews."

Further, Jesus proclaimed himself to be the Son of God (as described, for example, in John 5:17–23). He demonstrated extraordinary healing powers, used parables to teach moral and spiritual ways, and carried a message of God's redeeming love. All of this was convincing enough that many—mostly the common people—became disciples, or followers, of Christ during his lifetime.

Why was Jesus feared by the authorities?

There were two reasons Jesus (c. 6 B.C.–c. A.D. 30) was feared by Palestine's leaders. First, as an advocate of the poor, he was an outspoken critic of the privileged as well as of Palestine's oppressive Roman rulers. (Palestine was part of the Roman province of Syria during the lifetime of Jesus.) Jesus openly accused the ruling class of hypocrisy and injustices. Second, some feared that if Jesus was the Messiah, he would lead a revolution. So the governors viewed him and his teachings as a threat to their authority.

Who were **the apostles**?

The apostles were 12 men chosen by Jesus Christ (c. 6 B.C.–c. A.D. 30) to be his close followers. The apostles helped spread the word that they believed Jesus to be the Son of God. Matthew 10:1 explains that Jesus gave the 12 authority to drive out unclean spirits and to cure every kind of illness.

In Matthew 10:2–4, the names of the 12 apostles are given as Simon Peter (who is later simply called Peter), Andrew, James, John, Philip, Bartholomew, Thomas, Matthew, James the son of Alphaeus, Thaddaeus, Simon the Zealot, and Judas Iscariot. But the lists of apostles found in Luke 6:13–16 and in Acts 1:13 differ from that found in Matthew. While both Luke and Acts cite (Simon) Peter, Andrew, James, John, Philip, Bartholomew, Thomas, Matthew, James the son of Alphaeus, Simon the Zealot, and Judas Iscariot, they do not name Thaddaeus, but rather Judas the son of James. In other words, the lists agree on 11 of the 12 names.

After Judas Iscariot betrayed Jesus Christ, Matthias was chosen by the apostles to take his place (this is described in Acts 1:21–26). He was considered eligible since, like the 11 remaining apostles, he accompanied Jesus from the time of Jesus's baptism until "the day he was taken up from us."

Are **apostles and disciples** the same thing?

No. The apostles were the 12 followers chosen by Jesus (c. 6 B.C.–c. A.D. 30), as described in the New Testament of the Bible. The term *apostle* is also used to identify other early missionaries, such as Barnabas and Paul (Saul), who in Acts 13:2–3 are singled out by the Holy Spirit to be missionaries. A disciple is any follower of Jesus. Therefore, all the apostles were disciples. But not all disciples were apostles.

What is **Golgotha**?

Golgotha is the Hebrew name for Calvary, the site on a hillside outside ancient Jerusalem where Jesus Christ (c. 6 B.C.–c. A.D. 30) was crucified. Though the exact site is unknown, it is believed to be where the church of the Holy Sepulcher is located.

What are the **Four Horsemen of the Apocalypse**?

The Four Horsemen of the Apocalypse are allegorical figures mentioned in chapter six of Revelation (also called the Revelation of St. John the Divine). The chapter describes a scroll, which contains seven seals and is held in God's right hand. When the first four of the seals are opened, the four horsemen appear, each on a different colored horse. There are various interpretations of these allegorical figures, but the rider on the white horse is believed to represent conquest (or the return of Christ); the rider on the red horse is believed to represent war; the rider on the black horse, famine; and the pale horse, death. Some believe these hardships to be signs of the end of the world. The symbol of the four horsemen has appeared throughout art and literature.

How old is **Islam**?

Islam, one of the world's largest religions, originated with the teachings of the prophet Muhammad (c. 570–632) during the early 600s. Muhammad was born in Mecca (in present-day Saudi Arabia) and was orphaned at the age of six. He was raised by relatives, who trained him as a merchant. When he was 25 years old he married a wealthy widow, Khadijah, who bore him several children. In about 610, Muhammad began having visions in which he was called upon by God (Allah). More than 600 of these visions were written down, becoming the sacred text known as the Koran (or Qur'an). By 613 Muhammad had attracted followers with his messages of one God, Allah's power, the duty of worship and generosity, and the doctrine of the last judgment. Followers of this new religion became known as Muslims, an Arabic word meaning "those who submit" (to Allah); and the religion itself became known as Islam (meaning submission).

Today, there are Muslims in every part of the world, but the largest Muslim communities are in the Middle East, North Africa, Indonesia, Bangladesh, Pakistan, India, and central Asia. Additionally, most of the people of Turkey and Albania are Muslim. It is the world's second-largest religion (after Christianity).

Who wrote the **Koran**?

The Koran (or Qur'an) contains the holy scriptures of Islam and was written by the followers of the prophet Muhammad (c. 570–632). It is not known whether these texts were written down during Muhammad's lifetime or after his death. It is known that the texts were codified (organized into a body) between 644 and 656. Muslims believe the angel Gabriel revealed the book to Muhammad, beginning in 610 and continuing until the prophet's death in 632. The Koran, meaning "recitation," consists of 114 verses (*ayas*) that are organized in chapters (*suras*).

Muslims believe the beautiful prose of the Koran to be the words of God himself, who spoke through Muhammad. Further, it is believed to be only a copy of an eternal book, which is kept by Allah. The Koran is also held up by Muslims as proof that Muhammad was indeed a prophet since no human is capable of composing such text. Among the most widely read texts today, the Koran is also taught orally so that even Muslims who are illiterate may know and be able to recite verses.

When did the **Sunni and Shiah sects of Islam** form?

It was during the 600s, not long after Muhammad's death, when Muslims split into two main divisions: Sunni and Shiah.

Sunnite Muslims, who account for most of the Islamic world today, believe that Islamic leadership passes to caliphs (temporal and spiritual leaders) who are selected from the prophet Muhammad's tribe.

The Shiites believe, however, that the true leaders of Islam descend from Ali (c. 600–661), Muhammad's cousin and the husband of Muhammad's daughter, Fatimah (called the Shining One; c. 616–633). Ali, who was the fourth caliph (656–61), is revered by Shiites as the rightful successor to the prophet Muhammad and are led by his descendants.

Shiites form the largest subgroup, but there are other sects within Islam as well: the Wahhabi Muslims are a puritanical sect; the Baha'is emerged from the Shiites; and the Ismaili Khoja Muslims have been in existence almost from the beginning of Islam. While Islamic practices may vary somewhat among the sects, all Islamic people uphold the Five Pillars of Faith.

What are Islam's **Five Pillars of Faith**?

Muslims practice adherence to the Five Pillars of Faith: 1) belief in Allah as the only God and Muhammad as his prophet; 2) prayer five times daily—at dawn, at noon, in the afternoon, in the evening, and at nightfall; 3) giving alms to the poor; 4) fasting from dawn until dusk during the holy month of Ramadan); and 5) making the pilgrimage (*hajj*) to the holy city of Mecca at least once during their lifetime.

Hajj pilgrims pray during the annual Muslim pilgrimage to the holy city of Mecca, 2002.

Why is **Mecca** a holy city for Muslims?

Mecca, in western Saudi Arabia, is the birthplace of the prophet Muhammad (c. 570) and was his home until the year 622, when those who opposed him forced him to flee to Medina (a city about 200 miles north of Mecca). Muhammad later returned to Mecca and died there in 632.

Mecca is also the site of the Great Mosque, which is situated in the heart of the city. The outside of the mosque is an arcade, made up of a series of arches enclosing a courtyard. In that courtyard is the most sacred shrine of Islam, the Kaaba (or Caaba), a small stone building that contains the Black Stone, which Muslims believe was sent from heaven by Allah (God).When Muslims pray (five times a day, according to the Five Pillars of Faith), they face the Kaaba. It is also the destination of the *hajj,* or pilgrimage.

What were **the Crusades**?

The Crusades were a series of nine Christian military expeditions that took place during the end of the eleventh century and throughout the twelfth and thirteenth centuries. The stated goal of the Crusades was to recover from the Muslims the Holy Land of Palestine, where Jesus Christ (c. 6 B.C.–A.D. 30) lived. The word crusade

Why do Jews, Christians, and Muslims all claim the same Holy Land?

Palestine, in southwest Asia, at the eastern end of the Mediterranean Sea, is the Holy Land of Jews, since it was there that Moses led the Israelites after he led them out of slavery in Egypt (c. twelfth century B.C.) and where they subsequently established their homeland. It is the Holy Land of Christians because it was where Jesus Christ was born, lived, and died. And it was the Holy Land of Muslims, because the Arab people conquered Palestine in the seventh century and, except for a brief period during the Crusades, it was ruled by various Muslim dynasties until 1516 (when it became part of the Ottoman Empire). Palestine, which covers an area of just more than 10,000 square miles, is roughly the size of Maryland.

Palestine's capital, Jerusalem, is also claimed as a holy city by all three religions. Jews call it the City of David (or the City of the Great King) since it was made the capital of the ancient kingdom of Israel in about 1000 B.C. Christians regard it as holy because Jesus traveled with his disciples to Jerusalem, where he observed Passover. It is the site of the Last Supper, and just outside the city, at Golgotha, Jesus was crucified (c. 30 A.D.). Muslim Arabs captured the city in 638 A.D. (just after Muhammad's death), and, like the rest of Palestine, it has a long history of Muslim Arab rule. Jerusalem, which is now part of the modern state of Israel, is home to numerous synagogues, churches, and mosques. It has also been the site of numerous religious conflicts throughout history.

comes from the Latin word *crux* meaning "cross," and Crusaders were said to have "taken up the cross."

The Crusades began with an impassioned sermon given by Pope Urban II (c. 1035–1099) at Clermont, France, in November 1095. Earlier that year Byzantine emperor Alexius I Comnenus (1048–1118) had appealed to Urban for aid in fighting back the fierce Seljuk Turks. (The Seljuk Turks preceded the Ottoman Turks; the Seljuks were named for their traditional founder, Seljuq). Seeing the expansion of the Turks, who were Muslim, as a threat to Christianity, the Pope agreed to help. Not only did Urban rally support for the Byzantines in staving off the further advances of the Turks, he also advocated that the Holy Lands should be recovered from them. While the Arab Muslims who had previously controlled the Holy Land had allowed Christians to visit there, the Turks tolerated no such thing. Urban feared that if Palestine were not recovered, Christians would lose access to their holy places altogether.

But Urban also viewed the Crusades as a way of unifying western Europe: The feudal nobility there had long fought against each other. He believed a foreign war would unite them behind a common cause as Christians. Further, he hoped the Crusades would unite western with eastern (Byzantine) Europe behind one goal. If successful, the expeditions would also expand the pope's moral authority across a greater region.

En route from Clermont to Constantinople (present-day Istanbul, Turkey), where the Crusade was set to begin in August 1096, Urban continued to preach his message—at Limoges, Poitiers, Tours, Aquitaine, and Toulouse, France. The message found broad appeal, even if it appealed to something other than the people's religious sensibilities. Some of those who answered Urban's call took up arms not for the Christian cause, but for their own personal gain such as acquiring more land, expanding trade, or recovering religious relics. Many peasants "took up the cross" to escape hardships— in 1094 northern France and the Rhineland had been the site of flooding and pestilence, which was followed in 1095 by drought and famine.

The First Crusade actually turned into two. A Peasants' Crusade (which had never been Urban's intent) had gone ahead of the official expedition, and many lives were lost. It ended in failure. But the planned expedition, called the Crusade of Princes, ultimately succeeded in capturing Jerusalem in 1099. Western Christian feudal states were established at Edessa, Antioch, and Tripoli—all of which were placed under the authority of the Kingdom of Jerusalem. But Urban did not live to see the recovery of the Holy Land. And the Christian hold on Palestine was not to last, as the Muslims refused to give up the fight for control of lands they too considered to be holy. The Second (1147–49), Third (1189–92), Fourth (1202–04), Fifth (1217–21), Sixth (1228–29), Seventh (1248–54), and Eighth (1270) Crusades were prompted by a mix of religious, political, and social circumstances. The Crusades ended in 1291—almost 200 years after they had started—when the city of Acre, the last Christian stronghold in Palestine, fell to the Muslims, ending Christian rule in the East.

Yet another crusade, in 1212, was particularly tragic: Called the Children's Crusade, the expedition was led by a young visionary who had rallied French and German children to believe they could recover Jerusalem—since, as poor and faithful servants, they would have God on their side. As the children marched south across Europe, many of them died even before reaching the Mediterranean coast. Some believe that the Crusade was sabotaged, resulting in the children being sold into slavery in the East.

What impact did the **Crusades** have on **western Europe**?

The goals of the Crusades were not accomplished. The Holy Land had been recovered, but Christians were unable to keep control of it. And while the western Europeans had joined with the Eastern (Byzantine) Christians in their fight against the Muslims, the two groups remained bitter toward each other, which likely contributed to the fall of Byzantium to the Ottoman Turks in 1453. Nevertheless, the Crusades had a lasting effect on the European economy: During the expeditions, trade routes were established, new markets opened, and shipbuilding was improved. Having fortified themselves for the fight, the Christian monarchies in western Europe emerged from the Crusades in 1291 as strong as—if not stronger than—before Pope Urban had first rallied the troops in 1095.

How does the Catholic Church determine **who will be pope**?

Present-day procedures for electing a pope, who serves the church until his death, vary from those of earlier times. Before the 300s, the clergy of Rome and the outlying areas cast votes in what was essentially a local papal election. Throughout history, some of Europe's powerful rulers tried to influence the outcome of papal elections in order to establish a pope who would be favorable to their leadership. This interference in the process sometimes resulted in disputes over who was the rightful pope, with more than one pope claiming authority based on the support of emperors or factions within the church. (Those claiming to be popes but who were considered illegitimate are called antipopes.)

The process observed by the Roman Catholic Church today was established over time. The first important decision came in 1059 when Pope Nicholas II (c. 980–1061) declared that papal electors must be cardinals. In 1179 the Third Lateran Council established that all cardinals have an equal vote and that popes may only be elected by a two-thirds majority; this came after Pope Alexander III (c. 1105–1181) had been opposed by three antipopes. Pope Gregory X (1210–1276), after being elected to the papacy following a three-year vacancy, assembled the Council of Lyon in 1274, which decreed that the cardinals must meet within 10 days of a pope's death to determine the papacy. Further, the Council required that the cardinals remain together in strict seclusion until they have elected a new pope.

Today, when a pope dies, the dean of the Sacred College notifies all cardinals of the vacancy. The cardinals convene at Vatican City, in a meeting that is required to begin within 20 days of the pope's death. Any of three electoral processes may be used: ballot (which is most common), unanimous voice vote, or the unanimous decision of a committee of 9 to 15 delegates. Four votes are taken a day (two in the morning and two in the afternoon) until there is a two-thirds majority vote. As soon as a decision is reached, the dean asks the pope-elect if he accepts the position. Once he accepts, he is considered pope and has full authority over the church. He then proceeds to select a name, which he announces to the cardinals. Following these private meetings, the results of which are eagerly awaited by devout Roman Catholics, white smoke is sent up a chimney of the Vatican palace, signaling to the crowd in St. Peter's Square that the election is complete. (The smoke is created by burning ballots; ballots from elections in which a two-thirds majority was not achieved are also burned but in a way that creates black smoke. In this way, the public is apprised of the cardinals' decision-making process.) The oldest cardinal has the honor of making the official announcement to the people in St. Peter's Square. Then the people are blessed by the new pope in his first official act. The coronation of the pope is held later.

How many popes have there been?

The number given by the Vatican is 265, including Pope Benedict XVI, the former German cardinal who was elected on April 19, 2005, to succeed John Paul II (1920–2005). (Other lists cite 266 popes; the discrepancy arises around Stephen II, who died in 752 after he was elected but before he could be consecrated.) Except for a few brief interruptions when the papacy was vacant, the Roman Catholic Church has been led by the pope as its visible head (and Jesus Christ as its invisible head) since Jesus said to the apostle Peter: "And I say also unto thee, That thou art Peter, and upon / this rock I will build my church; and the gates of hell / shall not prevail against it" (Matthew 16:18).

The apostle Peter—who was earlier called Simon and is also called Simon Peter—became the leader of the Christian community after the crucifixion of Christ, and he made Jerusalem the headquarters of his preaching in Palestine. According to second-century sources, Peter traveled to Rome about 55 A.D. and became the city's first bishop. During the persecution of Christians under Roman emperor Nero (37–68 A.D.), Peter was crucified on Vatican Hill in the year 64. He died a martyr and was canonized. St. Peter's Church, the principal church of the Christian world, is said to have been built over Peter's burial place. These events in Rome during St. Peter's time long after gave the city special status within the church. It further established the site of the papal palace in Vatican City, which is an independent state that lies within the city of Rome. And the majority of popes (all but 18) have been Italian: When John Paul II, who was born in Poland, was elected pope in 1978, he was the first non-Italian pope since 1523.

Where does **the word "pope"** come from?

The word *pope* is derived from the Greek word *pappas,* meaning "papa" (father). The pope is also referred to as the Holy Father, the Vicar of Christ, and pontiff.

What was the controversy with **Thomas Becket**?

Archbishop of Canterbury Thomas Becket (c. 1118–1170) was killed by knights in the service of England's King Henry II (1133–1189); he had refused to be subservient to the monarch. In the long struggle between church and state, the story of St. Thomas Becket is a dramatic chapter. Born in London in about 1118, when he was in his twenties Becket entered into service for the archbishop of Canterbury, the spiritual head of the Church of England. He subsequently held various church offices, including archdeacon. When Henry II was coronated in 1154, becoming the worldly leader of the Church of England, he found in Becket one of his most vigorous champions. In 1162 Henry made him archbishop of Canterbury. But a transformation soon occurred in Thomas Becket, who put his spiritual duties first and began defending the church against the king's power. Henry, eager to increase royal authority, was determined to regain control over the church. A bitter struggle ensued between the two former friends. At one point, Becket fled the country because he was in fear for his life. When he returned to England six years later (in 1170), he renewed his opposition to the king, but nevertheless forced a reconciliation with him. Henry was still irked by Becket's open defiance to his authority, and he suggested to his knights that one among them might be brave enough to do away with him, ending the king's troubles. Four knights took the king at his word, and on December 29, 1170, they found Becket in Canterbury Cathedral and killed him as he made his evening prayers. Henry later did penance for the crime; Becket was canonized three years later.

What is an **infidel**?

Derived from the Latin *infidelis,* meaning "unbelieving" or "unfaithful," the term was used by the Catholic Church during the Middle Ages (500–1350) to describe a threat posed by Muslims. As the Moors (Muslims from North Africa) moved into Spain early in the eighth century and the Seljuk Turks conquered much of Asia Minor during the eleventh century, medieval Christians (of which there were a growing number) became increasingly fearful of growing Muslim influence. Not only were people of the Islamic faith occupying lands that were formerly Christian, they soon prevented Christian pilgrims from entering their Holy Land in the Middle East.

The church responded to the so-called infidels by inspiring western Europeans to take up arms in the Crusades, which began in 1095 and ended unsuccessfully in 1291. In another effort to drive back Muslim expansion, in 1231 Pope Gregory IX (c. 1170–1241) issued a papal bull creating the Inquisition, the system by which heretics were discovered and punished. Many of them were burned at the stake.

The Moorish dominance of the Iberian Peninsula (present-day Spain and Portugal) lasted hundreds of years before the North Africans were driven out by armies of the Christian states. Thereafter, during the reign of King Ferdinand and Queen Isabella, Spain embarked on a period of profound suspicion. It conducted the Spanish Inquisition, by which anyone thought to be an infidel (the definition now broadened to include Jews) was discovered and punished.

What caused the **East-West schism in the Catholic Church**?

During the Middle Ages (500–1350) cultural, geographical, and even political differences caused an increasingly wide divide between East (the Catholic churches in eastern and southeastern Europe, as well as in parts of western Asia) and West (the Catholic churches of western Europe). In the 800s a series of theological disputes began between the highest authority of the Eastern (Byzantine) churches, called the patriarch of Constantinople (also called the ecumenical patriarch), and the pope—particularly about the pope's authority over Christians in the East. Finally, in 1054, Pope Leo IX (1002–1054) issued an anathema (a formal curse) against the patriarch of Constantinople, Michael Cerularius (c. 1000–1059), excommunicating him and his followers from the Roman Catholic Church. The church had officially split. Thereafter the Eastern Orthodox churches would accept the patriarch of Constantinople as the highest church authority (in other words, they did not acknowledge the primacy of the pope) and they would follow the Byzantine rite (ceremonies); in the West, Roman Catholics followed the Latin rite and continued to regard the pope as the Holy Father.

When the Ottoman Empire captured Constantinople (present-day Istanbul, Turkey) in 1453, Orthodox Christians in the East came under Muslim rule; this lasted into the 1800s. Though there are still differences between the Eastern Orthodox churches (the Greek Orthodox church, the Russian Orthodox church, etc.) and the Roman Catholic Church today, the rift between them was healed in 1964 when Pope Paul VI (1897–1978) met with Ecumenical Patriarch Athenagoras I (1886–1972) in Jerusalem. The following year, the two religious leaders lifted the mutual anathemas between their churches.

When was the **Bible translated** into English?

The earliest English-language translations of any part of the Bible first appeared around the eighth century. The Vespasian Psalter was a primitive English translation of the Latin text of the book of Psalms. But the Latin translation of the Bible, called the Vulgate, continued to prevail for several centuries as the version authorized by the Roman Catholic Church. It was not until the 1300s that the first comprehensive translation into English was undertaken by the controversial religious reformer John Wycliffe (c. 1330–1384). Wycliffe, an English theologian and professor at Oxford University, was a forerunner of the Protestant Reformation. He believed that all authority—religious and secular—can be derived only from God, and he attacked the worldliness of the medieval church. In 1377 Pope Gregory XI (1329–1378) accused Wycliffe of heresy, but the reformer escaped trial. Nevertheless, he spent the rest of his life under the watchful eye of the church, which did everything in its power to curtail Wycliffe's activities, eventually forcing him to retire. He managed to initiate the first complete translation of the Bible, which was finished by his followers.

As an early leader of the Reformation, the Anglican priest William Tyndale (c. 1494–1536) believed that the Scriptures needed to be available to everyone, not just the scholarly or learned. So he began translating the New Testament and the Pentateuch (the first five books of the Old Testament: Genesis, Exodus, Leviticus, Numbers, and Deuteronomy) from Greek and Hebrew into English; he also referred to his friend Martin Luther's (1483–1546) German translation of the Bible. But Tyndale was unable to get his English translation published at home in England, where the Anglican church was determined to seize the work. In 1524 he fled the country for Germany, where he worked for the next five years to get his translations into print. Copies of it were smuggled back into England. Eventually he was arrested in Antwerp, Belgium, and condemned for heresy by the Roman Catholic Church. Tyndale was burned at the stake in 1536. His version later became the chief basis for the King James Version of the Bible (1611).

What was the **Reformation**?

It was a religious movement in Europe during the sixteenth and seventeenth centuries. It fomented inside the Catholic Church as people began questioning the church's doctrines, practices, and authority. While the movement was preceded by a swelling dissatisfaction with the church, the Reformation was officially, and some would say abruptly, begun in October 1517 when German monk and theology professor Martin Luther (1483–1546) nailed his Ninety-Five Theses to the door of the Castle Church at Wittenberg (Saxony, Germany), launching an attack on the church. The movement continued through the Thirty Years' War (1618–48). And though the resolution to that conflict brought about a measure of religious stability in Europe, the force of the Reformation did not end there. Both the freedom of dissent and the Protestantism people know today are the byproducts of the movement.

What **caused the Reformation**?

The religious movement during the sixteenth century also had political and cultural causes. As more and more people were converted to Christianity during the Middle Ages (500–1350), the pope's sphere of influence gradually increased—giving him greater authority than many secular rulers. This supremacy was defended by Pope Innocent III (c. 1160–1216), one of the most prominent figures of the Middle Ages, who asserted that the church should rightly retain its full power—in both secular and spiritual matters. But in western Europe, the monarchs became increasingly powerful as peasants began moving away from their farms and villages and to the emerging cities, which were protected by kings and emperors. The European monarchs often opposed the pope, regarding him as a leader of a foreign state. This conflict continued for centuries.

In 1309 Pope Clement V (c. 1260–1314) did something that would later divide the Roman Catholic Church when, being French himself, he moved the papacy from Rome to Avignon, France, where it stayed for 70 years. When Pope Gregory XI (1329–1378) moved it back to Rome in 1377, some French cardinals objected and elected a pope of their own (an antipope), installing him at Avignon. This resulted in two popes claiming supremacy, and in 1409 the situation grew more complicated when a third pope was added in Pisa, Italy. Not only had the power struggle divided the church, it had created tremendous confusion for the people, who further perceived that there were corrupt practices at work in the church. These included the selling of church positions and indulgences (pardons for sins) as well as the lavish lifestyle enjoyed by the bishops and the pope (which was on a par with that of royalty). While these injustices were decried by critics, the abuses did not stop. As a remedy some began to think that the church should be led not by the pope but by church councils.

Dissatisfaction with the church also extended to its message, which had turned away from God's mercy and the teachings of Jesus Christ, instead focusing on a life of good works as the way to salvation.

Europe was also in the midst of the Renaissance (1350 to 1600), which saw a proliferation in the number of universities, the circulation of printed materials (thanks to the advent of the movable-type printing press), and broader study of classic texts as well as of the Holy Scriptures—in their original languages instead of how they had been handed down in translation by the church. Before long, a middle class of educated people emerged in Europe.

These circumstances combined to bring about a period of religious reform that lasted until 1648. The movement itself, however, continued to exert influence through its emphasis on personal responsibility, individual freedom, and the secularization of society.

What was the importance of Martin Luther's **Ninety-Five Theses**?

The Reformation as a movement began on October 31, 1517, when the German monk and theology professor Martin Luther (1483–1546) nailed his Ninety-Five Theses to the door of the Castle Church at Wittenburg (Saxony, Germany). The theses (which are arguments or assertions) questioned the value of indulgences (the pardons that were disseminated by the church) and condemned the sale of them. Luther had already begun to preach the doctrine of salvation by faith rather than by works, and during 1518 he went on to publicly defend his beliefs, which were in direct opposition to the church. The following year he expanded his argument against the church by denying the supremacy of the pope.

In 1521 Pope Leo X (1475–1521) declared Luther a heretic and excommunicated him. Ordered to appear before the Diet of Worms in April 1521, Luther refused to retract his statements of his beliefs, saying, "Unless I am convinced by the testimony of the Scriptures or by clear reason…I am bound by the Scriptures I have quoted and my conscience is captive to the Word of God." The following month, Holy Roman Emperor Charles V (1500–1558) issued the Edict of Worms, declaring Luther to be an outlaw and authorizing his death. But the Prince of Saxony, known by history as Frederick the Wise (1463–1525), saw fit to protect Luther, whom he had appointed as a faculty member at the University of Wittenberg (founded by Frederick the Wise in 1502). There Luther translated the New Testament into German and undertook a translation of the entire Bible. Luther continued the Protestant movement until his death in 1546.

What was the **Counter Reformation**?

Also called the Catholic Reformation, it was the Roman Catholic Church's response to the Protestant Reformation of the sixteenth century. Some parishioners and members of the Roman Catholic clergy had already been calling for reforms within the church

What was the Diet of Worms?

In 1521 a diet, which is another word for an imperial council or an assembly of princes, was convened at Worms, situated on the Rhine River in Germany. This particular meeting had been called by Holy Roman Emperor Charles V (1500–1558) to consider the crisis that had been brought on by the Reformation. Since Martin Luther (1483–1546) was instrumental in both igniting the movement and furthering its causes, he was called before the diet to testify and defend himself. But the religious leader refused to yield his stance; he vowed to continue to make his argument against the church. He closed his statement with these famous words: "Here I stand. I cannot do otherwise. God help me. Amen."

The result of this was that on May 25 the Edict of Worms was issued, declaring Martin Luther an outlaw. However, Luther sought refuge in the protection of Frederick the Wise, and the edict itself served only to galvanize the cause of the reformers.

for more than two centuries when in 1517 German monk and theology professor Martin Luther (1483–1546) nailed his Ninety-Five Theses to the door of the Castle Church at Wittenberg (Saxony, Germany). His theses attacked the doctrines and authority of the church, sparking the Reformation. The movement's leaders, called Protestants because they protested against the Catholic Church, changed the religious landscape of Europe by creating new Christian churches. But a movement to make changes inside the Catholic Church also began. The turning point came in 1534 when Paul III (1468–1549) became pope. Realizing that the church must respond to what it viewed as a religious crisis, Pope Paul convened the Council of Trent (in Italy), which was charged with reviewing all aspects of religious life. The ecumenical group met from 1545 to 1547, 1551 to 1552, and 1562 to 1563, and out of those deliberations emerged the modern Catholic Church. The Counter Reformation was aided by a group of priests and brothers known as the Jesuits, members of the Society of Jesus, a religious order of the Roman Catholic Church. The Jesuits were instrumental in spreading the word of the reforms and in promoting a new spirit within the Catholic Church.

Who are the **Jesuits**?

The Jesuits are members of a Roman Catholic religious order called the Society of Jesus. The group was founded by Saint Ignatius of Loyola (1491–1556). Born into nobility as Iñigo de Oñaz y Loyola, the Spaniard became a knight in 1517. In that capacity he fought against the French in their siege 1521 of Pamplona, northeast of Madrid. But he was seriously wounded in the battle and retreated to a commune in northeast Spain from 1522 to 1523. There Ignatius heard a religious calling and subsequently undertook a pilgrimage to Jerusalem (1523 to 1524). Committed to a religious life, he embarked on a program of disciplined writing and study in Spain and in Paris, France. Even before he was ordained in 1537, Ignatius had gained followers—the Spanish missionary Francis Xavier (1506–1552) among them. With his companions Ignatius founded the Society of Jesus in 1539; the religious order was approved by Pope Paul III (1468–1549) the following year. Jesuits are known for leading structured lives and for their self-discipline, commitment to the pope, and missionary work. They have a profound belief in education, and as such have long been leaders in learning and in the sciences. The order was suppressed in 1773 but restored in 1814.

What was the **St. Bartholomew's Day Massacre**?

It was a mass murder in 1572, which began on August 24, the Roman Catholic feast day of St. Bartholomew. The slaughter, which eventually claimed an estimated 70,000 lives, was part of decades of civil wars between Roman Catholics and Protestants (Huguenots) in France.

The Protestant Reformation had begun in 1517, and it spread through Europe over the following decades. By the 1560s as many as one-third of the French people were Protestants. The growth of a reformed Christian faith threatened the Roman Catholic Church and eroded its power, both religious and political. In 1560 Charles IX (1550–1574), a Roman Catholic, rose to power in France. But since he was just 10 years old at the time, the country was ruled by a regent, his mother, Catherine de Medicis (1519–1589). Catherine was a descendant of the powerful Medici family of Florence and, as the wife of King Henry II (1519–1559), the queen of France. Even after Charles IX reached the age of majority, she continued to dominate him.

By 1572 Catherine feared the growing influence that another person, Admiral Gaspard de Coligny (1519–1572), had on her son, the king. Coligny was a leader of the Huguenots, the French Protestants, and, in fact, he did hold sway over Charles IX. To remove this threat, Catherine authorized a plot by Roman Catholic nobles to assassinate Coligny. On August 18, Catherine's daughter, Margaret of Valois (1553– 1615), was married to Henry of Navarre (1553–1610), a Huguenot. The royal event brought many Huguenot nobles to Paris. On August 22, an attempt was made on Coligny's life, but he survived. Catherine feared that an investigation into the attack on Coligny would reveal her role in it. She therefore gave approval to a wider plan to exterminate Huguenots, and she convinced her son, Charles IX, to order it. According to French statesman and historian Jacques-Auguste de Thou (1553–1617), the order stated that "it was the will of the king that, according to God's will, they should take vengeance on the band of rebels [the Huguenots] while they had the beasts in the toils."

The massacre began with the toll of a palace bell on August 24. Those who carried out the killings on the Huguenot leaders identified themselves with white armbands and a white cross on their hats. Coligny was among the first victims, but Protestants throughout Paris were targeted. As citizens became involved in the mayhem, several thousand perished in the city, causing the waters of the River Seine to run red. Throughout the provinces over the next several weeks, thousands more died, the victims of mob attacks on Protestants. News of the attacks was greeted with approval by Roman Catholic leaders; Protestants throughout Europe were horrified. The plot carried out on St. Bartholomew's Day had launched a mass extermination, fueling the hatred between Catholics and Protestants.

What were the **results of the Reformation**?

The emergence of the Protestants (who got their name for protesting against the Catholic Church) was officially recognized by Holy Roman Emperor Charles V (1500–1558) with the Peace of Augsburg (1555), which granted the people the right to worship as Lutherans (the church named for reformer Martin Luther [1483–1546]). But the hostility between Catholic and Protestant countries erupted in 1618 with the Thirty Years' War. That series of conflicts, which had become increasingly political as it raged, was ended with the Peace of Westphalia, which, among other things, stipulated that Lutheranism and Calvinism (or Presbyterianism, founded by Frenchman John Calvin [1509–64]) be given the same due as Catholicism.

Through acts of state, both Catholicism and Protestantism took hold in Europe, with the northern countries, including those of Scandinavia, turning toward the new churches and the southern countries remaining Catholic. For the most part, the Reformation fostered an attitude of religious tolerance among Christians. However, conflict would continue (to the present day) in the British Isles: After Queen Elizabeth I (1533–1603) adopted a moderate form of Protestantism (called Anglicanism) as the official religion of England, English Protestants colonized Ulster (Northern) Ireland and in so doing gave rise to hostilities with their Irish-Catholic countrymen to the south.

For its part, the Catholic Church, too, underwent a period of reform (called the Counter Reformation), which rid the church of many of its pre-Reformation problems to emerge as a stronger religious body.

The Reformation had without question brought about greater religious freedom than had been known before. Among the churches that emerged during the Reformation are the Lutheran, the Anabaptist (ancestors to the Amish and Mennonite churches), the Presbyterian, the Episcopal, and the Congregational and Unitarian (formerly Puritan)—all of which have strong followings today, both in Europe and in North America, where they were established by the colonists.

What was the **Peace of Augsburg**?

The Peace of Augsburg of 1555 came as a result of the Reformation and effectively carved up Europe between the Roman Catholic Church and the new Lutheran (Protestant) Church. Charles V (1500–1558) was Holy Roman Emperor at the time, and though he hated to concede lands to Protestantism, he also wished to end the religious divisions in the empire. Princes who had themselves converted to the new faith convinced Charles to allow each prince to choose the religion for his own land. Thus, the Peace of Augsburg officially recognized the Lutheran church and the right of the people to worship as Protestants.

Who were the **Puritans**?

The Puritans were members of a religious movement that began in England in the 1500s and lasted into the first half of the 1600s, when it spread to America as well. Influenced by the teachings of religious reformers John Wycliffe (c. 1330–1384) and John Calvin (1509–1564), the Puritans were so named because they wanted to "purify" the Anglican church (also known as the Church of England). They believed too much power rested with the church hierarchy (its priests, bishops, and cardinals), that the people (called the laity or lay members) should have more involvement in church matters, and that the ceremonies ought to be simplified to stress Bible reading and individual prayer. Further, they defied the authority of the pope, believing that each church congregation should have control of its own affairs, which should be guided by a church council (called a presbytery) made up of lay members.

These ideas are familiar to Americans today since they provided the basis not only for many Protestant churches but also influenced the formulation of U.S. government. When the Puritans faced persecution at home, they became religious pilgrims, traveling to the New World, where they established both their religious and social belief system.

What was the **Great Awakening**?

The Great Awakening was an American religious movement that began in New England in the mid-1730s. At its center were the fire-and-brimstone sermons delivered by charismatic preachers such as Congregational minister Jonathan Edwards (1703–1758) and Anglican missionary George Whitefield (1714–1770). Revivals were another cornerstone of the movement: These were evangelistic meetings that moved around the countryside, from Maine to Georgia, converting (or awakening) people to Christianity—not through the doctrines of the church, but rather through the individual's own experience. The theology of the Great Awakening was Calvinist, stressing the depravity of man and the sovereignty of God and promoting the belief that faith, and not conduct, is the road to salvation. In its emphasis on the individual and its espousal that the individual is the final arbiter of truth, the movement had a profound effect on the spiritual and political character of what would soon become the United States. Since many vehemently opposed the movement, it also served to divide churches between the revivalists and the traditionalists. Thus, it diversified American religious life and promoted religious tolerance.

Mother Teresa, as head of the Missionaries of Charity order, cradles an armless baby girl in Calcutta, India, 1978.

What is the **Protestant ethic**?

The Protestant ethic is a term describing a set of attitudes fostered by the leaders of the Reformation: Martin Luther (1483–1546), John Calvin (1509–1564), John Knox (1513–1572), Huldrych Zwingli (1484–1531), Conrad Grebel (c. 1498–1526), and their Protestant successors such as Methodist church founders John Wesley (1703–1791) and Charles Wesley (1707–1788). These church leaders stressed the holiness of a person's daily life, the importance of pastors to lead family lives (versus the celibacy of Catholic monks and nuns), education and study, and personal responsibility. According to these beliefs, the person who is hardworking, thrifty, and honest is a good person of value to their community and to God.

From 1904 to 1905, German sociologist Max Weber (1864–1920) wrote an essay called "The Protestant Ethic and the Spirit of Capitalism," asserting that Protestant principles contributed to the growth of industry and commerce during the 1700s and 1800s since the hard work, investment, and savings of individuals help build a capitalist economy.

How is **someone sainted**?

Criteria for sainthood (also called canonization) are leading a holy life, conducting miracles, and suffering or even dying because of one's faith (martyrdom). The Catholic Church keeps a list of saints, from which certain names were dropped in 1969 since their inclusion could not be justified by history.

How did **Mother Teresa** begin her life's work?

Born Agnes Gonxha Bojaxhiu in August 1910 in Skopje (in present-day Macedonia), the woman the world knew as Mother Teresa had by age 12 realized that she would spend her life aiding the poor. At age 18 she left her family to pursue that mission, joining a community of Irish nuns who were missionaries in Calcutta, India. There she took the name Sister Teresa and began teaching at St. Mary's High School, which she would continue to do for the next 17 years. She took her final vows as a nun in 1937.

In 1946 she became ill and was believed to have contracted tuberculosis. Sent to Darjeeling in northeast India to recover, she was on a train when she "heard the call to give up all and follow him to the slums to serve him among the poorest of the poor." In 1948 Pope Pius XII (1876–1958) allowed Sister Teresa to leave her order and pursue this mission. In 1950, after receiving some medical training, she founded the order of Missionaries of Charity in Calcutta. Two years later she opened a home for the dying poor (it was called Nirmal Hriday or "Pure Heart"). One year after that she opened her first orphanage. It was the children there who called her Mother Teresa, or sometimes simply "mother."

She spent her life helping the sick and the outcast, who she described as "Christ in distressing disguise." Small in stature, she was frail and in poor health in her final years, but she continued her work nevertheless. It was not until March 1997, just months before her death, that she finally stepped down as head of her order. Having started with only 12 members, the Missionaries of Charity had grown to include more than 4,000 nuns who continued to run orphanages and hospices around the world. Mother Teresa died on September 5 of that year. She was 87.

Will **Mother Teresa** be **sainted**?

As of the early 2000s Mother Teresa of Calcutta was on the road to sainthood. On October 19, 2003, she was beatified, a key step in the process that began just two years after her death. Her worldwide reputation of holiness prompted Pope John Paul II (1920–2005) to waive the customary five-year waiting period for the "cause of canonization" to begin.

Mother Teresa's "heroic virtues," a requirement for sainthood, were well known long before she died. The modern martyr spent nearly 70 years working as a missionary among the poor, the last 50 of them in outreach to society's most downtrodden—the impoverished sick and dying. She lived modestly, dressed simply (usually in a plain white sari, which she felt identified her with the poor), and devoted herself wholly to helping those society had forgotten. The so-called "saint of the gutter" received worldwide recognition for her work. In 1979 she was awarded the Nobel peace prize. Though she came into contact with some of the world's most influential people, she was unchanged by the attention. When the pope gave her a Lincoln Continental for her own personal transportation, she auctioned it off to raise needed funds for her works of charity. Reportedly, when she visited Britain's Princess Diana (1961–1997), the nun looked on the large rooms in the royal palace and uttered something to the effect of "just think how many people could live in these rooms.... "

John Paul II, pictured waving to the faithful in St. Peter's Square in 2002, had the third-longest papacy in history.

Another requirement of sainthood is involvement in miracles. In October 2002 it was reported that Pope John Paul's office had attributed a miracle to Mother Teresa: A young Indian woman was cured of a stomach tumor after praying to her. The Vatican, however, found no scientific explanation for the woman's recovery. On December 20, 2002, the pope approved "the decrees of her heroic virtues and miracles," paving the way for Mother Teresa's beatification in 2003. Her sainthood appeared to be imminent.

Why was **John Paul II** called the people's pope?

Because during his 27-year tenure, he dramatically changed the public perception of the pope. Polish cardinal Karol Wojtyla (1920–2005) was named Pope John Paul II on October 16, 1978, becoming the first non-Italian head of the Roman Catholic Church in 455 years. From the first moments of his service, it was clear that this was a different kind of pope. Upon his election, he greeted the cardinals of the conclave—his "brothers"—standing rather than seated, which was the tradition. A few weeks after his election, he leaned out the windows of the Vatican palace to sing carols with 50,000 children gathered in St. Peter's Square to celebrate Christmas. Instead of limiting his concerns to the administration of the church, he traveled far and wide to carry the message of Christianity to the people. Crowds, often numbering in the hundreds of thousands to millions, gathered to see him around the globe. In 1979 he made his first trip to the United States, after which *Time* magazine ran a cover story with the headline, "John Paul, Superstar." Pope John Paul fought for freedom of religion everywhere, even challenging his Communist homeland. His call for solidarity contributed to the downfall of communism in Poland and across the Eastern bloc. He published regularly—memoirs as well as books of prayers, lessons, meditations, and poetry. Despite his active ministry on the world stage, he remained a traditionalist, never wavering from the ages-old teachings of the Catholic Church. When he died on April 2, 2005, he was hailed both as a holy man and a man of peace by Christians and non-Christians alike.

EXPLORATION AND SETTLEMENT

When did people first **migrate to the Western Hemisphere**?

From Europe's discovery of the American Indian at the end of the fifteenth century to the present, the questions of who the native American populations are and how they came to the Western Hemisphere (North and South America and the surrounding waters) have intrigued scholars, clergymen, and laymen. The advancement of anthropology has yielded some answers: Since no skeletal remains of a human physical type earlier than *Homo sapiens* have yet been found in the Americas, researchers have concluded that the continents were settled through migration. Many scholars believe that Asians came to America during two periods: the first, between 50,000 and 40,000 B.C.; and the second, between 26,000 and 8000 B.C. They are believed to have come by way of a great land or ice bridge (Beringia) over the Bering Strait, between Asia (Russia) and North America (Alaska). (This causeway was covered by water from about 40,000 to 26,000 B.C. because of a period of melting, which would have prevented passage.)

Most scholars also agree that there were several discrete, and perhaps isolated, movements of various peoples from Asia to the Americas. The migrations might have been prompted by population increases in the tribes of central Asia, which impelled some to move eastward in quest of food sources—animals. As game moved across the Bering Strait, hunters followed.

Over time, population growth caused early man to continue southward through the Americas so that by 8000 B.C. there were primitive hunters even in Tierra del Fuego, which forms the southernmost part of South America. Around 5000 B.C. the disappearance of large game animals in both North and South America produced a series of regional developments, culminating in the emergence of several great civilizations, including the Inca, Maya, and Aztec.

How did the earliest **peoples arrive in North America**?

Long before the arrival of the Europeans in the Western Hemisphere in the late 1400s and early 1500s, Asian peoples are believed to have migrated over Beringia—a land bridge that is thought to have existed over the Bering Strait, the waterway that separates Asia (Russia) from North America (Alaska). Scholars believe that during the late Ice Age (known as the Pleistocene glacial epoch, which ended about 10,000 B.C.), a natural bridge was formed across the strait either by ice or by dropping sea levels that exposed landmasses. Asian peoples, who were hunters, are believed to have migrated over Beringia as they pursued large game, arriving in North America as early as 50,000 B.C. These people, called Paleo-Indians, were the first inhabitants of the Western Hemisphere. After their arrival, they spread out across North and South America. The many American Indian groups that were encountered by the Europeans upon their arrival were descendants of the migratory Paleo-Indians.

WEST MEETS EAST

When did **Marco Polo** travel to the Far East?

Marco Polo (1254–1324) was only in his teens when he left Venice in about 1270 with his father, Niccolò, and his uncle Maffeo, traveling an overland route to the East. The Polo brothers had made such a trip once before—in 1260 they had traveled as far as Beijing, China, but upon their return home, they learned that Niccolò's wife (Marco Polo's mother) had died. So when the pair of adventurers set out again, they took the young Marco Polo with them.

The Polos traveled from Acre, Israel, to Sivas, Turkey, then through Mosul and Baghdad (in Iraq) to Ormuz, a bustling trade center on the Persian Gulf, where they intended to take a ship for the East. Seeing the ships, the travelers determined they weren't reliable transport, so they opted to continue on land, heading north to Khorasan (in Iran), through Afghanistan, and to the Pamirs, a high plateau range in central Asia. It took the Polos 40 days to transverse the high-altitude range, finally reaching the garden city of Kashgar (China). From there, the Polos followed a path skirting the Taklamakan Desert and then rested before crossing the Gobi Desert, which they did in 30 days' time, covering some 300 miles. Stopping in Tun-hwang, the center of Buddhism in China, the European travelers then followed a southeast path that would have paralleled the Great Wall (constructed in the third century B.C.). After following the Yellow (Huanghe) River, the Polos were met by emissaries of Kublai Khan (1215–1294). They continued with their guides on a 40-day trip to Xanadu (Shang-tu), China, 300 miles north of Beijing, where they were received by Kublai Khan himself, founder and ruler of the Mongol dynasty and grandson of Genghis Khan (c. 1162–1227). It was May 1275.

Kublai Khan, who was an ardent Buddhist and a patron of the arts, took a liking to the young Marco Polo, who entered into diplomatic service for the ruler. In that capacity Marco Polo traveled to India and visited the Kingdom of Champa (what is now Vietnam), Thailand, the Malay Peninsula, Sumatra, Sri Lanka, and India. The Polos, European courtiers who were well liked by the Great Khan, stayed in China until 1292, finally returning home by way of Sumatra, India, and Persia (present-day Iran). In 1295 they arrived back in Venice, which they found at war with long-time rival Genoa. The Polos carried with them many riches, including ivory, jade, jewels, porcelain, and silk. Marco Polo was now a man in his forties and had spent most of his life thus far in the Far East.

Are the **adventures of Marco Polo** true?

Most of the tales are accepted as true and accurate by modern scholars. It is only those accounts that deal with places where it is not known that Marco Polo traveled, such as Africa, that are seen as legend rather than fact.

Upon his return to Venice in 1295, Marco Polo (1254–1324) took up the family occupation and worked as a merchant. Three years later, he was on board a ship that was captured by a rival Genoese ship. He was subsequently imprisoned in the port city of Genoa, where he met a writer named Rustichello (or Rusticiano), from the Italian city of Pisa. Polo recounted his stories to Rustichello, who wrote them down and published them as the *Divisament dou monde* (Description of the World). The book was an immediate popular success and became one of the most important sources of Western knowledge of the East. Readers today know the stories as *The Travels of Marco Polo.*

DISCOVERY OF THE NEW WORLD

Were the **Vikings** the first Europeans to reach North America?

It is believed that the sea-faring Norsemen, who are alternately called the Vikings, were in fact the first Europeans to see the Western Hemisphere (North and South America and the surrounding waters). Norwegian-born Leif Ericsson (c. 970–c. 1020) is generally credited with having been the first European to set foot on North American soil. Ericsson was the son of navigator Erik the Red, who founded a Norse settlement in Greenland and moved his family there in 985 or 986. About that same time another Norseman, Bjarni Herjolfsson, who was driven off course on his way from Iceland to Greenland, became the first European to sight North America, but he did not go ashore. It is believed that Ericsson decided he would follow up on Herjolfsson's discovery. About 1001 Ericsson set out from Greenland with a crew of 35 men and probably landed on the southern end of Baffin Island, due north of the province of Quebec.

The expedition likely made it to Labrador, Newfoundland (on the northeastern North American mainland), and later landed on the coast of what is today Nova Scotia or Newfoundland, Canada; this landfall may have been at L'Anse aux Meadows (on Newfoundland Island). Ericsson and his crew spent the winter of 1001–02 at a place he called Vinland, which was described as well wooded and produced fruit, especially grapes. He returned to Greenland in the spring of 1002.

The first authenticated European landing in North America was in 1500 when Portuguese navigator Gaspar de Côrte-Real (c. 1450–c. 1501) explored the coast of Labrador and Newfoundland. A year later, he made a second trip to North America but never returned home. In 1502 his brother Miguel went out in search of him; neither returned.

Why did Spain authorize **Columbus's expedition** in search of a westward route to the Indies?

When Christopher Columbus (born Cristoforo Colombo; 1451–1506), who was Italian, became convinced that Earth was round (he had been studying the writings of Ptolemy) and that he could, therefore, reach the East by traveling due west across the ocean, he first took the idea to the king of Portugal to seek his financial aid. This was about 1483. The move was a natural one: He had settled in Portugal at the age of 25, married a Portuguese woman (who bore him one son before she died), and Portugal was the leading seafaring nation of Europe at that time, carrying out southbound voyages with the intent of rounding Africa and reaching the Indies to the east. But Columbus was rebuffed by the Portuguese monarch. When in 1484 he took his plan to the Spanish monarchs, King Ferdinand (1452–1516) and Queen Isabella (1451–1504), they too refused to back him. But Columbus persisted, and in 1492 the Spanish king and queen agreed to sponsor the explorer's plan. There were two reasons for the decision: The overland trade route to the Indies (India and its adjacent lands and islands in the Far East) had long been cut off by the Turks, and the western Europeans found themselves in need of finding a new trade route to the Far East. Further, Ferdinand and Isabella were devout Christians, as was Columbus, and they all shared a desire to advance the Christian religion. In short, the monarchs saw that there were both material and religious advantages for backing Columbus's expedition.

Where did Christopher Columbus first land in the **New World**?

Columbus (1451–1506) set sail from Palos, in southwest Spain, on August 3, 1492, and he sighted land on October 12 that year. Going ashore, he named it San Salvador; alternately called Watlings Island (today it is one of the Bahamas). With his fleet of three vessels, the *Nina,* the *Pinta,* and the *Santa Maria,* Columbus then continued west and south, sailing along the north coast of Cuba and Haiti (which he named Hispaniola). When the *Santa Maria* ran aground, Columbus left a colony of about 40 men

In 1492 Christopher Columbus named his New World landing place San Salvador.

on the Haitian coast where they built a fort, which, being Christmastime, they named La Navidad (*Navidad* means "Christmas" in Spanish).

In January 1493 he set sail for home, arriving back in Palos on March 15 with a few "Indians" (native Americans) as well as some belts, aprons, bracelets, and gold on board. News of his successful voyage spread rapidly, and Columbus journeyed to Barcelona, Spain, where he was triumphantly received by Ferdinand and Isabella.

On his second voyage, which he undertook on September 25, 1493, he sailed with a fleet of 17 ships and some 1,500 men. In November he reached Dominica, Guadeloupe, Puerto Rico, and the Virgin Islands. Upon returning to Haiti (Hispaniola), Columbus found the colony at La Navidad had been destroyed by natives. In December 1493 he made a new settlement at Isabella (present-day Dominican Republic, the eastern end of Hispaniola), which became the first European town in the New World. Before returning to Spain in 1496, Columbus also landed at Jamaica.

On his third voyage, which he began in May 1498, Columbus reached Trinidad, just off the coast of South America. On his fourth and last trip he found the island of Martinique before arriving on the North American mainland at Honduras (in Central America). It was also on this voyage, in May 1502, that he sailed down to the Isthmus of Panama—finally believing himself to be near China. But Columbus suffered many difficulties, and in November 1504 he returned to Spain for good. He died two years later in poverty and neglect. He had, of course, never found the westward sea passage to the Indies in the Far East. Nevertheless, the Caribbean islands he discovered came to be known collectively as the West Indies. And the native peoples of North and South America came to be known collectively as "Indians."

Why does **controversy surround Christopher Columbus**?

It is due in part to the fact that history wrongly billed Columbus (1451–1506) as "the discoverer" of the New World. The native peoples living in the Americas before the arrival of Christopher Columbus were the true discoverers of these lands. These peoples had migrated thousands of years before: It is believed that as early as 50,000 B.C. they came across the Bering Strait from Asia, and migrated southward, throughout North and South America, reaching Tierra del Fuego by 8000 B.C. Therefore, it is correct to say that Columbus was the first European to discover the New World, and there he encountered its native peoples.

But it was for his treatment of these native peoples that Columbus is a controversial figure. Columbus was called back from the New World twice (on his second and third voyages) to be investigated for his dealings with the native Americans, including charges of cruelty. The first inquiry (1496–97) turned out favorably for the explorer: His case was heard before the Spanish king, and charges were dismissed. However, troublesome rumors continued to follow Columbus, and in 1500 he and two of his brothers (Bartholomeo and Diego) were arrested and sent back to Spain in chains. Though later released and allowed to continue his explorations (he would make one final trip to the New World), Christopher Columbus never regained his former stature, lost all honor, and died in poverty in the Spanish city of Valladolid in 1506.

Who were the ***conquistadors***?

Conquistador is the Spanish word for "conqueror." The Spaniards who arrived in North and South America in the late 1400s and early 1500s were just that—conquerors of the American Indians and their lands. In many cases, the Spaniards were the first Europeans to arrive in these lands, where they encountered native inhabitants including the Aztec of Mexico, the Maya of southern Mexico and Central America, and the Inca of western South America. By the mid-1500s these native peoples had been conquered, their populations decimated by the Spanish conquistadors. The conquest happened in two ways: First, the Spaniards rode on horseback and carried guns, while their native opponents were on foot and carried crude weapons such as spears and knives; and second, the European adventurers brought illnesses (such as smallpox and measles) to which the native populations of the Americas had no immunities, causing the people to become sick and die.

By 1535 conquistadors such as Francisco Pizarro (c. 1475–1541), Hernán Cortés (1485–1547), and Vasco Núñez de Balboa (1475–1519) had claimed the southwestern United States, Mexico, Central America, and much of the West Indies (islands of the Caribbean) for Spain.

Did the **Europeans introduce anything** besides disease **to the Americas**?

Yes, in addition to certain diseases, which were accidentally imported into the New World by the early Europeans, the explorers brought with them many things that were previously unknown in the Americas. When Christopher Columbus (1451–1506) landed at Hispaniola (present-day Haiti and Dominican Republic in the Caribbean) in 1492, he carried with him horses and cattle. These were the first seen in the Western Hemisphere; the American Indians had no beasts of burden prior to the arrival of the Europeans. In subsequent trips, Europeans introduced horses and livestock (including cattle, sheep, pigs, goats, and chickens) throughout South and North America. They later carried plants from Europe and the East back to the Americas, where they took hold. These included rice, sugar, indigo, wheat, and citrus fruits—all of which became established in the Western Hemisphere and became important crops during colonial times. With the exception of indigo (which was used as a fabric dye), these nonindigenous crops remain important to the countries of North and South America.

Besides the native peoples, what **discoveries did the Europeans make** in the New World?

The new lands of the Western Hemisphere yielded many plants unknown in Europe. On Christopher Columbus's first voyage in 1492 he became the first European to discover maize (corn), sweet potatoes, capsicums (peppers), plantains, and pineapples. Subsequent expeditions found potatoes, wild rice, squash, tomatoes, cacao (chocolate beans), peanuts, cashews, and tobacco. These plants, many of which had been developed and cultivated by the American Indians, were carried back to Europe, and their cultivation spread to suitable climates throughout the world.

How did **America get its name**?

America is derived from the name of Italian navigator Amerigo Vespucci (1454–1512), who took part in several early voyages to the New World. Vespucci had been a merchant in service of the Medici family in Florence and later moved to Spain where he worked for the company that outfitted the ships for Christopher Columbus's (1451–1506) second and third voyages. He sailed with the Spaniards on several expeditions (in 1497, 1499, 1501, and 1503). Though scholars today question his role as an explorer, in a work by German geographer Martin Waldseemüller (c. 1470–c. 1520) published in 1507, the author credited Vespucci with realizing that he had actually arrived in a New World—not

Spanish explorer Hernán Cortés and his men kneel before Aztec leader Montezuma.

in the Far East as other explorers (including Columbus) had believed. Thus, Waldseemüller suggested the new lands be named America after Amerigo Vespucci. For his part, Waldseemüller was led to believe this by Vespucci himself who had written to Lorenzo de Medici in 1502 or 1503, relaying his discovery of a new continent and vividly describing it. About a year later, the letter was published under the title *Mundus Novus* (New World), which was translated and published in future editions.

The designation *America* was used again in 1538 by Flemish cartographer Gerardus Mercator (Gerhard Kremer; 1512–1594). Today the term in the singular refers to either continent in the Western Hemisphere and sometimes specifically to the United States. In the plural, it refers to all of the lands of the Western Hemisphere, including North and South America and the West Indies.

How was **Cortés** able to claim Mexico for the Spaniards?

On behalf of the European power, Hernán Cortés (1485–1547) claimed Mexico after conquering its native peoples. In 1519 Cortés landed on the eastern coast of Mexico and founded the city of Veracruz. From there he marched inland, making an alliance with the Tlaxcalan Indians (who had fought wars against the powerful Aztecs in central Mexico). On November 8 of that year, Cortés marched into Mexico City (then named Tenochtitlán) and took the Aztec leader, Montezuma, hostage. Cortés then continued to Mexico's west coast. When he returned to Mexico City in the central part of the country in 1520, he found the Aztecs in revolt against the Spaniards. Fierce fighting ensued, and by the end of June Montezuma was dead. This period of warfare is still remembered today by Mexicans as *la noche triste* (the sorrowful night). It was not until the following year, in August 1521, that Cortés, after a four-month battle, claimed Mexico City, and the land came under control of the Spanish.

Who was **Montezuma**?

Montezuma (or Moctezuma) was the name of two rulers of the Aztec Indians in Mexico. Montezuma I ruled an area that extended from the Atlantic to the Pacific. He is credited with enlarging Tenochtitlán (the Aztec capital, which is today Mexico City). He died in 1469, but three years earlier was succeeded by his nephew, Montezuma II (1466–1520), who is the Montezuma most people are familiar with since it was during his reign that the Aztecs came into contact with—and were eventually conquered by—the Spaniards under explorer Hernán Cortés (1485–1547). In 1519 Montezuma tried to persuade Cortés not to come to Mexico City, but Cortés and his troops marched inland anyway. The Aztecs rose up to fight the Spaniards, but Montezuma was wounded as he addressed his warriors. He died a few days later, on June 30, 1520, one year before Mexico City fell to the Spaniards.

Who was **Cabeza de Vaca**?

Álvar Núñez Cabeza de Vaca (c. 1490–c. 1560) was a Spanish explorer who in 1527 joined an expedition to the New World, and because of his reports that he believed the area north of Mexico might be rich in precious metals, other Spanish explorers were later inspired to explore the region.

After landing in Florida in 1527, Cabeza de Vaca and a few others, including the black explorer Estevanico, became separated from the ships. The men built a barge and sailed across the Gulf of Mexico from northern Florida to the islands off the Texas coast, where their ship was wrecked. In 1528 Cabeza de Vaca was imprisoned by the Indians there. He escaped by 1530 and set out on foot through northern Mexico, exploring the region for some five years. Proceeding south, in 1536 Cabeza de Vaca reached Mexico City, which was then the capital of Spain's North American holdings. He returned to Spain the following year.

In 1541 the explorer led an expedition to the Río de la Plata region (of southern South America) and reached Asunción (in present-day Paraguay) in 1542. Cabeza de Vaca was appointed the Spanish colonial governor of Paraguay, a position he held for two years. He proved to be an inept leader, was deposed by the colonists, and returned to Spain. There he was banished to service in Africa.

What were the **Seven Cities of Cíbola**?

The reference is to an area in present-day northern New Mexico that was thought by early Spanish explorers to contain vast treasures. One expedition in search of these legendary golden cities was that led by Francisco Vásquez de Coronado (c. 1510–1554), who sought to claim the riches for Spain. In 1540 he set out from North Galicia (a province northwest of Mexico City) with some 300 Spanish troops as well as some Indians. They made it into the region where Arizona and New Mexico lie today. There they encountered Zuni Indian settlements and believed these to be Cíbola. The Spaniards captured the Zuni, who were the descendants of the Anasazi cliff dwellers that had settled the Southwest as early as 10,000 B.C. The Spaniards found no gold at the Zuni settlement. Separate expeditions set out, still hoping to locate riches in the area. They did not find any precious metals, but they did make some discoveries: They were the first Europeans to see the Grand Canyon, to travel up the Rio Grande Valley, and to encounter several native peoples living in the region. In 1546 Coronado was accused of cruelty in his treatment of these peoples.

What areas of the United States did **Hernando de Soto** explore?

Spanish explorer Hernando de Soto (c. 1500–1542) ventured throughout the Southeast before he caught fever and died along the banks of the Mississippi River.

Having been part of a brutal expedition that crushed the Inca Empire (in what is now Peru), in 1536 de Soto returned to Spain a hero. But he sought to go back to the New World and got his wish when King Charles I (1500–1558) appointed him governor of Cuba and authorized him to conquer and colonize the region that is now the southeastern United States.

Arriving in Florida in the winter of 1539, de Soto and an army of about 600 men headed north during the following spring and summer. In search of gold and silver, they traveled through present-day Georgia, through North and South Carolina, through the Great Smoky Mountains, and into Tennessee, Georgia, and Alabama. After defeating the Choctaw leader Tuscaloosa in October 1540 in south-central Alabama, the Spaniards headed north and west into Mississippi. They crossed the Mississippi River on May 21, 1540, and de Soto died later that same day. Since he had shown no mercy in his conquests of the native peoples, de Soto's troops sunk his body in the river so that it would not be discovered and desecrated by the Indians. Then his army continued on without him; under the direction of Luis de Moscoso they reached Mexico in 1541.

Were the **Spaniards the first Europeans to reach North America** after the Vikings?

No, that distinction goes to explorer John Cabot (c. 1451–1498), who in 1497 sailed westward from Bristol, England, in search of a trade route to the East. Cabot's story began in 1493, when Christopher Columbus (1451–1506) returned to Spain from his New World voyage, claiming to have reached Asia. From the accounts of the trip, Cabot, who was himself a navigator, believed it was unlikely that Columbus had traveled that far. He did, however, believe it was possible (as did subsequent explorers) to find a route—a northwest passage—that ran north of the landmass Columbus had discovered, and by which Asia could be reached. In 1495 the Italian Cabot, born Giovanni Caboto, took his family to England, and the following year, in March, appealed to King Henry VII (1457–1509) for his endorsement in pursuing the plan. For his part, the king, who was well aware of the claims made by the Spanish and Portuguese who had sponsored their own explorations, was eager to find new lands to rule. And so he granted a patent, authorizing Cabot's expedition.

Later that year, 1496, Cabot set sail, but problems on board the ship and foul weather forced him to turn back. The following spring, on May 20, 1497, he sailed again, in a small ship that had been christened *Matthew.* The crew of 20 included his son, Sebastian. On June 24, they sighted land and Cabot went ashore. While he saw signs of human habitation, he encountered no one. From reports of the trip, scholars believe Cabot had reached the coasts of present-day Maine, Nova Scotia, and probably Newfoundland. He then sailed home, returning to England on August 6. He reported to the king six days later and was given a reward, as well as authorization for a more sizeable expedition, which he undertook in May 1498. This time Cabot set sail with five ships in his command. But the expedition was not heard from again.

What became of John Cabot's son, **Sebastian**?

Sebastian (c. 1476–1557), who was born in Bristol, England, and sailed with his father on his successful expedition to North America the summer of 1497, did not take part in his father's ill-fated venture the following year. Had he done so, the world would have lost another great adventurer, since that expedition was never heard from again. Instead, Sebastian stayed behind and pursued his father's cause and that of other merchant-navigators who were determined to find overseas trade routes to the East.

During his lifetime, Sebastian Cabot drew up maps for both the English and Spanish royalty and from 1525 to 1528 led a Spanish expedition that reached South America's Río de la Plata, and sailed into the Paraná and Paraguay Rivers. In 1544 Cabot published an engraved map of the world. And seven years later, under a pension from King Edward VI (1537–1553), he founded the Merchant Adventurers of London. This group sponsored expeditions seeking a northeast passage (around Europe) to establish a trading route to the East. In so doing, the group effected trade with Russia.

CIRCUMNAVIGATION OF THE GLOBE

How long was it before **someone reached the East by sailing west**?

It was not until 1520 that a route was found. Portuguese navigator Ferdinand Magellan (c. 1480–1521) was on an expedition for Spain when he found a southwest passage, which took him around the southern tip of South America, through a winding waterway that still bears his name, the Strait of Magellan. Having set out from Spain in September 1519, it was a full year later before Magellan (born Fernão de Magalhães, and known in Spain as Fernando de Magallanes) reached this point, south of the South American mainland and north of the Tierra del Fuego island chain (today these islands are part of Argentina and Chile). And this was only after he had crushed a mutiny. Nevertheless, Magellan had found a connection between the Atlantic and the Pacific Oceans.

He sailed on from there, reaching the island we know as Guam on March 6, 1521. Ten days later, he discovered the Philippines. On the Philippine island of Cebú, he made an alliance with a treacherous native sovereign for whom he undertook an expedition to the nearby island of Mactan. It was there that Magellan met with his death in April 1521. His expedition continued without him, under the direction of Juan Sebastián de Elcano (c. 1476–1526), who in 1522 returned to Seville, Spain, along with 18 other survivors of the Magellan expedition. Their cargo, aboard the ship *Vittoria,* included valuable spices—which more than paid for the expense of the expedition.

Who was **the first Englishman to circumnavigate** the globe?

It was English admiral Sir Francis Drake (1540 or 1543–1596), who set out in 1577 to explore the Strait of Magellan. He did so, investigating the coast of South America (he and his crew plundered coastal Chile and Peru in the process), before continuing into the South Pacific and heading northward. He eventually reached the coast of present-day California, which he named New Albion (a name that did not stick), and claimed it for Queen Elizabeth I (1533–1603). He continued sailing northward and is believed to have reached Vancouver—still in search of the Northwest Passage. Not finding it (he was much too far south, explorers would later learn), he sailed westward. He reached the so-called Spice Islands (today known as the Moluccas) in east Indonesia in 1579. Drake also found the Indonesian island of Java, before continuing west through the Indian Ocean, rounding the southern tip of South Africa at the Cape of Good Hope and skirting the western coast of Africa northward to Sierra Leone. From there Drake returned home to Plymouth, England, where he landed in 1580, the first Englishman to travel around the world. He was knighted by the queen one year later.

It is also Drake who, along with his fellow countryman Sir Walter Raleigh (1554–1618), bears the dubious honor of introducing tobacco to his homeland. In 1586 Drake returned from another expedition to North and South America, where he did battle with Spanish fleets for control of lands. He then picked up colonists in Vir-

Who was the first to go around the world?

By sea, the first to circumnavigate the globe was the Basque navigator Juan Sebastián de Elcano (c. 1476–1526), though 18 sailors who made the trip with him also claim the distinction. The trip was completed in 1522 and had taken nearly three years. In 1519, Elcano had set out with Ferdinand Magellan (c. 1480–1521) on a Spanish-sponsored expedition that became the first one successful in finding a western route to the East. Having rounded the southernmost point of mainland South America in 1520, and entering into the South Pacific, the expedition reached the Philippines in 1521. When Magellan was killed there, it was Elcano who took leadership of the crew and guided the expedition westward, returning to Spain as the first sea captain to go around the world.

ginia, who carried with them potatoes and the materials and implements for tobacco smoking. Drake remained in the service of the queen for his whole life, going on to fight and defeat the Spanish Armada in 1588. On a mission to the West Indies in 1596, Drake died on board his own ship.

Who was **the first woman to circumnavigate** the globe?

It was a young French woman named Jeanne Baret. In 1766 Louis-Antoine de Bougainville (1729–1811), a French naval officer, undertook an around-the-world expedition, which was successful and returned to France in 1769. But the crew made an interesting discovery en route: When the French arrived in Tahiti, the Tahitians immediately noticed something the crew had not—that one of the servants on the expedition was a woman. "Jean" Baret had been hired in France by one of the ship's officers, Commercon, who also served as botanist for the expedition. Commercon did not know Baret was a woman. Her secret discovered by the Tahitians, she confessed, revealing that she was an orphan who had first disguised herself as a boy to get employment as a valet. When she learned about Bougainville's expedition, she decided to continue the disguise in order to carry out an adventure that would have been impossible for a woman in that day. She was the first woman known to have circled the globe.

What was **Thomas Cavendish**'s claim to fame?

English navigator Thomas Cavendish (c. 1560–1592) followed in Sir Francis Drake's (1540 or 1543–1596) footsteps. Seeing Drake return from his exploits at sea and against the Spanish, Cavendish was inspired. And it was for good reason: Drake had earned himself fame, wealth, and the honor of being knighted. So in 1586 Cavendish set out with three ships for Brazil, made it through the Strait of Magellan and then proceeded to capture Spanish treasure—including their prized ship, the *Santa Ana*. The Kings of Spain later mourned the loss and the fact that the ship had been taken by "an English youth…with 40 or 50 companions."

Cavendish, now in the Pacific, continued his voyage, which took him to the Philippines, Moluccas, and Java before he rounded the Cape of Good Hope (Africa) and returned home. The journey had taken 2 years and 50 days, cost him two of his own ships, and made him the third person to circumnavigate the globe.

But his welcome in England was not what he expected: Cavendish was received with acclaim, but was not knighted by the queen. The fame and fortune that had come his way quickly vanished; he spent most of his new money, and his renown soon faded. By 1590 Cavendish thought he would try the journey again. Setting sail with five ships in August 1591, the fleet was headed for trouble. Having made it to South America, heavy storms separated the ships as they attempted to make their way through the Strait of Magellan. The ship Cavendish captained turned back toward Brazil, attempting to make landfall. But Cavendish himself never made it. He died en route, believing he had been deserted by his mates.

Who was **Hawaii-Loa**?

He was a Polynesian chief who sailed some 2,400 miles of open water from the Marquesas Islands, near Tahiti, to discover the Hawaiian Islands in the A.D. 400s. The islands were first discovered by Europeans in 1778 when British navigator Captain James Cook (1728–1779) landed on the island of Kauai and named the islands after John Montagu (c. 1718–1792), who was the fourth earl of Sandwich and first lord of the admiralty. Captain Cook died there at the hand of the natives in a skirmish over a stolen boat.

What were **Captain Cook**'s discoveries?

British navigator Captain James Cook (1728–1779) was one of the world's greatest explorers, commanding three voyages to the Pacific Ocean and sailing around the world twice. From 1768 to 1771, aboard the ship *Endeavor,* Cook conducted an expedition to the South Pacific, where he landed in Tahiti, and made the first European discovery of the coasts of New Zealand, Australia, and New Guinea, which he also charted. In 1772 Cook set out to find the great southern continent that was believed to exist. He spent three years on this voyage, which edged along the ice fields of Antarctica. On his last voyage, which he undertook in 1776 on a mission to find a passage around North America from the Pacific, Cook charted the Pacific coast of North America as far north as the Bering Strait. He met his death in 1778 on the Hawaiian Islands. Cook's voyages led to the establishment of Pacific Ocean colonies by several European nations.

Who was the first European to traverse the **Bering Strait**?

The Bering Strait, which connects the Arctic Ocean and the Bering Sea is 53 miles across at its most narrow point. The first European to traverse it was Danish navigator Vitus Bering (1681–1741) in 1728. The explorer, for whom the strait and the sea were named, had been employed by Tsar Peter the Great (1672–1725) of Russia to determine whether Asia and North America are connected.

COLONIAL AMERICA

What were the **Spanish holdings in the New World**?

New Spain comprised much of the Spanish possessions in the New World during the colonial period. At its height, New Spain included what are today the southwestern United States, all of Mexico, Central America to the Isthmus of Panama, Florida, much of the West Indies (islands in the Caribbean), as well as the Philippines in the Pacific Ocean. The viceroyalty (province governed by a representative of the monarch) was governed from the capital at Mexico City beginning in 1535. In 1821 a Mexican rebellion ended Spanish rule there, and the colonial empire of New Spain was dissolved. By 1898 Spain had relinquished all its possessions in North America. Its last holdings were the islands of Cuba, Puerto Rico, Guam, and the Philippines, which were ceded to the United States after Spain lost the Spanish-American War (1898).

During the colonial period, Spain also claimed other territories in the New World—in northern and western South America. Most of these holdings fell under the viceroyalty of Peru, which was administered separately from the viceroyalty of New Spain. These possessions were also lost by Spain by the end of the 1800s.

What were the **French holdings** in the New World?

The French possessions in North America, called New France, consisted of the colonies of Canada, Acadia, and Louisiana. The first land claims were made in 1534 by French explorer Jacques Cartier (1491–1557) as he sailed the St. Lawrence River in eastern Canada. In 1604 Sieur de Monts (Pierre du Gua; c. 1568–c. 1630) established a settlement at Acadia (in present-day Nova Scotia, Canada), and French claims later extended the region to include what are today the province of New Brunswick and the eastern part of the state of Maine. After founding Quebec in 1608, explorer Samuel de Champlain (c. 1567–1635) penetrated the interior (present-day Ontario) as far as Georgian Bay on Lake Huron, extending French land claims westward. In 1672 French-Canadian explorer Louis Jolliet (1645–1700) and French missionary Jacques Marquette (1637–1675) became the first Europeans to discover the upper part of the

Mississippi River. Ten years later, French explorer Sieur de La Salle (1643–1687) followed the Mississippi to the Gulf of Mexico, claiming the river valley for France and naming it Louisiana. While the French expanded their North American claims, the majority of French settlers lived in Canada. France lost Canada to Great Britain in the Seven Years' War (1756–63). Louisiana changed hands numerous times before it was finally sold to the United States in 1803 as part of the Louisiana Purchase; it was France's last claim on the North American mainland. French culture and influence in these areas remains prevalent today.

In 1635 the French also claimed the West Indies islands of Martinique and Guadeloupe (and its small surrounding islands, including Saint Barthélemy). In 1946 the French government changed the status of these islands from colonies to "overseas departments."

What was the **Lost Colony**?

It was the second English colony established in America: Set up in 1587 on Roanoke Island, off the coast of North Carolina, by 1590 it had disappeared without a trace. Theories surround the disappearance, though it is not known for sure what happened.

Roanoke Island had also been the site of the first English colony, set up in 1585 by about 100 men who were sent there by Sir Walter Raleigh (1554–1618). Raleigh had perceived the island to be a good spot for English warships (that were then fighting the Spanish) to be repaired and loaded with new supplies. But the plan was not a success: The land wasn't fertile enough to support both the colonists and the Indians living nearby, and ships could not get close enough to the island since the surrounding sea proved too shallow. The colonists returned to England the following year. Meantime, Raleigh had dispatched another group of colonists from England. They arrived at Roanoke days after the original settlers left. Seeing that the site had been abandoned, all but 15 of the colonists opted to return to England.

In spring 1587 Raleigh sent yet another group of colonists to America, but these ships were headed for areas near Chesapeake Bay, farther north (in present-day Virginia). Reaching the Outer Banks in July, the ships' commander refused to take the colonists to their destination and instead left them at Roanoke Island. The colonists' leader, John White, who had also been among the first settlers at Roanoke, returned to England for supplies in August 1587. However, the ongoing war between England and Spain prevented him from returning to the colony until three years later. Arriving back at Roanoke in August 1590, expecting to be met by family members and the 100 or so settlers (including some women and children). Instead he discovered that the colony was abandoned.

The only clue that White found was the word *Croatoan,* which had been engraved on a tree. The Croatoan, or Hatteras, were friendly Indians who lived on an island south

A native American greets the Pilgrims, who in 1620 voyaged to America seeking religious freedom and self-government. (Original painting by Henry Sargent.)

of Roanoke Island. White set out to see if the colonists had joined the Hatteras Indians, but weather prevented the search and his expedition returned to England instead.

Two theories explain what might have become of the lost colonists: Since the shore of Chesapeake Bay was their original destination, the colonists might have moved there but, encountering resistance, perished at the hands of the Indians. Other evidence suggests that the colonists became integrated with several Indian tribes living in North Carolina. Either way, they were never seen again by Europeans.

Were the **Pilgrims** explorers?

The Pilgrims were early settlers who sought religious freedom and self-government in the New World. Since theirs was a religious journey, they described themselves as pilgrims. In fact, they were Separatists, Protestants who separated from the Anglican Church to set up their own church. In 1609 they fled their home in Scrooby, England, and settled in Holland. Fearing their children would lose contact with their own culture (becoming assimilated into the Dutch culture), the group decided to voyage to America to establish their own community. In 1620 they arrived on the rocky western shore of Cape Cod Bay, Massachusetts. Their transatlantic crossing had taken 66 days aboard the *Mayflower.* Two babies were born during the passage, bringing the number of settlers to 102, only some 35 of whom were Pilgrims; the rest were merchants. On November 21 the Pilgrims drafted the Mayflower Compact, an agreement by which the 41 signatories (the men aboard the *Mayflower*) formed a body politic that was authorized to enact and enforce laws for the community. Religious leader John Carver (1576–1621) was voted governor. Though their colonial charter from the London Company specified they were to settle in Virginia, they decided to establish their colony at Cape Cod, well outside the company's jurisdiction. By December 25 the Pilgrims had chosen the site for their settlement and began building at New Plymouth.

The first year was difficult and the Pilgrims faced many hardships: 35 more colonists arrived aboard the *Fortune,* putting a strain on already limited resources; sicknesses such as pneumonia, tuberculosis, and scurvy claimed many lives, including that of Governor Carver; and the merchants in the group challenged the purity of the settlement. Having secured a new patent from the Council of New England in June 1621, the lands of New Plymouth Colony were held in common by both the Pilgrims and the merchants. But this communal system of agriculture proved unsuccessful, and in 1624 William Bradford (1590–1657), who had succeeded Carver as governor, granted each family its own parcel of land. The Wampanoag Indians, who had previously occupied the land settled by the Pilgrims, proved friendly and were helpful advisers in agricultural matters. In 1626 the Pilgrims bought out the merchants' shares and claimed the colony for themselves. Though they were inexperienced at government before arriving in America and had not been formally educated, the Pilgrims successfully governed themselves according to the Scriptures, and Plymouth Colony remained independent until 1691, when it became part of Massachusetts Bay Colony—founded by the Puritans.

How were **the Puritans** different from the Pilgrims?

The Puritans who founded Massachusetts Bay Colony were, like the Pilgrims, religious Protestants (both sects "protested" the Anglican church). But while the Pilgrims separated from the church, the Puritans wished to purify it. Their religious movement began in England during the 1500s, and they were influenced by the teachings of reformer John Calvin (1509–1564). They also had strong feelings about government, maintaining that people can only be governed by contract (such as a constitution), which limits the power of a ruler. When King James I (1566–1625) ascended the throne of England in 1603, he was the first ruler of the house (royal family) of Stuart. The Stuart monarchs, particularly James's successor, King Charles I (1600–1649), tried to enforce absolute adherence to the High Church of Anglicanism and viewed the Puritan agitators as a threat to the authority of the crown.

Persecuted by the throne, groups of Puritans fled England for the New World. One group was granted a corporate charter for the Massachusetts Bay Company (1629).

Unlike other such contracts, which provided the framework for establishing colonies in America, this one did not require the stockholders to hold their meetings in England. Stockholders who made the voyage across the Atlantic would become voting citizens in their own settlement; the board of directors would form the legislative assembly; and the company president, Puritan leader John Winthrop (1588–1649), would become the governor. In 1630 they settled in what is today Boston and Salem, Massachusetts, establishing a Puritan Commonwealth. By 1643 more than 20,000 Puritans arrived in Massachusetts, in what is called the Great Migration. Puritans also settled in Rhode Island, Connecticut, and Virginia during the colonial period.

What were the **Dutch colonial holdings**?

New Netherlands was the only Dutch colony on the North American mainland. It consisted of lands surrounding the Hudson River (in present-day New York) and, later, the lower Delaware River (in New Jersey and Delaware). Explorers from the Netherlands first settled the area in about 1610. In 1624 the colony of New Netherlands was officially founded by the Dutch West India Company. On behalf of the company, in 1626 Dutch colonial official Peter Minuit (1580–1638) purchased the island of Manhattan from the American Indians for an estimated $24 in trinkets. The colonial capital of New Amsterdam (present-day New York City) was established there. The Dutch held the colony until 1664 when it was conquered by the English under the direction of the Duke of York (James II; 1633–1701), the king's brother. The English sought the territory since New Netherlands separated its American holdings. Under British control the area was divided into two colonies: New Jersey and New York.

During the colonial period the Netherlands also claimed the West Indies islands of Aruba, Bonaire, and Curacao (called the Netherlands Antilles), which were administered separately from New Netherlands on the North American mainland.

What were the **Swedish colonial holdings**?

The Swedish possessions consisted of a small colony called New Sweden, established in 1638 at Fort Christina (present-day Wilmington), Delaware. The Swedes gradually extended the settlement from the mouth of the Delaware Bay (south of Wilmington) northward along the Delaware River as far as present-day Trenton, New Jersey. The settlers were mostly fur traders, though there was farming in the colony as well. In 1655 the territory was taken by the Dutch in a military expedition led by director general of New Netherlands Peter Stuyvesant (c. 1610–1672). For nine years the territory was part of the Dutch colonial claims called New Netherlands. In 1664 the English claimed it and the rest of New Netherlands. Delaware was set up as a British proprietary colony, which it remained until the outbreak of the American Revolution (1775–83). New Sweden was the only Swedish colony in the Americas.

What are the **origins of slavery** in America?

The roots of slavery in North America date back to about 1400, when the Europeans arrived in Africa. At first, the result of African contact with Europeans was positive, opening trade routes and expanding markets. Europeans profited from Africa's rich mineral and agricultural resources and for a while abided by local laws governing their trade; Africans benefited from new technologies and products brought by the Europeans. But the relationship between the two cultures soon turned disastrous as the Europeans cast their attention on a decidedly different African resource—the people themselves. As the Portuguese in West and East Africa began trading in human lives and the Dutch in South Africa clashed with the native people who, once displaced by the wars, became servants and slaves, other Europeans began calculating the profits that could be made in the slave trade.

By the end of the 1400s Europeans had landed in the New World. Soon Europe's established and emerging powers vied for control of territories in the new lands of North and South America and the West Indies. The Spaniards, Portuguese, Dutch, English, and Swedes all made claims in the Western Hemisphere and began setting up colonies.

By the mid-1600s triangular patterns of trade emerged: The most common route began on Africa's west coast, where ships picked up slaves. The second stop was the Caribbean islands—predominately the British and French West Indies—where the slaves were sold to plantation owners and traders used the profits to purchase sugar, molasses, tobacco, and coffee. These raw materials were then transported north to the third stop, New England, where a rum industry was thriving. There ships were loaded with the spirits before traders made the last leg of their journey back across the Atlantic to Africa's west coast, where the process began again. Other trade routes operated as follows: 1) manufactured goods were transported from Europe to the African coast; slaves to the West Indies; and sugar, tobacco, and coffee transported back to Europe, where the route began again; and 2) lumber, cotton, and meat were transported from the colonies to southern Europe; wine and fruits to England; and manufactured goods to the colonies, where the route began again. There were as many possible routes as there were ports and demand for goods.

The tragic result of the triangular trade was the transport of an estimated 10 million black Africans. Sold into slavery, these human beings were often chained below deck and allowed only brief if any periods of exercise during the transatlantic crossing, which came to be called the Middle Passage. Conditions for the slaves were brutal and improved only slightly when traders realized that should slaves perish during the long journey across the ocean, it would adversely affect their profits upon arrival in the West Indies. After economies in the islands of the Caribbean crashed at the end of the 1600s, many slaves were sold to plantation owners on the North American mainland, initiating another tragic trade route. The slave trade was abolished in the 1800s, putting an end to the forced migration of Africans to the Western Hemisphere.

When did **the first Africans** arrive in the British colonies of North America?

In 1619 a Dutch ship carrying 20 Africans landed at Jamestown, Virginia. They were put to work as servants, not as slaves. Though they had fewer rights than their white counterparts, they were able to gain their freedom and acquire property, which prompted the development of a small class of "free Negroes" in colonial Virginia. For example, there is record of an Anthony Johnson arriving in Virginia in 1621 as a servant. He was freed one year later, and about thirty years after that he imported five servants himself, receiving from Virginia 250 acres of land for so doing.

THE EARLY REPUBLIC OF THE UNITED STATES

What was the goal of the **Lewis and Clark expedition**?

The expedition, which began in 1804 and took more than two years to complete, had three purposes: to chart a route that would be part of a passage between the Atlantic and Pacific Oceans; to trace the boundaries of the territory obtained in the Louisiana Purchase; and to lay claim to the Oregon Territory.

Thomas Jefferson (1743–1826) was president of the United States at the time, and he believed that a route could be found between St. Louis and the West Coast. As early as 1801, Jefferson had conceived of the idea that the Missouri and Columbia Rivers might be followed west, leading to the Pacific. The journey would also be a reconnaissance mission; information would be collected about the vast region and communications would be set up with its inhabitants. On April 30, 1803, the United States bought the Louisiana Territory from France. The purchase extended from the Mississippi River in the east to the Rocky Mountains in the west, and from the Gulf of Mexico in the south to British America (Canada) in the north. Jefferson soon picked his private secretary, Virginia-born Meriwether Lewis (1774–1809), to lead the westward expedition. Lewis then chose as his co-leader William Clark (1770–1838), who, as a lieutenant in the U.S. Army, had served General Anthony Wayne on the frontier (1792–96). Beginning in the summer of 1803, Lewis and Clark undertook the necessary preparations for the overland journey. These included studying the classification of plants and animals, learning how to determine geographical position by observing the stars, and recruiting qualified men (mostly hunters and soldiers) for the expedition.

On May 14, 1804, the Lewis and Clark expedition left St. Louis and headed up the Missouri River to its source. They then crossed the Great Divide and followed the Columbia River to its mouth (in present-day Oregon) at the Pacific Ocean, where they arrived in November 1805—one and a half years after they had set out. They arrived back in St.

Lewis and Clark, who headed an early nineteenth-century expedition to explore the western United States, address members of an Indian tribe. (Illustration by Patrick Glass.)

Louis on September 23, 1806, having gathered valuable information on natural features of the country, including its flora, fauna, and the Indian tribes who lived there.

The expedition had been helped by the addition, in what is now North Dakota, of a Shoshone Indian woman named Sacagawea (c. 1786–1812). Lewis and Clark had hired her husband, French-Canadian trader Toussaint Charbonneau, as an interpreter during the winter of 1804–05. Lewis and Clark thought that Sacagawea would be able to help them communicate with the Shoshone living in the Rocky Mountains, which she later did: Her brother was their chief.

After the expedition, Lewis was made governor of Louisiana Territory, a post he served from 1807 to 1809. Clark resigned from the army in 1807 and became brigadier general of the militia and superintendent of Indian affairs for Louisiana Territory. In 1813 he became governor of the Missouri Territory (the Louisiana Territory less the state of Louisiana, which was organized as a state admitted into the Union in 1812), a post he held until 1821.

What was the **Trail of Tears**?

The Trail of Tears was the government-enforced western migration of the American Indians, which began March 25, 1838. As an increasing number of white settlers

What did Manifest Destiny have to do with the expansion of the United States?

The doctrine of Manifest Destiny emerged in the United States in the early 1800s and by the 1840s had taken firm hold. Adherents believed that Americans had a God-given right and duty to expand their territory and influence throughout North America. Manifest Destiny was a rallying cry for expansionism and prompted rapid U.S. acquisition of territory during the 1800s. The acquisitions began in 1803 with the purchase of Louisiana Territory from France; in 1819 Florida and the southern strip of Alabama and Mississippi (collectively called the Old Southwest) were acquired from Spain in the Adams-Onís Treaty; in 1845 Texas was annexed after white settlers fought for and declared freedom from Mexico, forming the Republic of Texas and petitioning the Union for statehood; in 1846 the western border between Canada and the United States was determined to lie at 49 degrees north latitude, the northern boundary of what is today Washington State; in 1848 by the Treaty of Guadalupe Hidalgo, the United States secured New Mexico and California after winning the Mexican War (1846–48); and in 1853 southern Arizona was acquired from Mexico in the Gadsden Purchase. With the 1853 agreement the United States had completed the acquisition of territory that makes up the contiguous states.

The expansionist doctrine of Manifest Destiny was again invoked as justification for the Spanish-American War (1898), which was fought over the issue of freeing Cuba from Spain. Spain lost the war, dissolving its empire. Cuba achieved independence (though it was occupied by U.S. troops for three years). By the close of the nineteenth century, Manifest Destiny had resulted in U.S. acquisition of the outlying territories of Alaska, Hawaiian Islands, Midway Islands, the Philippines, Puerto Rico, Guam, Wake Island, American Samoa, Panama Canal Zone, and U.S. Virgin Islands.

moved inland from the coastal areas, they laid claim to Indian homelands; conflicts ensued. The government's solution was to relocate the Indians to make room for the pioneers. As many as 17,000 members of the Cherokee Nation were forced from tribal lands in Georgia, Alabama, and Tennessee and were escorted west by federal troops under the command of General Winfield Scott (1786–1866) along an 800-mile trail that followed the Tennessee, Ohio, Mississippi, and Arkansas Rivers to Indian territory in Oklahoma, north of the Red River.

The journey took between 93 and 139 days, and the movement westward was called the Trail of Tears not only because it was a journey the native people did not wish to make, to a place where they did not wish to go, but because an estimated 4,000 people—mostly infants, children, and the elderly—died en route. The deaths were caused by sickness, including measles, whooping cough, pneumonia, and tuberculosis.

Escorted in waves, it was a full year (spring of 1839) before the Cherokee had been relocated; some 1,000 had refused to leave their tribal lands in the southeast. The forced migration thus resulted in the fragmentation and weakening of the tribe.

Who were the **expansionists**?

Not long after the colonies won the American Revolution (1775–83), founding the United States of America, a nationalistic (superpatriotic) spirit emerged in the hearts of many citizens of the new country. Eager to spread American ideals, many looked westward, northward, and southward to expand the territory of the Union beyond the original 13 states. These people were called expansionists. Not only did they favor the settlement of the frontier, but some advocated seizure of the southwest (from Spain and later from Mexico), Florida (from Spain), the Louisiana Territory (from France), and the Northwest Territories and even Canada (from Britain). By the 1840s the doctrine of Manifest Destiny—stating that the United States had a God-given right and duty to expand its territory and influence throughout North America—took hold.

The fires of expansionism were fanned by population growth during the 1800s. Pioneer settlement of the Plains and the Old Northwest (the present-day states of Ohio, Michigan, Indiana, Illinois, Wisconsin, and part of Minnesota) resulted in an increase in farmland and overall crop production; Yankee ingenuity resulted in inventions such as the cotton gin (1793) and the McCormick reaper (1831), which improved the processing and harvesting of raw materials such as cotton and grain; and a continuous influx of immigrants from Europe supplied labor for the factories that had popped up across New England and the mid-Atlantic states. All these factors combined to create a rapid population growth. In the two decades between 1840 and 1860 alone, the population of the United States more than doubled, increasing from just over 17 million to more than 38 million. Though the eastern seaboard cities grew, a system of new canals, steamboats, roads, and railroads opened up the interior to increased settlement. By 1850 almost half the population lived outside the original 13 states.

Though Canada, of course, remained in the hands of the British, the spirit of expansionism resulted in the United States' relatively speedy acquisition of North American territories that had belonged to Spain, Mexico, France, and the British: By 1853 the United States owned all the territory of the present-day contiguous states, and by the end of the century, it owned all the territory of its present-day states—including Alaska (purchased from Russia in 1867) and Hawaii (annexed in 1898).

AMERICAN IMMIGRATION

What prompted widespread **Irish emigration** in the mid-1800s?

In 1845 Ireland's potato crop failed. Though crop failures in Europe were widespread at the time, the blight in Ireland was particularly hard hitting because of the reliance on a single crop, potatoes, as the primary source of sustenance. The Great Famine that resulted lasted until 1848. The effect was a drastic decline in the Irish population—due both to deaths and to emigration. Between 700,000 and 1 million people died in Ireland during the Great Famine. And between 1846 and 1854 alone, 1.75 million left the country in search of a better life elsewhere. Three-quarters of those were headed to the United States.

In the years after the Great Famine, experts determined that the blight of the late 1840s was caused by a fungus that was probably introduced to Ireland by a ship from North America, where there had been crop failures in the early 1840s. Sad irony that the very conduit that had likely borne the blight from North America to Ireland also carried immigrating Irish back to North American shores.

What was the **Great Famine**?

Typically the term refers to the Great Irish Famine, which began in 1845. That year crops failed across Europe, resulting in widespread hunger and disease, claiming 2.5 million lives. The famine was especially severe in Ireland, where many peasants exported their grain and millet and depended on potatoes for their own sustenance. The failure of the potato crop in 1845 was the twentieth in Ireland since 1727, and it marked the beginning of the Great Famine—which lasted into 1848 due to more disastrous crop failures each successive year. The crops had failed due to blight (previously unknown to Ireland), which was caused by a microscopic organism (fungus) believed to have been introduced by a ship from North America. British charity and government relief did little to alleviate the suffering. The resulting famine was responsible for a drastic decline in the Irish population—due both to death and emigration. Between 700,000 and 1 million people died in Ireland, and nearly 2 million people left the country in search of a better life elsewhere.

Other severe, or "great," famines include one in 1769 in Bengal; the disaster claimed the lives of 10 million Asian Indians, one-third of the population. In 1878 up to 20 million Chinese people died as a two-year drought held Asia in its grips, causing widespread crop failures. It is still considered the worst famine in history.

When were the **major waves of immigration** to the United States?

The first wave of immigration was during colonial times when most new arrivals to North America were from England. But other European countries were represented as well, including France, Germany, Ireland, Italy, the Netherlands, Sweden, and Wales. By 1700 roughly a quarter million people lived in the American colonies. By the beginning of the American Revolution (1775–83), the number had climbed to 700,000. Some of these new arrivals had been encouraged to immigrate by Virginia's "headright system": Englishmen who could pay their own Atlantic crossing were granted 50 acres of land; each of their sons and servants were also granted an additional 50 acres. Other colonies also adopted the headright system, with the land amounts varying in each. Other immigrants in this first wave were poor and could not afford the price of the transatlantic passage; by signing a contract agreeing to work as an indentured servant for a period (of typically three to seven years), their fare was paid by their future master. At the end of this period, the servant became a freeman and was usually granted land, tools, or money by the former master. During the American Revolution and for several decades after, the flow of immigrants into the new country slowed.

A second wave of immigration began in 1820. During the next 50 years, nearly 7.5 million newcomers arrived in the United States. Many were Irish who escaped the effects of the Great Famine back home, settling the cities along the eastern U.S. seaboard. An equal number (roughly a third) were German, who settled the nation's interior farmlands, particularly the Midwest. An economic depression in the 1870s stemmed the tide of immigrants, but only for a short time.

Between 1881 and 1920 a third wave brought more than 23 million immigrants to American shores. These new arrivals were largely from eastern and southern Europe. German immigration reached its peak in 1882. In 1883 the United States saw the peak of immigration from Denmark, Norway, Sweden, Switzerland, the Netherlands, and China. Just after the turn of the century, in 1902, U.S. immigration set new records as people from Italy, Austro-Hungary, and Russia made the transatlantic journey.

Between 1920 and 1965 immigration slowed. In the last three and a half decades of the twentieth century and into the twenty-first, a fourth wave of immigration has taken place. In spring 1998 the U.S. Census Bureau released a report citing that 9.6 percent of American residents are foreign-born, or roughly one in every ten. This is the highest percentage reported since the 1930s, when 11.6 percent of U.S. residents were natives of another country—a result of the third wave of immigration (1881–1920). However, the origin countries have shifted: Latin Americans now account for about half of all new arrivals, one-fourth are Asian-born, and one-fifth are European.

How many immigrants arrived at **Ellis Island**?

More than 12 million people first entered the United States through Ellis Island; their descendants account for an estimated 40 percent of the nation's current population. The majority of new arrivals were European, but immigrants also came from

Hundreds of immigrants in the Great Hall at Ellis Island await possible entry to the United States (undated photo).

the West Indies, Asia, and the Middle East. More men than women arrived at the immigration depot.

Originally a three-acre landmass, the island is situated in the New York Harbor, off the southern tip of Manhattan. It was named for Samuel Ellis, a merchant and farmer who owned the island during the late 1700s. New York acquired the land, and in 1808 sold it to the federal government. The site served as a fort, and later, an arsenal. By the end of the century, record numbers of immigrants prompted the federal government to establish a bureau to process the new arrivals, the vast majority of whom entered the country at its largest port, New York City.

On January 1, 1892, the Federal Immigration Station opened on the island—in the shadows of the Statue of Liberty (dedicated 1886 on nearby Bedloe Island). The Ellis Island facility, which by 1901 consisted of 35 buildings, was the country's chief immigration station. Its heaviest use was in processing the influx of immigrants who arrived between 1892 and 1924. The facility was closed on November 29, 1954, when immigration quotas had drastically reduced the number of incoming people and the mass-processing center was no longer needed. On May 11, 1965, Ellis Island was designated a national historic site. During the 1980s it was extensively restored so that visitors to the park and museum are afforded a glimpse of what their ancestors experienced upon arriving in this new land.

What did **immigrants experience** at Ellis Island?

The Ellis Island immigration depot was a processing center for third-class ship passengers arriving in New York Harbor. (Most first- and second-class passengers were processed by immigration officials on board their ships.) The new arrivals were ferried from their transatlantic vessels to Ellis Island, where they disembarked and were guided in groups into registration areas in the Great Hall, a room 200 feet long and 100 feet wide. There they were questioned by government officials who determined their eligibility to land. Upon completing the registration process, newcomers were ushered into rooms where they were examined by doctors. The processing was extremely businesslike—to the point of being dehumanizing. Processing typically took between three and five hours. An estimated 98 percent of those arriving at Ellis Island were allowed into the country. The remaining 2 percent were turned back for medical reasons (as U.S. health officials tried to keep out infectious diseases) or for reasons of insanity or criminal record. Other facilities at the Ellis Island Immigration Station included showers that could accommodate as many as 8,000 bathers a day, restaurants, railroad-ticket offices, a laundry, and a hospital. At its peak, the Ellis Island station processed some 5,000 immigrants and nonimmigrating aliens (visitors) daily.

EXPLORATION IN THE TWENTIETH CENTURY

What is the **Northwest Passage**?

The Northwest Passage is the circuitous sea passage between the Atlantic and the Pacific Oceans; it was long sought after by explorers. Though it was eventually found through a series of discoveries, it was not completely navigated until 1903 to 1906 by Norwegian explorer Roald Amundsen (1872–1928).

Convinced of the existence of such a passage, numerous navigators attempted to find it during the early years of European westward sea exploration. Their determination led to the discovery of the St. Lawrence River, between Canada and the United States, by French sailor and explorer Jacques Cartier (1491–1557) in 1534 to 1535; of Frobisher Bay, off the coast of Baffin Island, north of Quebec, by English commander Sir Martin Frobisher (1535–1594) in 1576; of Davis Strait, between Baffin Island and Greenland, by English navigator John Davis (c. 1550–1605) in 1587; and of the Hudson River, in eastern New York State, and Hudson Bay, the inland sea of east-central Canada, by English navigator Henry Hudson (?–1611) in 1609 to 1611.

After centuries of efforts, Roald Amundsen finally completed the first successful navigation of the Northwest Passage in September 1906; it was after a journey that lasted more than three years. The Norwegian adventurer had left the harbor at Oslo at midnight on June 16, 1903, aboard the *Gjöa,* a ship so small it required only a crew of six; Amundsen had bought the vessel with borrowed money. After a harrowing adventure that saw the *Gjöa* and her crew survive a shipboard fire, a collision with a reef, fierce winter storms, and ice that hemmed them in, Amundsen arrived in Nome, Alaska, where the entire town turned out to greet him and his crew. The nephew of Norwegian explorer Otto Sverdrup (1855–1930), who was there at the time, played the Norwegian national anthem as the ship pulled into the dock. Amundsen is said to have broken into tears.

Who was the first person to reach the **North Pole**?

There has been some dispute over this one: The credit usually goes to American explorer and former naval officer Robert E. Peary (1856–1920), who, after several tries, reached the North Pole by dogsled on April 6, 1909, along with Matthew A. Henson and three Inuit companions. Unbeknownst to him, five days before this achievement, another American explorer, Dr. Frederick Cook (1865–1940), claimed that he had reached the North Pole a year earlier. Peary and Cook knew each other: Cook had been the surgeon on the Peary Arctic expedition of 1891 to 1892, which reached Greenland. And for his part, Cook's claim was investigated by scientists, but the evidence he supplied did not substantiate the claim. Thus, Peary was recognized as the first to reach the northern extremity of Earth's axis.

Who was the first person to reach the **South Pole**?

Norwegian explorer Roald Amundsen (1872–1928) was first to reach the South Pole, in December 1911. Before earning this distinction, he had achieved another first—sailing the Northwest Passage (from 1903 to 1906).

Amundsen's desire to be an Arctic explorer had been with him almost his entire life. As a teen, he is said to have slept with his bedroom windows open year-round in order to become accustomed to the cold. When he was a young man of 21, he turned his attention away from the study of medicine to making an Arctic passage. He recognized that many of the previous (and failed) attempts to travel to the Arctic shared a common characteristic: "The commanders of these expeditions had not always been ships' captains." He resolved to become an experienced navigator and soon took jobs as a deck hand on various ships.

In 1897 Amundsen was chosen as the first mate on the *Belgica,* the ship that would carry the first Belgian Antarctic expedition under the command of Adrien Gerlache de Gomery (1866–1934). Also on board was the American Dr. Frederick Cook (1865–1940), who had been on one of Robert E. Peary's (1856–1920) earlier Arctic expeditions and who would, in 1909, dispute Peary's claim that he was the first to reach the North Pole. This was the same news that Amundsen would hear as he was preparing to make the North Pole. Upon learning of the success of Peary's 1909 expedition, Amundsen shifted his sights to reaching the South Pole instead, and quietly began to lay plans to do so. In fact, it was not until his expedition, which left Oslo in September 1910, was under way that he telegraphed his announcement back to Norway that he was in fact headed to the South, not the North, Pole. As it turned out, a race was on between the Norwegians and the British: Shortly after Amundsen had set sail, naval officer Robert Falcon Scott (1868–1912) had left England at the head of an expedition to reach the South Pole.

The Norwegians landed at Ross Ice Shelf, Antarctica, on February 10, 1911. It was not until 10 months later, on December 14, 1911, on a sunny afternoon, that they raised their country's flag at the spot their calculations told them was the South Pole. Before heading north again, they celebrated their achievement with double rations. When British naval officer Robert Falcon Scott's expedition arrived at the South Pole on the morning of January 18, 1912, they found the Norwegian flag flying over it. On their way back the crew died due to bad weather and insufficient food supplies. Amundsens's Norwegian expedition arrived safely at their base camp on January 25, 1912.

Who was the first to reach **Mount Everest**'s summit?

New Zealander Sir Edmund Hillary (1919–) was the first person to climb to the summit of Mount Everest, the highest mountain in the world. Everest, in the Himalaya Mountains, between Nepal and Tibet, rises nearly five and a half miles (29,028 feet) above sea level.

After numerous climbers made attempts on Everest between 1921 and 1952, Hillary reached the top on May 29, 1953, as part of a British-led expedition; he was followed by fellow climber Tenzing Norgay (1914–1986), a Nepalese Sherpa. Hillary took a picture of Tenzing at the summit, but Tenzing did not know how to work the camera so there is no picture of Hillary. The "Sir" was added to Hillary's name by Queen Elizabeth II (1926–), who took great pleasure in the fact that the triumph on Everest had been achieved by a British expedition. Having been crowned on June 2, 1953, it was one of her first official acts as queen. The mountain was named for another Briton, Sir George Everest (1790–1866), who served as a British surveyor-general of India from 1830 to 1843. (Tibetans call the mountain Chomolungma, and the Nepalese call it Sagarmatha.)

When was the first solo nonstop **transatlantic flight** made?

It was in 1927: On May 21, at 10:24 P.M., American Charles A. Lindbergh (1902–1974) landed his single-engine monoplane, the *Spirit of St. Louis,* at Le Bourget Air Field, Paris, after completing the first solo nonstop transatlantic flight. Lindbergh, declining to take a radio in order to save weight for an additional 90 gallons of gasoline, had taken off in the rain from Roosevelt Field, Long Island, New York, at 7:55 A.M. on May

20. The plane was so heavy with gasoline (a total of 451 gallons of it) that the *Spirit of St. Louis* had barely cleared telephone wires upon takeoff. Lindbergh covered 3,600 miles (about a third of it through snow and sleet) in 33 hours, 29 minutes. He won a $25,000 prize, which was offered in 1919, and became a world hero—hailed as the "Lone Eagle." His autobiography, titled *The Spirit of St. Louis* (published 1953), won the aviator the Pulitzer prize.

When was **Amelia Earhart** last heard from?

American aviator Amelia Earhart (1897–1937), the first woman to fly solo across the Atlantic Ocean, was last seen on July 1, 1937, and was last heard from on July 2, as she and navigator Fred Noonan (1893–1937) attempted to make an around-the-world flight along the equator.

The Kansas-born Earhart first became interested in aviation, which was very new at the time, during the early 1920s and began taking flying lessons. In 1928 she was invited to be the only woman on board a transatlantic flight, which departed from Newfoundland and landed in Wales. The trip made her famous as the first woman to cross the Atlantic Ocean by air. She followed that accomplishment in 1932 with a solo transatlantic flight: Earhart took off from Harbor Grace, Newfoundland, Canada, on the evening of May 20, 1932. Her destination was Paris. Within hours problems began for the aviator: She encountered a violent electrical storm, the altimeter failed, the wings iced up, and finally, the exhaust manifold caught on fire. Earhart decided to land in Ireland rather than attempting Paris. After a 15-hour flight, she touched down in a pasture outside of Londonderry in Northern Ireland. Again, fame and acclaim were hers, as the first woman to cross the Atlantic in a solo flight. She went on to set speed and distance records for aviation and soon conceived of the idea of flying around the world along the equator.

On May 20, 1937, Earhart and Noonan took off from Oakland, California. Reaching Miami, Florida, they stopped for repairs. On June 1, 1937, they departed Miami and headed for Brazil. From there, they flew across the Atlantic to Africa and then across the Red Sea to the Arabian Peninsula. Then it was on to Karachi, Pakistan; Calcutta, India; and Burma (present-day Myanmar). Earhart reached New Guinea on June 30, and she and Noonan prepared for the most difficult leg of the journey: to Howland Island, a tiny speck of land only two and a half miles long in the middle of the vast Pacific Ocean. The next day, July 1, they left New Guinea and began the 2,600-mile flight to Howland Island. On July 2 a U.S. Navy vessel picked up radio messages from Earhart that indicated reports of empty fuel tanks. Efforts to make radio contact with her failed. Though an extensive search effort ensued, no trace of the plane or two-person crew was found, and no one knows for certain what happened. Speculation surrounds the disappearance. One theory was that Earhart's true mission in making the around-the-world flight was to spy on the Japanese-occupied Pacific islands. However, this has never been substantiated, and, given the circumstances under which they were flying, the likelihood is that the plane crashed into the ocean, claiming Earhart's and Noonan's lives.

SPACE: THE FINAL FRONTIER

When did the **exploration of space** begin?

The "space age" began on October 4, 1957, when the Soviet Union launched *Sputnik* (later referred to as *Sputnik 1*), the first artificial satellite. The world reacted to the news of *Sputnik,* which took pictures of the far side of the moon, with a mix of shock and respect. Premier Nikita Khrushchev (1894–1971) of the Soviet Union immediately approved funding for follow-up projects. And leaders in the West, not to be outdone by the Soviets in exploring the last frontier, also vowed to support space programs. Four months later, the United States launched its first satellite, *Explorer 1,* on January 31, 1958. Not only had the launch of *Sputnik* initiated the space age, it had also started a "space race": The Soviet and American programs would continue to rival each other, with one accomplishment leap-frogging the other, for about the next three decades.

What was the first **animal sent into orbit**?

The Soviets immediately followed the success of *Sputnik 1* (launched October 4, 1957) by sending the first animal into space: a dog named Laika. The female Russian Samoyed traveled in a pressurized cabin aboard *Sputnik 2,* which was launched November 3, 1957, making her the first living creature to go into orbit. The trip ended badly for Laika, however; she died a few days into the journey. Before sending humans into orbit, both the Soviets and the Americans needed to prove that animals could survive in outer space. While the Soviets experimented with dogs traveling in space, by the end of 1958 the United States would send a monkey into space (but not into orbit). The following spring (May 28, 1959), two female monkeys, Able and Baker, were launched into orbit in a U.S. spacecraft and were recovered alive. They had traveled 300 miles aboard *Jupiter.*

Who was the **first person in space**?

The first person in space was Soviet cosmonaut Yuri A. Gagarin (1934–1968), who orbited the Earth in the spaceship *Vostok 1,* launched April 12, 1961. The flight lasted one hour and 48 minutes. The achievement made Gagarin an international hero. U.S. president John F. Kennedy (1917–1963) announced later that year, on November 25, that the United States would land a man on the moon before the end of the decade. The

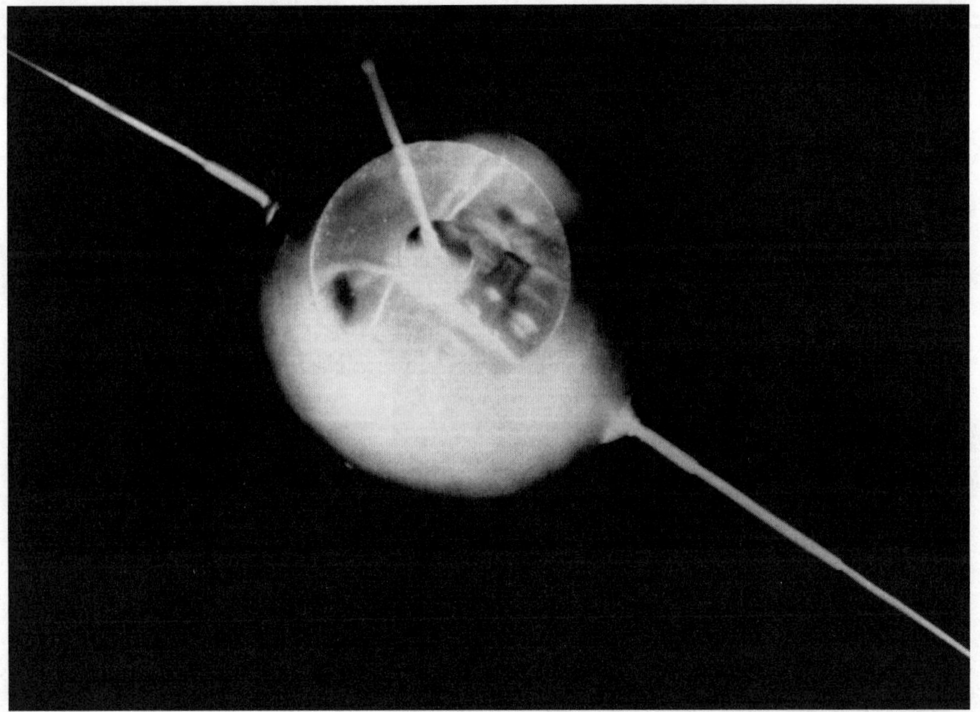

A model of *Sputnik 1* on display at the Prague, Czechoslovakia, exhibition. The *Sputnik 1* space capsule was launched by the Soviet Union in 1957, starting the "space race" with the United States.

first step toward reaching that goal was made by putting the first American into space: On May 5, 1961, Alan Shepard Jr. (1923–1998) piloted the first U.S. spaceflight, aboard the *Freedom 7.* The suborbital flight reached an altitude of 116.5 miles. Just more than nine months later, the United States put a man into orbit: On February 20, 1962, astronaut John Glenn Jr. (1921–) orbited Earth three times in the spaceship *Friendship 7.*

Who was the **first man** to walk **on the moon**?

It was American astronaut Neil Armstrong (1930–) who, on July 20, 1969, stepped out of the lunar module from *Apollo 11* and walked on the moon. Armstrong, who was joined by astronaut Edwin "Buzz" Aldrin Jr. (1930–), uttered the famous words, "That's one small step for a man, one giant leap for mankind." (The live voice transmission had dropped the "a" before "man," but it was added in later.)

Who was the **first woman in space**?

The Soviets claim the distinction of putting the first woman into space: On June 16, 1963, cosmonaut Valentina Tereshkova-Nikolaeva (1937–) was launched into space

Astronaut "Buzz" Aldrin walks on the moon during the *Apollo 11* mission, July 20, 1969.

aboard the *Vostok 6*. She spent three days circling Earth. It was 20 years before the United States would match the accomplishment: On June 18, 1983, Sally K. Ride (1951–) and four other crew members were launched into space aboard the space shuttle *Challenger*. Ride remained with the National Aeronautics and Space Administration (NASA) for four years after completing the mission. She served on the presidential commission that investigated the 1986 *Challenger* disaster.

What impact did the *Challenger* disaster have on the U.S. space program?

The immediate effect was that all scheduled launches were scratched, pending the outcome of the government investigation into the disaster. President Ronald Reagan (1911–2004) acted quickly to establish a presidential commission to look into the January 28, 1986, accident, appointing former secretary of state William P. Rogers (1913–2001) as the chair. Rogers conducted a thorough investigation involving public and private hearings, more than 6,000 people, 15,000 pages of testimony, 170,000 pages of documents, hundreds of photographs, and reports of independent technical studies. Additionally, the commission reviewed flight records, film evidence, and the recovered debris.

On June 6, 1986, the commission released its report, citing the cause of the disaster as the failure of the O-ring seals "that are intended to prevent hot gases from leaking through the joint during the propellant burn." The commission had learned that although both the National Aeronautics and Space Administration (NASA) and the O-ring manufacturer Morton Thiokol were concerned about the seals (which had also been used on other shuttles), they had come to regard them as an acceptable risk. The commission went on to say that "the decision to launch the *Challenger* was flawed." The U.S. House of Representatives' Committee on Science and Technology, which had spent two months conducting its own hearings, also concluded that the disaster could have been prevented, citing that "meeting flight schedules and cutting costs were given a higher priority than flight safety."

With the blame for the disaster placed on NASA's doorstep, public confidence in the agency plummeted and NASA's own astronauts became concerned that their lives had been put at risk. However, the commission also made nine recommendations to NASA, including redesigning the solid rocket booster joints, giving astronauts and engineers a greater role in approving launches, reviewing the astronaut escape systems, regulating the rate of shuttle flights to maximize safety, and reforming the shuttle program's management structure.

A turnover of personnel, which included some astronauts, resulted at the space agency, which spent almost three years rebuilding. It was not until September 29, 1988, that an American shuttle again flew in space.

Why was the **Mars *Pathfinder*** project important?

The historical space project yielded a tremendous amount of data—even more than scientists had planned or hoped—about the planet, totally reshaping many long-held notions about Mars, the fourth planet in the solar system. *Pathfinder* landed on Mars on July 4, 1997, and deployed a two-foot-long robotic rover called *Sojourner.* The small rover collected what project scientists described as a staggering amount of data: 2.3 billion bits of information about the planet; 8.5 million measurements of temperature and pressure; 16 chemical analyses of rocks, soil, and other surface materials; and 16,500 pictures. Among other things, the new data caused scientists to rethink the planet's color: long believed to be red, the color is now described as a deep amber or butterscotch. The evidence gathered by *Sojourner* supports the notion that there may have been life on Mars. Future projects aimed at studying the planet's properties will further explore the theory that there was once life on Mars. With the success of *Pathfinder,* project scientists turn their attention to improving the data-gathering capabilities of future rovers—including the addition of brushes to the robot, so that dust can be collected from rocks on the surface of Mars, which is a distance of 155 million miles from Earth.

What is the **Mars Rover**?

The Mars Rover is a National Aeronautics and Space Administration (NASA) project launched in June 2003 and that landed on the red planet in January 2004. Two rovers, named *Spirit* and *Opportunity,* were designed to be robotic geologists, collecting data

The International Space Station, pictured under construction in April 2002, is a scientific laboratory orbiting about 250 statute miles above Earth.

as they roam the surface of Mars. Each robot has features that NASA described as similar to what any living creature would need during exploration: a body, or structure, to protect the rover's vital organs; brains, or computers, to process information; temperature controls, including internal heaters and insulation to protect the structure from the elements; a neck and head, or a mast for the cameras to give the rovers a human-scale view; eyes and other "senses," or cameras and instruments that give the rovers information about their environment; wheels and legs for mobility; arms for reaching; and the ability to communicate, via antennas for "speaking" and "listening." The rovers are powered by batteries and solar panels.

Since the first data began to be returned to Earth in early March 2004, the rovers have provided scientists with data and impressive images for research and study. A dramatic finding from the twin rovers was the discovery that the arid planet had once been wet and possibly habitable: Based on data collected by *Opportunity,* researchers concluded that "liquid water was once intermittently present at the Martian surface." Observers can track the progress of *Spirit* and *Opportunity,* each exploring different areas of the red planet, at the NASA Web site (http://marsrovers.jpl.nasa.gov/home/index.html).

What is the **International Space Station**?

It is a scientific laboratory orbiting about 250 statute miles above Earth. The International Space Station (ISS) is a cooperative project among 16 nations, led by the United States; the other partners in what is called the most complex science project in history are Russia, Canada, Japan, the 11 nations of the European Space Agency, and Brazil. When completed, the ISS will have a mass of about 1,040,000 pounds, will measure 356 feet across and 290 feet long, and will have almost an acre of solar panels to provide electrical power to six state-of-the-art laboratories.

The ISS was built through a series of missions, the first of which was the U.S.-designed, Russian-built space module *Zarya* (Sunrise), launched November 20, 1998. The original plan called for completion of the ISS in 2004, but construction was extended into 2006. As of April 30, 2005, the ISS had been in orbit, and a work in progress, for 2,353 days, with a cumulative crew time in space of 1,640 days. (At that time, the United States had sent 11 expeditions to the ISS.) It then measured 240 feet across and 146 feet long, and weighed just more than 404,000 pounds.

Scientists believe that experiments conducted on the ISS will lead to discoveries in medicine and to the development of materials and new science that will benefit people around the world. The space station is also considered a monumental first step in preparing for future human space exploration.

Observers can track the ISS in orbit by using tools at NASA's Web site (http://spaceflight.nasa.gov/realdata/tracking/index.html). Construction updates and a timeline can be found at the Discovery Channel Web site (http://www.discovery.com/stories/science/iss/iss.html).

THE CLASSICAL AGE

What was the **Trojan War**?

According to Greek legend, the Trojan War was a 10-year siege, c. 1200 B.C., on the ancient city of Troy (also called Ilium). The Greek poet Homer (c. 850–? B.C.) chronicled the Trojan War in his epic the *Iliad.* According to Homer, the (Greek) Mycenaeans, under their great king Agamemnon, set out to conquer the city of Troy, situated on the Turkish coast at the southwestern part of the Hellespont. The Hellespont, now called the Dardanelles, and the Bosporus are narrow straits that connect the Aegean Sea with the Black Sea. As such, they are the gateway between Europe and Asia, and in Mycenaean times they held the key to control of the profitable Black Sea trade. In hopes of taking over that trade, Agamemnon's army attacked the powerful city of Troy, launching the decade-long campaign. The mythological war likely reflected a real war that was fought over the Dardanelles about 1200 B.C.

What is a **Trojan horse**?

Thanks to the Greek poet Homer (c. 850–? B.C.), a Trojan horse has come to symbolize anything that looks good but is actually subversive. According to his Trojan war epic the *Iliad,* after nearly 10 years of fighting the Trojans for control of their city, the Greek Mycenaean army built a huge wooden horse on wheels and offered it as a gift to their enemy. Leaving the "peace offering" outside the city walls of Troy, the Mycenaean army then departed. Despite warnings (including one from the Trojan princess Cassandra), the Trojans accepted the gift, and they opened the gates and wheeled the huge wooden horse into the city. It was a naive move: Once the horse was inside the city gates, Mycenaean soldiers who were hidden inside the wooden structure took Troy by storm, ending the decade-long campaign and taking control of the lucrative Black Sea trade.

THE PELOPONNESIAN WAR

What was the **Peloponnesian War**?

It was the war fought between the Greek city-states of Athens and Sparta between 431 and 404 B.C.; it left Athens ruined. The beginning of the war signaled the end of the golden age of Greece.

As the city-states (which were self-ruling regions made up of a city and the surrounding territory) developed, an intense rivalry grew between Athens and Sparta. The Spartans recruited allies into the Peloponnesian League (the Peloponnese Peninsula forms the southern part of mainland Greece), and together they attacked the Athenian Empire, which had been gaining in power.

The war consisted of three stages: The first was the Archidamian War (431–421 B.C.), named for Archidamus, the Spartan king who led the unsuccessful attacks on fortified Athens. In 421 the so-called Peace of Nicias (421 to 413 B.C.) began, which was negotiated by Athenian politician Nicias. But this truce was broken when an Athenian commander promoted counterattacks on Athens's aggressors in 418 and 415 B.C. The attacks on the Peloponnesian League were unsuccessful, and so the Ionian War broke out (413–404 B.C.). After years of fighting, the Ionian War finally ended in victory for Sparta, after the Peloponnesian League had not only gained the support of Persia to defeat Athens but had successfully encouraged Athens's own subjects to revolt. Athens surrendered to Sparta, ending the Peloponnesian War.

What happened to **Athens after the Peloponnesian War**?

At the same time that Athens was under attack from Sparta, the city-state also suffered a terrible plague, beginning in 430 B.C. The plague killed about a third of Athens's citizens. This contributed to Athens's eventual defeat at the hands of its Spartan rivals during the Peloponnesian War: After the plague had claimed so many lives, Athens simply couldn't muster the military leadership and strength that it needed to defeat the increasingly powerful Peloponnesian League. At the end of the fighting, in 404 B.C., Athens was in ruins. Sparta became the dominant city-state in the Greek world. But conflicts continued among the city-states, and Thebes eventually defeated Sparta in 371 B.C. As a result of the ongoing warfare, the Greek economic conditions declined, and the gap between rich and poor widened. The public spirit that had been the hallmark of the golden age of Greece disappeared. In short, the city-states were no longer the glorious entities they had once been.

While the Greek city-states were in decline, Greece's neighbor to the north, Macedonia, was growing more powerful. In 353 B.C. Macedonian king Philip II (382–336 B.C.) launched an attack on Greece. The war that resulted did not end until 338 B.C. when Greece was finally conquered. The Macedonian victory was only the first part of Philip's plans; he believed a combined Macedonian-Greek army could defeat the powerful Persians. But he was not to see this happen. When Philip was killed by a Macedonian in 336 B.C., he was succeeded by his 20-year-old son, Alexander (356–323 B.C.), a man history would come to call "the Great." Alexander carried out his father's plan to invade Persia. And two years after Philip's death, Alexander began a 10-year campaign that ultimately conquered the Persian Empire. By the time he was 30 years old, Alexander had conquered much of the known world, expanding his empire from Egypt to India.

PAN-HELLENISM

Why was Macedonian king **Alexander** known as "the Great"?

The Macedonian king Alexander the Great (356–323 B.C.) has passed through history as a legendary figure, a reputation attributable to the fact that he conquered virtually all of the known world in his day. In effect, he was king of the world, though his tenure in that role was relatively brief.

Born in the Greek city-state of Macedonia, Alexander was the son of King Philip II (382–336 B.C.), who had risen to power three years earlier. He had an upbringing and education befitting a young Greek prince: Alexander was tutored by Greek philosopher Aristotle (384–322 B.C.) and was trained in athletics and war. His studies of Greek literature and art would later combine with his skill as a warrior to produce a formidable conqueror: Inspired by Greek culture, he also possessed the military prowess and cunning necessary to spread Hellenism (Greek culture) throughout the known world.

At the age of 16, Alexander began running the government of Macedonia while his father, King Philip, waged military campaigns to protect and expand his kingdom. At the age of 17, Alexander joined his father on the battlefield, where he commanded a section of the army to defeat Thebes. When Philip was assassinated in 336 B.C., Alexander acted quickly to assert his claim to the throne. Having done so by 335 B.C., he continued to carry out his father's campaigns, securing Greece and the Balkan Peninsula by the fall of that year. He followed this with an all-out offensive on the Persian Empire, long the enemy of Greece. Supremely courageous and confident in his own abilities as well as in his troops (which numbered in the tens of thousands), by the fall of 331 B.C. Alexander had defeated the Persian army and along the way claimed Egypt. The centuries-old Persian Empire crumbled, and the young Macedonian king proclaimed himself "Lord of Asia." Still, he pressed on, claiming Afghanistan (where he solidified his claim by marrying a young princess) and then India. He was poised to

Alexander the Great in battle. The Macedonian king conquered virtually all of the known world in his day.

take the Arabian peninsula in 324 B.C. when he died of fever. He was 33. His vast kingdom, which he had ruled leniently but nevertheless authoritatively, was divided among his former generals.

The Roman Conquest

How was **Rome** able to conquer Greece?

After Alexander the Great died in 323 B.C., his generals divided his empire into successor states, but Greece remained under Macedonian control. Though the Greeks would fight the armies of the Macedonian kings into the 200s B.C., they would not achieve independence, and instead associations of Greek city-states again fought each other.

Meantime, just to the west, Rome had been conquering lands to become a formidable power in the Mediterranean and soon began to look eastward to expand its authority. When Rome conquered Macedonia in 197 B.C., Greece was liberated. Fifty years later, in 146 B.C., Greece was conquered by Rome and was divided into provinces. While the city-states had no military or political power, they nevertheless flourished under Roman rule. And the Romans, who had first started borrowing from Greek thought and culture around 300 B.C., were soon spreading Greek ideas, art, and religion throughout their empire, giving rise to the Greco-Roman culture inherited by modern western civilization.

What were the **Punic Wars**?

The Punic Wars were three major campaigns that Rome waged to expand its empire. Messina (a present-day province of Sicily, Italy) was the site of the First Punic War, which began in 264 B.C. when warring factions in the trade and transportation center called for assistance from both Carthage and Rome. The Carthaginians arrived first and secured the city. But the Romans, who had girded their navy for the battle, arrived and drove the Carthaginians out (241 B.C.), conquering Sicily. Messina became a free city but was allied with Rome.

The rivalry between Rome and Carthage did not end there: The Second Punic War (218–201 B.C.) was largely fought over control of Spain. When the great Carthaginian general Hannibal (247–183 B.C.) captured the Roman-allied city of Sagunto, Spain, in 218, he then crossed the Alps and invaded Italy, where he was met by and defeated the Roman armies. The deciding battle in the Second Punic War was fought in the North African town of Zama (southwest of Carthage) in 202 B.C. It was there that the Romans under general Scipio Africanus (236–183 B.C.) crushed the Carthaginians under Hannibal. Rome exacted payments from Carthage, and Carthage was also forced to surrender its claims in Spain. In 201 B.C. the two powers signed a peace treaty, which held for five decades.

The Third Punic War erupted in 149 B.C. when the Carthaginians rebelled against Roman rule. By 146 B.C. Carthage, which had been richer and more powerful than Rome when the Punic Wars began, was completely destroyed in this third and final conflict with the Roman army.

The Sacking of Rome

If the **Roman Empire** was so powerful, how could it have fallen?

One could argue that the Roman Empire collapsed under its own weight: It had become too vast to be effectively controlled by any one ruler.

By the close of the Punic Wars in 146 B.C., Greece, Macedonia, and the Mediterranean coasts of Spain and Africa had been brought under Roman control. Within a century, Rome again began to expand overseas. Under the Roman general Pompey (106–48 B.C.), eastern Asia Minor, Syria, and Judea (Palestine) were conquered. Next, Gaul was conquered by Pompey's rival, Julius Caesar (100–44 B.C.), adding the territo-

Who said "Veni, Vidi, Vici," and what does it mean?

The famous words were written by Roman statesman and general Julius Caesar (100–44 B.C.) as he announced the victory of his army in Asia Minor in early August 47 B.C. The extraordinarily concise message, which Caesar dispatched to Rome, means simply "I came, I saw, I conquered." The general had defeated Pharnaces II (63–47 B.C.) in a fight for control of Pontus, an ancient kingdom in northeast Asia Minor. The brief but decisive battle took place near Zela, in present-day Turkey.

ry west of Europe's Rhine River to the Roman world. In 31 B.C., in the Battle of Actium, Octavian (63 B.C.–A.D.14; Julius Caesar's adopted son and heir) defeated the forces of Marc Antony (c. 83–30 B.C.) and Cleopatra (69–30 B.C.), queen of Egypt, and in 30 B.C. Egypt became a Roman province.

In 27 B.C. Octavian became the first Roman emperor and was known as Augustus, meaning "exalted." Though Octavian's rule marked the beginning of the long period of stability called the Pax Romana, the Roman Empire had become so large—stretching across Europe and parts of Africa and the Middle East—that only a strong, central authority could govern it. During the 200 years of the Pax Romana, Rome's emperors gradually grew more powerful, to the point that after death an emperor was worshiped by the people.

But soon there were threats to this central control, not the least of which was the spread of Christianity, as well as invasions from the Germanic Goths and the Persians. Theodosius I (379–395) was the last emperor to rule the entire Roman Empire. When he died in A.D. 395, the empire was split into the West Roman Empire and the East Roman Empire, setting the stage for the decline of the Romans.

The West Roman Empire came under a series of attacks by various Germanic tribes including the Vandals and the Visigoths (the western division of the Goths), who invaded Spain, Gaul (in western Europe), and northern Africa. These assaults eventually led to the disintegration of the West Roman Empire by 476.

The East Roman Empire remained more or less intact, but it became known as the Byzantine Empire and was predominately a Greek-oriented culture from 476 until 1453, when it fell to the Turks.

How was **Rome "sacked"**?

After the split of the Roman Empire in 395, the West Roman Empire continued to weaken and Rome became subject to a series of brutal attacks by Germanic tribes. In

410 the Visigoths moved into Italy and looted Rome; in 455 the Vandals thoroughly ravaged the city; finally, in 476 the city fell when the Germanic chieftain Odoacer (433–493) forced Romulus Augustulus (c. 450–?), the last ruler of the empire, from the throne. By this time, however, Germanic chiefs had already begun claiming Roman lands and dividing them into several smaller kingdoms. The year 476 marks the official collapse of the West Roman Empire.

ATTILA THE HUN

Was **Attila the Hun** really a savage?

While Attila (c. 406–453) may have possessed some of the worthwhile qualities of a military leader, the king of the Huns was no doubt a ruthless and fierce figure. He is believed to have ascended through the ranks of the Hun army, coming to power as the leader of the nomadic group in 434. By this time, the Huns (who originated in central Asia) had occupied the Volga River valley in the area of present-day western Russia.

At first, like his predecessors, he was wholly occupied with fighting other barbarian tribes for control of lands. But under Attila's leadership, the Huns began to extend their power into central Europe. He waged battles with the eastern Roman armies, and, after murdering his older brother and co-ruler Bleda in 445, went on to trample the countries of the Balkan Peninsula and northern Greece—causing terrible destruction along the way. As Attila continued westward with his bloody campaigns, which each Hun fought using his own weapons and his own savage technique, he nearly destroyed the foundations of Christianity.

But the combined armies of the Romans and the Visigoths defeated Attila and the Huns at Châlons (in northeastern France) in June 451, which is known as one of the most decisive battles of all time. From there, Attila and his men moved into Italy, devastating the countryside before Pope Leo I (c. 400–461) succeeded in persuading the brutal leader to spare Rome. (For this and other reasons, Leo was later canonized, becoming St. Leo.) Attila died suddenly—and of natural causes—in 453, just as he was again preparing to cross the Alps and invade Italy anew.

THE MIDDLE AGES

What was the **Norman Conquest**?

The Norman Conquest is the brief but critical period in British history that began when the French duke William of Normandy (c. 1028–1087) sailed across the English

Channel in 1066 and invaded England. This was upon the death of what would turn out to be England's last Anglo-Saxon king, Edward the Confessor (c. 1003–1066). While William became known as William the Conqueror (and he did conduct a brutal conquest of Anglo-Saxon England), he might have had reason to believe he could claim the English throne upon King Edward's death: The named successor, Harold (c. 1022–1066), of the powerful Wessex family, had two years earlier become shipwrecked off the coast of France, where he reportedly took an oath that he would, upon King Edward's death, support William of Normandy (who was King Edward's distant cousin) as heir.

Hearing of Edward's death, William and his army set sail for England, where Harold had already assumed the throne as King Harold II. But Harold had previously quarreled with his brother Tostig, and the noble Wessex family was divided and engaged in a power struggle. Tostig was joined in his fight by the Norwegians who, at the same time that William was landing on England's southern coast, invaded from the north. Thus, William and his troops entered England without opposition (since Harold was focusing his efforts elsewhere). Though the king defeated the Norwegians and Tostig (who was slain in battle), he would not emerge the victor in his subsequent battle with William: On October 14 the two met in battle at Hastings, near the entrance to the Strait of Dover. Though he fought valiantly, Harold was killed.

William was crowned at Westminster Abbey on Christmas Day 1066. Within a few years, by 1070, he had killed many Anglo-Saxon nobles and the rest he deprived of their land. In the 21 years of his reign, William imposed Norman aristocracy on England, required that French be spoken at court, and drew England closer to Europe. He ruled until his death in 1087, after which the Norman nobility mixed with what was left of the Anglo-Saxons. It is this intermingling that produced the English language—from the German tongue of the Anglo-Saxons combined with the Norman French. William's descendants (albeit distantly so) have ruled England ever since his takeover in 1066.

Who was **Genghis Khan**?

He was a Mongol conqueror who rose to power in the early thirteenth century to rule over one of the greatest continental empires the world has seen. Born Temüjin (c. 1167–1227), he was named Genghis Khan, meaning "universal ruler," in 1206. He was a fearless military leader, a brilliant strategist, and a ruthless subjugator, known for his brutal methods.

Temuüjin was the first-born son of the leader of a small nomadic clan. When he was a young boy, his father was killed by a neighboring tribe (Tatars) and thus he rose to the status of chief. But instead of allowing a boy to lead them, clan members abandoned Temüjin and his family. He survived the hard-scrabbled youth of a destitute nomad. But by all accounts, he seemed destined to become a great leader.

By the time he was 20 years old, Temüjin had managed to forge alliances with various tribal leaders and claimed the leadership of a small clan. By 1189 he united two Mongol tribes, which he organized to conquer the rival Tatars by the year 1202. At a conference of Mongol leaders in 1206, Temüjin was pronounced the Great Ruler, or Genghis Khan, of the Unified Mongolian State. He began a transformation of the Mongol tribes, dividing them into military units, each one supported by a number of households. He imposed law and order, promoted education, and stimulated economic prosperity. Within five years, Mongol society was changed from a nomadic-tribal to a military-feudal system. Thus organized, Genghis Khan prepared his troops to expand the Mongolian empire.

Genghis Khan's armies embarked on a series of military campaigns, claiming land and subjugating peoples—sometimes using barbaric methods. By 1213 he controlled northern China to the Great Wall. By 1219 he controlled most of China and began campaigns into the Muslim world. When he died in the field in 1227, Genghis Khan commanded the vast territory from China to the Caspian Sea. He was succeeded by his sons, who continued to expand the Mongol holdings. His grandson was Kublai Khan (1215–1294), under whose leadership the Mongolian empire reached its pinnacle.

Who was **Tamerlane**?

Tamerlane (1336–1405) was a central Asian conqueror who gained power in the late 1300s. His Islamic name was Timur; Tamerlane is the English version. He was a barbaric warrior and a brilliant military leader whose fearsome tactics earned him the name Tamerlane the Terrible. By 1370 he was a powerful warlord whose government was centered in the province of Samarkand, in present-day Uzbekistan. In 1383 he launched a series of conquests that lasted more than 20 years and gained him control of a vast region including Iraq, Armenia, Mesopotamia, Georgia, Russia, and parts of India. He died in 1405, on an expedition to conquer China. His body was entombed in an elaborate mausoleum, which is considered a treasure of Islamic art. After his death, his sons and grandsons fought for control of his dynasty, which remained intact for another hundred years. Tamerlane and his heirs built Samarkand into a great city; in its day it was a center for culture and scholarship in central Asia.

Who was **Eleanor of Aquitaine**?

Eleanor of Aquitaine (c. 1122–1204) was among the most powerful figures of the Middle Ages (500–1350). Her father was William X (1099–1137), the duke of Aquitaine—one of the largest and wealthiest regions that became modern France (it extended from the Loire River to the Pyrenees Mountains). Eleanor inherited her father's vast holdings when he died in 1137; she was a teenager at the time. The same year, she married Louis VII (c. 1120–1180), who, within a month, succeeded his father as the king of France. As queen, Eleanor took an active role in her husband's business, even accompanying him on the Second Crusade (1147–49), purportedly so that she and her numerous female attendants could care for the wounded. The expedition ended in failure, largely because of a disagreement between the king and queen about the objective. Thereafter women were prohibited from joining the crusades. Though Eleanor and Louis had two children, both daughters, they were granted an annulment of their unhappy marriage in 1152 on the grounds that they were distantly related. According to feudal law, Eleanor regained Aquitaine. Within a year she married again, to Henry Plantagenet (1133–1189) of England, who was some ten years her junior. In 1154 he became King Henry II, and Eleanor was again queen, this time of England.

Though the couple had eight children (five sons and three daughters), their marriage was stormy. Henry had liaisons with other women, the result of which was an illegitimate line of heirs to the throne. Eleanor also resented him for his attempts to control Aquitaine, which she felt was rightly hers. In 1173, when Henry and Eleanor's three surviving sons attempted to depose their father, Eleanor threw her support behind her sons. The rebellion failed, and Eleanor fled for France. In 1174 she was captured and returned to England, where she was put in semi-confinement by her husband, the king, until his death in 1189. (Henry died while attempting to put down another rebellion by his sons, Richard the Lionhearted [1157–1199] and John Lackland [1167–1216]).

After King Henry's death, Eleanor again became a powerful force in European politics. Her son, Richard "the Lionhearted," inherited the English throne, and she took an active role in his administration. She even stepped in for him when he left England to fight in the crusades. When Richard died in 1199, his brother John ascended the throne. Eleanor, then in her late seventies, continued to wield power and influence. In an effort to ensure the future of the Plantagenets and to reconcile the English and French aristocracies, she arranged for one of her granddaughters to marry the son of the French king. She also managed to secure the family's land holdings in France for King John. She retired from her active political life in 1202 and died two years later.

What was the **Hundred Years' War**?

The term refers to a succession of wars between England and France. The fighting began in 1337 and did not end until 1453. However, the period was not one of constant warfare: truces and treaties brought about breaks in the military action between the countries. The reasons for the conflicts were many: England was trying to hang onto its provinces on the European continent; the French threw their support behind the Scots, who had their own battles with the English; the French wished to control the commercial center of Flanders (present-day Belgium), where the English had set up a profitable wool trade; and finally, the two countries disagreed about who should control the English Channel, the body of water that lies between them.

To further complicate matters, marriages between the English and French aristocracy meant that heirs to either throne could find themselves with a foreign relative, allowing them to lay claim to authority over the other country as well. When the first war broke out in 1337, King Edward III (1312–1377) of England claimed the French throne on the basis of the fact that his mother, Isabella, was the daughter of France's King Philip IV (called Philip the Fair; 1268–1314) and the sister of three French kings. Over the course of the next century, even though England won most of the battles and for a brief time controlled France (1420–22), it was the French who ultimately won the war in 1453. England lost all its territory on the continent, except Calais, which was also later taken by the French (in 1558).

Joan of Arc is surrounded by a crowd after leading the French into victory over the English in the Battle for Orleans in 1429. (Original painting by J.J. Scherrer.)

How did **Joan of Arc** become a warrior?

Joan of Arc (called the Maid of Orleans; c. 1412–1431) gained fame for leading the French into victory over the English in the Battle for Orleans in 1429. A year before the battle, the English forces had invaded northern France and took possession of an area that included the city of Reims, where all of France's kings were crowned in the cathedral. Thus, Charles VII (1403–1461), whom France recognized as their king, had never had a proper coronation. It is said that Joan, an extraordinarily devoted Catholic who was then just a teen, appealed to Charles to allow her to go into battle against the English who were besieging Orleans. Though he was skeptical of her at first (she claimed to have heard the voices of saints), he eventually conceded. In the battlefield, Joan also overcame the doubts of the French troops and their leaders, who were understandably hesitant to follow the young girl's lead. She proved to them that she was not only capable but also successful. In April 1429, in just 10 days' time, Joan led the French to victory over the English, who fled Orleans.

Still determined to see Charles properly crowned, Joan proceeded to lead a military escort for the king into Reims, where he was at last coronated on July 17, 1429—with Joan of Arc standing beside him. Next she determined that Charles should authorize her to try to free Paris from English control. Again the king acquiesced, but this time with dire results: She was captured by the French Burgundians (English sympathizers and loyalists), who turned her over to the English. Believing she was a heretic (by all reports Joan of Arc was clairvoyant), the English burned her at the stake in Rouen, France, on May 30, 1431. She is still considered a national hero of France. Recognizing Joan of Arc for her unswerving faith and for having valiantly pursued what she believed her mission to be, the Catholic Church canonized Joan of Arc in 1920. The feast day of St. Joan of Arc is celebrated on May 30.

What was the **impact of the Hundred Years' War**?

After waging war with each other for more than a century, in 1453, both England and France emerged as stronger, centralized governments. As the governments had gained strength, the nobility in both countries found themselves with less power and influence than they had enjoyed previously, and the system of feudalism, which before the war had been necessary in the absence of a larger, protective entity, was on the decline. In their strategies against each other, both countries had developed new military tactics. And though England had fewer resources than did France, it still managed to assert itself at sea, marking the beginning of that country's naval prowess.

What were the **Hussite Wars**?

The Hussite Wars were fought in the former kingdom of Bohemia (part of present-day Czech Republic) between 1420 and 1433. The country was plunged into the 13-year war by festering hostilities between papal forces and Bohemian peasants (the Hussites, named such since they were followers of religious reformer Jan Huss, c. 1373–1415). In the Four Articles of Prague (1420), the Hussites called for freedom of preaching, limits to church property holding, and civil punishment of mortal sin. The fighting ended in 1433 with the compromise of the Counsel of Basel and three years later was officially ended by the Compact of Iglau (1436), in which a faction of the Hussites agreed to accept the Holy Roman Emperor as king. The peasants had succeeded nevertheless in asserting Bohemian nationalism and had severed the country's ties to Germany.

What was the **Peasants' War**?

Fought in 1524 and 1525, the war was in part a religious one that came during the Reformation. It was the greatest mass uprising in German history. In 1517 the German monk Martin Luther (1483–1546) had begun questioning the authority of the Roman Catholic Church. He soon had followers—nobles and peasants alike—and his reform movement spread, giving birth to Protestantism (the Christian beliefs practiced by those who protested against the Catholic Church). While many Protestants were sincere in their faith, some had their own motives for following the movement. German peasants looked to the Reformation to end their oppression at the hands of the noble lords. When the peasants revolted at the end of 1524, they were forcibly suppressed. Some 100,000 peasants died. Prior to the uprising, they had aimed to get Martin Luther's endorsement, but he declined to give it.

RENAISSANCE AND THE ENLIGHTENMENT

What was the **Gunpowder Plot**?

In 1605 a group of 12 men, who believed the English government to be hostile to Roman Catholics, laid plans to blow up the Houses of Parliament—with King James I (1566–1625) and government officials in it. Their scheme was discovered, however, and all of the conspirators were put to death. The event is remembered today in two ways: On the night of November 4, which was when the plot was supposed to be carried out, the English hold a festival in which Guy Fawkes (1570–1606), the conspirator who originated the Gunpowder Plot, is burned in effigy. Further, the vaults beneath the Houses of Parliament are searched before each new session.

What was the **Thirty Years' War**?

The Thirty Years' War (1618–48), like the Hundred Years' War, was actually a series of related conflicts, rather than one long campaign. The conflict in Europe began as a religious one, with hostility between Roman Catholics and Protestants; but it eventually turned political before it was ended with the Peace of Westphalia. The war had four periods: the Bohemian (1618–24), the Danish (1625–29), the Swedish (1630–34), and the Swedish-French (1635–48).

In Bohemia (part of present-day Czech Republic) the trouble began in the capital of Prague when the archbishop authorized the destruction of a Protestant church. The act angered Bohemian Protestants and those elsewhere in Europe, who believed it was their right—granted by the Peace of Augsburg (1555)—to worship as Lutherans. When Holy Roman Emperor Matthias (1557–1619) failed to intervene on behalf of the Protestants, Prague became the scene of mayhem in May 1618. Disorder continued in

What was the Defenestration of Prague?

The event occurred at the beginning of what came to be known as the Thirty Years' War (1618–48) when Bohemian Protestants, angered at the destruction of their church, appealed to Holy Roman Emperor Matthias (1557–1619) to intervene on their behalf. When the appeal proved unsuccessful, in May 1618, rioting broke out in the Bohemian city of Prague and two of the emperor's officials were thrown out of windows; in other words, they were defenestrated.

Bohemia even as a new emperor ascended to the throne of the Holy Roman Empire. King Ferdinand II (1578–1637), a Habsburg, wielded an enormous amount of power, and in 1620 he squelched the Bohemian rebellion, which cost the Bohemians their independence. Further, Catholicism was reinstated as the state religion. These events caused other Protestant lands within the Holy Roman Empire to take notice. Soon the kings of Denmark, Sweden, and France entered into their own campaigns fighting King Ferdinand II for control of German lands. But the conflicts weren't strictly about religious freedom: Reducing the authority of the powerful Habsburg family became a primary objective as well.

What was the **Peace of Westphalia**?

In 1644, with Europe torn by the Thirty Years' War (1618–48), a peace conference was convened in Westphalia, Germany. But the negotiations were four long years in the making: the fighting continued until 1648, when the Peace of Westphalia was finally signed. Under this treaty, France and Sweden received some German lands. The agreement also made important allowances for Europe's religions: Not only was Lutheranism given the same due as Catholicism, but Calvinism, the religious movement begun by Frenchman John Calvin (1509–1564), was also was given the official nod. In short, the treaty not only ended the religious warfare in Europe, but it provided for some measure of religious tolerance.

Since the pact recognized the sovereignty of all the states of the Holy Roman Empire, it effectively dissolved the empire. Therefore, historians view the Peace of Westphalia as the beginning of Europe's modern state system.

What was the **Great Northern War**?

It was a war undertaken at the beginning of the eighteenth century that challenged Sweden's absolute monarchy and imperialism. During the seventeenth century, Sweden had become a power in the Baltic region, gradually bringing more and more territory under its control. Even the Peace of Westphalia (1648) had granted some German lands to Sweden. But much of Sweden's prosperity and expansion during this period had been under the rule of Charles XI (1655–1697). When he was succeeded by his young son, Charles XII (1682–1718) in 1697, the tides were about to turn for Sweden.

In 1700 an alliance formed by Denmark, Russia, Poland, and Saxony (part of present-day Germany) attacked Sweden, beginning the Great Northern War. Sweden readily defeated Denmark and the Russians that same year. But Poland and Saxony proved to be more formidable foes, and Charles XII spent almost seven years fighting—and eventually defeating—them. But the Russian army was to have another chance at the Swedish, and this time they were successful, defeating Charles XII's forces in 1709 at Poltava (Ukraine). Charles fled the country as the war continued and did not return until 1714. Four years after that, the monarch was killed as he observed a battle (in what is present-day Norway). Much of the country's lands in the Baltic were surrendered, and Sweden's period of absolute monarchy came to an end.

What was the **Seven Years' War**?

It was a worldwide conflict that began in 1756 between Prussia and Austria, who fought over control of Germany—and over who would be the supreme power in Europe. Great Britain threw its support behind King Frederick II (the Great; 1712–1786) of Prussia. But by the following year Austria was supported not only by France, but also by Sweden, most of the German states, and, very importantly, Russia. (Spain joined the fighting on the side of Austria also, but it was not until late in the game—after 1762.) With such alliances forged among the European states, the conflict soon spilled over, manifesting itself in the colonies of North America and India, where the French and British fought each other for control.

In order to assert his authority in Europe, Prussia's King Frederick had launched many military initiatives and was in a weakened state by the time his army faced the Austrians in 1756. He was spared certain defeat only by an event in Russia in 1762: Upon the death of Czarina Elizabeth Petrovna (1709–1762), who had feared King Frederick, Peter III (1728–1762) ascended the Russian throne. Peter, unlike his predecessor, held Frederick in high esteem, and he proceeded to withdraw Russia's support from Austria and reach a peace agreement with Austria's enemy, Prussia. He died that same year.

The Seven Years' War ended the next year, on February 15, 1763, with the signing of a peace agreement in Saxony, Germany. The area that Austria had fought to control remained, for the most part, under Prussian rule, positioning Prussia as a leading European power. There were no other territorial changes in Europe as a result of the war. In North America and India, Britain emerged from the conflicts as the victor—and as *the* colonial power.

COLONIALISM

How did the **British come to control** much of **North America** during colonial times?

British and French explorers laid claim to many parts of what is now the United States. During the late 1600s and into the mid-1700s, the two European powers fought a series of four wars in their struggle for control of territory in North America. Three of the wars broke out in Europe before they spread to America, where British and French colonists fought King William's War (1689–97), Queen Anne's War (1702–13), and King George's War (1744–48). King William's War saw no gains for either side. After Queen Anne's War, however, both sides signed the Treaty of Utrecht, in which France ceded Newfoundland, Acadia, and the Hudson Bay territory to Britain.

The struggle between England and France was not settled until a fourth war, the French and Indian War (1754–63), from which Britain emerged the victor.

What was the **French and Indian War**?

The French and Indian War (1754–63) was the last major conflict in North America before the Revolutionary War (1775–83). During colonial times, both Britain and France had steadily expanded their territories into the Ohio River valley. Since the fur trade prospered in this region, both countries wished to control it. But as the French encroached on their territory, the British colonists sent an ultimatum to them. This message was delivered by none other than George Washington (1732–1799), who had been sent by the British governor. But the French made it clear that they did not intend to back down. So in 1754 Washington (now a lieutenant colonel) and 150 troops established a British outpost at present-day Pittsburgh, not far from where the French had installed themselves at Fort Duquesne. That spring and summer fighting broke out.

Washington met the French, and though he and his troops mustered a strong resistance, there were early losses for the British. But a reinvigorated British force, under the leadership of Britain's secretary of state William Pitt (1759–1806), took French forts along the Allegheny River in western Pennsylvania and met French troops in battle at Quebec. In 1755 Washington was made colonel and led the Virginia troops in defending the frontier from French and Indian attacks. Though the British finally succeeded in occupying Fort Duquesne in 1758 (at which time Washington temporarily retired to his farm in Virginia), fighting continued until 1763 when the Treaty of Paris ended the war.

The British won the spoils, gaining control of all French lands in Canada as well as French territories east of the Mississippi River, with the exception of New Orleans. (The city was ceded to Spain, along with its holdings west of the Mississippi; Spain had become an ally to France late in the war, in 1762.) In exchange for Havana, Cuba, Spain turned over Florida to the British. France, which had once controlled a vast region of America, retained only two small islands off the coast of Newfoundland, Canada, and the two Caribbean islands of Martinique and Guadeloupe.

REVOLUTION AND THE MODERN ERA

AMERICAN REVOLUTION

Why is **Paul Revere's ride** so well known?

The April 18, 1775, event was famous in its own right but was memorialized by American writer Henry Wadsworth Longfellow (1807–1882) in his poem, "Paul Revere's Ride." The verse contains an error (or perhaps Longfellow simply took literary license) about the night that the American Revolution (1775–83) began: The light signal that was to be flashed from Boston's Old North Church (one light if the British were approaching the patriots by land and two if the approach was by sea) was sent not to Revere; it was received by Revere's compatriots in Charlestown (now part of Boston proper). However, Revere did ride that night—on a borrowed horse. He left Boston at about 10:00 P.M. and arrived in Lexington at midnight to warn Samuel Adams and John Hancock, who were wanted for treason, that the British were coming. The next day, April 19, the battles of Lexington and Concord were fought, starting the Revolutionary War in America.

As an American patriot, Revere (1735–1818) was known for his service as a special messenger, so much so that by 1773 he had already been mentioned in London newspapers. Revere also participated in the Boston Tea Party in 1773.

Did the **American colonies have any allies** in their fight against the British?

Yes. France—which was, of course, a longtime rival of Great Britain's—was a key ally to the Americans, supplying them with some 90 percent of their gunpowder.

Why was **Bunker Hill** important in the American Revolution?

The June 1775 battle on the hills outside Boston proved to be the bloodiest battle of the war. After the fighting in April at Lexington and Concord, more British troops arrived in Boston in late May. The Americans fortified Breed's Hill, near Bunker Hill,

Paul Revere's 1775 ride to Lexington to warn that the British were coming.

and on June 17, the British were ordered to attack the Americans there. The patriots, who needed to conserve ammunition, were given the famous direction not to fire until they saw the whites of their enemies' eyes. The patriots succeeded in driving the British back on their first two charges. But on the third charge, the patriots fled. The Battle of Bunker Hill resulted in more than 1,000 injured or dead British soldiers and 400 American soldiers killed or wounded.

Were all the battles of the American Revolution waged in **the Northeast**?

No, there was also fighting in the southern colonies. But the struggle between the American colonists and the British was further complicated in the South by the presence of slaves. Landowners feared that any fighting in the vicinity would inspire slaves to revolt against them. Knowing this, the British believed they could regain control of the southern colonies more readily than those in the north. In November 1775 the British governor of Virginia offered to free any slaves who would fight for the British. As many as 2,000 black slaves accepted the offer and took up arms. But there were also patriots in the South: It was Virginian Patrick Henry (1736–1799) who uttered the famous words, "Give me liberty or give me death."

In late February 1776 patriot forces confronted and defeated pro-British colonists near Wilmington, North Carolina. The British troops who were sailing from Boston, Massachusetts, to North Carolina to join the loyal colonists arrived too late to help. They instead sailed on to Charleston, South Carolina, which was also the scene of fighting that summer.

French Revolution

What was the **Oath of the Tennis Court**?

It was the oath taken in June 1789 by a group of representatives of France's third estate who, having been rejected by King Louis XVI (1754–1793) and the first and the second

Who were the minutemen?

They were volunteer soldiers, ready to take up arms at a minute's notice, who fought for the American colonies against the British during the Revolutionary War. The minutemen, who were trained and organized into the militia, are most well known for the battles at Lexington and Concord, on April 19, 1775, which marked the beginning of the war. They had been alerted to the approach of the British troops, or "redcoats," by the patriot Paul Revere (1735–1818). After the fighting at Lexington and Concord, which left 250 British killed or wounded and about 90 Americans dead, word spread quickly of the fighting, and the Revolutionary War had begun.

estates, vowed to form a French national assembly and write their own constitution. The pledge set off a string of events that began the French Revolution (1789–99).

French society had long been divided into three classes, called "estates": members of the clergy were the first estate, nobles comprised the second, and everyone else made up the third. When philosophers such as Jean-Jacques Rousseau (1712–1778) came along and challenged the king's supreme authority by promoting the idea that the right to rule came not from God but from the people, it fueled the discontent felt by the long-suffering peasants and the prosperous middle class who paid most of the taxes to run the government but who had no voice in it. In short these people were the disenfranchised third estate.

A government financial crisis brought on by the expense of war forced King Louis XVI to reluctantly call a meeting of the representatives of all three estates, called the Estates General, which had last convened in 1614. During the May 5, 1789, meeting at Versailles, the third estate attempted to seize power from the nobility, the clergy, and the king by insisting that the three estates be combined to form a national assembly in which each member had one vote; since the third estate had as many representatives as the other two combined, the people would at last have a voice. When the attempt failed, the representatives of the third estate gathered on a Versailles tennis court, where they vowed to change the government. Louis XVI began assembling troops to break up the meeting. Meantime, an armed resistance movement had begun to organize. The situation came to a head on July 14, 1789, with the storming of the Bastille in Paris.

What was the **Reign of Terror**?

It refers to the short but bloody period in French history that began in 1793 and ended July 1794. During this time revolutionary leader Maximilien Robespierre (1758–1794) led a tribunal that arrested, tried, and put to death more than 17,000 people—most of them by guillotine.

In the reforms that followed the 1789 Oath of the Tennis Court and the capture of the Bastille, France was transformed into a constitutional state, and French subjects became French citizens. An elected legislature (the Constituent Assembly) was given control of the government. Robespierre was elected first deputy from Paris and was the leader of the radical popular party. In this new era, those who had been associated with the old regime or those who opposed the French Revolution became the subjects of persecution. In January 1793 King Louis XVI (1754–1793) and his wife, Marie Antoinette (1755–1793), were executed, beginning the Reign of Terror that saw thousands more (mostly those who had made up the powerful first and second estates) suffer a similar fate at the hands of the revolutionaries. To escape certain death, many fled the country; this included top-ranking military officials, which made room for the rapid advancement of young military officers such as Napoleon Bonaparte (1769–1821).

The Reign of Terror ended on July 28, 1794, when Robespierre himself was put to death. As he gained power and influence, the revolutionary leader also had become increasingly paranoid, even putting two of his friends to death in 1794. He was overthrown on July 27 by the Revolution of 9th Thermidor and the next day died by guillotine.

How long did the **French Revolution** last?

The Revolution lasted some 10 years, and it grew increasingly violent as it progressed. It began in mid-1789 when the government found itself nearly bankrupt, and due to festering discontent among the commoners (the prosperous middle class included), that crisis quickly grew into a movement of reform. The Revolution ended in 1799 when French general Napoleon Bonaparte (1769–1821) seized control of the government. Democracy had not been established in France, but the Revolution had ended the supreme authority of the king, had strengthened the middle class, and had sent the message across Europe that the tenets of liberty and equality are not to be ignored.

What was the **Brumaire Coup d'État**?

It was the overthrow on November 9, 1799, of the French revolutionary government. The coup put Napoleon Bonaparte (1769–1821) in power as one of three counsels intended to head the government.

While Napoleon was in Egypt and Syria waging what were for the most part successful military campaigns on behalf of the French government, there was growing discontent back home with the Directory, the group of five men who had governed France since 1795. His army stranded in the Middle East, Napoleon received word that France might soon be under attack by the Second Coalition (the second in a series of six alliances that formed in Europe in order to stave off French domination). Leaving

Scene of rioting during the French Revolution, a tumultuous decade that ended when General Napoleon Bonaparte seized control of the government.

another man in command of his troops, Napoleon hurried home where he was welcomed as a hero. Aided by his brother, Lucien Bonaparte (1775–1840), and the French revolutionary leader Emmanuel-Joseph Sièyes (1748–1836), Napoleon carried out a coup d'état, overthrowing the Directory. A consulate was formed, with the young Napoleon becoming first consul; the other counsels had little influence, acting primarily as advisors to the ambitious Napoleon.

The coup marked the end of the French Revolution: After the chaos and violence of the previous decade, the French people looked to Napoleon as a strong leader who could bring order to the country. They did not know that the 30-year-old possessed a seemingly insatiable hunger for power, which would soon transform the government into a dictatorship. After a brief peace, Napoleon declared himself emperor of France on December 2, 1804, by which time he had already begun to wage a series of wars to gain himself more power in Europe.

When did the **Napoleonic Wars** begin and end?

The Napoleonic Wars began shortly after Napoleon Bonaparte (1769–1821) took power and lasted until 1815, when he was finally defeated at the Battle of Waterloo. Ever the general, Napoleon used his power to keep France at war throughout his reign.

After the Coup d'État of 18th Brumaire (November 1799), which had put Napoleon in power, at first he effected peace: In May 1800 he marched across the Alps to defeat the Austrians, ending the war with them that had begun eight years earlier. Britain, fearing a growing European power on the continent, had declared war on France in 1793; by 1802, having grown tired of battle, the country agreed to peace with Napoleon in the Treaty of Amiens. But the calm in Europe was not to last. By 1803 the diminutive but power-hungry Napoleon (nicknamed the Little Corporal) had begun to plot an invasion of Britain. Declaring himself emperor in 1804, he initiated a series of campaigns across Europe, and by 1806 most of the continent was under his control. He remained, of course, unable to beat the British, whose superior navy gave them supremacy at sea.

But the various alliances (called coalitions) formed by European countries against Napoleon eventually broke him. After he had been defeated in Russia in 1812, the European powers that had long been held in submission by Napoleon formed a sixth and final coalition against him: Great Britain, Russia, Sweden, Prussia, and Austria met Napoleon's army at the momentous Battle of the Nations at Leipzig, Poland, from October 16 to 19, 1813. Napoleon was defeated there in what is sometimes called the War of Liberation, and he retreated to France. The following March, the allies making up the Sixth Coalition took Paris; Napoleon's generals were defeated. He abdicated the throne on April 6. However, that was not the end of the Napoleonic era: Exiled to the Mediterranean island of Elba, Napoleon returned to Paris on March 20, 1815, believing he could recover power in the unstable atmosphere that followed his abdication. Three months later he was defeated at the Battle of Waterloo, on June 18. It was the last battle of the Napoleonic Wars. He was exiled to St. Helena island, where he died in 1821.

What happened at **Trafalgar**?

Cape Trafalgar, on the southwest coast of Spain, was in 1805 the scene of a decisive victory for Great Britain over Napoleon Bonaparte's (1769–1821) navy. Other than a one-year respite in 1802, France and Britain had been at war with each other since 1793. Napoleon remained determined to conquer Britain, just as he had most of continental Europe. But when his fleets met those of decorated English Admiral Horatio Nelson (1758–1805) off the coast of Spain, the certain defeat and destruction of Napoleon's navy ended the emperor's hopes of invading England.

The confrontation at Trafalgar was the culmination of a two-year game of cat and mouse between Nelson's fleets and the French under the direction of Admiral Villeneuve (1763–1806), whose sole objective it was to invade Britain. To prevent this from happening, in 1803 Nelson began a two-year blockade of Villeneuve and the French navy at Toulon, France (on the Mediterranean coast). When the French fleets escaped Toulon, attempting to lure the British out to sea, Nelson chased them all the way across the Atlantic—to the West Indies and back—before the showdown off Spain's coast, where the French were joined by Spanish fleets. Meantime, the coast of England remained protected by the British navy, leaving no opportunity for invasion by the French.

On October 21, 1805, seeing the enemy sailing out of Trafalgar, Nelson formed his fleet of 28 ships into two columns, intending to divide and conquer the combined French and Spanish force of 33 ships. About noon that day, as they prepared for the confrontation, Admiral Nelson sent out one of the most famous commands of naval history: "England expects that every man will do his duty." While the British prevailed, destroying Napoleon's fleet in less than four hours' time, Nelson was fatally wounded by a sharpshooter, and the English navy hero died just as victory was his. The brave Nelson had seen fate coming: The night before Trafalgar, he had revised his will, and just before the battle had begun, he told Captain Henry Blackwood, "God bless you, Blackwood, I shall never speak to you again." Nevertheless, Nelson died knowing that he had won, uttering the still famous words, "Thank God, I have done my duty."

Why did the **Russians burn Moscow**?

The September 14, 1812, torching of their own city was directed by Tsar Alexander I (1777–1825), who wished to prevent Napoleon Bonaparte (1769–1821) and his invading armies from reaping the benefits of anything Russian. Through a series of wars, Napoleon had dominated most of Europe by 1805. The authority of Alexander was certainly threatened by the French emperor. In 1805 and 1807 Russia suffered major losses in battles with Napoleon's armies. In the face of these defeats, what Alexander did next was a stroke of genius, though he had many detractors at the time: Napoleon's forces, though victorious, were weary from fighting and were unable to pursue the Russian armies further. So, Alexander made peace with the emperor in the Treaty of Tilsit (1807). The Russian ruler vowed support of Napoleon, and for his part, Napoleon believed Alexander had extended him a hand of friendship. Instead, the cunning Russian ruler had bought himself and his country the time they needed to gird themselves against powerful Napoleon.

By 1812 Russia, its economy dependent on exports, resumed trade with Great Britain, Napoleon's archenemy. This prompted the return of Napoleon's troops to Russia: Later that year the French emperor marched into Russia with a force of as many as 600,000 men, but the Russians still delivered Napoleon a crushing defeat. The Russian army had relied on guerrilla warfare tactics, including burning their own countryside. Napoleon returned to Paris in defeat by the end of the year.

WAR OF 1812

What caused the **War of 1812**?

The war between the young United States and powerful Great Britain largely came about because of France. After the French navy was crushed by the British under Admiral Nelson (1758–1805) at the Battle of Trafalgar, Napoleon turned to economic warfare in his long struggle with the British: He directed all countries under French control not to trade with Great Britain. Its economy dependent on trade, Britain struck back by imposing a naval blockade on France, which soon interfered with U.S. shipping. Ever since the struggle between the two European powers began in 1793, the United States tried to remain neutral. But the interruption of shipping to and from the continent and the search and seizure of ships posed significant problems to the American export business: In 1807 Great Britain had issued an Order in Council that required even neutral vessels destined for a continental port to stop first in England; Napoleon countered with the Milan Decree, which stated that any neutral vessel that had submitted to British search be seized.

Back in America, the people of New England, the region most dependent on shipping, nevertheless vehemently opposed entering into war with the British. But the country's economy was depressed as a result of the interruption of exports, and the U.S. Congress declared war on June 18, 1812. In these days before telegraph and radio, the United States did not know that two days before, on June 16, Britain had withdrawn its Orders in Council, lifting its policy of shipping interference, which had been the chief reason for the war declaration. Thus the two countries engaged in fighting for the next two and a half years. On December 24, 1814, the Treaty of Ghent officially ended the war. But once again, poor communication led to fighting: Two weeks after the treaty was signed, troops in New Orleans, unaware of this event, fought for control over the Mississippi River in the worst battle of the entire conflict. Though both the United States and Great Britain claimed victory in the War of 1812, neither side had gained anything.

Who were the **War Hawks**?

The War Hawks were a group of Republicans in the U.S. Congress who advocated war with Great Britain. Elected in 1810, the congressmen took office in 1811, the failure of the Erskine agreement fresh in their memories: That bit of 1809 diplomacy, arranged by British minister to the United States George Erskine and the then U.S. secretary of state James Madison (1751–1836), would have provided for the suspension of Britain's maritime practices that interfered with U.S. shipping, but the agreement fell apart when Erskine was recalled from office. The relationship between the United States and Great Britain, tenuous since 1807 due to trade embargoes and the impressment of American sailors into British service, deteriorated. The newly elected congressmen were tired of the failure of diplomacy to resolve maritime problems with the British; they further felt that the British were challenging the young United States through their policies, which purportedly included British aid to American Indians in the Northwest. War Hawk leader Henry Clay (1777–1852) was named Speaker of the House, and Congress soon passed a series of resolutions to strengthen the army and navy. When Congress was called upon by President James Madison to declare war on the British in June of 1812, it was the War Hawks who swung the close vote. Thus the

War of 1812 was declared. Some historians believe the true motive behind the War Hawks was not resolution of the shipping problems, but rather the desire to annex parts of southern Canada to the United States.

Who said, **"We have met the enemy and they are ours"**?

It was Captain Oliver Hazard Perry (1785–1819) who wrote the famous words in a letter to General William Henry Harrison (1773–1841) after defeating the British at the Battle of Lake Erie on September 10, 1813. An improvised U.S. squadron commanded by Captain Perry, 28 years old, achieved the victory in the War of 1812 battle. The message he sent to Harrison (who later went on to become the ninth U.S. president) was: "We have met the enemy and they are ours: two ships, two brigs, one schooner, and one sloop." Perry received a gold medal and thanks from Congress for the victory.

Greek War of Independence

What was the **London Protocol**?

The 1830 decree recognized an independent Greek nation after the eight-year Greek War for Independence (1821–29). The London Protocol officially ended Ottoman-Turk rule of Greece, which had begun almost 400 years earlier.

In 1453 the Ottoman Turks conquered Constantinople (present-day Istanbul, Turkey), and they soon moved westward to bring the Greek peninsula under their control as well: By 1456 most of Greece had been absorbed into the Ottoman Empire. Hundreds of years later, in 1770, the Greeks tried to overthrow the Turks and were aided in this effort by Russian Tsarina Catherine the Great (1729–1796), whose aim it was to replace Muslim rule with Orthodox Christian rule throughout the Near East. But the effort was unsuccessful, and it was 50 years before the Greeks would rise again to assert their independence. On March 25, 1821, the Greeks, led by the archbishop of Patras, proclaimed a war of independence against the Turks. Soon, Egypt had thrown its military support behind the Turks, but even the combined force could neither defeat the Greeks nor squelch the revolution.

In 1827 Britain, France, and Russia, all sympathetic to the Greek cause, came to their aid. In October of that year a combined fleet of the three European powers defeated the Turk and Egyptian fleet in the Battle of Navarino, off the Peloponnese Peninsula. But the deciding moment came when Russia declared war on the Ottoman Empire in 1828, and the Ottoman Turks turned their attention to fighting the Russians. The following year, the Egyptians withdrew from Greece. In March 1830 the London Protocol was signed by Britain, France, and Russia, recognizing an independent Greece. Weary from the fighting, the Ottoman Turks accepted the terms of the proclamation later that year.

Texas War of Independence

What does **"Remember the Alamo"** mean?

The saying was a rallying cry for Texans in their war for independence from Mexico. The movement for independence had begun in the winter of 1835–36 when the people of Texas decided to cut off relations with Mexico, and soon turned into a war when the Mexican government sent a force of some 4,000 troops, under the command of General Antonio Lopez de Santa Anna (1794–1876), to squelch the rebellion. As the Mexican army approached, the force of about 150 men who were determined to defend the city of San Antonio retreated to the Alamo, a Spanish mission built in the previous century. There they were joined by another 50 men but were still no match for the Mexicans, who kept the Alamo under siege for 13 days—from February 23 to March 6, 1836. The Texans, low on ammunition, ceased to return fire. On the morning of March 6, Santa Anna's troops seized the Alamo. The fierce frontiersmen, Davy Crockett (1786–1836) among them, are believed to have fought using the butts of their rifles. All the Texans who fought that day at the Alamo died.

Meantime, General Sam Houston (1793–1863) had assembled his forces, and with the rallying cry "remember the Alamo" (and their fellow Texans who had bravely fought and died there), he set out to face the Mexican army and secure independence. This he did, at San Jacinto, Texas, on April 21, 1836, in a quick and decisive battle that had caught Santa Anna's troops by surprise. The following day the Mexican general was captured and made to sign a treaty giving Texas independence.

Mexican War

What caused the **Mexican War**?

The two-year war (1846–48) was fought over the United States' annexation of Texas. The events that led up to the conflict began in 1837 when President Andrew Jackson (1767–1845) recognized Texas as independent (this was just after Texas had won its war with Mexico). Republic of Texas president Sam Houston (1793–1863) felt that protection against a Mexican invasion may be necessary, so he eyed annexation to the United States. In the meantime, Mexican president Antonio Lopez de Santa Anna (1794–1876) warned that such an action on the part of the United States would be "equivalent to a declaration of war against the Mexican Republic." In June 1844 the U.S. Senate rejected a proposed annexation treaty. But later that year Democratic

Party nominee James K. Polk (1795–1849), an ardent expansionist, was elected president. Because the annexation of Texas had figured prominently in his campaign platform, outgoing President John Tyler (1790–1862) viewed Polk's victory as a public mandate for annexation, and he recommended that Congress pass a joint resolution to invite Texas into the union. Congress did so in February, and President Tyler signed the resolution on March 1, 1845, three days before leaving office.

Mexico responded by breaking off diplomatic relations with the United States. A border dispute made the situation increasingly tenuous: Texas claimed that its southern border was the Rio Grande River, while Mexico insisted it was the Nueces River, situated farther north. In June President Polk ordered Brigadier General Zachary Taylor (1784–1850) to move his forces into the disputed area. In November the U.S. government received word that Mexico was prepared to talk. Polk dispatched Congressman John Slidell (1793–1871) to Mexico to discuss three other outstanding issues: the purchase of California (for $25 million), the purchase of New Mexico (for $5 million), and the payment of damages to American nationals for losses incurred in Mexican revolutions. This last point was critical to the negotiations, as Polk was prepared to have the United States assume payment of damages to its own citizens in exchange for Mexico's recognition of the Rio Grande as the southern border of Texas.

But upon arrival in Mexico City, Slidell was refused the meeting—President José Joaquín Herrera (1792–1854) had bowed to pressure, opposing discussions with the United States. When Polk received news of the scuttled talks, he authorized General Taylor to advance through the disputed territory to the Rio Grande. Meanwhile, Mexico overthrew President Herrera, putting into office the fervent nationalist General Mariano Paredes y Arrillaga (1797–1849), who reaffirmed Mexico's claim to Texas and pledged to defend Mexican territory.

While Polk worked through Slidell to get an audience with the Mexican government, the attempts failed, and on May 9 the Cabinet met and approved the president's recommendation to ask Congress to declare war. The next day news arrived in Washington that on April 25 a sizeable Mexican force had crossed the Rio Grande and surrounded a smaller American reconnaissance party. Eleven Americans were killed and the rest were wounded or captured. On May 11 Polk delivered a message to Congress, concluding, "Mexico has...shed American blood upon the American soil....War exists...by the act of Mexico herself." By the time the war was officially declared on May 13, just more than one year after Polk had been sworn into office, General Taylor had already fought and won key battles against the Mexicans and had occupied the northern Mexico city of Matamoros.

What did the United States gain from the **Mexican War**?

The Mexican War (1846–48) was officially ended when the U.S. Senate ratified the Treaty of Guadalupe Hidalgo on March 10, 1848. By the treaty, Mexico relinquished roughly half its territory—New Mexico and California—to the United States. Mexico also recognized the Rio Grande as its border with Texas.

Mexico received payments in the millions from the United States, which also assumed the payment of claims of its citizens. Five years later under the terms of the Gadsden Purchase, the United States purchased a small portion of land from Mexico for another $10 million, which was widely regarded as further compensation for the land lost in the war. The territory the United States gained was in present-day Arizona and New Mexico, south of the Gila River.

What is **privateering**?

Privateering is the hiring of privately owned ships and their crews to fight during battle. The practice, dating back to the 1400s, continued well into the 1800s, eventually replaced by the development of strong navies. Privateers were, essentially, gunboats for hire. They played a crucial role in the American Revolutionary War (1775– 83) after the Second Continental Congress authorized their use on March 18, 1776, enabling the colonists to capture about 600 British ships. The Americans would again employ privateers in the War of 1812 (1812–14).

But during times of peace, some privateers turned to pirating, which at least in part prompted European nations to sign the Treaty of Paris of 1856, which ended the Crimean War (1853–56) and outlawed privateering. Since the United States had relied on privateers in the past and had yet to develop its own navy, the Americans did not sign the treaty. While there was some privateering during the American Civil War (1861–65), the need for them soon subsided as navies developed—by enlistment and draft. Privateering has not been used in more than 100 years.

Crimean War

What was the **Crimean War**?

The Crimean War was fought from 1853 to 1856 between Russian forces and the allied armies of Britain, France, the Ottoman Empire (present-day Turkey), and Sardinia (part of present-day Italy). The Crimean Peninsula, which juts out into the Black Sea and is today part of Ukraine, was the setting for many of the battles. The source of the conflict was Russia's continued expansion into the Black Sea region—which, if left unchecked, would have resulted in strategic and commercial advantages for Russia. But Russia was unable to muster the strength it needed to combat the powerful alliance formed by the European countries and the Ottoman Empire. The war was ended with the signing of the Treaty of Paris (1856), which required Russia to surrender lands it had taken from the Ottoman Empire and abolished Russian navy and military presence in the Black Sea region. It was the first conflict that was covered by newspaper reporters at the front.

War of Reform

What was the **War of Reform**?

It was the period in Mexican history from 1858 to 1861 when the federalist government collapsed and civil war ensued. In 1858 President Ignacio Comonfort (1812–1863), who had become Mexico's leader when he helped overthrow President Antonio Lopez de Santa Anna (1794–1876) in 1855, felt political pressure and fled the country. Benito Júarez (1806–1872), who had served as minister of justice and minister of the interior, assumed the presidency. His position was immediately opposed by centralists who rallied around rebellious army forces. Under this pressure, the federalists, led by Júarez, withdrew from Mexico City and set up the capital at Veracruz, on the Gulf Coast. There they had control over customs receipts, which allowed them to purchase arms and finance their government. Eventually they defeated the centralists and reentered Mexico City in January 1861. Júarez was elected president later that year, but his authority was challenged again with the arrival of the French, who quickly put Maximilian (1832–1867) in power as emperor of Mexico. Júarez led the country in a successful campaign against the French, who were expelled in 1867 when Júarez resumed the presidency. He died in office in 1872.

What does **"Fifty-four forty or fight"** mean?

The slogan refers to a dispute between the United States and Great Britain over Oregon Country, which an 1818 treaty allowed both nations to occupy. This was the territory that began at 42 degrees north latitude (the southern boundary of present-day Oregon) and extended north to 54 degrees 40 minutes north latitude (in present-day British Columbia). During the 1830s and early 1840s American expansionists insisted that U.S. rights to the Oregon Country extended north to latitude 54 degrees 40 minutes, which was then the recognized southern boundary of Russian America (roughly present-day Alaska).

The eleventh president of the United States, James K. Polk (1795–1849), used the slogan in his political campaign of 1844. After he was elected, Polk settled the dispute with Great Britain (in 1846), and the boundary was set at 49 degrees north, the northern boundary of what is today Washington State and the border between the United States and Canada. This agreement—reached without the fight threatened in the slogan—gave the United States the territory lying between 42 and 49 degrees north latitude and Great Britain the territory between 49 degrees and 54 degrees 40 minutes north latitude as well as Vancouver Island. The United States' portion is present-day Washington, Oregon, and Idaho as well as parts of Montana and Wyoming.

Was the Civil War fought because of **slavery**?

For years, American schoolchildren learned that the question of slavery was the only cause of the Civil War (1861–65): With 19 free states and 15 slave states making up the Union, Abraham Lincoln (1809–1865) had called the country "a house divided" even before he became president. While slavery was central to the conflict, many believe the bloody four-year war had other causes as well.

By the mid-1800s important differences had developed between the South and the North—and many maintain these differences, or vestiges of them, are still with the country today. The economy in the South was based on agriculture while the North was industrialized; the ideals and lifestyles of each region reflected these economic realities. Southerners believed their agrarian lifestyle was dependent on the labor of slaves. For a long time, slavery was viewed by some as a necessary evil. But by the early 1800s the view that slavery is morally wrong was beginning to take hold. Northern abolitionists had begun a movement to end slavery in the states. But, except for a small antislavery faction, these views were not shared in the South.

There were other factors that contributed to the declaration of secession and the formation of the Confederacy, although some still argue these factors were merely smoke screens for the defense of slavery. Disputes between the federal government and the states had limited the power of the states, and this policy was called into question by Southerners. Further, the political party system was in disarray in mid-1850s America. The disorder prompted feelings of distrust for the elected politicians who set national policy. Before the 1860 presidential election, Southern leaders urged that the South secede from the Union if Lincoln, who had publicly taken a stand against slavery, won.

How did the **Civil War begin and end**?

Unhappy with the outcome of the 1860 presidential election, in which Abraham Lincoln (1809–1865) was elected, and fearing a loss of their agrarian way of life, the Southern states began to make good on their promise to secede if Lincoln won the presidency: South Carolina was the first (in December of that year). In January 1861 five more states followed: Mississippi, Florida, Alabama, Georgia, and Louisiana. When representatives from the six states met the next month in Montgomery, Alabama, they established the Confederate States of America and elected Jefferson Davis (1808–1889) president. Two days before Lincoln's inauguration, Texas joined the Confederacy. (Virginia, Arkansas, North Carolina, and Tennessee joined in April, shortly after the Civil War had already begun.)

The Civil War, also called the War of Secession and the War between the States, began on April 12, 1861, when Southern troops fired on Fort Sumter, a U.S. military post in Charleston, South Carolina. Brutal fighting continued for four years. On April

The 1863 Battle of Gettysburg is considered the turning point in the American Civil War.

9, 1865, General Robert E. Lee (1807–1870) surrendered his ragged Confederate troops to General Ulysses S. Grant (1822–1885) of the Union at old Appomattox Court House, Virginia. The war had not only been between the states, it had also been between brothers: the conflict divided the nation. The Civil War took more American lives than any other war in history.

Why did **Andrew Johnson** vow he would burn Nashville before surrendering it?

It seems a strange thing for a politician to say about his home state's capital city. Andrew Johnson had served Tennessee in both the U.S. House of Representatives (1843–53) and the U.S. Senate (1857–62); he had also been governor (1853–57). But after the Civil War broke out, the Southern democrat made a surprising move: he sided with the Union. This show of allegiance was largely owing to Johnson's strongly held belief that the South's secession was unconstitutional. Having thus made his stand, President Abraham Lincoln (1809–1865) saw fit to appoint Johnson as the Union's military governor in Tennessee. When rebel forces surrounded Nashville and seemed poised to take it, Johnson proclaimed he would sooner burn the city before surrendering it. But he was forced to do neither: In mid-December 1864 Union forces used the hand-cranked Gatling gun (invented in 1861) to help defeat the Confederate forces.

Why was the **battle at Gettysburg** important?

The 1863 battle, fought when the two sides met accidentally in the southern Pennsylvania town, was a turning point in the Civil War. From July 1 to 3 General George Meade (1815–1872) led his troops (about 90,000 strong) to defeat the advancing Confederate troops (numbering some 75,000) under General Robert E. Lee (1807–1870). The Union win effectively stopped Lee's invasion of the North.

The following November 19 President Abraham Lincoln (1809–1865) made the historical address at Gettysburg, as he dedicated part of the battlefield as a national cemetery. Beginning with the now-famous words "Four score and seven years ago our fathers brought forth upon this continent a new nation, conceived in Liberty, and dedicated to the proposition that all men are created equal," the short speech (which Lincoln rewrote many times) closed by issuing a rallying cry for the nation as a whole, saying, "we here highly resolve that the dead shall not have died in vain—that the nation shall, under God, have a new birth of freedom—and that governments of the people, by the people, and for the people, shall not perish from the earth."

Indian Wars

What was the Battle of **Tippecanoe**?

The battle near present-day Lafayette, Indiana, took place on November 7, 1811, when U.S. troops under the command of General William Henry Harrison (1773–1841) defeated the Shawnee Indians while their leader, Tecumseh (1768–1813), was away.

Before the November battle, the Shawnee Indians had been steadily pushed back from their ancestral lands along the Cumberland River in Kentucky. Once they settled in the Ohio River valley, they formed a wall of resistance to further pressure. Resistance fighting began in 1763 and did not end until 30 years later (1793) when the American forces of General Anthony Wayne (1745–1796) defeated the Shawnee at Fallen Timbers in northwest Ohio. Tecumseh was 25 years old at the time of this critical defeat of his people. He became determined, despite the counsel of the tribe elders, to halt the westward movement of the Shawnee. Tecumseh and his brother, Tenskwatawa (c. 1768–1834), built a settlement on Tippecanoe Creek near the Wabash River, Indiana. Called Prophet's Town, it became a rallying point for those Indians who resisted displacement by colonists. Soon the Shawnee who had settled at Prophet's Town were joined by bands of Wyandot, Potawatomi, Miami, and Delaware Indians. The British became aware of Tecumseh's defiance and soon supplied Prophet's Town Indians with arms to use against the Americans.

In 1810 the governor of the Indiana Territory, William Henry Harrison (1773–1841), requested a meeting with Tecumseh. But both were suspicious of each other and could reach no agreement for peace. Settlers who were aware of Tecumseh's armed resistance movement pressured President James Madison (1751–1836) to take action,

U.S. General William Henry Harrison leads his troops against the Shawnee Indians at the Battle of Tippecanoe, 1811.

which he did: Under orders from the president, Harrison set out in the fall of 1811 with a force of 1,000 men. Arriving at Tippecanoe on November 6, he met with the Indians under a flag of truce and then made camp about a mile away from Prophet's Town. Tecumseh was visiting the Creek Indians in Alabama at the time, trying to rally their support in the white resistance movement. But his brother, Tenskwatawa, rallied the men at Prophet's Town, and at the break of day, they attacked Harrison's camp. Hand-to-hand fighting ensued in an icy drizzle. The Indian alliance retreated and though Harrison took heavy losses, he had managed to destroy Prophet's Town.

The battle undermined Tecumseh's initiative and launched the career of Harrison. In 1840, after successful military duty and service in the U.S. Congress, Harrison rode a wave of public support all the way to the White House to become the ninth president. His campaign slogan was "Tippecanoe and Tyler, too" (John Tyler was his running mate). Eventually the Shawnee were pushed into Kansas. Tecumseh became an ally of the British in the War of 1812 (1812–14) and was killed in the fighting.

What was the **Sioux uprising**?

The uprising took place in August and September 1862 in southwestern Minnesota when the Sioux there suddenly had been made to give up half their reservation lands.

Their situation was made worse by crop failures. While the government debated over whether it would make the payments it owed to Indian nations in gold or in paper currency, the Sioux were also without money. The U.S. agent at the Sioux reservation refused to give out any food to the Indians until their money arrived from Washington. The Sioux people were hungry and angry, and white observers could see there was trouble coming and warned the government. But the situation soon erupted in August when four young men having a shooting contest suddenly fired into a party of whites, killing five people. The Sioux refused to surrender the four men to the authorities, and, under the leadership of Chief Little Crow (c. 1820–1863), they raided white settlements in the Minnesota River valley. A small U.S. military force sent out against the Sioux was annihilated. Meantime, white settlers fled the region in panic.

On September 23 Minnesota sent out 1,400 men who defeated Little Crow at the Battle of Wood Lake. The raids had already claimed the lives of 490 white civilians. Thirty-three Sioux were killed in the fighting with the military. While most of the Indians who had taken up arms fled to the Dakotas, the government began to round up native men who were suspected of participating in the campaign against white settlers. More than 300 men were tried and sentenced to death, many of them on flimsy evidence. Episcopal bishop Henry Whipple (1822–1901) interceded in their behalf, making a personal plea to President Abraham Lincoln (1809–1865). The bishop was able to get 265 death sentences reduced to prison terms. But 38 Sioux men, accused of murder or rape, were hung in a public ceremony on December 26, 1862. The Sioux reservation lands were broken up, and the remaining Sioux were dispersed. Minnesota nevertheless continued to man military posts in that part of the state for years to come.

The events during and after the uprising were brutal for both sides, but many observers had seen that the treatment of the Sioux was going to lead to conflict: One missionary, after witnessing the harsh way the policy with the Indians had been carried out, wrote to Bishop Whipple, saying, "If I were an Indian I would never lay down the war club while I lived."

What was **"Custer's Last Stand"**?

The term refers to the defeat of General George A. Custer (1839–1876) at the Battle of Little Bighorn on June 25, 1876. Custer had a national reputation as a Civil War general and Indian fighter in the west, and when he and his troops were outnumbered and badly beaten by the Sioux led by Sitting Bull (c. 1831–1890)—just as the country was about to celebrate the hundredth anniversary of the Declaration of Independence—the result was a stunning reversal in the national mood.

Little Bighorn was part of a series of campaigns known collectively as the Sioux War. Several events led to the conflict that became Custer's Last Stand. The Sioux were nontreaty Indians, which means they had refused to accept the white-dictated limits on their territory. They were outraged at the repeated violation of their lands by the onrush of miners to new gold strikes in the Black Hills of South Dakota. Further, there had been eight attacks by the Sioux on the Crow who were living on reservation land. Finally, Sitting Bull, the chief of the Hunkpapa band of Sioux, refused government demands that he and his people return to reservation lands. Meanwhile, unbeknownst to the government's military strategists, by spring 1876 Sitting Bull had been joined in his cause by other groups of northern Plains tribes, including the Cheyenne led by Crazy Horse (c. 1842–1877). With the government ready to use force to return Sitting Bull and his band of Sioux to reservations, the stage was set for a conflict—bigger than any Washington official had imagined.

On June 25 Custer rode into Montana territory with his Seventh Cavalry to meet the Sioux. Despite orders to simply contain the Indians and prevent their escape, he attacked. While historians remain divided on how Custer could have been defeated on that fateful June day, one thing remains certain: Custer and his men were badly outnumbered. Having divided his regiment into three parts, Custer rode with about 225 men against a force of at least 2,000—the largest gathering of Indian warriors in Western history. Custer and his soldiers all died. The fighting continued into the next day, with those Indians that remained finally disbanding and returning to their designated territory. Meantime, Sitting Bull and his band retreated into Canada. Returning to the United States five years later, in 1890, Sitting Bull was killed by authorities.

The battle became the subject of countless movies, books, and songs. It's remembered by some Native Americans as a galvanizing force—proof that brave men who fight for what they believe in can win.

BOER WARS

What were the **Boer Wars**?

They were conflicts between the British and the Afrikaners (or Boers, who were Dutch descendants living in South Africa) at the end of the nineteenth century in what is today South Africa. The first war, a Boer rebellion, broke out in 1880 when the British and the Afrikaners fought over the Kimberley area (Griqualand West), where a diamond field had been discovered. The fighting lasted a year, at which time the South African Republic (established in 1856) was restored. But the stability would not last long: In 1886 gold was discovered in the Transvaal, and though the Afrikaner region was too strong for the British to attempt to annex it, they blocked the Afrikaners' access to the sea. In 1899 the Afrikaner republics of the Orange Free State and the Transvaal joined forces in a war against Britain. The fighting raged until 1902, when the Afrikaners surrendered. For a time after the Boer War (also called the South African War), the Transvaal became a British crown colony. In 1910 the British government combined its holdings in southern Africa into the Union of South Africa.

What caused the **Spanish-American War**?

The 1898 war, which lasted only a matter of months (late April to mid-August), was fought over the liberation of Cuba. During the 1870s the Cuban people rebelled against Spanish rule. But once that long rebellion had been put down, peace on the Caribbean island did not hold: worsening economic conditions prompted revolution in 1895. American leaders, the bloody American Civil War (1861–65) still in their memories, feared that while the Cuban rebels could not win their battle against the Spanish, neither were the Spanish strong enough to fully put down the insurrection. Meanwhile, the American public, fed by a steady stream of newspaper accounts reporting oppressive conditions on the island, increasingly supported U.S. intervention in the Cuban conflict.

In November 1897 President William McKinley (1843–1901) did intervene, but it was through political, rather than military, pressure. As a result, Spain granted Cuba limited self-government within the Spanish empire. However, the move did not satisfy the Cuban rebels, who were determined to achieve independence from Spain; the fighting continued. Rioting broke out in Havana, and in order to protect Americans living there, the United States sent the battleship *Maine* to the port on January 25, 1898. On February 15, an explosion blew up the *Maine,* killing more than 200 people. Blame for the blast was promptly—and history would later conclude wrongly—assigned to Spain. While President McKinley again made several attempts to pressure Spain into granting Cuba full independence, it was to no avail. Nevertheless, on April 19, the U.S. Congress passed a joint resolution recognizing an independent Cuba, disclaiming American intention to acquire the island, and authorizing the use of the American army and navy to force Spanish withdrawal. On April 25, the United States formally declared that the country was at war with Spain.

In the months that followed, American forces battled the Spanish and Spanish loyalists in Cuba and the Spanish-controlled Philippines. There was also military activity on Puerto Rico, however, the American forces there met little resistance. Once Santiago, Cuba, was surrendered by the Spanish after the battle at San Juan Hill in July 1898, it would only be a matter of weeks before a cease-fire was called and an armistice was signed (on August 12), ending the brief war.

What was the charge up **San Juan Hill**?

On July 1, 1898, during the Spanish-American War, Colonel Theodore Roosevelt (1858–1919) led his American troops, known as the Rough Riders, on an attack of the Spanish blockhouse (a small fort) on San Juan Hill, near Santiago, Cuba. Newspaper reports

Colonel Theodore Roosevelt (standing just left of center, wearing white suspenders) and his Rough Riders after fighting the Spanish at San Juan Hill, near Santiago, Cuba, 1898.

made Roosevelt and the Rough Riders into celebrities, and even after he became a U.S. president, Teddy Roosevelt remarked that "San Juan was the great day of my life."

San Juan Hill was part of a two-pronged assault on Santiago. While the Rough Riders regiment attacked the Spanish defenses at San Juan and Kettle Hills, another American division, led by General Henry Lawton (1843–1899), captured the Spanish fort at El Caney. The success of the two initiatives on July 1 combined to give the Americans command over the ridges surrounding Santiago. By July 3 the American forces had destroyed the Spanish fleet under the command of Admiral Pascual Cervera y Topete (1839–1909). On July 17 the Spanish surrendered the city.

Though the victory was critical to the outcome of the war, the assault on Kettle Hill and San Juan Hill had come at a high price: 1,600 American lives were lost, in a battle that had seen American troops—black and white—fight the Spanish shoulder to shoulder.

What is the **Treaty of Paris**?

There has been more than one Treaty of Paris. The following international agreements were signed in the French capitol:

In 1763 representatives of Great Britain, France, and Spain signed a treaty, which, along with the Treaty of Hubertusburg (February 15, 1763), ended the Seven Years' War (1756–63).

On September 3, 1783, the Treaty of Paris, which had been under negotiation since 1782, was signed by the British and the Americans, represented by statesmen Benjamin Franklin (1706–1790), John Adams (1735–1826), and John Jay (1745–1829). The agreement officially ended the American Revolution (1775–83), establishing the United States as an independent country and drawing the boundaries of the new nation—which extended west to the Mississippi River, north to Canada, east to the Atlantic Ocean, and south as far as Florida, which was given to Spain.

In 1814 and 1815 treaties were signed ending the Napoleonic Wars, which the French ruler Napoleon Bonaparte (1769–1821) had begun shortly after taking power in 1799.

In 1856 European nations signed a treaty in Paris ending the Crimean War (1853–56) and outlawing the wartime practice of privateering.

The Treaty of Paris that was signed December 10, 1898, settled the conflict that had resulted in the Spanish-American War (1898). This treaty provided for Cuba's full independence from Spain. It also granted control of Guam and Puerto Rico to the United States. The pact further stipulated that the United States would pay Spain $20 million for the Philippine Islands.

In 1951, in the wake of World War II (1939–45), Belgium, France, Italy, Luxembourg, the Netherlands, and West Germany signed the Treaty of Paris, which established the European Coal and Steel Community (ECSC). The desire was to bring about economic and political unity among the democratic nations of Europe. This agreement paved the way for the European Union effected by the Maastricht Treaty, an economic agreement signed by representatives of 12 European countries in the Netherlands in 1992.

What happened to the **Philippines after they were ceded by Spain**?

After the United States gained control of the Philippines in the Treaty of Paris (1898), the Pacific archipelago was soon embroiled in a conflict similar to the one in Cuba, which had developed into the Spanish-American War: Filipinos, determined to achieve independence, revolted in an uprising that lasted from 1899 to 1901. A civil government was established on the Philippines in 1901, and in November 1935, the Commonwealth of the Philippines was officially established. However, the islands would continue to be the site of conflict in the coming decades, as the United States again struggled with a foreign power—this time the Japanese—for control of the islands during World War II (1939–45).

Mexican Revolution

What does "¡Viva Zapata!" mean?

It was the cry that went up in support of the rebel general Emiliano Zapata (1879–1919), whose chief concern during the Mexican Revolution (1910–20) was the distribution of land to the people.

An advocate of Mexico's lower classes, Zapata began revolutionary activities against the government of Porfirio Díaz (1830–1915) as early as 1897. Zapata rose to prominence in helping the liberal and idealistic Francisco Madero (1873–1913) overthrow Díaz in 1911. With Madero placed in power, Zapata promptly began pressing his co-conspirator for a program to distribute the hacienda (large estate) lands to the peasants. Rebuffed by Madero that same year, Zapata drafted the agrarian Plan of Ayala and renewed the revolution. Madero's government never achieved stability and proved to be ineffective, prompting a second overthrow in 1913: Victoriano Huerta (1854–1916) seized power from Madero, whom he had helped put into office, and in the chaos surrounding the coup, Madero was shot and killed.

But Zapata refused to support Huerta and remained a leader of the revolution, continuing his crusade for the people—who supported him with cheers of "¡Viva Zapata!," meaning "long live Zapata!" The bitter fighting of the revolution continued and soon those who had supported the slain Madero—including Zapata and Pancho Villa (1878–1923)—threw their backing behind another revolutionary, Venustiano Carranza (1859–1920). In 1914 Carranza's forces occupied Mexico City and forced Huerta to leave the country. No sooner had Carranza taken office than the revolutionaries began fighting among themselves. Zapata and Pancho Villa demanded dramatic reforms and together they attacked Mexico City in 1914. Five years later, and one year before the end of the revolution, Carranza's army ambushed and assassinated Zapata in his home state of Morelos.

Why did the U.S. government send troops after **Pancho Villa**?

Pancho Villa (1878–1923) was sought by the U.S. government because in 1916 he and his followers attacked Americans on both sides of the border. In 1915 the United States decided it would back the acting chief of Mexico, Venustiano Carranza (1859–1920), even as he faced attacks from two of his fellow revolutionaries, Emiliano Zapata (1879–1919) and Pancho Villa. Four years earlier, Villa had himself sought to control Mexico after the fall of President Porfirio Díaz. When the United States cut off the flow of ammunition to the rebels, Villa, who was a fierce fighter, earned himself a reputation as a bandit, seeking revenge on Americans in Mexico by stopping trains and shooting the passengers. In 1916 Villa raided the small New Mexico village of Columbus, where he killed 18 people. The attack prompted President Woodrow Wilson (1856–1924) to send U.S. soldiers to hunt Villa down and capture him. Though thousands of men were put on the initiative under General John Pershing (1860–1940), they never caught up with the bandit. Wilson withdrew the forces from Mexico after the government there expressed resentment for the U.S. effort—which the Mexican people, President Carranza included, viewed as a meddlesome American interference in the Mexican Revolution (1910–20). The revolution ended three years later, after 10 years of fighting and disorder.

WARS IN ASIA

What was the **Chinese-Japanese War**?

It was a war fought in 1894 to 1895 over control of Korea, which was a vassal state of China. When an uprising broke out in Korea in 1894, China sent troops in to suppress it. Korea's ports had been open to Japan since 1876 and in order to protect its interests there, Japan, too, sent troops to the island nation when trouble broke out. But once the rebellion had been put down, the Japanese troops refused to withdraw. In July 1894 fighting broke out between Japan and China, with Japan emerging as the victor, having crushed China's navy. A peace treaty signed on April 17, 1895, provided for an independent Korea (which only lasted until 1910, when Japan took possession) and for China to turn over to Japan the island of Taiwan and the Liaodong Peninsula (the peninsula was later returned to China for a fee after Russia, Germany, and France forced Japan to do so). The war, though relatively brief, seriously weakened China, and in the imperialist years that followed, the European powers scrambled for land concessions there.

What caused the **Russo-Japanese War**?

From 1904 to 1905 the war was fought by Russia and Japan over their interests in China (particularly Manchuria) and Korea—areas of strategic importance to each country. Before fighting broke out Japan moved to settle the conflict, but the overture was rejected by Tsar Nicholas II (1868–1918), and Japan soon severed all diplomatic relations (on February 6, 1904) with Russia. Two days later the Japanese issued a surprise attack on Russian ships at Lushun (Port Arthur), Manchuria. On February 10 Japan officially declared itself at war with Russia. The battles—both on land and at sea—went badly for the Russian forces, which could not be adequately reinforced or supplied to meet the powerful and disciplined Japanese. Early in 1905, the war effort already unpopular back home, revolution broke out in Russia, further weakening the country's resolve.

After an eight-month siege at Lushun, it became clear that Russia could no longer muster a fight. Further, the war was expensive for Japan, which sought the intervention of the United States in settling the conflict. President Theodore Roosevelt (1858–1919) became involved in mediating the dispute; a peace treaty was signed Sep-

What was the Boxer Rebellion?

It was a Chinese uprising in 1900, which was put down through the combined forces of eight foreign countries including Germany, Italy, Japan, Russia, and the United States. The Chinese-Japanese War (1894–95) had seriously weakened China, and in 1898 the country agreed to lease its Kiaochow region to Germany. Soon other European countries followed suit and before long Western influence was being felt in China. This angered many Chinese, including members of a secret society that opposed the Manchu government for having allowed the foreign incursions. This secret society was made up of athletic young men, and so they were called Boxers by China's Westerners.

Between June 21 and August 14, 1900, Boxers rebelled against anything foreign and began a raid of the country that was intended to drive out all foreign influence. The uprising was aimed at not only Westerners and foreign diplomats, but missionaries, Chinese Christians, and any Chinese who were thought to support Western ideas. Houses, schools, and churches were burned. Much of the destruction took place in Beijing, and when foreign diplomats there called for help, it arrived from eight countries. The Manchu government did not welcome this interference in its affairs and promptly declared war on the eight nations.

On September 7, 1901, the Manchu government signed a peace settlement with the foreign countries. The Boxer Protocol called for China to punish officials who had been involved in the rebellion and pay damages in the hundreds of millions of dollars. The United States, Britain, and Japan later returned part of the money to China, specifying that it be used for educational purposes.

tember 5, 1905, at a shipyard in Portsmouth, New Hampshire, following one month of deliberations. The terms of the treaty were these: both nations agreed to evacuate Manchuria; Russia ceded to Japan the southern half of Sakhalin Island, which lies between the two countries (the island was ceded back to Russia after World War II); Korea became a Japanese protectorate; and Russia transferred to Japan the lease of China's Liaodong Peninsula. And so Japan emerged as a power onto the world scene.

Russian Revolution

What was **Bloody Sunday**?

The January 22, 1905, event, which is also known as Red Sunday, signaled the beginning of revolutionary activity in Russia that would not end until 1917. On that winter day the young Russian Orthodox priest Georgi Gapon (1870–1906), carrying a cross over his shoulder, led what was intended to be a peaceful workers' demonstration in front of the Winter Palace at St. Petersburg. But, as the *London Times* correspondent reported that day, when the crowd was refused entry into the common gathering ground of the Palace Square, "the passions of the mob broke loose like a bursting dam." Despite Father Gabon's thinking that they too would join the workers in protest, the Cossack guards and troops, still loyal to the Romanov tsar Nicholas II (1868–1918), shot into the crowd of demonstrators, killing about 150 people—children, women, and young people among them.

Father Gapon, who had intended to deliver to the tsar a petition on behalf of the workers, was injured during the day's events and later fled in exile. His thinking that the palace guards would come over to the workers' side was not his only miscalculation: Tsar Nicholas was not even at the palace that Sunday, having left days earlier. But Nicholas's reign was threatened by his troops' response to the gathering crowd: So horrific was the bloodshed that the snow-covered streets of St. Petersburg were stained in red, and the correspondent for the French newspaper *Le Matin* reported that the Cossacks had opened fire "as if they were playing at bloodshed." The event sent shockwaves through the country, where hostilities had been mounting against Nicholas's ineffective government. It also stirred up unrest elsewhere, including in Moscow—where, in a related event, the tsar's uncle, Grand Duke Serge, was killed in early February. The death was a sure sign that popular anger had been focused on the tsar and his family. In the countryside, the peasants rose up against their landlords, seizing land, crops, and livestock.

The events foreshadowed the downfall of tsarist Russia: Though the outbreak in 1905 was unsuccessful in effecting any change and Nicholas remained in power for 12 more years, he was the last tsar to rule Russia.

Who was **Rasputin**?

Grigory Rasputin (1872–1916) was a Russian mystic and quasi-holy man who rose from peasant farmer to become adviser to Tsar Nicholas II (1868–1918) and his wife, Tsarina Alexandra (1872–1918). Sometime in 1905 or shortly thereafter, Alexandra had come into contact with Rasputin, and, showing he was able to effectively treat Nicholas's and Alexandra's severely hemophiliac son Alexis (1904–1918), Rasputin quickly gained favor with the Russian rulers. But the prime minister and members of the legislative assembly, the Duma, could see Rasputin was a disreputable character, and they feared his influence on the tsar. They even tried to exile Rasputin, but to no avail.

By 1913, one year before the outbreak of World War I (1914–18), the Russian people had become acutely aware of Tsar Nicholas's weaknesses as a ruler—not only was his government subject to the influence of a pretender like Rasputin, but the events of Bloody Sunday had irreversibly marred the tsar's reputation. That year the Romanov dynasty was marking its 300th anniversary: members of the royal family had ruled

Russia since 1613. But public celebrations, intended to be jubilant affairs, were instead ominous, as the crowds greeted Nicholas's public appearances with silence.

Russia's entry into World War I proved to be the beginning of the end for Nicholas, with Rasputin at the front and center of the controversy that swirled around the royal court. During the first year of fighting against Germany, Russia suffered one military catastrophe after another. These losses did further damage to the tsar and his ministers. In the fall of 1915, urged on by his wife, Nicholas left St. Petersburg and headed to the front to lead the Russian troops in battle himself. With Alexandra left in charge of government affairs, Rasputin's influence became more dangerous than ever. But in December 1916, a group of aristocrats put an end to it once and for all when, during a palace party, they laced Rasputin's wine with cyanide. Though the poison failed to kill Rasputin, the noblemen shot him and deposited his body in a river later that night. Nevertheless, the damage to Nicholas and Alexandra had already been done: By that time virtually all educated Russians opposed the tsar, who had removed many capable officials from government office, only to replace them with the weak and incompetent executives favored by Rasputin. The stage had been set for revolution.

What was the **Bolshevik Revolution**?

It was the November 1917 revolution in which the Bolsheviks, an extremist faction within the Russian Social Democratic Labor Party (later renamed the Russian Communist Party), seized control of the government, ushering in the Soviet age. The event is also known as the October Revolution since by the old Russian calendar (in use until 1918), the government takeover happened on October 25.

The Bolshevik Revolution was the culmination of a series of events in 1917. In March, with Russia still in the midst of World War I (1914–18), the country faced hardship. Shortages of food and fuel made conditions miserable. The people had lost faith in the war effort and were loathe to support it by sending any more young men into battle. In the Russian capital of Petrograd (which had been known as St. Petersburg until 1914), workers went on strike and rioting broke out. In the chaos (called the March Revolution), Tsar Nicholas II (1868–1918) ordered the legislative body, the Duma, to disband; instead, the representatives set up a provisional government. Having lost all political influence, Nicholas abdicated the throne on March 15. He and his family were imprisoned and are believed to have been killed in July of the following year.

Hearing of Nicholas's abdication, longtime political exile Vladimir Lenin (1870–1924) returned from Europe to Petrograd, where he led the Bolsheviks in rallying the Russian people with calls for peace, land reform, and worker empowerment; their slogan was "Land, Peace, and Bread." The Bolsheviks grew in numbers and became increasingly radical, in spite of efforts by the provisional government headed by revolutionary Alexander Kerensky (1881–1970) to curb the Bolsheviks' influence. The only socialist member of the first provisional government, Kerensky's government proved

The Bolsheviks lead workers and disgruntled soldiers in a takeover of Petrograd's Winter Palace, Russia, 1917.

ineffective and failed to meet the demands of the people. He also failed to end the country's involvement in World War I, which the Bolsheviks viewed as an imperialistic war.

On November 7 the Bolsheviks led workers and disgruntled soldiers and sailors in a takeover of Petrograd's Winter Palace, the scene of Bloody Sunday in 1905, and which had become the headquarters of Kerensky's provisional government. By November 8 the provisional government had fallen.

What was the **Red Terror**?

The Red Terror was the brutal coercion used by the Communists during the tumultuous years of civil unrest that followed the Bolshevik Revolution of November 1917. After the revolution, the Bolsheviks, now called Communists, put their leader, Vladimir Lenin (1870–1924), into power. Delivering on the Bolshevik promise to end the country's involvement in World War I (1914–18), Lenin immediately called for peace talks with Germany, ending the fighting on the eastern front. (Germany and the other Central Powers would be prevented from victory on the western front by the entry of the United States into the war that same year.) But the Brest-Litovsk Treaty, signed March 3, 1918, dictated harsh—and many believed humiliating—terms to Russia, which was forced to give up vast territories including Finland, Poland, Belarus, Ukraine, Moldavia, and the Baltic states of Estonia, Latvia, and Lithuania.

Meantime, Russians had elected officials to a parliamentary assembly. But when the results were unfavorable to Lenin (of the 703 deputies chosen, only 168 were Communists), he ordered his troops to bar the deputies from convening, and so the assembly was permanently disbanded. In its place, Lenin established a dictatorship based on Communist secret police, the Cheka. Further, the radical social reforms he had promised took the form of government takeover of Russia's industries and the seizure of farm products from the peasants. Lenin's hard-handed tactics created opposition to the Communists—colloquially known as the Reds. The opposition organized their White army, and civil war ensued. In September 1918 Lenin was nearly assassinated by a political opponent, prompting Lenin's supporters to organize the retaliative initiative that came to be known as the Red Terror. Though thousands of Communist opponents were killed as a result, the unrest in Russia would not end until 1920. And some believe the ruthless repression of the Red Terror lasted into 1924.

WORLD WAR I

How did **World War I begin**?

Though the Great War, as it was called until World War II (1939–45), was sparked by the June 28, 1914, assassination of Archduke Francis Ferdinand of Austria-Hungary (1863–1914), the war in Europe had been precipitated by several developments. National pride had been growing among Europeans; nations increased their armed forces through drafts; and colonialism continued to be a focus of the European powers, as they competed with each other for control of lands in far-off places. Meantime, weapons and other implements of war had been improved by industry, making them deadlier than ever. So on that June day in the city of Sarajevo (then the capital of Austria-Hungary's province of Bosnia and Herzegovina), when a gunman named Gavrilo Princip (1894–1918) shot down Archduke Ferdinand, it is not surprising that Austria-Hungary responded with force. Since Princip was known to have ties to a Serbian terrorist organization, Austria-Hungary declared war on Serbia. Both sides, however, believed that the battle would be decided quickly. But instead fighting would spread, involving more countries. Four years of fighting—aided by the airplane, the submarine, tanks, and machine guns—would cause greater destruction than any other war to that date.

What **alliances** were **forged during World War I**?

In its declaration of war against Serbia in late July 1914, Austria-Hungary was joined in early August by its ally Germany, which together formed the Central Powers. In October 1914 Bulgaria and the Ottoman Empire joined the Central Powers.

When the fighting began, France, Britain, and Russia threw their support behind Serbia, and together were known as the Allies. The Allies declared war on the Ottoman Empire in November 1914, after Turkish ships bombarded Russian ports on the Black Sea and Turkish troops invaded Russia. Eventually, 20 more nations joined the Allies, but not all of them sent troops to the front. Belgium, Montenegro, and Japan joined the Allies in August 1914, with Japan declaring war on Germany and invading several Pacific islands to drive out the Germans. In 1915 Italy and San Marino joined; as fighting wore on, in 1916, Romania and Portugal became Allied nations; and 1917 saw the entry of eight countries, most notably the United States and China, but also Liberia, Greece, Siam, Panama, Cuba, and Brazil. Before the war ended in 1918, Guatemala, Haiti, Honduras, Costa Rica, and Nicaragua all became supporters of the Allies.

What did the ***Lusitania*** have to do with World War I?

World War I (1914–18) was already under way when in May of 1915 a German U-boat sank a British passenger ship, the SS *Lusitania,* off the coast of Ireland. The ship had been launched in 1907 by Britain's Cunard Line to become the largest passenger ship afloat. When she was downed in the North Atlantic, 1,200 civilians, including 128 American travelers, were killed. President Woodrow Wilson (1856–1924) warned Germany that another such incident would force the United States into entering the war. Germany heeded the warning only for a time.

Why did the **United States** get involved **in World War I**?

When war broke out in Europe in August 1914, Americans opposed the involvement of U.S. troops, and President Woodrow Wilson (1856–1924) declared the country's neutrality. But as the fighting continued and the German tactics threatened civilian lives, Americans began siding with the Allies.

After the sinking of the passenger liner SS *Lusitania,* Germany adopted restricted submarine warfare. But early in 1917 Germany again began attacking unarmed ships, this time American cargo boats, goading the United States into the war. Meantime, German U-boats were positioning to cut off shipping to and from Britain, in an effort to force the power to surrender. Tensions between the United States and Germany peaked when the British intercepted, decoded, and turned over to President Wilson a telegram Germany had sent to its ambassador in Mexico. The so-called "Zimmermann note," which originated in the office of German foreign minister Arthur Zimmermann (1864–1940), urged the German officials in Mexico to persuade the Mexican government into war with the United States—in order to reconquer lost territory in Texas, New Mexico, and Arizona. The message was published in the United States in early March. One month later, on April 6, 1917, the U.S. Congress declared war on Germany after President Wilson had asserted that "the world must be made safe for democracy."

The SS *Lusitania* was torpedoed by a German U-boat in May 1915; the event forced U.S. involvement in World War I.

How did **World War I end**?

Though the United States had been little prepared to enter the war, the American government mobilized quickly to rally the troops—and the citizens—behind the war effort: In April 1917 the U.S. Regular Army was comprised of just more than 100,000 men; by the end of the war, the American armed forces stood some 5 million strong. It was the arrival of the U.S. troops that gave the Allies the manpower they needed to win the war. After continued fighting in the trenches of Europe, which had left almost 10 million dead, in November 1918, Germany agreed to an armistice and the Central Powers finally surrendered. In January 1919 Allied representatives gathered in Paris to draw up the peace settlement.

Who were the **Big Four**?

Though the Paris Peace Conference, which began in January 1919, was attended by representatives of all the Allied nations, the decisions were made by four heads of government, called the Big Four: President Woodrow Wilson (1856–1924) of the United States, Prime Minister David Lloyd George (1863–1945) of Great Britain, Premier Georges Clemenceau (1841–1929) of France, and Premier Vittorio Orlando (1860–

German soldiers ready a machine gun as they watch enemy positions during World War I.

1952) of Italy. Other representatives formed committees to work out the details of the treaties that were drawn up with each of the countries that had made up World War I's Central Powers: the Treaty of Versailles was signed with Germany, the Treaty of St. Germain was signed with Austria, the Treaty of Neuilly was made with Bulgaria, the Treaty of Trianon was made with Hungary, and the Treaty of Sevres was signed with the Ottoman Empire.

How were **Europe's lines redrawn** as a result of World War I?

The treaties that came out of the Paris Peace Conference (1919–20) redrew Europe's boundaries, carving new nations out of the defeated powers. The Treaty of Versailles forced Germany to give up territory to Belgium, Czechoslovakia, Denmark, France, and Poland. Germany also forfeited all of its overseas colonies and turned over coal fields to France for the next 15 years. The treaties of St. Germain and Trianon toppled the former empire of Austria-Hungary (whose archduke had been assassinated in 1914, triggering the war) so that the separate nations of Austria and Hungary were formed, each occupying less than a third of their former area. Their former territory was divided among Italy, Romania, and the countries newly recognized by the treaties: Czechoslovakia, Poland, and the kingdom that later became Yugoslavia. The Treaty of

Sevres took Mesopotamia (present-day Iraq), Palestine, and Syria away from the Ottoman Empire, which three years later became the Republic of Turkey. Finally, Bulgaria lost territory to Greece and Romania. However, these new borders would serve to heighten tensions between some countries, as the territorial claims of the newly redrawn nations overlapped with each other.

How did the **Treaty of Versailles** pave the way for World War II?

In the aftermath of World War I (1914–18), Germany was severely punished: One clause in the Treaty of Versailles even stipulated that Germany take responsibility for causing the war. In addition to its territorial losses, Germany was also made to pay for an Allied military force that would occupy the west bank of the Rhine River, intended to keep Germany in check for the next 15 years. The treaty also limited the size of Germany's military. In 1921 Germany received a bill for reparations: It owed the Allies $33 million.

While the postwar German government had been made to sign the Treaty of Versailles under the threat of more fighting from the Allies, the German people nevertheless faulted their leaders for accepting such strident terms. Not only was the German government weakened, but public resentment over the Treaty of Versailles soon developed into a strong nationalist movement—led by German chancellor and führer Adolf Hitler (1889–1945).

What was the **League of Nations**?

The League of Nations was the forerunner to the United Nations. It was an international organization established by the Treaty of Versailles at the end of World War I (1914–18). Since the United States never ratified that treaty, it was not a member.

The league was set up to handle disputes among countries and to avoid another major conflict such as the Great War (which is how World War I was referred to until the outbreak of World War II). But the organization proved to be ineffective; it was unable to intervene in such acts of aggression as Japan's invasion of Manchuria in 1931, Italy's conquest of Ethiopia during 1935 to 1936 and occupation of Albania in 1939, and Germany's takeover of Austria in 1938.

The League of Nations dissolved itself during World War II (1939–45). Though unsuccessful, the organization did establish a basic model for a permanent international organization.

THE ERA BETWEEN THE WORLD WARS

What was **Nazism**?

Short for national socialism, "Nazi" was a derisive abbreviation that held. The Nazi doctrine rests on three philosophies: extreme nationalism, anti-Semitism, and anti-communism. As one of the Central Powers, Germany's defeat in World War I (1914–18) resulted in severe punishment of that country and its seriously diminished role in Europe. The doctrines of Nazism took hold there, appealing to the masses with promises of a rebuilt Germany.

The "bible" of Nazism was Adolf Hitler's *Mein Kampf* (*My Struggle;* 1923), which asserted the superiority of a pure Aryan race (Aryans are non-Jewish Caucasians, particularly those of northern European descent), led by an infallible ruler (called "the führer"); the reestablishment of a German empire (the Third Reich); and the systematic annihilation of people who Nazis perceived to be Germany's worst enemies: Jews and Communists. Nazis ruled Germany from 1933, when Hitler rose to power as head of the National Socialist German Workers' Party. In their own country, they enforced their policies through a secret police (the Gestapo), storm troops (called the SS), and Hitler's bodyguard (called the SA). Elsewhere in Europe, the Nazis used sheer force in imposing their system. Their aggression and ruthlessness resulted in World War II (1939–45). During the Holocaust (1933–45), Nazi soldiers, led by "Hitler's henchmen," persecuted and exterminated upwards of 12 million people, at least half of whom were European Jews. Nazism ended in 1945, when Hitler killed himself and Germany lost the war. The doctrine, which demonstrated how detrimentally powerful a theory can be, was outlawed thereafter. Sadly, the late twentieth century saw a resurgence of "neo-Nazism" among extremists in Germany and the United States.

SINO–JAPANESE WAR

What was the **Sino-Japanese War**?

This dispute between China and Japan (who had not that long ago clashed in the Chinese-Japanese War of 1894–95) began in 1937 and was absorbed by World War II (1939–45). The trouble between the Asian powers began when Japan, having already taken Manchuria and the Jehol Province from China, attacked China again. Though China was in the midst of internal conflict—with the nationalist forces of Generalissimo Chiang Kai-shek (1887–1975) fighting the Communists under Mao Tse-tung (1893–1976)—China turned its attention to fighting the foreign aggressor. The fighting between the two countries continued into 1941 before war was officially declared by China. In so doing, China was at war not only with the Japanese but with Japan's

Adolf Hitler salutes his troops in Warsaw on October 5, 1939, after Germany's invasion of Poland.

Axis allies—Germany and Italy—as well. The conflict then became part of World War II, which ended with the surrender of Japan to the Allies in September 1945.

What was the **Nanking Massacre**?

One of the most brutal chapters in modern history, the Nanking Massacre, also called the Rape of Nanking, was a mass execution of hundreds of thousands of unarmed Chinese civilians by invading Japanese soldiers in December 1937 and January 1938. No one knows for certain how many people were murdered in the mass killings, but most estimates place the number at 300,000, with another 80,000 people raped and tortured, including women and children.

On December 13, 1937, the Japanese royal army swept into the eastern Chinese city of Nanking (today called Nanjing), which was then the capital of China. In the weeks that followed, the Japanese soldiers went on an orgy of violence. The atrocities were documented on film by the Japanese themselves as well as by helpless foreigners in the city at the time of the seizure. Surviving photos show unimaginable cruelties. It is believed that Japan's military had been trained to carry out the killings and atrocities in order to make an example of Nanking to other Chinese people, thereby facilitating Japan's intended occupation.

The horrific event was the source of recent controversy stirred by the 1997 publication of *The Rape of Nanking: The Forgotten Holocaust of World War II* by Iris Chang (1968–), a historian and journalist whose grandparents narrowly escaped the massacre. Unlike Germany, which accepted responsibility for the holocaust of Jews during World War II (1939–45) and whose Nazi leaders were tried in number at Nuremberg, Japan never acknowledged its crimes committed at Nanking. After World War II only a few of Japan's military leaders were tried and found guilty of war crimes related to the taking of Nanking. This chapter in Japan's national history has been largely denied by its officials, some of whom accused Chang of issuing propaganda. Chang stood by her research, which included interviews with survivors as well as with Japanese soldiers who participated in the violence. The massacre remains a deeply divisive event between the two nations and their people.

Spanish Civil War

What caused the **Spanish Civil War**?

From 1936 to 1939, two sides fought for control of Spain: the nationalists and the loyalists. The insurgent nationalists were aristocrats, military leaders, Roman Catholic clergy, and members of a political group called the Falange Party; they were supported by Nazi Germany (under Adolf Hitler [1889–1945]) and fascist Italy (under Benito Mussolini [1883–1945]) in their effort to wrest control. The loyalists were liberals, socialists, and communists; they were supported by the Soviet Union (under Joseph Stalin [1879–1953]). A number of non-Spanish idealists, who believed saving the republic from the fascist rebels was worth dying for, joined the ranks of the loyalists to form the International Brigade. (In his novel *For Whom the Bell Tolls,* Ernest Hemingway [1899–1961], who had covered the war as a correspondent, wrote about one young American man who took up arms in behalf of the loyalist effort.) The nationalists, under Generalissimo Francisco Franco (1892–1975), won the war when they captured Madrid in March 1939, beginning an era of harsh right-wing rule. And, as with any war, the fascist victory had come at a dear price: hundreds of thousands dead and massive destruction throughout the country.

Who was **Generalissimo Franco**?

Generalissimo Francisco Franco (1892–1975) was the fascist leader of Spain from 1939 until 1973. He rose to power in the Spanish Civil War (1936–39) as he led a rebel nationalist army against the loyalist forces. Capturing Madrid in 1939, Franco assumed the role of head of government. Though he and the nationalists had received considerable help from Nazi Germany and fascist Italy to win the civil war, when fighting broke out in World War II (1939–45), Spain stayed neutral (at least nominally so).

In 1947, with the fighting in Europe over, Franco declared himself monarch of Spain and ruled as an authoritative dictator. Two years before he died, he stepped down as head of state, though he retained the title generalissimo, meaning "commander in chief." Franco named as his successor Prince Juan Carlos (1938–). When Franco died in 1975, Juan Carlos I became the first Spanish monarch to control Spain since his grandfather, King Alfonso XIII (1886–1941), was deposed in 1931 to make way for the brief republic (which was later overthrown by Franco and the nationalists). King Juan Carlos played an important role in transforming Spain into a modern democracy.

WORLD WAR II

What was the **Munich Pact**?

It was a failed effort to appease the territory- and power-hungry German leader Adolf Hitler (1889–1945) in the days leading up to World War II (1939–45). After Germany annexed neighboring Austria in the *Anschluss* of March 1938, it became known that Hitler had designs on the Sudetenland, a heavily German region of Czechoslovakia. With World War I (1914–18) a fresh memory, and European nations still recovering from heavy losses, Europe's powers were eager to avoid another conflict. On September 29 and 30, 1938, British prime minister Neville Chamberlain met with Hitler in Munich; they were joined by Italian dictator Benito Mussolini, a German ally, and French Premier Edouard Daladier, a Czech ally. Czechoslovakia did not have any representatives at the conference. The leaders quickly worked out a plan for Germany to occupy the Sudetenland. Chamberlain considered Czechoslovakia's concession a reasonable price to pay for peace on the continent. But the effort to assuage Hitler was not successful: In March 1939 Germany moved to occupy the rest of Czechoslovakia; on September 1, Germany marched into Poland, and World War II began.

How did **World War II begin**?

The war began on September 1, 1939, when Germany invaded Poland, which was soon crushed by German chancellor and führer Adolf Hitler's (1889–1945) war machine. But while the Nazis moved in from the west, Poland was under attack by the Soviets from the north and east. The events in the Eastern European country had set the stage for a major conflict.

After Poland, the Germans moved into Denmark, Luxembourg, the Netherlands, Belgium, Norway, and France, taking control as they went. By June 1940 only Great Britain stood against Hitler, who was joined by Axis power Italy. Before long, fighting had spread into Greece and northern Africa.

In June 1941 Germany invaded the Soviet Union, enlarging the scope of the conflict again. With the world's focus on war-torn Europe, Japan executed a surprise

More than 2,300 people were killed in the 1941 bombing of Pearl Harbor; the attack drew the United States into World War II.

attack on the U.S. naval base at Pearl Harbor, Hawaii, in December 1941, which drew Americans into the war. The war would not end until 1945.

How many countries were part of the **Axis powers**?

The Axis, which was forged in 1936, included an alliance of three nations: Germany, Italy, and Japan. These major powers were joined by six smaller countries, the Axis satellites: Albania, Bulgaria, Finland, Hungary, Romania, and Thailand. But together these countries never comprised the unified front and strength that the Allied powers did.

Germany started the war on September 1, 1939, and was joined in June 1940 by Italy and Albania. In the middle of 1941 Bulgaria, Hungary, Romania, and Finland joined the Axis effort. Japan, which was already fighting with China (in the Sino-Japanese War), entered the fray on December 7, 1941, with its attack on the U.S. naval base at Pearl Harbor, Hawaii. Thailand was the last Axis country to enter the war, on January 25, 1942.

Which countries comprised **the Allies in World War II**?

The three major Allied powers were Great Britain, the United States, and Soviet Union. Their leaders, Winston Churchill (1874–1965), Franklin Roosevelt (1882–

1945), and Joseph Stalin (1879–1953), were referred to as the Big Three. They and their military advisors developed the strategy to defeat the Axis countries—though Stalin, for the most part, acted alone on the Soviet front. China, which had been at war with Asian rival Japan since 1937, also joined the Allies. Forty-six other countries became part of the Allied front before the war was over.

Germany invaded Poland on September 1, 1939, and within days Great Britain entered into fighting against Germany. Australia, New Zealand, India, France, South Africa, and Canada also allied with Great Britain, as did Norway, Denmark, Belgium, the Netherlands, and Luxembourg in 1940—all of them under siege by Nazi Germany. Greece entered the war later that year, as did Yugoslavia in the spring of 1941. On June 22, 1941, the Soviet Union entered the war. And in the days after the Japanese bombing of the U.S. naval base at Pearl Harbor, Hawaii, on December 7, 1941, 12 more Allied countries became involved in the war, chief among them, the United States and China. (The others, with the exception of Czechoslovakia, were all Caribbean and Latin American countries: Panama, Costa Rica, the Dominican Republic, Haiti, Nicaragua, El Salvador, Honduras, Cuba, and Guatemala.) The year 1942 saw three more countries join the Allies—Mexico in May, Brazil in August, and Ethiopia in December. In 1943 and 1944—in what were perhaps the darkest days of the war—Iraq, Bolivia, Iran, and Columbia signed on as Allied nations, followed by the tiny country of San Marino (significant since it is situated wholly within the boundaries of Axis power Italy), Colombia, and Liberia. February and March of 1945 saw another wave of nations siding with the Allies: the South American countries of Ecuador, Paraguay, Peru, Chile, Venezuela, Uruguay, and Argentina; along with the Middle Eastern countries of Egypt, Syria, Lebanon, and Saudi Arabia. Mongolia (the Mongolian People's Republic), in Central Asia, was the last to join the Allies, on August 9, 1945. Of course, the level of support each Allied nation lent to the war effort varied. But it was significant that the list of Allied nations grew longer with each year that the war was fought.

What was **Lend-Lease**?

Lend-Lease was a plan, developed and strongly supported by President Franklin D. Roosevelt (1882–1945), to extend material assistance to the Allied powers fighting the Axis powers in World War II (1939–45). In the days preceding U.S. involvement in the war, Roosevelt argued that it was imperative for the country to come to the aid of those fighting Germany and Italy—it was similar to helping your neighbor put out a fire in his house in order to prevent your own house from catching fire and burning.

Under Lend-Lease, which was passed by Congress on March 11, 1941, approximately $50 billion of aid in the form of food and supplies, weapons, machinery, and other equipment was provided to the Allied nations—primarily to Britain and the Commonwealth nations first, but later to all nations fighting against Hitler's war machine.

The return of the goods was not addressed until after the war had ended. At that time, most people felt the Allies had all contributed everything they had to the war effort, and that the sacrifices made by Allied Europe in the days prior to U.S. entry into the fighting were balanced by the contributions made under the Lend-Lease Act.

What was the **Atlantic Charter**?

On the eve of direct U.S. involvement in World War II (1939–45), President Franklin D. Roosevelt (1882–1945) met with British prime minister Winston Churchill (1874–1965) on board a ship off the coast of Newfoundland, Canada. There the two leaders drew up a program of peace objectives known as the Atlantic Charter, which they signed on August 14, 1941. In addition to other peacetime goals, the charter roughly contained Roosevelt's Four Freedoms, which he had outlined in his speech to Congress on January 6, 1941, as the legislative body considered passage of the Lend-Lease Act. Roosevelt believed that freedom of speech and expression, freedom of worship, freedom from want, and freedom from fear should prevail around the world.

Briefly, in the Atlantic Charter the two leaders stated that neither of their countries sought new territories; that they respected the right of the people of each country to choose their own form of government; that no country ("great or small, victor or vanquished") would be deprived access to the raw materials it needed for its own economic prosperity; that countries should cooperate to improve labor standards and social security; that after the "final destruction of the Nazi tyranny, all the men in all the lands may live out their lives in freedom from fear and want"; and that a "wider and permanent system of general security" would be necessary to ensure peace. (This last statement alludes to the future establishment of the United Nations.)

Why did the **Japanese attack Pearl Harbor**?

There is still disagreement among historians, military scholars, and investigators about why the island nation of Japan issued this surprise attack on the U.S. military installation at Pearl Harbor, Hawaii. Some believe that Japan had been baited into making the attack in order to marshal public opinion behind U.S. entry into World War II (1939–45); others maintain that the United States was unprepared for such an assault, or at least, the Japanese believed Americans to be in a state of unreadiness; and still others theorize that Pearl Harbor was an all-or-nothing gamble on the part of Japan to knock America's navy out of the war before it had even entered into the fray.

These are the facts: In 1941 Japanese troops had moved into the southern part of Indochina, prompting the United States to cut off its exports to Japan. In fall of that year, as General Hideki Tojo (1884–1948) became prime minister of Japan, the country's military leaders were laying plans to wage war on the United States. On December 7 Pearl Harbor, the hub of U.S. naval power in the Pacific, became the target of

Japanese attacks, as did the American military bases at Guam, Wake Island, and the Philippines. But it was the bombing of Pearl Harbor that became the rallying cry for Americans during the long days of World War II—since it was at this strategic naval station, which had been occupied under treaty by the U.S. military since 1908, that Americans had felt the impact of the conflict.

What happened at **Pearl Harbor**?

On the night before the attack, the Japanese moved a fleet of 33 ships to within 200 miles of the Hawaiian island of Oahu, where Pearl Harbor is situated. More than 300 planes took off from the Japanese carriers, dropping the first bombs on Pearl Harbor just before 8:00 A.M. on December 7, 1941. There were eight American battleships and more than 90 naval vessels in the harbor at the time. Twenty-one of these were destroyed or damaged, as were 300 planes. The biggest single loss of the day was the sinking of the battleship USS *Arizona,* which went down in less than nine minutes. More than half the fatalities at Pearl Harbor that infamous December day were due to the sinking of the *Arizona.* By the end of the raid, more than 2,300 people had been killed and about the same number were wounded.

Pearl Harbor forever changed the United States and its role in the world. When President Franklin D. Roosevelt (1882–1945) addressed Congress the next day, he called December 7 "a date which will live in infamy." The United States declared war against Japan, and on December 11 Germany and Italy—Japan's Axis allies—declared war on the United States. The events of December 7 had brought America into the war, a conflict from which it would emerge as the leader of the free world.

When did the **first U.S. troops** begin fighting in World War II?

Late in 1942 the United States sent its first troops across the Atlantic, making amphibious landings in North Africa, followed by Sicily and the Italian peninsula. The first allied landings were in Morocco (Casablanca) and Algeria (Oran and Algiers) on November 8 of that year. (Algiers became the Allied headquarters in North Africa for the duration of the war.) The combined forces of the initial landing included more than 100,000 troops, launching the American military effort in the Atlantic theater of conflict. One American newspaper headline announced: "Yanks Invade Africa."

Was the **U.S. mainland attacked** during World War II?

Yes, the continental United States was hit twice during the war, but with no casualties and only minimal damage. The first attack occurred at approximately 7:00 P.M. on February 23, 1942, when a Japanese submarine shelled an oil storage field at Ellwood Beach, California, about 12 miles north of Santa Barbara. The Japanese were trying to hit oil tanks there, evidently with the intent of producing a spectacular explosion. But after firing a reported 16 or 17 rounds, they had struck only a pier. Most of the shells fell into the sea, well short of their targets. U.S. planes gave chase, but the submarine got away. There were no injuries and only minimal damage, but the event put the nation on heightened alert to the possibility of more attacks.

The February 23 attack took place shortly after President Franklin Roosevelt (1882–1945) had begun his fireside chat, addressing the nation over the radio. He talked about how this war was different, since it was being waged on "every continent, every island, every sea, every air-lane in the world." He also said, "The broad oceans which have been heralded in the past as our protection from attack have become endless battlefields on which we are constantly being challenged by our enemies." The unsuccessful assault at Ellwood was the first attack on mainland U.S. soil since the War of 1812 (1812–14). The event stirred fears of conspiracy and rattled nerves up and down the West Coast.

There was one other strike on mainland soil during World War II: at Fort Stevens, Oregon, at the mouth of the Columbia River. On the evening of June 21, 1942, a Japanese submarine fired some 17 rounds of shells at the coastal military installation but caused no damage.

Why did the U.S. government order the **internment of Japanese Americans** during World War II?

After the attack on Pearl Harbor, American citizens of Japanese descent were viewed as threats to the nation's security. On February 19, 1942, President Franklin Roosevelt (1882– 1945) signed an Executive Order directing that they be moved to camps for containment for the duration of the war. More than 100,000 people, most of them from California and other West Coast states, were rounded up and ordered to live in secure camps. The action drew immediate criticism. With thousands of lives interrupted without cause, the chapter is one of the saddest in American history. In 1988 President Ronald Reagan signed the Civil Liberties Act, which made reparations to the victims of the Japanese internment; $20,000 was paid to "internees, evacuees, and persons of Japanese ancestry who lost liberty or property because of discriminatory action by the Federal government during World War II." It also established a $1.25-billion public education fund to teach children and the public about the internment period.

The War in Europe

What happened at **Anzio**?

Anzio, Italy, was the site of a four-month battle between Allied troops and the Germans during World War II (1939–45). On January 22, 1944, more than 36,000 Allied

Long lines of American soldiers move into Normandy, France, as part of the D-Day assault of June 6, 1944.

troops and thousands of vehicles made an amphibious landing at Anzio, which is situated on a peninsula jutting into the Tyrrhenian Sea. But German soldiers, led by Field Marshal Albert Kesserling (1885–1960), were able to surround the Allied forces, containing them along the shoreline into May of that year. Fighting was intense, with an estimated 60,000 casualties, about half on each side. On May 25, 1944, the Germans withdrew in defeat, enabling the Allies to march toward Rome (33 miles to the north-northwest). The taking of Anzio was a tactical surprise on the part of the U.S. and British, and their eventual victory there was a turning point for the Allies in the war.

What is **D day**?

The military uses the term *D day* to designate when an initiative is set to begin, counting all events out from that date for planning. For example, "D day minus two" would be a plan for what needs to happen two days before the beginning of the military operation. While the military planned and executed many D days during World War II (1939–45), most of them landings on enemy-held coasts, it was the June 6, 1944, invasion of Normandy that went down in history as *the* D Day.

What happened at **Normandy**?

Normandy, a region in northwestern France that lies along the English Channel, is known for the June 6, 1944, arrival of Allied troops, which proved to be a turning point in World War II (1939–45). Officially called Operation Overlord (but known historically as D day) and headed by General Dwight D. Eisenhower (1890–1969) of the United States, the initiative had been in the planning since 1943 and it constituted the largest seaborne invasion in history. After several delays due to poor weather, the Allied troops crossed the English Channel and arrived on the beaches of Normandy on the morning of June 6. Brutal fighting ensued that day, with heavy losses on both sides. At the end of the day, the Allied troops had taken hold of the beaches—a firm foothold that would allow them to march inland against the Nazis, eventually pushing them back to Germany. While it was a critical Allied victory (which history has treated as the beginning of the end for German chancellor and führer Adolf Hitler), the invasion at Normandy was still to be followed by 11 more months of bloody conflict; Germany would not surrender until May 7 of the following year.

What was the **Battle of the Bulge**?

The term refers to the December 16, 1944, German confrontation with the American forces in the Ardennes Mountains, a forested plateau range that extends from northern France into Belgium and Luxembourg. Even though Germany appeared to be beaten at this late point in the war, Hitler rallied his remaining forces and launched a surprise assault on the American soldiers in Belgium and Luxembourg. But Germany could not sustain the front, and within two weeks the Americans had halted the German advance near Belgium's Meuse River (south of Brussels). The offensive became known as the Battle of the Bulge because of the protruding shape of the battleground on a map.

The Ardennes were also the site of conflict earlier in World War II, in 1940, as well as in World War I, in 1914 and 1918.

THE WAR IN THE PACIFIC

Why did **General MacArthur** vow to return?

Two weeks after the Japanese bombing of the U.S. military bases at Pearl Harbor and the Philippines, Japan invaded the Philippine Islands. General Douglas MacArthur (1880–1964), the commander of the U.S. Army forces in the Far East, led the defense of the archipelago. He had begun to organize his troops around Manila Bay when, in March 1942, he received orders from the president to leave the islands. When he reached Australia, MacArthur said, "I shall return," in reference to the Philippines.

U.S. Army troops make their way across the coral reef in the South Pacific Mariana Islands in July 1944.

Under new commands, MacArthur directed the Allied forces' offensive against Japan throughout the Southwest Pacific Islands. After a string of successes, on October 20, 1944, MacArthur made good on his promise, landing on the Philippine island of Leyte, accompanied by a great invasion force. By July of the following year, the general had established practical control of the Philippines. When Japan surrendered in August, MacArthur was made the supreme commander of the Allies, and as such, he presided over the Japanese surrender aboard the USS *Missouri* on September 2. He received the Medal of Honor for his defense of the Philippines, but he wasn't the only hero in the MacArthur family: His father, Arthur MacArthur (1845–1912), had received the nation's highest military award during the Civil War (1861–65).

What was the **Bataan Death March**?

It was one of the most brutal chapters of World War II (1939–45). On April 9, 1942, American forces on the Bataan Peninsula, Philippines, surrendered to the Japanese. More than 75,000 American and Filipino troops became prisoners of war (POWs). On April 10, they were forced to begin a 65-mile march to a POW camp. Conditions were torturous—high temperatures, meager provisions, and gross maltreatment. The troops were denied food and water for days at a time; they were not allowed to rest in the shade; they were indiscriminately beaten; and those who fell behind were killed. On stretches where some troops were transported by train, the boxcars were packed so tightly that many POWs died of suffocation. The forced march lasted more than a week. Twenty thousand men died along the way.

But the end of the march was not the end of the horrors for the surviving POWs. About 56,000 men were held until the end of the war. They endured starvation, torture, and horrific cruelties; some were forced to work as slave laborers in Japanese industrial plants and some became subjects of medical experiments. In August 1945 their POW camp was liberated by the Allied forces, and the surviving troops were put on U.S. Navy vessels for the trip home. As part of the United States' 1951 peace treaty with Japan, surviving POWs were barred from seeking reparations from Japanese firms that had benefited from their slave labor. This injustice continued to be the subject of proposed Congressional legislation into the early 2000s, with no positive outcome for the veterans as of 2005.

Why was the **battle of Midway** important in World War II?

It was the turning point for the allied forces fighting the Japanese in the Pacific. The battle for Midway Island (actually two small islands situated about 1,300 miles west-northwest of Honolulu, Hawaii) began on June 4, 1942. The Japanese aimed to control Midway as a position from which its air force could launch further attacks on Hawaii. As the Japanese fleet approached the islands, which was home to a U.S. Navy base (established in 1941), U.S. forces attacked. Fighting continued until June 6. The Japanese were decisively defeated, losing four aircraft carriers; the United States lost one. The victory proved that Allied naval might could overcome Japan's.

What happened on **Iwo Jima**?

During the month of February 1945 Allied forces and the Japanese fought for control of Iwo Jima, a small island in the northwest Pacific Ocean, 759 miles south of Tokyo. Japan was using Iwo Jima as a base from which to launch air attacks on U.S. bombers in the Pacific. Capturing the island from the Japanese became a key objective for the United States. On February 19, 1945, the Fourth and Fifth U.S. Marine Divisions invaded the island. Fighting over the next several days claimed more than 6,000 U.S. troops. On the morning of February 23, after a rigorous climb to the top of Mount Suribachi (Iwo Jima's 550-foot inactive volcano), U.S. Marines planted an American flag. Though small, it was visible from around the island. Later that day, a larger flag was raised atop Mount Suribachi by five marines and a navy hospital corpsman. The moment was captured by American news photographer Joe Rosenthal. His famous photo became the inspiration for the U.S. Marine Corps Memorial (dedicated November 10, 1954) in Arlington, Virginia.

Who were the **code talkers**?

Code talkers were Navajo Indians who provided secure communications for the U.S. military's Pacific operations during World War II (1939–45). Serving in the marines, the Navajo servicemen were recruited because of their language, which is unwritten and extremely complex. Military officials believed the Japanese would not be able to decipher intercepted communiqués that were transmitted in the Navajo language. The first Navajo recruits attended boot camp in May 1942 and developed and memorized a dictionary of military terms to use in encrypting messages. The trained code talkers were deployed with marine units throughout the Pacific theater. They worked around the clock, with tremendous speed and accuracy, to transmit vital information about military tactics, battle orders, and troop movements. Their messages were sent over telephone and radio; because of the complexity of the language, the Japanese military was never able to break the code. About 400 Navajos served as code talkers during the war, contributing mightily to the success of U.S. military assaults at Guadalcanal, Iwo Jima, and other Pacific venues.

What was the ***Enola Gay***?

It was the American B-29 bomber that dropped the first atomic bomb ever used in warfare. On August 6, 1945, the *Enola Gay* flew over Hiroshima, Japan, to drop an A-bomb over the city. The explosion killed an estimated 80,000 people and leveled an area of about five square miles in Hiroshima, an important manufacturing and military center. Thousands more died later from radiation exposure.

How many were killed by the A-bomb that was dropped on **Nagasaki**?

The death toll from the explosion, on August 9, 1945, was about 40,000. But, as in Hiroshima, thousands more died later due to radiation exposure from the atomic bomb. American military strategists believed that the first A-bomb, on Hiroshima, would force Japan's leaders to surrender. But when they did not, the second bomb was dropped on Nagasaki, an important seaport and commercial city. That catastrophic attack on Japan brought an end to the war, on August 14. Japan signed the surrender agreement on September 2 aboard the U.S. battleship *Missouri,* in Tokyo Bay.

Who was **Tokyo Rose**?

Wartime radio broadcasts by a woman called Tokyo Rose were an Axis propaganda effort intended to weaken the resolve of American and Allied troops. The Japanese pressed into service Allied POWs and a woman named Iva Toguri d'Aquino (a U.S. citizen caught in wartime Japan) to read over the radio what were intended to be demoralizing messages to the Allied troops. Originating in Japan and heard by soldiers and sailors in the Pacific, the broadcasts were either disregarded by their intended audi-

New York's Times Square is packed with crowds celebrating the news of Germany's surrender in World War II, May 7, 1945.

ence or were found mildly amusing. Though d'Aquino was later convicted of treason, in 1977 President Gerald Ford issued an unconditional pardon in the case, which was built on "tainted" facts. The Axis powers did the same sort of broadcasts from Germany where "Axis Sally" aired messages that were heard throughout Europe.

WORLD WAR II ENDS

What are **V-E Day and V-J Day**?

V-E Day stands for Victory in Europe Day, and V-J Day stands for Victory over Japan Day. After the German surrender was signed in Reims, France (the headquarters of General Dwight D. Eisenhower [1890–1969]), in the wee hours of May 7, 1945, U.S. president Harry S. Truman (1884–1972) declared May 8 V-E Day—the end of the World War II fighting in Europe.

But it was not until the Japanese agreed to surrender on August 14, 1945, that World War II ended. September 2, 1945, was declared the official V-J Day since it was then that Japan signed the terms of surrender on the USS *Missouri* anchored in Tokyo Bay.

Were any countries besides Switzerland **neutral during World War II**?

Yes, in its official stance of neutrality, Switzerland was joined by Spain, Portugal, Sweden, Turkey, and Argentina. However, postwar findings indicated the neutrality of these countries—with the exception of Argentina—was not an absolute policy. (Some have described these countries as only nominally neutral.) A 1998 report released by U.S. Undersecretary of State Stuart Eizenstat indicated that the Swiss had converted Nazi gold to Swiss francs and that Germany had used that exchange to buy minerals from Spain, Portugal, Sweden, and Turkey. The report further pointed out that Sweden had allowed a quarter of a million Nazi troops to cross its country in order to reach neighboring Finland, where the Germans fought Soviet forces.

Eizenstat, who headed a U.S. government effort to determine where Nazi gold ended up, was assisted in his research by State Department historians. Although the investigation's reports were critical of these neutral countries, Eizenstat also pointed out that all the countries were in difficult positions during the conflict. For example, Switzerland was completely surrounded by German-occupied countries. Nevertheless, Jewish groups brought lawsuits against the Swiss government and three Swiss banks for their role in converting looted Nazi gold into currency during World War II.

How was it that **Anne Frank's diary** survived World War II?

The young German-Jewish girl's diaries, which chronicle 26 months of hiding from the German authorities in Amsterdam during World War II (1939–45), were given to her father, Otto Frank (1889–1980), when he returned to Holland from Auschwitz after the war. The notebooks and papers had been left behind by the secret police, and were found in the Frank family's hiding place by two Dutch women who had helped the fugitives survive. Otto Frank published his daughter's diaries in 1947. Twenty years later, the English-language edition, *The Diary of a Young Girl,* was released. It was subsequently made into a play and a film, both called *The Diary of Anne Frank.*

The diary is an astonishingly poignant account of the suffering—and heroism—during the Nazi occupation of the Netherlands. Anne Frank, born in 1929, had known religious persecution her entire life. She was only four years old when her family escaped Nazi Germany in 1933 and fled to the Netherlands. During the Nazi occupation of that country (beginning in 1942), the Frank family hid in a secret annex behind her father's office at 263 Prinsengracht in Amsterdam. It was there that Anne lived with seven others for just more than two years, recording her thoughts in three notebooks. But the family was finally found (on August 4, 1944) by the Nazis, who arrested them and put them in German concentration camps. In March 1945, two months before the German surrender, Anne Frank died of typhoid fever in the Bergen-Belsen concentration camp. Of the eight who had hidden in the secret annex, Anne's father, Otto, was the only survivor.

A group of blockaded Berliners watch a U.S. Air Force plane in 1948. For eleven months, West Berlin was supplied with food and fuel entirely by airplanes.

What was the **Berlin airlift**?

It was the response during 1948 and 1949 to the Soviet blockade of West Berlin. After World War II (1939–45), the German city had been divided into four occupation zones: American, British, French, and Soviet. But following the conclusion of the war, it did not take long for the Cold War (1947–89) between the Western powers and the Soviet Union to heat up. When the Americans, British, and French agreed to combine their three areas of Berlin into one economic entity, the Soviets responded by cutting the area off from all supply routes. In June 1948 all arteries—road, rail, and water—into West Berlin were blocked by Soviet troops. Since Berlin was completely surrounded by the Soviet occupation zone, the Soviets clearly believed the blockade would be an effective move that would prompt the Western countries to pull out. But the move failed: the Americans, British, and French set up a massive airlift. For the next eleven months, West Berlin was supplied with food and fuel entirely by airplanes. The Soviets lifted the blockade in May 1949, and the airlift ended by September.

COLD WAR

What is **NATO**?

NATO stands for the North Atlantic Treaty Organization, a military alliance formed on April 4, 1949, when 12 countries signed the North Atlantic Treaty in Washington, D.C. The original 12 NATO countries were Belgium, Canada, Denmark, France, Iceland, Italy, Luxembourg, the Netherlands, Norway, Portugal, the United Kingdom, and the United States. Each member nation agreed to treat attacks on any other member nation as if it were an attack on itself. In other words, any aggressor would have to face the entire alliance. This was NATO's policy of deterrence, a way of discouraging any attacks by the Soviet Union or other Eastern bloc countries. The organization had the further benefit of discouraging fighting among the member countries.

Three years after it was formed, the alliance was joined by Greece and Turkey (in 1952). West Germany followed three years after that, in 1955, and Spain joined in 1982. After the fall of communism and the reunification of East and West Germany (c. 1990), all of Germany joined the alliance. At this point, with the Cold War (1947–89) over, many wondered what purpose the organization could serve. After all, the Soviet threat was no longer existent. However, other conflicts loomed on the horizon, including those in Bosnia and Herzegovina and in the Albanian republic of Kosovo. Fearing the civil war in the former Yugoslav republic would spread, NATO sent in troops on the side of the Bosnian government. NATO also formed the Partnership for Peace in 1994: This program was joined by more than 20 countries, including former Eastern bloc nations, including Russia. Though these nations are not full members in the NATO alliance, the Partnership for Peace provides for joint military planning among signing nations. On March 12, 1999, three former Eastern bloc nations were given full membership in NATO: Poland, Czechoslovakia, and Hungary. Observers hailed the additions as evidence that Europe is becoming more unified. On April 23, 1999, the 50th anniversary of the alliance's founding, the 19 NATO member nations gathered in Washington, D.C., to commemorate the event, just after NATO had begun air-bombing Yugoslavia to pressure the government there to accept international terms aimed at bringing peace to the nation's Kosovo province, where ethnic conflicts between Serbs and Albanians had turned deadly.

NATO is governed by the North Atlantic Council, which is made up of the heads of government of member nations or their representatives. It was headquartered in Paris until 1967, at which time the offices were moved to Brussels, Belgium.

What was the **Warsaw Pact**?

The Warsaw Pact was the Eastern bloc countries' answer to the North Atlantic Treaty Organization (NATO). Seeing the Western nations form a strong alliance, in May 1955 the Soviet Union and its allies met in Warsaw, Poland, where they signed a treaty

Did the Cold War ever get "hot"?

After World War II (1939–45), an intense rivalry developed between so-called Eastern bloc countries (made up of the Communist Soviet Union and its allies) and the Western bloc countries (the United States and its democratic allies). Though mutual suspicion, distrust, and fear ran deep on both sides of the conflict, the Cold War (1947–89) never resulted in fighting: it never became "hot." But both sides nevertheless prepared for that possibility by strengthening military alliances and developing and stockpiling weapons. The Cuban Missile Crisis (1962) was probably the closest the Cold War came to getting hot.

agreeing that they, too, would mutually defend one another. The eight member nations were Albania (which withdrew in 1968), Bulgaria, Czechoslovakia, East Germany, Hungary, Poland, Romania, and the Soviet Union. The Warsaw Pact was headquartered in Moscow and, in addition to discouraging attacks from Western bloc/NATO countries, the organization also sought to quell any democratic uprisings in Warsaw Pact nations.

But in 1990 the pact and the Soviet Union's control of it weakened as democracy movements in member nations could not be put down. As the former Eastern bloc countries underwent relatively peaceful revolutions, Warsaw Pact members began announcing their intentions to withdraw from the organization. East Germany withdrew when it was reunified with West Germany, and the restored Germany joined NATO (in 1990). The Warsaw Pact was dissolved by the remaining member nations in 1991.

What was the **arms race**?

The arms race refers to the buildup of nuclear weapons by the Soviet Union and the United States during the Cold War. The "race" began on August 29, 1949, when the Soviet Union tested an atomic bomb. (Prior to this, only the United States had the knowledge to build the atomic bomb.)

What was the **Bay of Pigs**?

Bay of Pigs is the name of an unsuccessful 1961 invasion of Cuba, which was backed by the U.S. government. About 1,500 Cuban expatriates living in the United States had been supplied with arms and trained by the U.S. Central Intelligence Agency (CIA). On April 17, 1961, the group of men who opposed the regime of Cuba's Fidel Castro (1926–) landed at the Bahia de Cochinos (Bay of Pigs) in west-central Cuba. Most of the rebels were captured by the Cuban forces; others were killed. In order to secure the release of the more than 1,100 men who had been captured during the invasion, private donors in the United States accumulated $53 million in food and medicine, which was given to Castro's government in exchange for the rebels' release. The failed invasion came as a terrible embarrassment to the Kennedy administration, and many believe the Bay of Pigs incident directly led to the Cuban Missile Crisis.

What was the **Cuban Missile Crisis**?

The 1962 events, which happened very quickly, nevertheless constituted a major confrontation of the Cold War (1947–89). After the disastrous Bay of Pigs invasion, when the United States backed Cuban expatriates in an attempt to oust Fidel Castro (1926–), the Soviet Union quietly began building missile sites in Cuba. Since the island nation is situated just south of Florida, when U.S. reconnaissance flights detected the Soviet military construction projects there, it was an alarming discovery. On October 22, 1962, President John F. Kennedy (1917–1963) demanded that the Soviet Union withdraw its missiles from Cuba. Kennedy also ordered a naval blockade of the island. Six days later, the Soviets agreed to dismantle the sites, ending the crisis.

What was the impact of the **Soviet invasion of Afghanistan**?

When the Soviet Union invaded Afghanistan in December 1979 to bolster a pro-Communist government in the Middle Eastern nation, no one could have anticipated the far-reaching effects—effects that would be felt decades later and around the globe. What immediately followed was a 10-year civil war, in which Soviet troops fought Afghan guerrillas, or the *mujahideen.* The war in Afghanistan became a *jihad,* or holy war, and a rallying point for many Muslims, with the conflict drawing young men from across the Muslim world to fight on the side of the guerillas. According to *The 9/11 Commission Report,* "mosques, schools, and boarding houses served as recruiting stations in many parts of the world, including the United States." The war was a virtual stalemate for seven years. But a turning point came in 1986 after the United States and Great Britain supplied shoulder-fired surface-to-air missiles to the Afghan guerrillas. The weaponry gave the scrubby ground forces a fighting chance against Soviet air power. As *The 9/11 Commission Report* asserts, together with Saudi Arabia, the United States supplied billions of dollars worth of secret assistance to rebel Afghan groups resisting the Soviet occupation. Thus supported, in April 1988 the Afghans declared victory, and early the next year the Soviet troops began to withdraw.

The war was over, but it had fueled an extremist Islamic ideology (the *jihad* as holy war) and put into place an infrastructure out of which emerged a powerful and deadly terrorist network. Though most Muslims hold peaceful views, a minority of Muslims view any non-Muslims as unbelievers. It was from this minority, trained and financed as a result of the Afghan War, that the global network of terrorists called al Qaeda emerged.

WARS IN SOUTHEAST ASIA

Who were the **Khmer Rouge**?

The Khmer Rouge (or Red Khmer) were a group of Cambodian Communists led by radical Marxist leader Pol Pot (1925–1998). Between 1970 and 1975 the Khmer Rouge guerrilla force, supported by Communists from neighboring Vietnam, waged a war to topple the U.S.-supported government of Lon Nol (1913–1985). On April 16, 1975, Lon Nol's regime fell, and the next day the Khmer Rouge seized the Cambodian capital, Phnom Penh. The ruthless revolutionary leader Pol Pot became prime minister of a Communist Cambodia and instituted a reign of terror. In his attempt to turn Cambodia into an agriculture-based society, the Khmer Rouge systematically emptied the cities, forcibly moving the people onto collective farms where they performed hard labor. Anyone thought to be opposed to the Khmer Rouge was killed. An estimated 2 million people died—by execution, overwork, and starvation. Pol Pot's "experiment" had failed, and his efforts to revolutionize Cambodia amounted to nothing short of genocide.

A Vietnamese invasion ousted the Khmer Rouge in 1979 and installed a new leadership. But civil wars were fought throughout the 1980s. The warring factions, who had made various alliances among themselves, finally signed a peace treaty in 1991. Under the watchful eye of the United Nations, elections were held in 1993. The resulting constitution provided for a democratic government with a limited monarchy. At that point, the Cambodian leadership seemed to come full circle—with Norodom Sihanouk (1922–) being crowned king in 1993: In 1970 Sihanouk had been deposed by Lon Nol, whose regime later became the target of the Khmer Rouge. During his lifetime Sihanouk made strides in establishing Cambodia's independence, and he enjoyed great public support. Due to failing health, he abdicated the throne in November 2004 and was succeeded by his son, Norodom Sihamoni (1953–).

For decades after Pol Pot was deposed, he continued to lead a revolutionary force of the Khmer in Cambodia, though he remained out of public view. His own men turned against him in early 1998, and Pol Pot died in April of that year. In December the last main fighting force of the Khmer Rouge surrendered to the Cambodian government. The event was broadcast on national television. Though some Khmer leaders remained in hiding and small bands of guerrilla fighters were thought to still exist, the radical Marxist group, which had terrorized Cambodia, no longer presented a threat to the government.

What were the **"killing fields"**?

After Communist leader Pol Pot (1928?–1998), head of the Khmer Rouge, took over the Cambodian government in 1975, he ordered a collectivization drive, rounding up anyone who was believed to have been in collusion with or otherwise supported the former regime of Lon Nol (1913–1985). The government-instituted executions, forced labor (in so-called re-education camps), and famine combined to kill one in every five Cambodians, or an estimated 2 million people, during Pol Pot's reign. He was removed from power in the Vietnamese invasion of 1978 to 1979, and he died in hiding in 1998.

On December 29 of that year, two former Khmer Rouge leaders surrendered to authorities: Khieu Samphan, age 67, and Nuon Chea, 71. The two appeared in a televised news conference. Asked if he was sorry for the suffering that claimed the lives of millions of Cambodians, Khieu Samphan looked straight at the questioner and answered in English: "Yes, sorry, very sorry." Nuon Chea, said, "We are very sorry, not just for the human lives but also animal lives that were lost in the war." However, neither Samphan nor Chea accepted personal responsibility for the killing fields. While Samphan pled not to be tried for his crimes and Prime Minister Hun Sen (1950–) of Cambodia seemed inclined toward closing the book on this dark chapter in the country's history, there was public outcry to bring the former Khmer leaders to justice. Supporters of a trial assert that Cambodia will have no peace until someone is punished for the killing fields—for the Khmer's genocidal regime.

Korean War

Why did the United States get involved in the **Korean War**?

Americans became involved in the Korean conflict when the United Nations (UN), only five years old, called upon member countries to give military support to South Korea, which had been invaded by troops from Communist-ruled North Korea on June 25, 1950. The United Nations considered the invasion to be a violation of international peace and called on the Communists to withdraw. When they did not, 16 countries sent troops and some 40 countries sent supplies and military equipment to the aid of the South Korean armies. About 90 percent of the UN aid came from the United States. But North Korea received aid too—the Chinese sent troops and the Soviet Union provided equipment for them to sustain the war, which lasted until July 27, 1953. After three years of fighting, an armistice was called, but a formal peace treaty was never drawn up between the neighboring countries, prompting the United States to maintain military forces in South Korea in an effort to discourage any further acts of aggression from the north.

Vietnam War

What caused the **Vietnam War**?

In the simplest terms, the long conflict in Southeast Asia was fought over the unification of Communist North Vietnam and non-Communist South Vietnam. The two

U.S. military helicopters sweep in to rescue injured soldiers during the Vietnam War (c. late 1960s).

countries had been set up in 1954. Prior to that, all of Vietnam was part of the French colony of Indochina. But in 1946, the Vietnamese fought the French for control of their own country. The United States provided financial support to France, but the French were ultimately defeated in 1954. Once France had withdrawn its troops, an international conference was convened in Geneva to decide what should be done with Vietnam. The country was divided into two partitions, along the 17th parallel. This division of land was not intended to be permanent, but the elections that were supposed to reunite the partitions were never held. Vietnamese president Ho Chi Minh (1892–1969) took power in the north while Emperor Bao Dai (1913–1997), for a while, ruled the south.

But the Communist government in the north opposed the non-Communist government of South Vietnam and believed the country should still be united. The North Vietnamese supported antigovernment groups in the south and over time, stepped up aid to those groups. These Communist-trained South Vietnamese were known as the Viet Cong. Between 1957 and 1965, the Viet Cong struggled against the South Vietnamese government. But in the mid-1960s, North Vietnam initiated a large-scale troop infiltration into South Vietnam, and the fighting became a full-fledged war.

China and the Soviet Union provided the North Vietnamese with military equipment, but not manpower. The United States provided both equipment and troops to non-Communist South Vietnam in its struggle against the Viet Cong and North Vietnam. By 1969 there were more than half a million American troops in South Vietnam. This policy was controversial back in America, where protests to involvement in the Vietnam War continued until the last U.S. troops were brought home in 1973. In January of that year, the two sides had agreed to a cease-fire, but the fighting broke out again after the American ground troops left. On April 30, 1975, South Vietnam surrendered to North Vietnam and the war, which had lasted nearly two decades, ended. North Vietnam unified the countries as the Socialist Republic of Vietnam.

For its part, the North Vietnamese called the conflict a "war of national liberation": They viewed the long struggle as an extension of the earlier struggle with France. They also perceived the war to be another attempt by a foreign power (this time the United States) to rule Vietnam.

Why did the **United States** get involved **in Vietnam**?

The policy of involvement in the Vietnam conflict began in the mid-1950s when President Harry S. Truman (1884–1972) provided U.S. support to the French in their struggle to retain control of Vietnam, which was then part of French Indochina. In the Cold War era (1947–89), government leaders believed that the United States must come to the assistance of any country threatened by communism. Truman's successors in the White House, Presidents Dwight D. Eisenhower (1890–1969), John F. Kennedy (1917–1963), and Lyndon B. Johnson (1908–1973), also followed this school of thought, fearing a "domino effect" among neighboring nations—if one fell, they'd all fall.

What was the **Tet Offensive**?

The Tet Offensive was a turning point in the Vietnam War (1954–75). The assault began during Tet, a festival of the lunar new year, on January 30, 1968. Though a truce had been called for the holiday, North Vietnam and the Viet Cong issued a series of attacks on dozens of South Vietnamese cities, including the capital of Saigon, as well as military and air installations. American troops and the South Vietnamese struggled to regain control of the cities, in one case destroying a village (Ben Tre) in order to "save it" from the enemy. Fighting continued into February. Though the Communist North ultimately failed in its objective to hold any of the cities, the offensive was critical in the outcome of the war: As images of the fighting and destruction filled print and television media, Americans saw that the war was far from over, despite pre-Tet reports of progress in Vietnam. The Tet Offensive strengthened the public opinion that the war could not be won. It altered the course of the American war effort, with President Lyndon Johnson (1908–1973) scaling back U.S. commitment to defend South Vietnam.

What was the **My Lai Massacre**?

It was a horrific chapter in American military history, during which U.S. troops fighting in South Vietnam took the small village of My Lai on March 16, 1968. The incident did not come to light until more than a year later, after which time it became clear that the unit of 105 soldiers who entered My Lai that morning had faced no opposition from the villagers. Even so, at the end of the day as many as 500 civilians, including women and children, lay dead. Though charges were brought against some of the men, only the commander of the company, Lt. William Calley, was convicted. His sentence of life imprisonment for the murder of at least 22 people was later reduced to 20 years, and he was released on full parole in November 1974.

Why did so many Americans **protest U.S. involvement in the Vietnam War**?

The Vietnam War (1954–75) divided the American public: The antiwar movement maintained that the conflict in Southeast Asia did not pose a risk to U.S. security (contrary to the "domino effect" that Washington, D.C., foresaw), and in the absence of a threat to national security, protesters wondered, "What are we fighting for?" Meanwhile, President Lyndon B. Johnson (1908–1973) slowly stepped up the number of troops sent to Vietnam. Many never came home, and those who did came home changed. Mass protests were held, including the hallmark of the era, the sit-in. Protesters accused the U.S. government of not only involving Americans in a conflict in which the country had no part, but of supporting a corrupt, unpopular—and undemocratic—government in South Vietnam.

Those Americans who supported the nation's fight against Communism eventually became frustrated by the United States' inability to achieve a decisive victory in Vietnam. Even for the so-called hawks, who supported the war, the mounting costs of the war hit home when President Johnson requested new taxes. As the casualty count soared, public approval of U.S. participation in Vietnam dropped. By the end of the 1960s, under increasing public pressure, the government began to withdraw American troops from Vietnam. The evacuation of the ground troops was not complete until 1973. But even then, soldiers who were missing in action (MIAs) and prisoners of war (POWs) were left behind.

THE MIDDLE EAST

What is the basis of the conflict over the **Gaza Strip and the West Bank**?

The conflict is rooted in Jewish and Arab claims to the same lands in the Palestine region, which was under British control between 1917 and 1947. The Gaza Strip is a tiny piece of territory along the eastern Mediterranean Sea and adjacent to Egypt. After the nation of Israel was established and boundaries were determined by the United Nations (UN) in 1947, the Gaza Strip—bounded on two sides by the new Israel—came under Egyptian control. The Arab-Israeli war of 1967 resulted in Israeli takeover and occupation of Gaza. But unrest continued, and in 1987 and 1988 the region was the site of Arab uprisings known as the Intifada. A historic accord between the Palestine Liberation Organization (PLO) and Israel, signed in May 1994, provided for Palestinian self-rule in the Gaza Strip. This has been in effect since—though peace in the region remains elusive, as extremists on both sides of the conflict stage sporadic acts of violence.

The West Bank, which does *not* neighbor the Gaza Strip, is an area on the east of Israel, along the Jordan River and Dead Sea. The West Bank includes the towns of Jericho, Bethlehem, and Hebron. The holy city of Jerusalem is situated on the shared border between Israel and the West Bank. By the UN mandate that established the independent Jewish state of Israel in 1948, the West Bank area was supposed to become Palestinian. But Arabs who were unhappy with the UN agreement in the first place attacked Israel, and Israel responded by occupying the West Bank. A 1950 truce brought the West Bank under the control of neighboring Jordan; this situation lasted until 1967, when Israeli forces again occupied the region. Israelis soon began establishing settlements there, which provoked the resentment of Arabs. The Intifada uprisings that began in the Gaza Strip in 1987 soon spread to the West Bank. In 1988 Jordan relinquished its claim to the area, but fighting between the PLO and Israeli troops continued. Peace talks began in 1991, and the agreements that provided for Palestinian self-rule in the Gaza Strip also provided for the gradual return of West Bank lands to Palestinians. The city of Jericho was the first of these lands.

In August 2005 Israel began pulling out of the Gaza Strip after 38 years of occupation. Some Israeli settlers resisted Prime Minister Ariel Sharon's call for withdrawal; but Sharon insisted the move was a critical step toward peace and securing Israel's future.

What were the **Camp David Accords**?

Camp David Accords is the popular name for a 1979 peace treaty between Israel and Egypt. The name stuck since President Jimmy Carter (1924–) met with Israel's Menachem Begin (1913–1992) and Egypt's Anwar Sadat (1918–1981) at the presidential retreat at Camp David, Maryland. The treaty was actually signed on March 26, 1979, in Washington, D.C., with Carter as a witness to the agreement between the warring Middle Eastern nations.

The pact, which was denounced by Arab countries, provided for the return of the Sinai Peninsula to Egypt. The mountainous area, adjacent to Israel and at the north end of the Red Sea, had been the site of a major campaign during the Arab-Israeli War of 1967 and had been occupied by Israel since. The transfer of the peninsula back to

Egypt was completed in 1982. The Camp David Accords had also outlined that the two sides would negotiate Palestinian autonomy in the occupied West Bank and Gaza Strip. However, Sadat was assassinated in 1981, and this initiative saw no progress as a result of the Camp David Accords.

What is the **Wye Accord**?

Officially called the Wye River Memorandum, the accord outlined a limited and interim land-for-peace settlement between Israel and Palestine. It was signed October 23, 1998, by Israeli prime minister Benjamin Netanyahu (1949–) and Palestinian leader Yasser Arafat (1929–2004) at a summit held at Wye Mills, on the banks of Maryland's Wye River. The meeting was the follow-up to the 1993 Middle East Summit in Oslo, Norway. There, after months of talks, both sides agreed to an interim framework of Palestinian autonomy in the West Bank and Gaza Strip. The Wye meeting was the opportunity for both sides to make good on the promises made in Oslo.

The Wye Accord was brokered after a 21-hour bargaining session mediated by U.S. president Bill Clinton (1946–). The points of the agreement included developing a security plan to crackdown on terrorism; the withdrawal of Israeli troops from an additional 13 percent of the West Bank (along with a commitment for future additional withdrawals); a transfer of roughly 14 percent of the West Bank from joint Israeli-Palestinian control to Palestinian control; Palestinian agreement that anti-Israeli clauses in its national charter would be removed; Israel's guarantee that it would provide two corridors of safe passage between the Gaza Strip and the West Bank; Israeli release of 750 Palestinian prisoners; and the opening of a Palestinian airport in Gaza.

The Knesset, Israel's parliament, approved the accords on November 17, 1998. But by December Israel suspended its obligations in the Wye, citing Palestinian failure to comply with the accords. Benjamin Netanyahu's successor, Prime Minister Ehud Barak (1942–), pledged to resume implementing the Wye Accord but at the same time delayed its timetable, saying the measures should be included in a final peace agreement with the Palestinians. On September 4, 1999, the two sides met again at Sharm al-Sheikh, Egypt, where they agreed on a new timetable for the Wye. That document was signed by Barak and the Palestinian Authority's Arafat and was witnessed by diplomats from Egypt, Jordan, and the United States (Secretary of State Madeleine Albright). But both Barak and Arafat faced mounting political opposition at home, posing immediate challenges to the revised agreement, which stalled again.

What was **Camp David II**?

In July 2000 President Bill Clinton (1946–) invited Israeli and Palestinian Authority leaders to the presidential retreat at Camp David, Maryland, to hammer out a final peace agreement in the Middle East. In what could have been the major breakthrough in the conflict, Israeli prime minister Ehud Barak (1942–) agreed to a Palestinian state, including the West Bank and East Jerusalem, and the administration of all Jerusalem holy sites by a third party (i.e. neither Israel nor Palestine). In exchange, Palestinian leader Yasser Arafat (1929–2004) was asked to sign an "end of conflict" addendum to the final agreement, which would have required him to bring the militant Arab group Hamas under control and end all Palestinian attacks on Israelis. But Arafat refused the deal. The July 11–25 meeting ended without an agreement. Violence erupted again in Israel, beginning the Second Intifada.

What is the **intifada**?

Intifada means "uprising" in Arabic and refers specifically to two recent periods of intense and regular violence between Palestinians and Israelis in the disputed territories of the Middle East. The First Intifada began in 1987 and did not subside until a peace settlement was brokered between the two sides in Oslo, Norway, in 1993. The Second Intifada began in September 2000; the violence of that uprising continued into 2005 despite cease-fire vows from Arab-Palestinian and Israeli leaders. The second wave of violence was marked by almost daily suicide bombings of Israeli targets, including markets, restaurants, buses, and other public places.

Israelis view the intifadas as acts of terrorism, claiming the lives of thousands of Israeli citizens. Palestinians in general and the Palestine Liberation Organization (PLO) and the militant Islamic group Hamas in particular consider the tactics part of their war of national liberation—a decades-long struggle to establish an autonomous Palestinian homeland in Israel. In the waves of violence, casualties were high on both sides (more than 4,000 died in the Second Intifada alone). The cycle of conflict appeared endless: Palestinian acts of violence were met with retaliation from the Israelis, which incited further violence from the Palestinians.

Hopes for a lasting peace were raised again in February 2005 when Palestinian leader Mahmoud Abbas (1935–) and Israeli prime minister Ariel Sharon (1928–) held a summit meeting in Sharm al-Sheikh, Egypt. They declared a cease-fire agreement and their intentions to resume negotiations, as outlined in the Roadmap for Peace. But the militant Palestinian group Hamas immediately claimed the verbal agreement was nonbinding. Indeed, Hamas, which was largely responsible for the intifada, posed major roadblocks to an end of violence in the region. Before they would put down arms, the group demanded total Israeli withdrawal from the West Bank, Gaza, and East Jerusalem, as well as the right of return of Palestinian refugees; Hamas was also dissatisfied with Israel's agreement to release only 900 of the 8,000 Palestinians held prisoner by Israel.

What is the **Roadmap for Peace**?

It is a plan for lasting peace in the Palestinian-Israeli conflict. The roadmap was announced by the Bush administration on March 14, 2003, after several months of working with Russian, European Union, and United Nations (UN) officials to develop the plan, which calls for a permanent two-state solution. The roadmap outlines specific, actionable steps to be taken by each side in the conflict. The U.S. State Department described it as a "performance-based and goal-driven roadmap, with clear phases, timelines, target dates, and benchmarks aiming at progress through reciprocal steps by the two parties in the political, security, economic, humanitarian, and institution-building fields." In announcing the roadmap, President George W. Bush (1946–) called for an end to the recent wave of violence (intifada) in the Middle East, for authoritative Palestinian leadership, and for Israeli readiness to comply. He asked for Israeli and Palestinian leaders to contribute to and discuss the roadmap, and he repeated his call for "all parties in the Middle East to abandon old hatreds and to meet their responsibilities for peace." The plan, the last phase of which was to have been completed in 2004–05, foundered amid the daily violence of the Second Intifada, but was given new promise in May 2005 when Palestinian and Israeli leaders met in Egypt to discuss peace.

LATIN AMERICA & THE CARIBBEAN

Who is **Che Guevara**?

Ernesto "Che" Guevara (1928–1967) was an idealist who became involved in revolutionary movements in at least three countries. The Argentinian, who had earned his degree in medicine in Buenos Aires in 1953, believed that social change and the elimination of poverty could only come about through armed conflict. Guevara met Fidel Castro (1926–) in Mexico in 1954 and served his guerrilla forces as a physician and military commander during the Cuban Revolution (1956–59). Once in power, Castro appointed Guevara as president of the National Bank of Cuba. Between 1965 and 1967 Guevara became active in leftists movements in Congo and in Latin America. He was leading a force against Bolivian government when he was killed in 1967.

What was the **"dirty war"**?

This is the name that has been given to a troubling chapter in Mexico's recent history. In the late 1960s and early 1970s thousands of left-wing reformists were killed in Mexico, and hundreds more went missing and are presumed to have been killed. After the 2000 election of reform-minded President Vicente Fox (1942–), the human rights abuses of Mexico's dirty war era at last were being investigated. In June 2002 secret security files were released detailing the government's iron-fisted crackdown on its opponents during the late 1960s and into the 1970s. In July 2002 former Mexican president Luis Echeverria (1922–) was called before a special prosecutor, who quizzed the octogenarian about two massacres that occurred when he was a high-ranking government official

How old is guerrilla warfare?

Guerrilla warfare dates back to ancient times but got its name during the Peninsular War of 1809 to 1814 when Napoleon Bonaparte (1769–1821) fought for control of the Iberian Peninsula (Spain and Portugal). In Spanish, *guerrilla* means "small war." The resistance to Napoleon's troops employed tactics that are typical of what we know as guerrilla warfare—fighting in small bands, ambushes, sudden raids, and sabotage.

It is Chinese Communist leader Mao Tse-tung (1893–1976) who, in his 22-year fight against the Chinese nationalists (from 1927 to 1949), is believed to have developed the techniques of modern guerrilla warfare. Chairman Mao slowly but surely gained the support and sympathy of the common people—in particular those living in rural areas. Eventually, he had control of the masses who believed the reforms he would make once in office would be favorable to them. The people would provide the manpower and supplies that would sustain the fight. If any followers faulted in their loyalty to the cause, they would be punished.

Today, guerrillas rely on terrorist attacks against governments, goading the military into action, which, in turn, rallies the public in its outrage against government. In this way, guerrilla movements can gain popular support over time. Such movements are by no means limited to the countryside: Urban attacks include tactics such as kidnapping and assassination. Such guerrilla measures have led to the outbreak of civil wars.

and then president. In the two events, one in 1968 and the other in 1971, antigovernment protesters were beaten and shot. Echeverria and his presidential successor, Jose Lopez Portillo (1920–2004), both denied any involvement in the events or in the disappearance of radicals who actively protested their regimes. In late February 2005 Mexico's Supreme Court, citing a 30-year time limit, ruled that Echeverria could not be tried on alleged human rights abuses. The decision was considered a setback for Fox's reform efforts, which centered on punishing past crimes. But the special prosecutor in the case did not rule out bringing other charges against Echeverria, saying the former president ordered men loyal to him to attack student demonstrators.

What was the **Chiapas uprising**?

Also called the Zapatista uprising, it was a January 1994 revolt staged by Mayan rebels in Chiapas, Mexico's southernmost state. On New Year's Day, members of the Zapatista Army of National Liberation (EZLN) launched a coordinated attack on four municipal capitals and a Mexican army headquarters in the remote region. With the cry of *"tierra y libertad"* ("land and liberty"), the armed insurgents invoked the name and spirit of

Zapatistas carry machetes as they march on January 1, 2003, in Mexico's Chiapas state to commemorate the ninth anniversary of the rebel group's uprising against government oppression.

Emiliano Zapata (1879–1919), Mexico's early twentieth-century revolutionary leader. The EZLN, or "Zapatistas," destroyed government offices, burned land deeds, and freed prisoners. At least 135 people died in the rebellion.

On January 12, after 11 days of heavy fighting, a cease-fire was called. The next month peace talks began between EZLN representatives and the Mexican government. Negotiations between the two sides proved to be a frustrating and lengthy process. Seeking "democracy, liberty, and justice for all Mexicans," the EZLN called for government reforms, including local autonomy, as well as land redistribution and other measures to aid the region's impoverished indigenous population. In February 1996 the two sides signed the San Andrés Accords and agreed to more talks. But in August of that year, the dialog stalled; the EZLN said it would not return to the negotiating table until the government implemented the San Andrés Accords.

Meanwhile, progovernment paramilitary groups with ties to Mexico's ruling PRI party (Institutional Revolutionary Party) made their presence known in Chiapas. There were violent episodes, the most horrific of which occurred on December 23, 1997, when the progovernment paramilitary group Paz y Justicia (Peace and Justice) brutally attacked a group of unarmed indigenous people in the village of Acteal. A total of 45 people, mostly women and children, were slaughtered, and 25 more were injured.

The turbulence in Chiapas is fueled by deep-seated antigovernment feelings among the indigenous (Mayan) population. Despite the fact that it is rich in natural resources (including coffee, corn, timber, and oil), it is one of the poorest regions of Latin America; the wealth of Chiapas rests in the hands of a few. In 1990 half of the population in the state was malnourished, 42 percent had no access to clean water, 33 percent was without electricity, and 62 percent did not have a grade-school education. It was no coincidence that the 1994 uprising took place the same day that the North American Free Trade Agreement (NAFTA) went into effect: According to one Zapatista leader, NAFTA was the "death sentence" for Mexico's poor farmers, who would now have to compete with farmers north of the border. In 2005, more than 10 years after the Zapatistas burst onto the scene, the situation in Chiapas remained unresolved.

GRENADA

What happened in **Grenada**?

On October 24, 1983, about 3,000 U.S. Marines and U.S. Army rangers landed on the Caribbean island. The number included about 300 military personnel from neighboring Antigua, Barbados, Dominica, Jamaica, St. Kitts-Nevis, St. Lucia, and St. Vincent. The arrival followed the October 12 through 19 coup in which Prime Minister Maurice Bishop (1943–1983) was overthrown and killed by a hard-line Marxist military council headed by General Hudson Austin.

While the United Nations and friends of the United States condemned President Ronald Reagan (1911–2004) for the action, American troops detained General Austin and restored order on the island. Governor-general Sir Paul Scoon (1935–) formed an interim government to prepare for elections. Most of the U.S. military presence was withdrawn by December 1983, with nominal forces remaining on the island through most of 1985.

President Reagan justified the tactic by citing that the coup had put in danger American students on the island, but prevailing political conditions on tiny Grenada were more likely to have inspired the show of force: Increasingly stronger ties between Grenada and Cuba had made many American officials nervous; they feared the island would be used as a way-station for shipping Soviet and Cuban arms to Central America.

PERSIAN GULF WAR

What did **President George H.W. Bush** mean when he said the U.S. had to "draw a line in the sand"?

President George H.W. Bush (1924–) was reacting to Iraqi leader Saddam Hussein's (1937–) act of aggression when on August 2, 1990, his troops invaded neighboring

Kuwait. The United Nations (UN) gave Iraq until January 15, 1991, to withdraw from Kuwait. Iraq failed to comply. The "line in the sand" that Hussein crossed was soon defended: On January 16, 1991, Operation Desert Storm was launched to liberate the Arab nation of Kuwait from Iraq, whose military dictator had not only invaded Kuwait but proclaimed it a new Iraqi province. Bush averred, "This will not stand," and in order to protect U.S. oil supplies in the country, the president mobilized U.S. forces, which were joined by a coalition of 39 nations, to soundly and quickly defeat Iraq.

Why was the **Persian Gulf War** important?

The six-week war, telecast around the world from start to finish (February-April 1991), was significant because it was the first major international crisis to take place in the post–Cold War era. The United Nations proved to be effective in organizing the coalition against aggressor Iraq. Leading members of the coalition included Egypt, France, Great Britain, Saudi Arabia, Syria, and the United States. The conflict also tested the ability of the United States and the Soviet Union (then still in existence as such) to cooperate in world affairs.

WAR IN THE BALTICS

What caused the **Bosnian War**?

To understand the war in Bosnia (1992–95) it is important to review the history of Yugoslavia. Treaties at the end of World War I (1914–18) dissolved the Austria-Hungarian Empire, creating separate nations of Austria and Hungary and dividing their former territory into three new countries: Czechoslovakia, Poland, and the Kingdom of Serbs, Croats, and Slovenes.

The various factions within the Kingdom of Serbs, Croats, and Slovenes struggled for power. In 1929 King Alexander I (1888–1934), an ethnic Serbian, dismissed the national parliament, did away with the constitution (1921), and declared an absolute monarchy. He also changed the country's name to Yugoslavia. The government was then dominated by ethnic Serbs, who had settled in the region as early as the seventh century A.D. and were converted to Eastern Christianity (Orthodox Christianity) by the ninth century. But the Serbian authority was challenged by the nation's ethnic Croats, whose ancestors had settled in the region by the seventh century A.D. and were converted to Western Christianity (Roman Catholicism) by the Franks. To try to end the struggle, in 1939 Croats were given limited autonomy within Yugoslavia. The arrangement was short-lived: Yugoslavia was invaded by the Axis powers in April 1941 in World War II (1939–45).

Ethnic Albanian refugees escape the fighting in Macedonia and arrive in the village of Donje Ljubinje in southern Kosovo, March 2001.

The war over, in 1946 Yugoslavia was divided into six federated republics: Bosnia and Herzegovina, Croatia, Macedonia, Montenegro, Serbia, and Slovenia. But the lines of demarcation between these republics paid little regard to the ethnic boundaries of Serbs, Croats, and Muslims (who have been in the region since 1526, when it was invaded by Turks). Federal power was in the hands of Communist leader Josip Broz Tito (1892–1980). At first Tito tied his government to the Soviet Union; he directed the nationalization of land, industry, utilities, and natural resources. But after 1948 he pursued a policy of nonalignment.

In the 1980s Yugoslavia's economy weakened, exacerbating regional differences. Tensions among ethnic groups flared. In 1991, as communism fell across Eastern Europe, Yugoslavia began to break apart. By March 1992 four of its republics had declared independence: Croatia, Slovenia, Macedonia, and Bosnia and Herzegovina. What remained of Yugoslavia were Serbia and Montenegro.

Serbs living in Bosnia and Herzegovina objected to the declaration of independence, which had been approved by the republic's Croats and Muslims. Fighting broke out in Bosnia, centered around the capital city of Sarajevo. Troops from Serbia entered the region to back the ethnic uprising in Bosnia. As with many civil wars, the conflict divided families and friends. Evidence mounted that the Serbs, under the direction of leader Radovan Karadzic (1945–), were engaged in a program of ethnic cleansing, including the mass murder of tens of thousands of Muslim refugees. In May 1995, after the Serbian military in Bosnia refused to comply with a United Nations (UN) ultimatum, the North Atlantic Treaty Organization (NATO) began a campaign of strategic air strikes on Serbian targets. The NATO assaults weakened the Serbs and brought them to the negotiating table in November 1995, when U.S. mediators helped broker a peace agreement in Dayton, Ohio. A single state (Bosnia and Herzegovina) was re-established; it was to be governed through a power-sharing arrangement among Serbs, Croats, and Muslims. But the conflict in Yugoslavia was not over; by 1998 the region's ethnic disputes erupted into another civil war, this time in the Kosovo province.

In 2003 Yugoslavia was effectively dissolved with the establishment of the country of Serbia and Montenegro through a peace accord brokered by officials from the European Union (EU). The new arrangement gave greater autonomy to each republic.

What happened to **"the Butcher of Bosnia"**?

During the Bosnian war, Radovan Karadzic (1945–), the former president of the Serb Republic and commander of its armed forces, earned himself the ignominious nickname "the Butcher of Bosnia" for directing the massacres and mass victimization of enemies, many of them Muslims. Following the war, the United Nation's International Court Tribunal for the former Yugoslavia (ICTY) in The Hague issued two indictments of Karadzic, charging him with genocide, war crimes, and crimes against humanity. Karadzic disappeared in 1996 and was still at large in spring 2005. There were warrants for his arrest.

In early December 1998, North Atlantic Treaty Organization (NATO) forces arrested Karadzic henchman General-Major Radislav Krstic (1948–). The high-ranking Serbian official was believed to have taken part in the July 1995 massacre of as many as 8,000 Muslims in the eastern Bosnian town of Srebrenica. On August 2, 2001, Krstic was found guilty of genocide, persecutions for murders, cruel and inhumane treatment, terrorizing the civilian population, forcible transfer and destruction of personal property of Bosnian Muslim civilians, and murder as a violation of the Laws and Customs of War; he was sentenced to 46 years in prison. After appeals, in April 2004, Krstic's sentence was reduced to 35 years based on the court's belief that he had aided and abetted acts of genocide but had not instigated them. He was transferred to Great Britain to serve out his sentence.

Another high-ranking Serbian military leader who faced charges of genocide before the ICTY was General Ratko Mladic (1942–). As the former commander of the Bosnian Serb forces in Bosnia and Herzegovina, Mladic was considered responsible for the "serious breaches" of international humanitarian law committed by the Bosnian Serb forces between May 1992 and July 1995, including the massacre at Srebrenica. His indictment also included charges of war crimes and crimes against humanity. He was still wanted in spring 2005.

KOSOVO

What caused the **fighting in Kosovo**?

The conflict in Kosovo, like that in Bosnia, was ethnic-based. Kosovo is a province at the southern end of Serbia; it neighbors Albania and the former Yugoslav republic of Macedonia. Many Albanians still live in Kosovo and see themselves more closely aligned with Albania, to the southwest, than with Yugoslavia. Thus, a separatist movement began, which caused mounting tension between ethnic Albanians and Serbian authorities.

Early in 1998 Serbian forces and Yugoslav army units moved to suppress the Kosovo Liberation Army (KLA), the guerrilla force that sought independence for the province's Albanian population. In October Yugoslav President Slobodan Milosevic (1941–) agreed to end the crackdown—but this was only after the North Atlantic Treaty Organization (NATO) had repeatedly threatened air strikes. However, in the months that followed Milosevic's stated compliance, there was more violence against ethnic Albanians. By January 1999 hundreds had been killed and more than a quarter million people were displaced from their homes—many of them seeking shelter in makeshift huts in the forest. Victims included the elderly, women, and children. On March 23, 1999, Yugoslav's Serb parliament rejected NATO demands for autonomy in Kosovo as well as the plan to send NATO peacekeeping troops into the troubled province. The following day, NATO launched a campaign of air strikes against Yugoslavia, with the intent to weaken Milosevic and force him to comply with international demands to settle the conflict. After more than 50 days of air strikes, which included some controversial and deadly errors on NATO's part, it appeared that while NATO was winning the air campaign against Yugoslavia, the Serbian government of Yugoslavia was winning a ground campaign against the Kosovar Albanians: Of the estimated 1.8 million ethnic Albanians in Yugoslavia, Milosevic's forces had driven out or killed all but 130,000. Meantime, evidence of Serb atrocities toward ethnic Albanians mounted, as mass graves were discovered and survivors who fled Kosovo reported horrific tales of torture and rape at the hands of the Serbs.

Ultimately NATO's Operation Allied Force was successful. On June 10, 1999, after a 77-day air campaign, the bombing was temporarily suspended because Yugoslav forces had begun to fully withdraw from Kosovo. The withdrawal was in compliance with an agreement drawn up between NATO and the Federal Republic of Yugoslavia and Republic of Serbia on the evening of June 9. By June 20 the Serb withdrawal was complete and a multinational security force was established to keep the peace. That same day NATO announced that it had formally terminated the air campaign. NATO personnel were reassigned as peacekeepers and to move humanitarian aid, including food, water, tents, and medical supplies. Though the crisis in Kosovo was over, the aftermath was immense: It was estimated that by the end of May 1999, 1.5 million people, or 90 percent of the Kosovar population, had been expelled from their homes and a quarter million Kosovar men were missing. Further, there was mounting evidence that ethnic Albanians had been the victims of genocide. This evidence included the discoveries of mass graves; reports of mass executions, expulsions, and rape; and the systematic destruction of property and crops.

In 2003, in a peace accord brokered by the European Union, the nation of Serbia and Montenegro was established, with each republic receiving greater autonomy. The nation of Yugoslavia no longer existed.

What happened to **Slobodan Milosevic**?

Like other Serb leaders involved in the recent Baltic wars, former Yugoslav president Slobodan Milosevic (1941–) faced charges of genocide, war crimes, and crimes against humanity at the International Court Tribunal for the former Yugoslavia (ICTY) in The Hague. He was arrested in Belgrade on April 1, 2001, and transferred to The Hague in June. There he faced 66 counts of war crimes from the Balkan conflicts. Before the ICTY, he pled innocent of three indictments against him, one for each major war crimes scene: Kosovo, Croatia, and Bosnia. His trial began on February 12, 2002, and was marked by numerous idiosyncrasies, including Milosevic's attempts to represent himself (without the benefit of counsel), his frequent refusals to cooperate with the court, and numerous days lost to his various illnesses. The prosecution wrapped up its case against him on February 25, 2004, after hearing testimony from almost 300 witnesses. The defense portion of the trial began on August 31, 2004, and was slated to last 150 court days (not the same as calendar days). As of June 2005 the trial was still underway in The Hague.

CHECHNYA

What is the **conflict over Chechnya**?

Since the 1990s separatist factions have been fighting for the independence of the tiny Russian republic (at just more than 6,000 square miles in area, it is about the size of Hawaii).

Chechnya's population falls into three main ethnic groups: Chechens (the majority), Ingush, and Russian. The religion of the Chechen and the Ingush peoples is Islam, while the Russian population is mostly Orthodox Christian. After the breakup of the Soviet Union, the region remained with the new Russian Federation as the Chechen-Ingush republic. But dissent grew, and in 1991 a rebel faction led by Dzhokhar Dudayev (1944–1996) took control of the government. Chechnya separated from Ingushetia to form two separate republics in 1992.

In 1994 Russian troops invaded to reclaim the Chechen capital of Grozny, an oil and manufacturing center. By this time two factions existed among the rebels: one was a nationalist movement and the other a fundamentalist Islamic movement. In 1995 the conflict spilled over into neighboring regions, as Chechen militants began a series of terror attacks. In June of that year gunmen seized a hospital in Budyonnovsk, about 90 miles north of the Chechen border, in the Stavropol territory, and held 1,800 people hostage for six days. More than 100 people were killed and hundreds were injured. In 1996 Chechen terrorists seized another hospital, holding 2,000 hostages, this time in neighboring Dagestan; at least 23 people were killed. Russia's military strikes in Chechnya continued until a cease-fire was negotiated in 1996, by which time dissident leader Dudayev had been killed.

Even after a peace treaty was signed in 1997, ending the First Chechen War, the status of Russia's "breakaway republic" remained unclear. In 1999 the battle over Chechnya's status was taken straight to the Russian capital of Moscow, where five bombings in four weeks claimed 300 lives; Islamic militants in Chechnya were blamed. The Kremlin responded with force, leveling Grozny by early 2000 and displacing a quarter of a million people from their homes. A pro-Moscow administration was put into place.

Violence related to the Chechen conflict continued throughout Russia. Suicide bombings alone claimed more than 260 lives between 2002 and 2004. There were large-scale assaults as well: In October 2002 Chechen militants took more than 700 people hostage at a Moscow theater; after a two-day standoff, Russian Special Forces stormed the building, killing 41 Chechen fighters and 129 hostages. In August 2004 two airliners crashed within minutes of each other after taking off from the same airport; 90 people died in the crashes, which Russian president Vladimir Putin (1952–) labeled terrorism. And on September 1, 2004, 32 armed militants seized an elementary school in Beslan, North Ossetia, a region bordering Chechnya. The terrorists held some 1,200 hostages for 48 hours, at which time Russian forces stormed the building; 335 people died, most of them children. Leaders of the Chechen nationalist movement distanced themselves from the terrorist acts, the responsibility for which were claimed by a militant Chechen Muslim group.

Chechnya continued to be unstable, with Putin resolved to a hard-line approach to the breakaway republic. In elections held in March 2003, voters reportedly approved a new constitution that declared Chechnya to be part of Russia. But critics found the election results irregular, with almost 96 percent of voters expressing support for the referendum. Months later, pro-Moscow leadership was elected in the person of Deputy Prime Minister Ramzan Kadyrov; critics called it a puppet regime.

In March 2005 Chechen nationalist leader Aslan Maskhadov (1951–2005) was killed in a Russian assault. His death was seen as a victory for Moscow; Maskhadov, who had briefly been president (1997–1999) of an independent Chechnya, was generally viewed as a moderate who believed an honest dialogue between the two sides could bring an end to the decade-long conflict.

IRELAND

What is the "Irish Question"?

The Irish Question is an ages-old and very complicated problem that encompasses issues of land ownership, religion, and politics between Ireland and Britain. The chronology of events is:

Twelfth century: A feudal landowning system is imposed on Ireland by the British, creating an absentee landlord class and an impoverished Irish peasantry.

1700s: The English try to impose Protestantism on a largely Catholic Ireland; Irish rebellions flair.

1801: The Act of Union unites England and Ireland, forming the United Kingdom. Though Ireland has representation in Parliament, it is divided.

1800s: The British Crown begins to populate the six counties of the northeastern Irish province of Ulster with Scottish and British settlers, giving the area a decidedly Protestant character. The division deepens between the 26 counties of southern Ireland and Ulster: The north becomes increasingly industrialized and Protestant, while the south remains agricultural and Catholic.

1840s: Irish discontent with British rule heightens when a great famine strikes Ireland, resulting in widespread hunger, illness, and death. Many who survive emigrate to seek a better life elsewhere.

1858: A secret revolutionary society forms in Ireland and among Irish emigrants in the United States. Called the Fenian movement, the group's objective is to achieve Irish independence from England by force. Fenians stage rebellions, which are suppressed by the British.

1868: The head of the Liberal Party, William Gladstone (1809–1898), becomes British prime minister. He will be prime minister three more times during the last four decades of the century. Gladstone becomes an advocate for the peaceful settlement of the Irish Question.

1870: Parliament passes the First Land Act, encouraging British landlords to sell land and providing reduced-rate loans to Irish tenants wishing to buy land.

1875: Irish nationalist leader C. S. Parnell (1846–1891) enters British Parliament. He uses filibusters to prevent Parliament from discussing anything but the Irish Question.

1886: C. S. Parnell forms an alliance with Prime Minister William Gladstone. The First Home Rule Bill (providing for Irish self-government) is introduced in Parliament but fails to pass.

1893: The Second Home Rule Bill is introduced in Parliament. It is passed by the House of Commons but fails to be passed by the Lords.

1905: An Irish nationalistic movement called Sinn Fein (meaning "we ourselves") forms under the leadership of Arthur Griffith (1872–1922). The group seeks to establish an economically and politically independent Ireland.

1912: The Third Home Rule Bill is introduced in Parliament. Again, the House of Commons passes the legislation providing for Irish self-government. Fearing domination by Catholic southern Ireland in the event of Irish Home Rule, there is agitation—including threat of civil war— in Ulster (northern) Ireland. To prevent the outbreak of violence, the Lords exclude Ulster Ireland from the provisions of the Home Rule Bill. It does not take effect due to continued unrest.

1916: Refusing to accept the divided Ireland proscribed by the British parliament, revolutionary Sinn Fein member Michael Collins (1890–1922) organizes a guerrilla movement led by the Irish Republican Army (IRA). The nationalist organization becomes dedicated to the unification of Ireland.

1920: Parliament passes the Government of Ireland Act, establishing separate domestic legislatures for the north and south, as well as continued representation in British Parliament. The six northern counties of Ireland accept the act and become Northern Ireland. The 26 southern counties refuse to accept the legislation.

1921: The Anglo-Irish Treaty is signed, providing for the 26 counties of southern Ireland to become the Irish Free State (now the Republic of Ireland).

1922: The Irish Free State, headed by Sinn Fein leader Arthur Griffith, is officially declared. It gradually severs its ties to Britain.

1927: Sinn Fein ends as a movement, but some intransigent members join the IRA, and Sinn Fein becomes the political arm of the Irish Revolutionary Army.

1939–45: IRA violence and its pro-German stance cause both Irish governments to outlaw the organization, which goes underground.

1969: The IRA splits into an "official" majority, which disclaims violence, and a "provisional" wing, which stages attacks on British troops in Northern Ireland via random bombings and other terrorist acts. Still determined to forge a unified and independent Ireland, the IRA continues to stage acts of violence into the 1990s. At times, bombings become part of everyday life in Belfast, Northern Ireland. London is also the target of random IRA bomb attacks.

June 6, 1996: Britain and Ireland agree on an agenda for multiparty peace negotiations on Northern Ireland, but finding ways to disarm Northern Ireland's rival guerrilla groups remains an obstacle to political settlement.

April 10, 1998: A Good Friday peace accord is signed by Catholic and Protestant leaders who agree to form a multiparty administration by October. Disarmament will remain a sticking point to moving this agreement forward.

May 22, 1998: 71 percent of voters in Northern Ireland vote for an agreement on a power-sharing government, which is also backed by 94 percent of the voters in the Republic of Ireland. The peace agreement is designed to heal the divisions between Catholics and Protestants that have left 3,400 dead, 40,000 injured, and millions of dollars of property damage. Irish prime minister Bertie Ahern (1951–) has a simple message for those who still aimed to promote violence: "Forget it. The people on whose behalf you claim to act have spoken. Your ways are the ways of the past."

August 15, 1998: Violence is renewed when a car bomb in Omagh, Northern Ireland, kills 28 people and injures 220. The Real IRA, a dissident faction of the Irish Republican Army, claims responsibility.

October 16, 1998: The leaders of Northern Ireland's two main political parties win the Nobel prize for their efforts to end three decades of religious-inspired violence in the British province. John Hume (1937–), a Catholic, heads the Social Democratic and Labour Party (whose membership is predominately Roman Catholic); David Trimble (1944–), a Protestant, leads the Ulster Unionist Party (consisting of pro-British Protestants).

December 1998: British prime minister Tony Blair (1953–) tries to jumpstart the stalled Good Friday peace agreement: Catholic and Protestant politicians have failed to set up the multiparty administration to which they had agreed. Leaders have been hampered in their efforts to do so by the question of disarmament: Protestants demand exclusion of the Sinn Fein party from government until its military wing, the IRA, disarms. Blair also presses for north-south bodies, committees that will oversee cooperation between the Irish Republic and Northern Ireland on matters of industry, tourism, and transportation. Blair insists that his government will hold all parties to a May 2000 disarmament deadline.

2000: The disarmament deadline is not met.

October 23, 2001: The IRA promises to disarm in order to "save the peace process."

2002: The peace process is suspended after police discover that the IRA is operating a spy ring inside Northern Ireland government offices. Sporadic violence continues. Irish prime minister Bertie Ahern and other major political figures in the Republic of Ireland move to distance themselves from Sinn Fein, the IRA's political arm.

December 2004: Belfast's Northern Bank is robbed of $50 million the week before Christmas. The biggest heist in the history of British crime is believed to have been carried out by the IRA to fund its operations.

Early 2005: Despite the Good Friday agreement, the IRA has not yet disarmed. According to news reports, IRA gunmen have become an "increasingly mafia-like crime organization," involved in drug trafficking, extortion, money laundering, and the violence associated with those criminal activities. In a January 30 Belfast pub brawl, Robert McCartney, an "amiable" forklift operator, is murdered in what the perpetrators tell pub patrons is "IRA business." Though witnesses are intimidated, McCartney's sisters are not; the five women demand that the IRA be held accountable for the slaying. The McCartney sisters are hailed as heroes for their courage to speak out against the IRA and are embraced by Sinn Fein. But Sinn Fein's electoral prospects dimmed in the aftermath of the murder. The event provides another roadblock to peace.

July 2005: The IRA renounces the use of violence in its fight against British rule in Northern Ireland. The group vows to disarm, and to use only peaceful means in its ongoing efforts. The announcement is widely hailed.

AFRICA

What was **"Black Hawk Down"**?

Though the U.S. military uses the term to communicate any crash of one of its Black Hawk helicopters, the phrase is closely associated with events in Mogadishu, Somalia, on October 3, 1993. The term became synonymous with that day after American journalist Mark Bowden wrote a book, by the same title, describing a disastrous U.S. raid on a Mogadishu warlord. The book was turned into a movie in 2001.

The background is this: Somalia threw off its colonial constraints in 1960 to become an independent nation. But warring factions within the impoverished east African nation made a stable central government elusive. After staging a 1969 coup, Soviet-influenced army commander Mohammed Siad Barre (1919 or 1921–1995) established a military dictatorship in Somalia. His authoritarian rule, which was marked by human rights abuses, lasted until 1991 when he was deposed in a popular uprising (he died in exile four years later). The nation of about 8 million people was in chaos, and many were starving. International donations of food were hijacked and used by competing warlords to secure weapons from other nations, thus furthering civil strife. After a 1992 cease-fire, the United Nations sent peacekeepers to Somalia and launched a humanitarian relief operation. Outgoing U.S. President George H.W. Bush (1924–) supported the UN effort by approving a deployment of 25,000 American troops to Somalia to help secure trade routes over which badly needed food supplies could move. In 1993 the United States, then led by President Bill Clinton (1946–), reduced the number of troops to less than half the original deployment.

Trouble was ignited on June 5, 1993, when 24 Pakistani soldiers, in Somalia as part of the UN operation, were killed in an ambush. The warlord thought to be responsible for the massacre was Mohammed Farah Aidid. Somalia's government ordered Aidid's arrest. His capture was an imperative to peace: He and his followers were staging a violent rebellion against the provisional Somali government, led by Aidid rival Ali Mahdi. Over the next several months, UN and U.S. forces launched several attacks on what were believed to be Aidid clan strongholds, but Aidid himself remained an elusive target.

On October 3 U.S. elite forces launched an assault on a Mogadishu hotel believed to be an Aidid hideout. They were met with an ambush. Over the following 17 hours, U.S. troops, including a military mission to rescue downed Black Hawk helicopter crews, engaged in a battle with armed Somalis in the streets of Mogadishu. Eighteen American servicemen were killed; the bodies of some were dragged through the streets of the city. Another 84 American soldiers were wounded. Hundreds of Somalis were killed in the fighting. Video footage of the chaos was shown on international television. The Battle of Mogadishu, as it is officially called, was the most intense combat firefight experienced by U.S. troops since Vietnam. On October 7, President Clinton signed orders to withdraw all American troops from Somalia. The United States pulled out in 1994, and the UN peacekeepers followed in 1995. Even after a 2002 reconciliation conference, Somalis had not secured a central government by 2004. The country remained impoverished, strife-ridden, and lawless. The UN and other non-governmental organizations (NGOs) worked to provide much-needed humanitarian relief to Somalis.

Some military and foreign affairs experts point to the Battle of Mogadishu as a primary reason for American reluctance to engage troops in the world's hotspots in the 1990s.

What happened in the **Rwandan genocide**?

On April 6, 1994, the airplane carrying Rwandan president Juvenal Habyarimana (1937–1994), of Rwanda's majority Hutu ethnic group, was shot down by unknown attackers. The event touched off, or was used as an excuse for, what one journalist described as a "premeditated orgy of killing" in which ethnic Hutu extremists carried out a campaign of mass murder against minority Tutsis. Ten years after the horrific events, the *Chicago Tribune*'s Africa correspondent recounted how Rwanda's "Hutu

A worker arranges human bones and skulls in Kigali, Rwanda, in 2000, as part of a nationwide effort to retrieve corpses and arrange dignified reburials of the nearly one million victims of the 1994 genocide.

majority, equipped with machetes and called to action by government radio announcements, slaughtered neighbors, friends, co-workers. Priests killed parishioners who sought refuge in churches. Teachers murdered pupils. Hundreds of thousands of women were raped, children burned or drowned, bodies pushed into mass graves." The massacre continued for three months, ending only when Tutsi fighters managed to seize the capital at Kigali and take power.

In 2004 the Rwandan government released its official estimate of the death toll: 937,000 people were murdered, making it the worst ethnic cleansing the world has seen since the Holocaust during World War II (1939–45). But, unlike the Holocaust, which was carried out by a dictatorial military machine, the Rwandan genocide was carried out by the masses. In the aftermath, more than 150,000 people were accused of participating in the massive violence, though the Rwandan courts had only tried a small fraction of those—no more than 10 percent in 10 years. An estimated 2 million Hutus, many who probably feared retribution, had fled to neighboring countries after the Tutsis gained power. The genocide had happened at the hands of many.

In the decade since the Rwandan genocide, Rwandans, and the world, have grappled with difficult and perhaps unanswerable questions. How could so many people (Hutus) have participated in the mass cleansing? Experts point to Rwanda's deeply divided history; the rivalry between Hutus and Tutsis dates back hundreds of years—since the Tutsis first arrived in the central African region in the fourteenth century. How could the world body have "allowed" such an event to happen, particularly with the Nazi Holocaust a not-too-distant memory? Again, there is no easy answer. The United Nations withdrew its people when the violence began in April 1994, but some UN officials estimated later that perhaps as few as 5,000 troops might have been able to prevent the annihilation. The United States did not intervene in the Rwandan genocide; the recent images of dead American soldiers being dragged through the streets of Mogadishu, Somalia (in October 1993), had left the country reticent to involve itself in the world's hotspots. But, even with the lesson of Rwanda, experts wondered how the international response might differ today. Finally, how can Rwanda rebuild? The governing Tutsis have been able to enforce a calm on the tiny landlocked nation. But some believe it is an uneasy calm and wonder how long minority Tutsis can retain control over a country that is 85 percent Hutu. In Rwanda's efforts to serve justice, its jails and courts remained jammed 10 years after the genocide. And since so many had fled the country and others escaped prosecution, justice could not be served fully. Some legal experts thought that a general amnesty, issued by the Rwandan government, would help put the nation's problems behind them.

What was the crisis in **Darfur, Sudan**?

The brutal violence, really a genocide, in Darfur began in February 2003 and was still ongoing two years later as United Nations (UN) negotiators tried to broker a peace agreement. Sudanese government troops and government-backed Arab militia, called the janjaweed, were engaged in a violent campaign that targeted black African civilians from the Fur, Massalit, and Zaghawa ethnic groups in the western Sudanese state of Darfur (which is about the size of Texas). Reports from the Human Rights Watch and UN agencies working in Sudan indicated that large-scale bombing and burning campaigns had destroyed entire villages, tens of thousands of Darfurians had been raped and killed, and as many as 2 million had been displaced by early 2005. International relief efforts were hampered by ongoing violence, putting millions of women and children at risk of starvation and disease. Observers called for an international response to the crisis to avert "another Rwanda." In April 2005 the UN Security Council referred the Darfur situation to the International Criminal Court. But to date, no one had been able to stop the campaign of ethnic cleansing.

The Darfur crisis was part of a greater and older conflict in Sudan. The east African nation, whose neighbors include Egypt to the north and Ethiopia to the east, was in 2005 the scene of Africa's oldest civil war. The conflict had pit the Arab (Muslim) north against the Christian and animist south since 1984. But diplomatic efforts to resolve that decades-old conflict had so far failed in 2005: The Muslim regime based in the national capital of Khartoum was unwilling to share power and wealth. Thus, when Khartoum was hit with a rebellion in 2003, it retaliated by inciting the brutality in Darfur. Diplomats worked to resolve the north-south civil conflict, offering Sudan a package of debt relief and development aid that could reach $100 million. Critics of that plan said that incentives were not enough, and that sanctions should be levied against the nation to pressure the government to move toward ceasefire.

What is the incidence of **conflict in Africa**?

Since the end of World War II (1939–45), there have been at least 56 separate, identifiable conflicts in Africa, with about a dozen of them ongoing in 2005. The continent seemed to be in a state of turmoil, with conflicts ranging from regional skirmishes between warlords to civil war. The unimaginable horrors of the Rwandan genocide (1994), the decades-old civil unrest and ethnic cleansing in Sudan, and longstanding rebel violence in Uganda made news the world over. The war-ravaged image of Africa caused many observers to wonder if the continent was unusually prone to civil war. In a study published by the World Bank, researchers concluded in 2000 that "Africa has had a similar incidence of civil conflict to that of other developing regions, and that, with minor exceptions, its conflicts are consistent with the global pattern." The authors of that report acknowledged that the rising trend of conflict in Africa was due to its "atypically poor economic performance." While many could see only Africa as war-torn, others pointed to the fact that most of Africa's nations were peaceful, and that the trouble spots gave the international community an overstated impression of despair on the continent.

TERRORISM

What happened after the **1993 World Trade Center bombing**?

During the Federal Bureau of Investigation (FBI) investigation that followed the February 26, 1993, tragedy, it was learned that the World Trade Center was one of several intended targets of an Islamic extremist group. The bomb explosion in lower Manhattan killed six people and started a fire that sent black smoke through the 110-story twin towers, injuring hundreds and forcing 100,000 people to evacuate the premises.

Days later, on March 4, 25-year-old Mohammed A. Salameh, an illegal Jordanian immigrant, was arrested in Jersey City, New Jersey. Salameh was later found to be a follower of self-exiled Islamic fundamentalist leader Sheik Omar Abdel Rahman (1933–), who was wanted by Egypt for having incited antigovernment riots in 1989. In June investigators seized Arab terrorists they accused of plotting to blow up several New York City sites, including the United Nations headquarters and the Holland and Lincoln tunnels. U.S. authorities then arrested Rahman and imprisoned him on suspicion of complicity in the World Trade Center bombing. On October 1, 1995, a federal jury found Rahman and nine other militant Muslims guilty of conspiring to carry out a campaign of terrorist bombings and assassinations aimed at forcing Washington to abandon its support of Israel and Egypt.

The 1993 bombing foreshadowed the terrorist strikes of September 11, 2001, which destroyed the landmark twin towers of the World Trade Center in lower Manhattan and launched what came to be called a global war on terrorism.

What happened in the **Oklahoma City bombing**?

The attack took place at 9:02 A.M. on April 19, 1995, when a truck bomb exploded outside the Alfred P. Murrah Federal Building in Oklahoma City, Oklahoma. Of the 168 people who were killed, 19 were children. Another 500 people were injured. The blast, which investigators later learned had been caused by a bomb made of more than two tons of ammonium nitrate and fuel oil, sheered off the front half of the nine-story building and left a crater 8 feet deep and 30 feet wide. Nearby buildings were damaged or destroyed, including a YMCA day care center where many children were seriously injured. The force of the explosion shattered windows blocks away. Survivors of the blast and others in the vicinity began rescue efforts right away. Eventually more than 3,600 people from around the country participated in rescue operations, including police, firefighters, and members of the Federal Emergency Management Agency (FEMA).

Police and Federal Bureau of Investigation (FBI) agents arrested members of an American right-wing militant group who were suspected of wanting to avenge the April 19, 1993, FBI/ATF raid on the Branch Davidian religious compound in Waco, Texas. Former army buddies Timothy J. McVeigh, then 27, and Terry L. Nichols, then 40, were indicted on August 10, 1995, on 11 charges each. The two were tried separately and convicted in federal court. McVeigh was found guilty of murder and conspiracy in June 1997, and a federal jury sentenced him to death. Nichols was later found guilty of conspiracy and involuntary manslaughter, but in January 1998 the jury deadlocked on the sentence, which was then to be decided by the judge. In early June it was decided that Nichols would serve life in prison.

Officials believe the Murrah Federal Building was targeted in the antigovernment attack because it housed 15 federal agencies, including offices of the Social Security Administration (SSA), Housing and Urban Development (HUD), the Drug Enforcement Administration (DEA), and the Bureau of Alcohol, Tobacco, and Firearms (ATF), as well as several defense department offices and a government-run day care center.

Is the **bombing** of buildings a new phenomenon?

Though terrorist bombings seem a plague of recent decades, the history of such strikes goes back to a time before the word *terrorism* was part of everyday language. In

1920 a bomb explosion on September 16 ripped through the J. P. Morgan Bank Building in New York City, killing 39 people (30 of them instantly), injuring 300 more, and causing $2 million in property damage. According to eyewitness accounts, the bomb had been carried by a horse-drawn carriage into the heart of America's financial center just before midday, and it exploded as nearby church bells tolled noon. Among the victims were passersby in the street and people working at their desks, including some high-ranking personnel at J. P. Morgan. Suspicion centered around anarchists, some of whom were questioned by the police and Federal Bureau of Investigation (FBI). But no culprit was ever found. Until September 11, 2001, that event was the deadliest bombing in New York City history.

Further, the many bombings carried out by white supremacists during the civil rights movement in the 1960s are also testimony to the fact that targeted bombings are not a new phenomenon. Between 1962 and 1965, as the Council of Federated Organizations worked to register voters in Mississippi, racial extremists turned to violence. Among the tactics used—in addition to shootings, beatings, and lynches—were bombings. During Freedom Summer (1964) alone, more than 65 homes, churches, and other buildings were bombed in Mississippi.

Perhaps the most widely publicized of those bombings was the Sixteenth Street Baptist Church in Birmingham, Alabama, on September 15, 1963. Two hundred people were attending Sunday services when a bomb exploded. Four young African American girls were killed. But the sorrow of that day did not stop there: The bombing provoked racial riots, and police used dogs to control the crowds. Two black schoolboys were killed in the melee.

Was the **Madrid commuter train bombing** an act of terrorism?

Spanish officials concluded that the March 11, 2004, bombing of packed commuter trains in Madrid was an act of terrorism, likely motivated by Spain's arrest of dozens of al Qaeda suspects after the September 11, 2001, attacks on the United States. (At least three of those arrested in Spain were charged with helping organize the 9/11 attacks.) On March 11, 2004, 10 backpacks loaded with dynamite exploded on four trains at the height of morning rush hour, killing 191 people and injuring 1,800. In the investigation that followed, officials uncovered other terrorist plots, including suicide bombings and assassinations aimed at interrupting Spain's court system that tries terrorist suspects. The discoveries led officials to conclude that their nation had become a "crossroads" for Muslim extremists, in part because of Spain's proximity to northern Africa. In April 2004, as authorities closed in on the hideout of the suspected Madrid bombing ringleader, a Tunisian man, and his associates, the suspects blew up their apartment. A total of 62 suspects were arrested in 2004 in association with the train bombing.

Rescue workers line up bodies following explosions on passenger trains in Madrid, Spain. The March 11, 2004, bomb attacks killed more than 190 people.

9/11

What was the chronology of events on **September 11, 2001**?

The sequence of events related to the terrorist attacks on 9/11 were as follows (all times are eastern daylight time):

> 8:46 A.M.: A passenger plane crashes into the north tower of New York City's World Trade Center. At first it is assumed to be an accident. But within hours, it is learned that the plane was American Airlines Flight 11, and commandeered by hijackers shortly after takeoff from Boston headed to Los Angeles; it carried 81 passengers (including 5 hijackers) and 11 crewmembers.
>
> 9:03 A.M.: A second passenger plane slams into the south tower of the World Trade Center and explodes. United Airlines later announces that this was Flight 175, another Boston-Los Angeles plane; it carried 56 passengers (including 5 hijackers) and 9 crewmembers. Both towers of the World Trade Center are in flames.
>
> 9:21 A.M.: Bridges and tunnels leading into New York City are closed. Within the hour, Mayor Rudolph Giuliani (1944–) orders an evacuation of Manhattan south of Canal Street.

9:25 A.M.: All flights in the United States are grounded. It is the first time in history that the Federal Airlines Administration (FAA) halts all flights.

9:30 A.M.: President George W. Bush (1946–), scheduled to speak at a grade school in Sarasota, Florida, remarks that the nation has been the target of an "apparent terrorist attack."

9:45 A.M.: A third passenger plane crashes into the east wall of the Pentagon in Arlington, Virginia. The plane was American Airlines Flight 77, carrying 58 passengers (including 5 hijackers) and 6 crewmembers. It originated at Washington, D.C.'s Dulles International Airport and was headed to Los Angeles. After being hijacked, the plane made a u-turn over the Ohio-Kentucky border to return to the Washington, D.C.-area, where it hit its intended target.

9:45 A.M.: The White House is evacuated.

10:05 A.M.: The south tower of the World Trade Center collapses.

10:10 A.M.: A large section of the Pentagon collapses.

10:10 A.M.: A fourth passenger plane crashes in a field in Somerset County, in southwestern Pennsylvania (outside of Pittsburgh). It was United Airlines Flight 93, originating in Newark, New Jersey, and bound for San Francisco. After being hijacked, the plane made a u-turn over Ohio and was headed for a target in Washington, D.C. (later thought to be Camp David, the White House, or the U.S. Capitol building). After learning via cell phones of the other hijackings, a group of passengers stormed the cockpit in an effort to take the plane, which crashed in the countryside. The plane carried 37 passengers (including 4 hijackers) and a crew of 7. There were no survivors.

10:24 A.M.: The FAA reports that it has diverted all incoming transatlantic flights to Canada.

10:28 A.M.: The north tower of the World Trade Center falls. The collapse of each tower blankets lower Manhattan in smoke and ash, turning the brilliantly sunny day to darkness.

10:45 A.M.: All federal office buildings in the nation's capital are evacuated.

2:00 P.M.: Senior FBI sources reveal to the media that they are working under the assumption that the four airplanes that crashed were part of a terrorist attack. The FBI later learns the identities of all 19 hijackers (there were 5 on board each plane, except Flight 93, which had 4).

4:00 P.M.: Media reports indicate that U.S. officials have credible evidence tying the morning's attacks to Saudi militant Osama bin Laden (1957–).

5:20 P.M.: Building 7 (a 47-story structure) of the World Trade Center complex collapses. Nearby buildings are in flames.

A fireball erupts from the north tower of the World Trade Center after it was struck by the second of two hijacked airplanes on the morning of September 11, 2001.

Where was the **president on September 11**?

President George W. Bush (1946–) began the day at a Sarasota, Florida, elementary school, where he was expected to speak on his education program. At 9:00 A.M. he was notified of the first plane crash into the World Trade Center's north tower. He was preparing to make his planned speech when, a few moments later, he was told of the second crash. After consulting with advisers, he addressed the gathering at 9:30 A.M., saying there had been "an apparent terrorist attack on our country." At 9:57 A.M., Air Force One, with the president on board, took off from Florida. By 10:00 A.M. Bush had ordered the nation's military to high alert status around the world. At 11:40 A.M., the president's plane landed at Barksdale Air Force Base in Louisiana. There he remarked to the press, "Make no mistake, the United States will hunt down and punish those responsible for these cowardly acts." Shortly after 1:00 P.M., Air Force One departed Louisiana and headed to Nebraska's Offutt Air Force Base.

Throughout the day the president was in contact with his top advisers, including Vice President Dick Cheney (1941–), who was reported to be in a "secure location" in Washington, D.C. At 4:36 P.M. the president's plane took off from Offutt to make the trip back to Washington. Bush arrived at the White House aboard presidential helicopter Marine One after Air Force One made an escorted landing at Andrews Air Force Base in Maryland. First Lady Laura Bush (1946–) arrived back at the White House separately via motorcade; she, too, was reportedly in a "secure location" throughout the day. At 8:30 P.M. the president addressed the nation via a televised broadcast from the Oval Office. In part he said, "Thousands of lives were suddenly ended by evil, despicable acts of terror." He went on to say, "Terrorist attacks can shake the foundations of our biggest buildings, but they cannot touch the foundation of America. These acts shatter steel, but they cannot dent the steel of American resolve." Further, he promised that the search was underway for those responsible, and he said the nation would stand with its friends and allies to "win the war against terrorism." He ended his day with a National Security meeting, which lasted nearly two hours.

How many died in the September 11 attacks?

About 3,000 people died that day as a result of the terrorist attacks on the United States. The strikes on New York's World Trade Center claimed 2,602 lives, and another 125 died at the Pentagon, outside Washington, D.C. (The numbers include firefighters and police officers who died as part of the rescue effort.) The victims on board the hijacked flights, which were used as terrorist weapons, numbered an additional 87 on American Airlines Flight 11, which crashed into the North Tower of the World Trade Center; 60 on United Airlines Flight 175, which crashed into the South Tower of the World Trade Center; 59 on American Airlines Flight 77, which slammed into the Pentagon; and 40 on United Airlines Flight 93, which crashed in rural Shanksville, Pennsylvania.

In February 2005 officials announced that they had exhausted all possible methods of identifying the human remains recovered at Ground Zero. Of the 2,749 people reported missing at the World Trade Center, positive identification had been made on 1,161 people. The rest were presumed dead; 2,749 death certificates were issued related to the World Trade Center attack.

How many were injured in the September 11 attacks?

The exact number of injuries resulting from the attacks is not known. Data from the aftermath show that thousands of people were treated in hospitals after the attacks. Some sustained serious physical injuries such as burns, eye injuries, and respiratory problems. But exact figures on the number and extent of injuries could not be gathered, partly because some had chosen not to receive medical treatment or had received it later. According to a September 2004 report by the U.S. Government Accountability Office (GAO), "A multitude of physical and mental health effects have been reported in the years since the terrorist attack on the World Trade Center on September 11, 2001, but the full health impact of the attack is unknown. Concern about potential long-term effects on people affected by the attack remains." The report cites several injury categories including respiratory health, reproductive health, and mental health. Post-traumatic stress disorder (PTSD) alone was thought to have affected a great number of New Yorkers, particularly emergency responders and children. According to the report, "PTSD is an often debilitating and potentially chronic disorder that can develop after experiencing or witnessing a traumatic event. It includes such symptoms as difficulty sleeping, irritability or anger, detachment or estrangement, poor concentration, distressing dreams, intrusive memories and images, and avoidance of reminders of the trauma." Mental health specialists reported that people across the nation demonstrated symptoms consistent with PTSD after the 9/11 attacks.

What was the **9/11 Commission**?

It was the 10-member group created by a congressional act signed by the president on November 27, 2002. The bipartisan commission, consisting of five Republicans and five Democrats, was chosen by Congress to look into how the attacks of September 11, 2001, could have happened and how such a tragedy could be avoided in the future. During its investigation, the commissioners and their staff reviewed more than 2.5 million pages of documents, interviewed more than 1,200 people in 10 countries, held 19 days of hearings, and took public testimony from 160 witnesses. The commission wrapped up its work in all due speed, publishing a full report less than two years after it received its mandate. The 567-page report chronicles the events of 9/11, looks at the roots and growth of the "new terrorism," reviews the U.S. response to the attacks and to previous assaults (including the August 1998 U.S. embassy bombings in Kenya and

Tanzania), and recommends changes to prevent further terrorist strikes. The report was made available as a book and an online document, presented "to the American people for their consideration."

Among the key disclosures in the report were that the commission found "no credible evidence that Iraq and al Qaeda cooperated on the attacks against the United States." That finding was immediately dismissed by the White House in June 2004. The commission also reported that the original plan for the al Qaeda attacks on the U.S. homeland included a total of 10 hijacked airplanes, striking targets on the East and West coasts; the plan was dismissed by al Qaeda leader Osama bin Laden as too complex. In the aftermath of the 9/11 attacks and the U.S.-led retaliatory strikes on Afghanistan, the commission believed that al Qaeda had become more decentralized, with cell leaders assuming greater authority for decision-making.

Anthrax

Were the **anthrax attacks** acts of terrorism?

Yes, as U.S. attorney general John Ashcroft said, "When people send anthrax through the mail to hurt people and invoke terror, it's a terrorist act." Investigators believed the attacks, which killed 5 people and sickened 17 others, were acts of domestic terrorism. But whoever was responsible for mailing anthrax spores in the fall of 2001 remained unknown in mid-2005. Some observers believed the attacks were carried out to expose the nation's vulnerability to bio-terrorist attacks.

What happened in the **anthrax scare** of 2001?

The anthrax scare unfolded shortly after the September 11, 2001, terrorist attacks on the World Trade Center and the Pentagon. When news first broke that anthrax-laced letters were received at East Coast media outlets and U.S. congressional offices, there were fears that the spreading of the infectious bacterium was linked to the 9/11 attacks. But the investigation later pointed toward a domestic perpetrator, still unknown in mid-2005.

According to a report from the office of Homeland Security, the anthrax events began on October 2, 2001, when an infectious disease doctor in Palm Beach County, Florida, reported a suspected case of inhalation anthrax. That diagnosis was confirmed by the Centers for Disease Control (CDC) on October 4 and the information was released to the public, rattling the nerves of an already jittery nation. On October 5 the first victim of the anthrax attacks, the 63-year-old south Florida man who had been hospitalized, died. On October 7 investigators announced that evidence of the anthrax bacterium was found in his workplace, the Florida offices of American Media Inc. (AMI), publisher of the *National Enquirer, Sun,* and other tabloids.

Anthrax clean-up techniques are demonstrated during a news conference on October 30, 2001, in Washington, D.C. There were fears that the anthrax attacks were linked to 9/11, but the investigation pointed to a domestic perpetrator.

On October 12 New York mayor Rudolph Giuliani announced that an NBC News employee in the office of anchorman Tom Brokaw had tested positive for cutaneous (skin) anthrax, a less dangerous form of the bacterium. Three days later, it was reported that a letter containing suspicious powder was opened in the office of Senator Tom Daschle (South Dakota) in the Hart Senate Office Building in Washington, D.C. The next day the powder tested positive for anthrax.

October 16 brought the news that the infant son of an ABC News producer in New York City had tested positive for exposure to the skin form of anthrax; the baby had visited the ABC News offices weeks earlier, and officials believed exposure had occurred at that time. The same day, it was announced that a second AMI worker, a 73-year-old mailroom employee, was diagnosed with inhalation anthrax and was in intensive care. The announcements brought the number of confirmed anthrax cases to four, two inhalation and two cutaneous. Within days CDC officials had linked all four cases to "the intentional delivery of *B. anthracis* spores through mailed letters or packages." The FBI urged the public to handle any suspicious packages with care; to not open, smell, or taste them; and to call 911 immediately.

On October 18 a worker at CBS News in New York was reported to have been infected with skin anthrax. The next day the public learned that an editorial assistant in the Manhattan offices of the *New York Post* had also been diagnosed with the skin form of the disease. Also on October 18 and 19 the CDC confirmed diagnoses of both inhalation and cutaneous anthrax in New Jersey postal workers, who had begun showing symptoms of infection on October 13.

Between October 19 and 22 four workers from a Washington, D.C., postal facility were confirmed to have been infected with inhalation anthrax and were hospitalized. On October 22 two of them died. Two days later the U.S. postmaster general announced to the public, "There are no guarantees that the mail is safe," and he advised handwashing after handling the mail. On October 31 a New York hospital worker died of inhalation anthrax, becoming the fourth fatality. But investigators could not determine how the woman had been exposed to the bacterium.

By mid-November traces of anthrax had been discovered in mail facilities that supplied the CIA, the Senate and House office buildings, the Supreme Court, Walter Reed Army Institute of Research, the White House, Washington, D.C.'s Brentwood Mail Processing and Distribution Center, Washington, D.C.'s Southwest Postal Station, a Kansas City postal facility, the Pentagon's post office, and four more New Jersey postal facilities. Antibiotics were distributed to thousands of workers as a preventive measure.

On November 16 an anthrax-laced letter addressed to Senator Patrick Leahy (Vermont) was discovered in a batch of quarantined Capitol Hill mail. The handwriting, all uppercase, was similar to that in the letter addressed to Senator Daschle, and the two letters carried identical messages: "09-11-01 / you can not stop us. / we have the anthrax. / you die now. / are you afraid? / death to America. / death to Israel. / Allah is great." The letters sent to the senators were postmarked October 9, 2001, in Trenton, New Jersey. The message and handwriting in those letters were also similar to those in letters addressed to the editor of the *New York Post* and to Tom Brokaw at NBC News, both of which read: "09-11-01 / this is next / take penacilin [sic] now / death to America / death to Israel / Allah is great." The letters sent to the media outlets were postmarked September 18, 2001, also in Trenton, New Jersey. The "Franklin Park, NJ" return address was identical on all four letters; but it was a nonexistent location. Investigators believed that whoever sent out the deadly missives was familiar with the area.

On November 20 it was confirmed that an elderly Connecticut woman, who had died the day before, had been infected with anthrax. The 94-year-old was believed to have been exposed to a letter that had been cross-contaminated in the mail. All five anthrax deaths were caused by the inhalation form of the bacterium. In total, 22 cases of anthrax were identified between October 4 and November 20; 11 of them were the deadly inhalation form (6 people survived) and 11 were the less serious skin form. In all but two cases (the New York hospital worker and the elderly Connecticut woman), the victims were mail-handlers or were exposed to worksites that had been contaminated by mail.

Investigators concluded that all of the anthrax spores were of the same strain, called Ames, but that the letters contained different grades of the bacterium. By November 9,

2001, the FBI had issued a "behavioral/linguistic assessment" of the offender based on the known anthrax parcels: "The offender is believed to be an adult male who has access to a source of anthrax and possesses the knowledge and expertise to refine it." The FBI headed a multi-agency effort to identify the perpetrator of the deadly attacks.

THE WAR ON TERROR

What is the **War on Terror**?

In his remarks the evening of September 11, 2001, President George W. Bush (1946–) vowed that "America and our friends and allies join with all those who want peace and security in the world, and we stand together to win the war against terrorism." TV newscasts were soon emblazoned with the message, "America's War on Terror," or simply, "The War on Terror." The events of 9/11 catapulted the free world into a new era, in which conflicts no longer were limited to wars between nations. There was a new enemy, which knew no national boundaries, whose "army" was covert, and which mercilessly targeted civilians.

Acknowledging that the new threat could not be met by the United States alone, the Bush administration began forging an alliance of nations that together would use diplomacy, take military action, and coordinate intelligence and law enforcement efforts to combat terrorists around the globe. On September 12, 2001, Secretary of State Colin Powell (1937–) called for a "global coalition against terrorism." Eventually the Bush administration put together an alliance of 84 countries, called the Global Coalition Against Terrorism, "united against a common danger, and joined in a common purpose."

The strike on Afghanistan, called Operation Enduring Freedom, was the first military strategy in the new war. The next major initiative was the war in Iraq. While those operations were underway, terrorist strikes continued around the globe. In March 2004, following the commuter train blasts in Spain, President Bush remarked that "the murders in Madrid are a reminder that the civilized world is at war. And in this new kind of war, civilians find themselves suddenly on the front lines. In recent years, terrorists have struck from Spain, to Russia, to Israel, to East Africa, to Morocco, to the Philippines, and to the United States. They've targeted Arab states such as Saudi Arabia, Jordan, and Yemen. They've attacked Muslims in Indonesia, Turkey, Pakistan, Iraq, and Afghanistan. No nation or region is exempt from the terrorists' campaign of violence."

Why did **NATO** respond to the **9/11** attacks?

The North Atlantic Treaty Organization (NATO) responded to the terrorist attacks on the United States because its charter states that an attack on any member nation is considered an attack on the alliance. The language is contained in Article 5 of the NATO Treaty, signed April 4, 1949, in Washington, D.C.: "The Parties agree that an armed attack against one or more of them in Europe or North America shall be considered an attack against them all and consequently they agree that, if such an armed attack occurs, each of them, in exercise of the right of individual or collective self-defense...will assist the Party or Parties so attacked." It was the first time Article 5 had been invoked by NATO since its founding.

On September 12, 2001, NATO convened a special meeting in response to the attacks on American soil and afterward issued a statement saying, in part, that the United States could rely on the support and assistance of NATO if it was found that the attack was directed from abroad. The organization's secretary general, Lord Robertson (1946–), strongly condemned the attacks and called for the "international community and the members of the Alliance to unite their forces in fighting the scourge of terrorism."

The invocation of Article 5 was confirmed by NATO on October 2, after U.S. Ambassador-at-Large Frank Taylor briefed the organization's chief decision-making body on the investigations into the terrorist attacks. The North Atlantic Council determined that the information provided by Taylor confirmed that "that the individuals who carried out the attacks belonged to the world-wide terrorist network of Al-Qaida, headed by Osama bin Laden and protected by the Taleban regime in Afghanistan."

At a press conference held October 8, Secretary General Lord Robertson announced NATO's full support for the U.S.-led invasion of Afghanistan. The following day it was confirmed that NATO assets had been deployed to the eastern Mediterranean to establish a presence in the region. But the alliance did not take a lead role in the military effort to oust the Taliban from Afghanistan.

What is **al Qaeda**?

Al Qaeda (Arabic, meaning "the base") is a global network of terrorists who banded together during the 1990s and proclaimed to be carrying out a holy war on non-Islamic nations. The group knows no national boundaries, though certain nations, including Afghanistan, were known to be al Qaeda strongholds. Led by the elusive Osama bin Laden (1957–), a wealthy exiled Saudi, the group conducted terrorist training programs in several Muslim (mostly Middle Eastern) countries and was funded by loyalists around the world. One of the United States' first actions following the September 11, 2001, terrorist attacks on the World Trade Center and Pentagon (which were later confirmed to have been carried out by al Qaeda operatives) was to freeze bank accounts of persons and organizations with suspected ties to the terrorist group.

The roots of al Qaeda can be traced to the Soviet invasion of Afghanistan in 1979, when thousands of Muslims, including bin Laden, joined the Afghan resistance. The 10-year conflict was a rallying point for Islamic extremists. Bin Laden returned home to Saudi Arabia in 1989, determined to perpetuate a holy war (*jihad*) by maintaining

During a 1999 anti-American rally in Pakistan, a protester holds a poster of al Qaeda leader Osama bin Laden.

the funding, organization, and training that had made the Afghan resistance victorious against the Soviets. By the early 1990s he emerged as a leader in the Muslim world, proclaiming his goal to reinstate the Caliphate, a unified Muslim state. He also proclaimed the United States to be an enemy to Islam; he considered the nation responsible for all conflicts involving Muslims. The Saudi government rescinded his passport in 1994, and bin Laden fled his homeland. He eventually found safe harbor in Taliban-ruled Afghanistan. According to the report issued by the 9/11 Commission, bin Laden's declaration of war came in February 1998, when he and fugitive Egyptian physician Ayman al Zawahiri "arranged from their Afghanistan headquarters for an Arabic newspaper in London to publish what they termed a *fatwa* issued in the name of a 'World Islamic Front.'" The statement claimed that America had declared war against God and his messenger, and they called for retaliation.

Under bin Laden's direction, al Qaeda carried out several attacks on American targets, including the August 7, 1998, bombings of U.S. embassies in Kenya and Tanzania, which killed 258 and injured 5,000, and the September 11, 2001, attacks on the World Trade Center and Pentagon, which killed nearly 3,000 people. After the Global Coalition Against Terrorism, led by U.S. forces, launched its attack on Afghanistan in October 2001, bin Laden was believed to have fled for Pakistan. Capturing him and other al Qaeda leaders and operatives was the key objective of the United States in its efforts to dismantle the terrorist network.

What was the **Taliban**?

It was the ultraconservative faction that ruled Afghanistan from late 1996 until December 2001, when its government crumbled following a U.S.-led military campaign. The Persian word *taleban* means "students"; the group was made up of Afghan refugees who, during the Soviet invasion (1979–89), had fled their country for Pakistan, where they attended conservative Islamic religious schools. After the Soviet withdrawal from Afghanistan and amidst the unrest that ensued, the Taliban rose to prominence. They gained control of the nation region by region, eventually taking the capital of Kabul in 1996.

While in power, the group put into force strict laws based on a fundamental interpretation of Islam. The Taliban excluded women from Afghan society, and it allowed the nation to become a training ground for Islamic terror groups such as al Qaeda. Very few nations of the world recognized the Taliban government. Its human rights abuses, principally the complete disenfranchisement of women and girls, were decried by the international community. But the breaking point came after the September 11, 2001, terrorist attacks on the U.S. homeland: When the American government requested that the suspected mastermind of those attacks be extradited from Afghanistan to the United States, the Taliban refused. The group was forcibly ousted in the brief military campaign that followed. In December 2001 the United Nations (UN) convened a conference in Bonn, Germany, where leaders of anti-Taliban ethnic factions decided on a post-Taliban transitional government, led by Pashtun leader Hamid Karzai. (The Pashtuns are the dominant group in Afghanistan, representing about 42 percent of the population. The next largest group is the Tajik, which represents about 27 percent of the population in the highly fragmented nation of 28.5 million people.) Attendees also agreed to a UN-led peacekeeping operation. Though ethnic rivalries and sporadic conflicts continued under the transitional government, Afghanistan made strides in building a stable, democratic government. One encouraging sign of reform came in March 2005, with the appointment of Habiba Sarobi, the first woman to become a provincial governor in Afghanistan history. Sarobi was chosen by the president for her post, which she assumed in late March.

When did the **U.S.-led operation in Afghanistan** begin?

Joined principally by the United Kingdom, the U.S.-led military strikes on Afghanistan began on October 7, 2001—26 days after the terrorist attacks on the U.S. homeland. The goals were to weaken the Taliban government and root out terrorist cells in the Mideast nation. Al Qaeda leader Osama bin Laden (1957–), the suspected 9/11 mastermind, was also a target. The strikes on Afghanistan, which the United States called Operation Enduring Freedom (OEF), were the first in the new war on terror.

OEF's primary goal—ousting the ruling Taliban government—was achieved relatively quickly. By mid-November, the combined efforts of the U.S. military, British, and Afghanistan's Northern Alliance, managed to take control of major cities, including the capital of Kabul. In late December an interim government was established at a UN-convened conference in Bonn, Germany. In June 2002 Hamid Karzai (1957–) was overwhelmingly elected as transitional president by the *loya jirga,* the traditional Afghan assembly of ethnic leaders; the transitional government was to run Afghanistan until national elections could be held in 2004. Despite strides in establishing the new government, fighting continued in the rugged mountainous terrain of eastern Afghanistan; Operation Anaconda was the name given to the U.S. military's effort to combat Taliban pockets of resistance, root out terrorist cells, and capture key al Qaeda leaders, including Osama bin Laden (though he had reportedly fled to neighboring Pakistan). Violence also preceded the October 2004 election.

What were the **international contributions to the War on Terror**?

According to the U.S. Central Command, by early 2005 there were 70 nations supporting the global War on Terror, and 21 nations had deployed more than 16,000 troops to the U.S.-led operations in Afghanistan and Iraq. Coalition forces contributed in a variety of ways, including providing intelligence, personnel, security assistance, equipment, and other assets for use in ground, air, and sea operations. Coalition members also provided humanitarian aid to civilians.

What are **WMD**?

WMD are weapons of mass destruction: nuclear, biological, or chemical weapons that can cause extensive casualties. The term emerged during World War II (1939–45); the abbreviated "WMD" became part of everyday language in the late 1990s and early 2000s, as the world's superpowers and the United Nations turned their attention to serious threats posed by rogue states and terrorists in a post Cold War society.

Following the launch of the 2003 Iraq war, WMD were regularly in the news. The Bush administration and its chief ally, British prime minister Tony Blair (1953–), faced sharp public criticism when no weapons of mass destruction were found in Iraq. The presence of WMD in that rogue state had been the justification for the controversial invasion. In 2004 President Bush appointed a bipartisan commission to look into why U.S. intelligence agencies had concluded that Iraq possessed WMD. In late March 2005 the commission released to the public an unclassified version of its report. The conclusion: Intelligence errors had overstated Iraq's WMD programs. It stated, "The daily intelligence briefings...before the Iraq war were flawed.... This was a major intelligence error." The commission outlined 74 recommendations to improve intelligence-gathering among the United States's 15 spy agencies. The classified version of the report contained information on the intelligence community's assessments of the nuclear programs of many of the "world's most dangerous actors." It also provided more details on intelligence concerning the al Qaeda terrorist network.

In Britain, the intelligence failures concerning Iraq spurred a years-long controversy, which damaged Blair's approval ratings and posed tragic consequences. In addition to the loss of life in Iraq, British weapons inspector David Kelly took his own life after publicly accusing the government of overstating the need for war. As in the United States, Britain took steps to tighten controls on its intelligence-gathering to prevent errors in judgment.

Despite the fact that no WMD were found in Iraq, the White House stood firm on the decision to invade Iraq. On September 11, 2004, the Office of the Press Secretary released a fact sheet titled, "Three Years of Progress in the War on Terror." The document stated in part, "We were right to go into Iraq. We removed a declared enemy of America, who had defied the international community for 12 years, and who had the capability of producing weapons of mass murder, and could have passed that capability to the terrorists bent on acquiring them. Although we have not found stockpiles of weapons of mass destruction, in the world after September 11th, that was a risk we could not afford to take."

When did **Operation Iraqi Freedom begin**?

The U.S.-led multinational military campaign in Iraq began on March 19, 2003, with air strikes on the capital of Baghdad; ground forces moved into southern Iraq from neighboring Kuwait. After taking the southern city of Basra, U.S. marines and army infantry moved northward, toward the capital. U.S. troops took control of Baghdad on April 9, after which images of gleeful Iraqis dismantling statues and other symbols of Saddam Hussein's despotic rule flooded the American media. Coalition forces, American troops, and U.S.-backed Kurdish fighters then pressed into northern Iraq, including Tikrit, Saddam's hometown and a loyalist stronghold. On April 14 Tikrit fell. The war seemed to be near conclusion, but the hard combat had only begun—and continued for years, even after the end of major combat was declared.

When did the major combat phase of **Operation Iraqi Freedom end**?

President George W. Bush (1946–) declared an end to major combat on May 1, 2003. But the stabilization of Iraq was far from over; the fighting continued more than two years later, the result of an increasingly violent Iraqi insurgency. Faced with the ongoing resistance, in December 2004 the number of U.S. troops in the war-torn nation was increased from 130,000 to 150,000.

Most of the casualties occurred after the declared end of major combat: On April 8, 2005, the Pentagon reported that there had been 1,543 American fatalities in the war to date—1,174 in hostile actions and 369 in nonhostile actions, including accidents during routine maneuvers. Of the 1,543 U.S. military deaths, 1,404 died after the declared end of major combat, 1,065 of them from hostile action. More than 7,000 had been injured to date.

In addition to the American fatalities, the British military had reported 86 deaths as of early April 2005; Italy, 21; Ukraine, 18; Poland, 17; Spain, 11; Bulgaria, 8; Slovakia, 3; Estonia, Thailand, and the Netherlands, 2 each; and Denmark, El Salvador, Hungary, Kazakhstan, and Latvia, 1 each.

The figures fueled criticism for the lingering war, with some observers wondering if stabilization was possible in the fractious nation. There were several factors contributing to the growing lists of casualties and injuries: coalition forces were frequent-

A 30-foot statue of former Iraqi leader Saddam Hussein is toppled by plastic explosives in Tikrit, Iraq, 2003.

ly ambushed in attacks from resistance fighters and suicide bombers; U.S. troops faced continued combat in parts of Baghdad and its outskirts; the southern towns of Najaf and Kufa were holdouts of resistance; and there was intense fighting in the Sunni cities of Fallujah, Ramadi, and Samarra, which remained under insurgent control even after the transfer of political authority from the United States to the interim Iraqi authority on June 28, 2004.

When was **Saddam Hussein captured**?

The former Iraqi leader, known for his cruelty, was caught on December 14, 2003, eight months after the fall of Baghdad. Hussein (1937–) was found in Ad Dawr, about nine miles from his hometown of Tikrit. He was said to be hiding "like a rat," in a hole across the Tigris River from one of his palaces. His 6-to-8-foot bunker was equipped with a basic ventilation system and was camouflaged with bricks and dirt. A disheveled Hussein had in his possession about $750,000 as well as arms, which he did not use. The former dictator was taken into custody in what U.S. defense secretary Donald Rumsfeld called a "surprisingly peaceful manner." Despite that, Hussein reportedly remained defiant and unrepentant: When he was asked about the thousands of people killed and dumped into mass graves during his regime, he dismissed his victims as "thieves."

The news of his capture prompted jubilation in Baghdad and across the nation: Crowds of Iraqis flooded into the streets to celebrate the end of his brutal rule. But his hometown of Tikrit, considered a loyalist stronghold, remained quiet. The news of his capture was welcomed by leaders around the globe. In a short televised address from the White House, President Bush remarked that Hussein would "face the justice he denied to millions." Bush also reassured the Iraqi people that they would "not have to fear the rule of Saddam Hussein ever again." Hussein was held for questioning in Iraq. Upon the June 28, 2004, transfer of authority, Iraq was given legal custody of the former ruler, who became a criminal defendant instead of a prisoner of war.

Why did some observers draw **comparisons between the war in Iraq and the Vietnam War**?

Critics of the U.S.-led war in Iraq found several similarities to American military involvement in Vietnam. Both conflicts seemed to many people to be without ample cause. Critics of the U.S.-led invasion of Iraq charged that there was no compelling reason for Americans to go to war in the Middle Eastern nation; they called the operation hawkish, saying it was an unwarranted expansion of the War on Terror. These arguments were strengthened when no weapons of mass destruction were found in Iraq. Media images of protests against the invasion of Iraq harkened back to the antiwar demonstrations of the Vietnam era. And, like Vietnam, there seemed to be no clear exit strategy for the U.S. military from Iraq—these were extended operations, or "wars without end." Some observers also drew comparisons between Vietnam's My Lai Massacre and the Abu Ghraib prison scandal in Iraq: They pointed to military spokespersons who categorized the atrocities as isolated events, unrepresentative of American military policy.

The Bush administration stood by the decision to enter Iraq, ousting and capturing Saddam Hussein (1937–), saying it was a rogue state led by a despotic ruler who had the capability of aiding terrorist groups. The controversy over Iraq continued long after President George W. Bush (1946–) declared the end to major combat operations, as the casualty count climbed—the result of continued coalition fighting against pockets of resistance and attacks on U.S. installments in the unstable nation. In 2005, as U.S. officials slowly handed over security to Iraqi forces, there was not yet a timetable for U.S. withdrawal. But the newly elected interim president, Kurdish leader Jalal Talabani, vowed that his government would work to provide security so that U.S.-led coalition forces could return home.

What was the **Abu Ghraib** prison scandal?

News of the atrocities of Abu Ghraib, where Iraqi prisoners were being held by members of the U.S. military during Operation Iraqi Freedom, first surfaced in April 2004 when shocking photographs began appearing in the American media. The photos depicted an array of hideous abuses, all of them in violation of the Geneva Conventions for the treatment of prisoners of war.

The deep troubles first surfaced in spring 2003 when International Red Cross and human rights groups complained that American troops had been mistreating Iraqi prisoners. The U.S. Army launched an investigation into its prison system; a fact-finding mission was eventually led by Major-General Antonio Taguba, who completed his report in February 2004. It stated that one police company at Abu Ghraib had committed "sadistic, blatant, and wanton criminal abuses." The Taguba report also found that "supervisory omission was rampant" at the prison. More investigations and reports followed, taking into account other prisons under Central Command. Some of the reports tied prisoner deaths to the abuses.

The scandal sent shockwaves around the world and sparked an intense debate about how such abuses could have happened. The tortures were labeled by some as "unauthorized actions taken by a few individuals." But others believed they were a manifestation of policy that had gone wrong. The scandal soon widened: By May 2004 British newspapers began reporting abuses by British troops. While Prime Minister Tony Blair (1953–) apologized for any mistreatment Iraqi prisoners may have suffered at the hands of British troops, President Bush went on Arabic television to denounce the abuses as "abhorrent"; he stopped short of an apology, saying that the mistreatment "does not represent the America that I know." Both nations moved quickly to bring charges against several soldiers shown in the pictures. The guilty verdicts, and subsequent sentences, began being handed down in May 2004.

The prison scandals fueled criticism of the controversial war in Iraq. They also prompted a deadly backlash from the Arab world: Insurgents in Iraq began a series of kidnappings of Americans and British citizens working in Iraq, most of them ending in ghastly and widely publicized killings. Between spring 2004 and spring 2005 more than 200 foreigners were taken captive in Iraq; more than 30 of them were killed by their kidnappers.

GOVERNMENT AND POLITICS

What was the **first national government**?

It is believed to have been that of the first Egyptian king, Menes, who united Upper and Lower Egypt in 3110 B.C. and founded a central government at Memphis (near present-day Cairo). Ruling for 62 years, Menes established the first of what would eventually number 30 dynasties that ruled ancient Egypt for nearly 3,000 years—until 332 B.C.

By the time the Third Dynasty began around 2700 B.C., the central government was well established and strong—subjects believed their kings and queens to be half-human and half-god. The pharaohs lived in magnificent luxury: Palaces and temples were built for them and were filled with exotic goods from other lands. These treasures were even buried with the pharaohs in order to be enjoyed in the afterworld. It was during the Third Dynasty that the 500-year period known as the Old Kingdom or the Pyramid Age began; it would become the period that saw the building of gigantic pyramids for Egypt's kings.

EMPIRES, ROYALTY, AND DYNASTIES

Was **King Tut** the greatest ruler of ancient Egypt?

No, in fact, King Tut's reign was relatively unimportant in the vast history of ancient Egypt. A ruler of the Eighteenth Dynasty, Tutankhamen (c. 1370–1352 B.C.) was in power from age nine (1361 B.C.) until his death at the age of 18—a nine-year period that would be of little significance were it not for the November 1922 discovery of his tomb in the Valley of the Kings near ancient Thebes (present-day Luxor). Of the 27 pharaohs buried near Thebes, only the tomb of the minor king, Tutankhamen, was spared looting through the ages. Having not been opened since ancient times, the tomb still contained its treasures.

In the antechamber English archeologist Howard Carter (1873–1939) found more than 600 artifacts, including funerary bouquets, sandals, robes, cups and jars, a painted casket, life-size wooden statues of Tutankhamen, animal-sided couches, remnants of chariots, and a golden throne. In the burial chambers a team of archaeologists discovered four golden shrines and the golden coffin containing the royal mummy of Tutankhamen—complete with a golden mask covering his head and shoulders.

Earlier in his career, Carter had discovered the tombs of King Tut's predecessors, Queen Hatshepsut (c. 1520–c. 1468 B.C.) and King Thutmose IV (d. 1417 B.C.), both of whom were also rulers during Egypt's Eighteenth Dynasty.

The Mayan Empire

Was the **Mayan Empire** the most advanced early civilization?

In some regards, the Mayas were more advanced than other civilizations. Their development preceded that of the other agrarian civilizations in North and South America, principally the Aztec and the Inca.

The Mayas were an agricultural people who in about 1000 B.C. settled in southern Mexico and Central America. Their territory covered Mexico's Yucatan Peninsula, Belize, much of Guatemala, and parts of Honduras and El Salvador. They developed a civilization that was highly advanced: Not only did the Mayas produce remarkable architecture (including flat-topped pyramids, temples, and towers that are still visited by tourists today) and art (including sculpture, painting, and murals), but they developed their own writing system—probably the first in the Western Hemisphere. They used this system to record time, astronomical events, their history, and religion (they believed in more than 160 gods). They also developed an advanced mathematics as well as a 365-day calendar believed by some to be even more accurate than the Gregorian calendar in use today.

At its peak, the Mayan population numbered some 14 million. Their history is divided into three periods. The Pre-Classic period began about the time they originated (roughly 1000 B.C.) and extended into A.D. 300; this was the group's formative period. During the Classic period, 300 to 900, Mayan culture spread throughout the area and city-centers were developed at Copán (Honduras), Palenque, Uxmal, and Chichén Itzá (Mexico), and Piedras Negras, Uaxactún, and Tikal (Guatemala). Scholars believe that Tikal was home to some 50,000 people and was not only a center for government, education, economics, and science, but was also a spiritual mecca for the Maya.

It was in the second half of the Classic period that the Maya made their greatest accomplishments in art and science: Europe would not produce a superior system of mathematics for centuries to come. During the Post-Classic period (900–1546), they were invaded by the Toltecs. However, the Maya absorbed these people rather than being conquered by them. Nevertheless, by the time the Spaniards arrived in the mid-1500s, the Mayan civilization was in decline. Some historians attribute this to widespread famine or disease while others believe the decline was due to a rebellion of the people against the harsh government. Though they were conquered by the Spaniards and became assimilated into the larger culture that developed in the region, Maya Indians still survive in Mexico and Central America today.

Aztec Empire

How did the **Aztec and Inca Empires** compare with the Mayan?

While all were advanced civilizations that were eventually conquered by Spaniards, the Inca and Aztec cultures reached their peaks in the fifteenth century—just before the arrival of the Europeans in the New World. The Mayan civilization reached its zenith about 500 years earlier and was already in decline by the time of European incursion. Each group also occupied a different region of the Americas, where each carved out its own stronghold and flourished: The Aztecs settled in central Mexico, the Incas in western South America (primarily Peru), and the Mayas in the Yucatan Peninsula and Central America. There is evidence that they traded with each other as well as with American Indians to the north.

The Aztecs founded their central city of Tenochtitlán (the site of Mexico City) about 1325. A poor nomadic people before their arrival in Mexico's central region, the Aztecs believed the Lake Texcoco marsh was a prophetic place to settle. Before they built it into a great city, they first had to fill in the swampy area, which they did by creating artificial islands. In the 1500s, when the Spanish first saw the remarkable city—with its system of causeways, canals, bridges, and aqueducts—they called it the Venice of the New World. In addition to constructing the impressive trade and cultural center of Tenochtitán, the Aztecs were farmers, astronomers, mathematicians, and historians who recorded the events of their civilization. Their religion was pantheistic, meaning they worshiped many gods. Given that, it's not surprising that when the Spanish conquistadors arrived, at first the Maya believed they were gods (or at least, the heavenly hosts of their long-awaited god Quetzalcoatl), and even welcomed them with gifts. Later, the Aztec rose up against the Europeans, but under the leadership of Hernán Cortés (1485–1547), the Spaniards conquered the group, claiming Mexico in August 1521.

The Incas developed one of the most extensive empires in all the Americas. During the hundred years before the arrival of the Europeans, the Incas expanded their territory along the western coast of South America to include parts of present-day Peru, Ecuador, Colombia, Bolivia, Chile, and Argentina. Though it was a vast region, it was nevertheless a closely knit state ruled by a powerful emperor. The government was subdivided down to the local level, but because the emperor required total obedience from his subjects, local rulers were kept in check.

Like the Aztecs in Mexico, the Inca developed an infrastructure that included a network of roads, bridges, and ferries as well as irrigation systems. They, too, built impressive edifices, demonstrating their abilities as engineers. The magnificent city of Machu Picchu was modeled in clay before construction began. The Inca were also skilled craftspeople, working with gold, silver, and textiles. Like the Aztecs, the Incas worshiped many gods. And when the Spanish explorer Francisco Pizarro (c. 1475–1541) arrived in the region in 1532, he was welcomed as a god at first. However, by 1537 the Inca were brought under Spanish control.

The Roman Republic

Why was **Julius Caesar** murdered?

The Roman general and statesman Julius Caesar (100–44 B.C.) was stabbed to death in the senate house by a group of men, including some of his former friends, who viewed him as an ambitious tyrant and a threat to the Roman Republic. The date of the assassination, March 15, fell into relatively common usage thanks to William Shakespeare's tragedy, *Julius Caesar* (written in the late 1590s), which has a soothsayer warning the Roman general to "Beware the ides of March." After Caesar's death in 44 B.C., a triumvirate was formed to rule Rome, with Lepidus, Octavian (who would in 27 B.C. become Augustus, the Roman Empire's first ruler), and Marc Antony sharing power. It was Marc Antony (c. 83–30 B.C.), of "Antony and Cleopatra" fame, who aroused the mobs against Caesar's conspirators, driving them out of Rome.

The events illustrate the controversy about Julius Caesar: While some clearly viewed him as a demagogue who forced his way into power, others considered the patrician-born Caesar a man of noble character who defended the rights of the people in an oligarchic state—where the government was controlled by a few people who had only their own interests in mind. This divided opinion has followed Caesar throughout history.

While opinion is still divided on what kind of a ruler Caesar was, there can be no denying his contributions—both to Rome (which would soon emerge as the Roman Empire) and to modern civilization. In his battles, Caesar brought the provinces of Italy under control and defeated his former co-ruler, Pompey the Great (who had, along with Caesar and Crassus, formed the first triumvirate), effectively ending the oligarchy that had ruled Rome. In so doing, he had succeeded in ending the disorder that had plagued Rome for decades and laid the groundwork for the formation of the empire under his grandnephew Augustus in 27 B.C. While Caesar was in office, he planned and carried out several reforms, not the least of which was the implementation of the Julian calendar, which he introduced in 46 B.C. The Gregorian calendar we use today evolved from it.

Caesar also left a legacy of literature: He penned a total of 10 books on his battles in Gaul (c. 58–50 B.C.) and on the civil war, which he had more or less started in 49 B.C. These clear commentaries are still considered masterpieces of military history.

THE ROMAN EMPIRE

Who were the most important rulers of the **Roman Empire**?

The 500 years of the Roman Empire (27 B.C. to A.D. 476) gave history some of its most noteworthy—and most diabolical—leaders. The major emperors are names that are familiar to most every student of Western civilization: Augustus, Tiberius, Caligula, Claudius, Nero, Trajan, Marcus Aurelius, Diocletian, and Constantine I (called "the Great").

Augustus, Tiberius, Caligula, Claudius, and Nero were the first five emperors, a succession covering 75 years of Roman rule. Octavian (63 B.C.–A.D. 14), later known as Augustus, became Roman emperor when, after the assassination of his great-uncle, Julius Caesar, a power struggle ensued and he defeated Mark Antony and Cleopatra to take the throne. Under Augustus's rule from 27 B.C. to A.D. 14 began the 200 years of the Pax Romana, a period of relative peace. During this time no power emerged that was strong enough to sustain conflict with the Roman army. Consequently, Rome was able to turn its attention to the arts, literature, education, and trade.

As second emperor of Rome, Tiberius (42 B.C.–A.D. 37) came under the influence of Roman politician and conspirator Sejanus (d. A.D. 31). Tiberius was the adopted son of Emperor Augustus, and though he had been carefully schooled and groomed to take on the leadership role, ultimately he became a tyrannical ruler; the final years of his reign were marked by viciousness and cruelty. Upon Tiberius's death, his nephew Caligula (A.D. 12–41) ascended the throne. Born Gaius Caesar, he was nicknamed Caligula, meaning "Little Boots," since he was brought up in military camps and at an early age had been dressed as a soldier. For a short time Caligula ruled with moderation. But not long after he came to power, he fell ill and thereafter exhibited the erratic behavior for which he is well known. Most scholars agree that Caligula must have been crazy. He was murdered in A.D. 41, and Claudius, also nephew to Tiberius, was then proclaimed emperor.

Claudius (10 B.C.–A.D. 54) renewed the expansion of Rome, waging battle with Germany, Syria, and Mauritania (present-day Algeria and Morocco), and conquering half of Britain. Though his administration was reportedly well run, he had his enemies; among them was his niece, Agrippina the Younger (A.D. 15?–59), who is believed to have murdered him in 54, after securing her son, Nero, as successor to the throne.

In Nero (A.D. 37–68) the early Roman Empire had perhaps its most despotic ruler: Though his early years in power were marked by the efficient conduct of public affairs, in 59 he had his mother assassinated (she reportedly had tried to rule through her son), and Nero's legacy from that point forward is one of ruthless behavior. He was involved in murder plots, ordered the deaths of many Romans, instituted the persecution of Christians, and led an extravagant lifestyle that emptied the public coffers. He was declared a public enemy by the Roman Senate and in the year 68 took his own life.

With the exception of Augustus, the first century A.D. of the Roman Empire was marked by extreme rulers. The second century A.D. was marked by the leadership of soldiers and statesmen. Trajan (53–117), who ruled from the year 98 until his death 19 years later, is best known for his military campaigns, which expanded Rome's territory. He was also a builder—constructing bridges, roads, and many buildings. When Marcus Aurelius (121–80) ascended to emperor in 161, he had already been in public office for more than 20 years. A man of great experience, he was reportedly both learned and of gentle character. His generals put down revolting tribes, and, in addition to winning victories along the Danube River, his troops also fought barbarians in the north. Diocletian (245?–313?) had served as an army commander before becoming emperor in 284. In an effort to effectively rule the expansive territory, he divided it into four regions, each with its own ruler, though he himself remained the acknowledged chief. Two years before he abdicated the throne (305), he began the persecution of Christians—a surprising move since he had long been friendly toward them. Unlike his predecessors who died in office, Diocletian had a retirement, which he reportedly spent gardening.

Constantine the Great (who ruled from 306 until his death in 337) is notable for reuniting the regions that Diocletian had created, bringing them all under his rule by 324. He was also the first Roman emperor to convert to Christianity. Theodosius I (347–395), also called "the Great," is known to many since he was the last to rule the united Roman Empire (379–95).

THE BYZANTINE EMPIRE

Why was **Constantine I** called "the Great"?

Roman emperor Constantine the Great (c. 275–337) is credited with no less than beginning a new era in history. His father, Constantius, was ruler of the Roman Empire when he died in 306. Though Constantine was named emperor by Roman soldiers, a power struggle ensued. During a battle near Rome in 312, Constantine, who had always been sympathetic toward Christians, reportedly saw a vision of a flaming cross. He emerged from the conflict both converted and victorious. For the next 12 years, Constantine ruled the West Roman Empire while Licinius (also tolerant to Christians) ruled the East. But a struggle between the two emperors ended in death for Licinius and, beginning in 325, Constantine ruled as sole emperor.

During Constantine's reign, Christians regained freedom of worship and the Christian Church became legal. In 325 he convened the Council of Nicaea (from whence came the Nicene Creed so familiar to Christians today). In moving the capital of the Roman Empire to Byzantium (in 330), Constantine shifted the empire's focus from west to east and in so doing laid the foundation for the Byzantine Empire. The Eastern Orthodox Church regards Constantine as a saint.

How was the **Byzantine Empire formed**?

The Byzantine Empire was a continuation of the Roman Empire—its citizens even called themselves Romans. Two dates are given for the formation of the Byzantine Empire, which, though boundaries shifted constantly, was centered in Asia Minor and the Balkan Peninsula: In A.D. 395, upon the death of emperor Theodosius the Great, the Roman Empire was divided into two: East and West. In the years that followed, the West Roman Empire was subject to repeated attacks from nomadic barbarian groups, and Rome finally fell in 476. The East Roman Empire survived as the Byzantine Empire, which, after the fall of Rome, laid claim to much of the lands in the west.

However, many historians date the beginning of the Byzantine Empire earlier—at A.D. 330, when Roman emperor Constantine the Great moved the capital of the then united Roman Empire from Rome to Byzantium (present-day Istanbul, Turkey—subsequently known as Constantinople). By this definition of the empire, Constantine the Great was its first ruler. He was succeeded by nearly 100 rulers over the course of more than 1,000 years of Byzantine rule. At its height, during the sixth century reign of Justinian I from 483 to 565, the empire included parts of southern and eastern Europe, northern Africa, and the Middle East. The Byzantine Empire ended when the Ottoman Turks conquered Constantinople in 1453.

What is the **Byzantine Empire's role** in history?

The Byzantine Empire (330 or 395–1453) is considered a link between ancient and modern civilizations. Though the empire was constantly fighting off invaders, plagued by religious controversies, and marred by political strife, as the heir of the Roman world it allowed for the customs of Greco-Roman civilization to mix with those of the East and with Christianity. The fall of the Byzantine Empire in 1453 typically marks the beginning of the modern era.

The Tang Dynasty

What was the **Tang dynasty**?

The Tang (617–907) was the sixth-to-last Chinese dynasty. It's well known since the period saw great achievements not only in government and business, but in letters and the arts—principally lyrical poetry, formal prose, painting, sculpture, and porcelain pottery. The first published book, the *Diamond Sutra,* was produced during this time (in 868). Considered a golden age of Chinese civilization, the Tang was also an age of great expansion. At its height, the empire stretched from Turkmenistan in the west to Korea (which was a vassal state) in the east, and from Manchuria to northern India. As a result, trade prospered, with Chinese jade, porcelain, silk, rice, spices, and teas exported to India, the Middle East, and Europe. One historian called the Tang "the consummate Chinese dynasty...formidable, influential, and innovative."

One of the Tang's innovations was the balance of administrative power: government was separated into three main branches: the Imperial Secretariat (which organized the emperor's directives into policies), the Imperial Chancellery (which reviewed the policies and monitored bureaucracy), and the Department of States Affairs (which carried out the policies through the administration of six ministries). Add to this triumvirate a Board of Censors, which ensured that corruption was kept to a minimum. This form of government outlasted the Tang dynasty: Subsequent monarchies perpetuated the system into the twentieth century.

Yet another example of the forward thinking of the Tang dynasty was a civil service. Candidates for public service were trained in the Confucian principles before they took an exam that would qualify them for official duty.

The Ming Dynasty

What were the highlights of the **Ming dynasty**?

The focus on Chinese culture that was the hallmark of the Ming dynasty (1368–1644) was both its strength and its weakness. After the foreign Mongols (whose dynasty had been established by Kublai Khan in 1260) were overthrown as rulers of China in 1368, the Ming emperors returned their—and their subjects'—attention to those things that are distinctively Chinese. The focus on Chinese culture produced a flowering in the arts, evidenced by the name *Ming* itself, meaning "bright" or "brilliant." Architects working during this period produced the splendor of Beijing's Forbidden City. Ming porcelain, bronze, and lacquerware are coveted collectors' items today. Additionally, the novel and drama flourished.

And though the Ming rulers promoted this artistic renaissance and reinstated Confucianism and the program of civil service suspended by the Mongols, the rulers' myopia prevented them from seeing the threat of the nomadic Manchu people on the horizon. In 1644 the Manchus invaded from the north and conquered China, setting up the last dynastic period in Chinese history (it lasted until 1912). Nevertheless, it was the Ming, and not the Manchu, who formed the last great and truly Chinese dynasty.

CAROLINGIAN EMPIRE

What was the **Carolingian Empire**?

The empire, which united most of western Europe under a single leader from about 734 until 987, was named for the Carolingians, a family of Frankish kings.

After the decline of the Roman Empire (in 476), various Germanic tribes (including the Goths, the Vandals, the Franks, and the Anglo-Saxons) dominated western Europe—fighting each other as well as the advancing Muslims to protect and expand their territories. In 719 Charles Martel (c. 688–741) united the lands of all the Franks under his rule. (The Franks were the descendants of Germanic tribes who settled in the Rhine River region of western Europe.) He then went on to protect France from Arab incursion and campaigned against the Burgundians and the Frisians, eventually bringing them under his control. Upon his death in 741, Charles Martel was succeeded by his son, Pepin III (known better perhaps as Pepin the Short; c. 714–768). It was Pepin who established the Carolingian Empire and brought the Lombards into the empire. Upon his death, he was succeeded by his son Charlemagne, or Charles the Great (742–814).

Why is **Charlemagne** so well known?

Charlemagne's popularity with history students is due not only to the ruler's great accomplishments during his lifetime, but also to the fact that these accomplishments were documented: His biography, titled *Vita Caroli Magni* (*The Life of Charlemagne*) was written by a fellow named Einhard (c. 770–840), who was his adviser.

Charlemagne (742–814), or Charles the Great, became king of the western Franks when his father, Pepin the Short (c. 714–768) died in 768. Upon the death of his brother, Carloman, in 771, Charles became king of all the Franks. He then went on to conquer much of western Europe, including Saxony, Lombardy, northeastern Spain, and Bavaria. He sometimes employed brutal tactics in bringing people and regions under control: During the last two decades of the eighth century, he used mass executions to subdue Saxon rebellions. Charlemagne succeeded in uniting all of these areas under one empire, and on Christmas Day 800, he had Pope Leo III (c. 750–816) crown him Emperor of the West, thus initiating the Holy Roman Empire. As a patron of the arts, literature, and science, Charlemagne revived western Europe, which had been in decline since the fall of the West Roman Empire (in 476). He is credited with laying the foundation for the Holy Roman Empire and the European civilization that developed later in the Middle Ages. He ruled until his death in 814.

THE OTTOMAN EMPIRE

What was the **Ottoman Empire**?

It was a vast Turkish state founded in the thirteenth century by the Osmani Turks, Turks who were led by descendants of Osman I (1258–c. 1326). By the middle of the next century, the Ottoman Empire consisted roughly of modern-day Turkey (the terms "Turkey" and "Ottoman Empire" are used interchangeably). The empire was expanded further by conquests during the 1400s, including the conquest of the Byzantine Empire in 1453. At its height, the Byzantine Empire extended over an area that included the Balkan Peninsula (present-day Slovenia, Croatia, Bosnia and Herzegovina, Macedonia, Yugoslavia, Romania, Bulgaria, Albania, Greece, and Turkey), Syria, Egypt, Iraq, the northern coast of Africa, Palestine, and parts of Arabia, Russia, and Hungary. The capital was placed at Constantinople (present-day Istanbul, Turkey). Thus the Turks established a Muslim empire that would remain a formidable force and influence in the region and in Europe for the next three centuries.

During the 1500s and 1600s the Ottoman Empire was the most powerful in the world. It reached its most glorious heights during the reign of Süleyman the Magnificent (1494–1566), who ruled from 1520 to 1566: It was he who added parts of Hungary to the Ottoman territory. He also tried to take Vienna, but failed. He did succeed in strengthening the Ottoman navy, which dominated the Mediterranean Sea. Süleyman was not only an expansionist, but a patron of the arts and a builder. He ordered the construction of mosques (to spread the Islamic religion throughout the empire), bridges, and other public works.

But by the time World War I began in 1914, the Ottoman Empire had been in decline for some 300 years and only consisted of Asia Minor, parts of southwestern Asia, and part of the Balkan Peninsula. As one of the losing Central powers, the Ottoman Empire was dissolved in 1922 by the peace treaties that ended the war.

THE HOLY ROMAN EMPIRE

Who were the **Habsburgs**?

The Habsburgs were Europe's most powerful royal family. Even if one chose to argue the point of power, there can be no arguing the longevity of the house of Habsburg: They supplied Europe with a nearly uninterrupted stream of rulers for more than 600 years.

Also spelled Hapsburg (which is closer to the pronunciation, HAPS-berg), the name came from the castle of Habichtsburg (meaning "Hawk's Castle"), built during the early eleventh century in Switzerland. The first member of the family to bear the name was Count Werner I (who died in 1096). It was Werner's descendant, Rudolf I (1218–1291), who was elected king of Germany and the Holy Roman Empire in 1273. When Rudolf conquered Austria three years later, he established that country as the family's new home. Austria, Bohemia, Germany, Hungary, and Spain were among the European states ruled by the house of Habsburg. With only one exception, the Habsburg family also ruled the Holy Roman Empire from 1438 (when Albert II was elected) until 1806.

It was the reign of Emperor Charles V during the sixteenth century that the Habsburg influence reached its high-water mark. When in 1496 Spain's Philip I (called Philip the Handsome; 1478–1506) married Joan of Castile (1479–1555), it assured that their son Charles V (1500–1558) would inherit the crown of Spain, which he did in 1516. (Charles was grandson to Spain's King Ferdinand and Queen Isabella.) He also inherited the rest of what by then was a vast empire, and he ruled as Holy Roman Emperor from 1519 until 1556.

Charles V is considered the greatest of all the Habsburgs, though he did face problems. Chief among these were the Protestant Reformation; opposition from his lifelong rival, Francis I (1494–1547), king of France; and the Ottoman Turks, who were at the height of their power during his reign. Nevertheless, he was a successful ruler and his accomplishments included Spain's conquest of lands in the New World—Mexico (at the hands of Hernán Cortés, 1485–1547) and Peru (by Francisco Pizarro, c. 1475–1541).

In 1867 the Habsburg empire was reorganized as the Austro-Hungarian monarchy. That monarchy was dissolved in 1918, after World War I, with the Treaty of Versailles establishing new boundaries for the successor states.

Ferdinand and Isabella

Why were Spain's **King Ferdinand and Queen Isabella** so powerful?

The 1469 marriage of Ferdinand (1452–1516) and Isabella (1451–1504) brought previously separate Spanish kingdoms (Aragon and Castile) under their joint control. Together the monarchs went on to rule Spain and expand their realm of influence until Isabella's death in 1504. (Ferdinand ruled without his wife thereafter.) Theirs was a reign that seemed to have religion on its side: In 1496 Pope Alexander VI conferred upon each of them the title "Catholic," as in, "Ferdinand the Catholic" and "Isabella the Catholic." And for good reason, because the king's and queen's most well-known acts seemed to have been motivated by their beliefs.

It was Ferdinand and Isabella who in 1478 established the infamous Spanish Inquisition, a court that imprisoned or killed Catholics who were suspected of not fol-

Christopher Columbus, after returning from his first voyage to America, kneels before Spain's King Ferdinand and Queen Isabella. (Original lithograph by George Schlegel.)

lowing religious teachings. While the Inquisition was aimed at discovering and punishing Muslims and Jews who had converted to Catholicism but who were thought to be insincere, soon all Spaniards came to fear its power. In 1482 the monarchs undertook a war with the (Muslim) Moors, conquering the last Moorish stronghold at Granada in 1492, and forcing them back to Africa after four centuries of occupation—and influence—in the Iberian Peninsula (Spain and Portugal). The recovery of Iberia had been motivated by religion; when the king and queen expelled the Moors, they also believed they were expelling Islam from their kingdom.

That year, 1492, was a fateful one for the Spanish: Not only were the Moors driven out, but Ferdinand and Isabella also turned their attention to the Jewish "threat," expelling them, too. (Those who remained went underground with their faith; those Iberian Jews who migrated spread their division of Judaism, called Sephardim, to North Africa and the Middle East.)

Most students of history know 1492 best as the year that explorer Christopher Columbus (1451–1506) sailed to the New World. It was Ferdinand and Isabella who sponsored his voyage, believing that the conquered lands would not only add to their authority but would provide new territory for the spread of Catholicism. The Spaniards soon emerged as a formidable sea power in the Atlantic.

For all their fervor, Isabella and Ferdinand were also interested in education and the arts, and they sponsored advances in both areas during their reign. Their legacy included their grandson Charles V (1500–1558), who, through marriage, became Holy Roman Emperor and ruled from 1519 to 1558 as one of the all-powerful Habsburgs.

The Romanov Dynasty

Who were the **Romanovs**?

The Romanov family ruled Russia from 1613 until 1917, when Nicholas II (1868–1918) was overthrown by the Russian Revolution (1905–17). The dynasty was established by Michael Romanov, grandnephew to Ivan the Terrible, who ruled from 1533 to 1584. There were 18 Romanov rulers, including the much-studied Peter the Great, who ruled from 1682 to 1725, and Catherine the Great, who ruled from 1762 to 1796.

As the last tsar of Russia, Nicholas II, who ruled from 1894 to 1917, likely suffered not only the recrimination that was due him, but the public hostility that had accumulated over centuries of ruthless Romanov leadership. Nicholas's difficulties came to a head when he got Russia involved in World War I (1914–18), which produced serious hardships for the Russian people and for which there was little public support. Once Tsar Nicholas was overthrown (and later killed) in the Russian Revolution (1905–17), Bolshevik leader Vladimir Lenin (1870–1924) set about extracting Russia from the conflict by agreeing to sever concessions to Germany. Oddly enough, the Romanov family had, in the fourteenth century, originated with a German nobleman, Andrew Kobyla, who had emigrated to Russia.

Why were **tsars Peter and Catherine** known as "the Great"?

The epithet "the Great" can be misleading: While Romanov tsars Peter the Great, who ruled from 1682 to 1725, and Catherine the Great, who ruled from 1762 to 1796, are among the best-known of the Romanov dynasty and both had many accomplishments during their reigns, they are also known for having increased their power at the expense of others.

Peter is recognized for introducing western European civilization to Russia and for elevating Russia to the status of a great European power. But he also relied on the serfs (the peasants who were little more than indentured servants to the lords) not only to provide the bulk of the funding he needed to fight almost continuous wars, but for the manpower as well: most soldiers were serfs. The man responsible for establishing schools (including the Academy of Sciences), reforming the calendar, and simplifying the alphabet also carried out ruthless reforms. Peter's most vain-glorious act was, perhaps, to move the capital from Moscow to the city he had built for himself on the swampy lands ceded by Sweden: St. Petersburg (known as Petrograd 1914–24, as Leningrad 1924–91). As his "window on Europe," Peter succeeded in making the city into a brilliant cultural center.

For her part, Catherine the Great may well be acknowledged as a patron of the arts and literature (one who corresponded with the likes of French writer Voltaire, 1694–1778), but she, too, increased the privileges of the nobility while making the lives of the serfs even more miserable. Her true colors were shown by how she ascended to power in the first place. In 1744 she married Peter (III), who became tsar of Russia in 1762. That same year, Catherine conspired with her husband's enemies to depose him. He was later killed. And so Catherine came to power, proclaiming herself tsarina. She began her reign by attempting reforms, but a peasant uprising (1773–74) and the French Revolution (which began in 1789) prompted her only to strengthen and protect her absolute authority. Like Peter the Great, she, too, extended the frontiers of the empire through a series of conquests. By the end of her reign, in 1796, Catherine had reduced even the free peasants to the level of serfdom.

Rulers of Great Britain

What was the **Domesday Book**?

It is an important document surviving from the reign of England's William I (c. 1028–1087), a Norman who had conquered England in 1066 to become king. He ordered the survey so that he could have a complete record of England's lands, property owners, and resources. He used this information to his advantage, even taking possession of some properties thereafter. The census is considered an excellent record of Europe's Middle Ages (500–1350).

What are **England's royal "houses"**?

England's royal houses are simply families, including ancestors, descendants, and kin. Since 1066 England's rulers have come from a series of 10 royal houses: Normandy (ruled 1066–1135), Blois (1135–54), Plantagenet (1154–1399), Lancaster (1399–1471), York (1471–85), Tudor (1485–1603), Stuart (1603–49, restored 1660–1714), Hanover (1714–1901), Saxe-Coborg (1901–10), and Windsor (1910 to present).

Prior to the establishment of the House of Normandy, England had been ruled by Saxons and Danes since 802. The first king of the House of Normandy was William I (also known as William the Conqueror [c. 1028–1087]), who was the son of the French Duke of Normandy. William invaded England in 1066 on the death of Edward the Confessor and the ascension of Harold II (c. 1022–1066). Ousting Harold, William was coronated at Westminster Abbey on Christmas Day. William's grandson, Stephen, was all that consisted of the short-lived reign of the House of Blois (named such since Stephen was the Count of Blois and Chartres, though he was raised in the court of his uncle, King Henry I, whom he succeeded).

The House of Plantagenet, also called the House of Anjou, included the 10-year reign (1189–99) of Richard I, or Richard the Lionhearted (1157–1099), who fought his father, Henry II (1068–1135), and his brothers for control of the throne. Richard's military prowess made him the hero of romantic legends. Thereafter, two contending branches of the House of Plantagenet—the houses of Lancaster and York—vied for the crown in the infamous War of the Roses (1455–85). The struggle finally ended when Henry VII (a Lancaster) ascended the throne and married into the House of York, reuniting the two sides of the family under the newly minted House of Tudor.

The Tudors were a famous lot, remembered for the reigns of Henry VIII (1509–47) and his daughters, Mary I (1553 to 1558) and Elizabeth I (1558–1603). The Tudors were followed by the Stuarts, whose reign was interrupted by the establishment of the Commonwealth and Protectorate under Oliver Cromwell (1599–1658). The house was restored to power, giving history the eight-year period known as the Restoration, in 1660. It was King William III (1650–1702), a Stuart, and Queen Mary II (1662–1694), who began in 1689 to rule England in a more modern fashion—through Parliament.

A Stuart descendant, George I (1660–1727), established the House of Hanover, which originated in Germany. Queen Victoria (1819–1901), who presided over the Victorian Age (1837–1901), was of the House of Hanover. She was succeeded by her son, Edward VII (1841–1910), who established the House of Saxe-Coburg. Technically, this is the royal house still at the helm today—the name was changed to Windsor during World War I (1914–18).

The Tudors

How did the **House of Tudor** originate?

The name of a royal family that ruled England for well over a century, from 1485 to 1603, came from a Welshman, Owen Tudor, who sometime after 1422 married Catherine of Valois (1401–1437), the widow of Henry V, who ruled from 1413 to 1422. The family did not come to power until Henry VII ascended the throne. Ruling from 1485 until his death in 1509, Henry VII (1457–1509) ended the bitter, 30-year War of the Roses—during which two noble families, the houses of Lancaster and York, had struggled against each other for control of the throne. (The conflict earned its name since the badge of each house depicted a rose, one red and the other white.) In taking power, Henry VII became the head of the House of Lancaster, and in 1486 he married into the House of York, thus uniting the two former enemies and founding his own Tudor dynasty. (He's known as Henry Tudor.)

Because of papal conflicts and England's break with the Roman Catholic Church, the reign of Henry VIII is perhaps the most well-known Tudor monarchy. (Original painting by Hans Holbein II.)

Why is **Henry VIII** so famous?

The reign of Henry VIII, from 1509 to 1547, is perhaps the most well-known Tudor monarchy. It was marked by papal conflicts and England's subsequent break with the Roman Catholic Church. When Henry's wife, Catherine of Aragon (1485–1536), failed to produce a male heir, he appealed to the pope to grant him a divorce. The request was of course denied. Though Henry went on to have his marriage to Catherine declared invalid (on the grounds that she was his brother's widow) and he secretly married Anne Boleyn (c. 1507–1536) in 1533, his troubles with the church continued. In 1534 he set up the Church of England, declaring the monarch as its head. He went to extreme measures to ensure the act was upheld—even executing his appointed chancellor, Sir Thomas More (1478–1535), for his refusal to acknowledge royal supremacy.

Henry VIII was eventually successful in procuring a male heir to the throne—but it required a third marriage, to Jane Seymour (c. 1509–1537): His son, Edward VI (1537–53), succeeded him in 1547. Nevertheless, it was Henry VIII's daughters who went on to make history. Mary I, who ruled England and Ireland from 1553 to 1558, was the daughter of Henry and his first wife, Catherine of Aragon. In 1554 Mary wed Spain's Philip II (1527–1598), forming a temporary alliance between the two powers. The following year she realigned England with the Catholic Church, undertaking the persecution of Protestants and earning herself the name "Bloody Mary."

Why did **Queen Elizabeth I** have an entire age named after her?

Ascending the throne in 1558, Queen Elizabeth (1533–1603) remained in power until 1603, a 45-year period during which England dominated the seas to become a European power; colonization began, with Sir Walter Raleigh (1554–1618) and others establishing British settlements in North America; culture flourished, with the likes of William Shakespeare (1564–1616), Christopher Marlowe (1564–1593), and Sir Edmund Spenser (1552–1599) producing literary masterpieces; and industry and commerce boomed. This was the Elizabethan Age, one of England's high-water marks.

One of the most well-known events of Elizabeth's reign was England's defeat of the Spanish Armada in July–August 1588. This event has an interesting twist. The only child of Henry VIII (1491–1547) and his second wife, Anne Boleyn (c. 1507–1536), Elizabeth had succeeded her half-sister Mary Tudor (1516–58) as queen. In the summer of 1588, the "invincible" Spanish fleet, which was headed for invasion of England, had been dispatched by none other than Spain's Philip II (1527–1598), Mary Tudor's former husband. Though the English were severely outnumbered (having only 34 ships to Spain's 132), they were aided by weather and defeated the Spanish Armada on August 8. This victory at sea opened the world to English trade and colonization.

Queen Elizabeth was the last of the Tudor rulers, who are credited with strengthening the monarchy in England. They were succeeded by the House of Stuart in 1603, when James I (1566–1625), who ruled from 1603 to 1625, ascended the throne. It is that James who gave the world the King James version of the Bible. He is also notable for having been the son of Mary, Queen of Scots (1542–1587), whom Elizabeth I had reluctantly put to death.

HOUSE OF STUART

Who was **Mary, Queen of Scots**?

Mary Stuart was born in 1542 to James V of Scotland, who ruled from 1513 to 1542, and his wife, Mary of Guise. When Mary was just six days old, her father died, making the infant a queen. Her mother ruled the country as a regent until 1561, when Mary officially took on her duties. She was, by all reports, a beautiful and charming young woman whose courage and mettle would be tested by time. When she ascended the throne, she inherited her mother's struggle with the Protestants, who were led by John Knox (1513–1572), a former Catholic priest who was involved in the Reformation. As a Roman Catholic, Mary was subject to harsh verbal attacks issued by Knox, who denounced the pope's authority and the practices of the church. But this was not the worst of her troubles: In 1565 Mary wed her English cousin, Lord Darnley (1545–1567), in an attempt to secure her claim to the English throne as successor to Elizabeth I (1533–1603), also her cousin.

But Mary's ambitions would be her undoing. She quickly grew to dislike her husband, who became aligned with her Protestant opponents and successfully carried out a plot to murder—in her presence—Mary's adviser, David Rizzio (c. 1533–1566). Surprisingly, Mary and Darnley reconciled shortly thereafter (a politically savvy move on her part), and she conceived a child, James, who was born in 1566. Darnley had enemies of his own, and one year later he was murdered. Mary promptly married the Earl of Bothwell (1536–1578), with whom she had fallen in love well before becoming a widow. Bothwell was accused of Darnley's murder, and though he was acquitted, his marriage to the queen shocked Scotland. The people took up arms, forcing Mary to abdicate the throne in 1567. She was 25 years old.

Fleeing to England, Mary, Queen of Scots, was given refuge by Elizabeth I (1533–1603). Though she was technically a prisoner, Mary nevertheless was able to conspire with Elizabeth's enemies—including English Catholics and the Spanish—in attempts to kill her. When one such plot was discovered in 1586, Mary was charged for her involvement in it and was put on trial. Found guilty, she was put to death in 1587, though Elizabeth hesitated to take such action.

Meantime, Mary's son, James VI of Scotland (1566–1625), had taken on his responsibilities in 1583—after Scotland had been ruled on his behalf by regents since Mary's abdication some 16 years prior. He had promptly formed an alliance with Elizabeth I and, in 1587, accepted with resignation his mother's execution. In 1603 James succeeded Elizabeth, becoming James I, King of England, uniting Scotland and England under one throne. The union was made official about a century later (in 1707) when Parliament passed the Act of Union.

The Windsors

How did the **House of Windsor** originate?

The origins of Windsor, the family name of the royal house of Great Britain, can be traced to the 1840 marriage of Queen Victoria, who ruled from 1837 to 1901, to her first cousin Albert, the son of the Duke of Saxe-Coburg-Gotha (in present-day Germany). As a foreigner, Prince Albert had to overcome the distrust of the British public, which he did by proving himself to be a devoted husband to Queen Victoria and by demonstrating his genuine concern in Britain's national affairs. Victoria and Albert had nine children. Their oldest son, Albert Edward (1841–1910), became King Edward VII upon Victoria's death in 1901. But Edward's reign lasted only until 1910, when he died and his son, George V (1865–1936), ascended the throne. George was king during World War I (1914–18), and, in 1917, with Britain and Germany bitter enemies, he denounced his ties and claims to Germany, superseding his grandfather's (Prince Albert's) family name of Wettin and establishing the House of Windsor.

Was King George III really insane?

King George III (1738–1820) suffered from bouts of mental illness during his 60-year reign. Today many believe that he was ill with porphyria, a metabolic disorder that results in excessive amounts of porphyrins (a basic substance in body tissue, blood, and urine). He ascended the throne in 1760, upon the death of his grandfather, King George II (1683–1760). Five years later he suffered his first attack of mental illness; others followed in 1788 and 1789 and in 1803 and 1804. A final—and devastating—attack came in 1811; it left him deranged and blind.

As the popular play and movie *The Madness of King George* attests, the monarch's illness presented England with a difficult problem: What do you do when a ruler becomes irrational? When the king became ill in 1788, his prime minister, William Pitt the Younger (1759–1806), and the queen ran the government, dutifully protecting the king's interests—even while the king's son George, the Prince of Wales (1762–1830), openly associated with his father's parliamentary opposition and himself plotted to take control. Once King George III had recovered (in 1789), he reduced his official activities and grew increasingly reliant on Pitt, who proved to be an effective national leader. Pitt resigned in 1801 and was replaced by Henry Addington (1757–1844) until 1804 when the king, feeling that invasion by France was imminent, asked Pitt to prepare a new government. Pitt again assumed the role of prime minister, a role he retained until his death in 1806. Thereafter a coalition government, sometimes called the All-the-Talents administration, was headed by William Grenville (1759–1834) until 1807. Following that, the Duke of Portland (William Henry Bentinck; 1738–1809) became prime minister, until his death in 1809. That year, King George III participated in forming a new government for the last time. After 1811 the Prince of Wales (who would become King George IV) ruled the country as regent.

Despite his health problems King George III led England during one of the most crucial and trying periods in British history. His reign saw the American colonies fight for and win independence from Britain; the French Revolution challenge and oust royal authority in a neighboring country; Britain challenged by Napoleon's Grand Army; and the dawning of the Industrial Revolution. Further, Ireland was brought under England by the 1801 Act of Union, forming the United Kingdom.

Thus, George V was the first ruling member of the House of Windsor. The others were Edward VIII (1894–1972), who abdicated the throne in 1936 so that he could marry American socialite Wallis Simpson (1896–1986); George VI (1895–1952), who became king upon his brother Edward's abdication and who would work tirelessly during World War II (1939–45) to keep up the morale of the British people; and Elizabeth

II (1926–), George VI's elder daughter, who still reigns as Queen of England today. Elizabeth, who ascended the throne in 1952, proclaimed that she and all her descendants who bear the title prince or princess are to be known as Windsor.

Queen Victoria was England's longest-reigning monarch; her rule began in 1837 and ended with her death in 1901.

Which **British monarch has ruled the longest**?

Victoria (1819–1901) claims the distinction of having remained on the throne of England the longest: She became queen in June 1837 upon the death of her uncle, King William IV (1765–1837). Since she was but a teenager at the time, during the early years of her reign she relied on the guidance of the prime minister, the tactful Lord Melbourne (William Lamb; 1779–1848), whom she counted as a friend. After marrying Prince Albert (1819–1861) in 1840, she also sought out advice from him and his adviser. But by 1850 she had grown confident in her abilities, and she managed the country's affairs with authority. Later in her reign, prime ministers Benjamin Disraeli (1804–1881) and William Gladstone (1809–1898) gradually secured increased authority for that office—without alienating the queen. It is for this reason that she is often seen as the first modern monarch, of which she remains a lasting symbol. Queen Victoria celebrated her Diamond Jubilee in 1897 amidst an outpouring of public support. She remained on the throne four more years, until her death in 1901.

During her 64-year reign Queen Victoria presided over the rise of industrialization in Great Britain as well as British imperialism abroad. Architecture, art, and literature flourished, in part due to the influence and interests of Prince Albert. It was the prince who in 1851 sponsored the forward-looking Great Exhibition at London's Crystal Palace. It was the first international exposition—a world fair. The grand Crystal Palace remained a symbol of the Victorian Age until it was destroyed by fire in 1936.

Why is the **Magna Carta** important?

The Magna Carta, arguably the most famous document in British history, has had many interpreters since it was signed by King John (1167–1216), who was under pressure to do so, on June 15, 1215, "in the meadow which is called Runnymede, between Windsor and Staines," in Surrey, England. Drawn up by English barons who were angered by the king's encroachment on their rights, the charter has been credited with no less than insuring personal liberty and putting forth the rights of the individual, which include the guarantee of a trial by jury: "No freeman shall be arrested and imprisoned, or dispossessed, or outlawed, or banished, or in any way molested; nor will be set forth against him, nor send against him, unless by the lawful judgment of his peers, and by the law of the land."

King John, who ruled from 1199 to 1216 and was also called John Lackland, had a long history of abuse of power. While his brother Richard I (or Richard the Lionhearted, 1157–1199) was still king, John tried to wrest the throne from him. Though he failed in his effort, when Richard died in 1199, John did ascend to king, inheriting four French duchies (which he soon lost). Upon his refusal to recognize the new archbishop of Canterbury, John was excommunicated. In order to regain favor with the pope, he was forced to give up his kingdom (1213) and receive it back as a papal fief. He was further required to pay an annual tribute to the pope. It was in raising funds that John ran into trouble with England's powerful barons, who were outraged at and tired of his interference in their affairs. The barons drafted the 63 chapters of the Magna Carta, writing it in the king's voice, and met John along the banks of the Thames River as he returned in 1215 from an unsuccessful invasion of France. The document, to which John was forced to put his seal, asserted the rights of the barons, churchmen, and townspeople, and provided for the king's assurance that he would not encroach on their privileges. In short, the Magna Carta stipulated that the king, too, was subject to the laws of the land.

In that the Magna Carta made a provision for a Great Council, to be comprised of nobles and clergy who would approve the actions of the king vis-a-vis his subjects and ensure the tenets set forth in the charter were upheld, it is credited with laying the foundation for a parliamentary government in England.

After signing it, John immediately appealed to Pope Innocent III (1160 or 1161–1216), who issued an annulment of the charter. Nevertheless, John died before he could fight it, and the Magna Carta was later upheld as the basis of English feudal justice. It is still considered by many to be the cornerstone of constitutional government.

THE BIRTH OF MODERN GOVERNMENT

BRITAIN

How old is the **British parliament**?

The legislative assembly of Great Britain has roots dating back to the Middle Ages (500–1350) when a great council, known as the Curia Regis, advised the king. This body was made up of nobility and clergy. The body evolved over time and progressively gained more power to govern.

Today Parliament consists of the House of Commons, a democratically elected body (roughly equivalent to the U.S. House of Representatives); the House of Lords, which consists of noblemen (dukes, marquesses, earls, viscounts, and barons) as well as high-ranking Anglican clergy (bishops and abbots); and the monarch (king or queen). Since 1911 the power of the House of Lords has been negligible, with the House of Commons charged with electing the prime minister (who must be a member of Commons). The prime minister—not the monarch—is the executive head of government.

Why was **Oliver Cromwell** important to British history?

English soldier and statesman Oliver Cromwell (1599–1658) was a key player in a chain of events that shaped modern British government. The events began when Charles I (1600–1649), House of Stuart, ascended the throne in 1625 and shortly thereafter married a French Catholic princess, immediately raising the ire of his Protestant subjects. This was not the end of England's problems with King Charles: After repeated struggles with the primarily Puritan Parliament, Charles dismissed the legislative body in 1629 and went on to rule without it for 11 years. During this period, religious and civil liberties were seriously diminished, and political and religious strife prevailed. Fearing the king's growing power, Parliament moved to raise an army, and soon civil war broke out (1642–48). Like the French Revolution (1789–99), fought 150 years later, the struggle in England was largely one between a king who claimed to rule by divine right and a government body (in this case Parliament) that claimed the right to govern the nation in behalf of the people.

Enter military leader Oliver Cromwell: After two years of indecisive battles in the English civil war, Cromwell led his parliamentary army troops to victories at Marston Moor (1644) and Naseby (1645), which resulted in Charles's surrender. But when Charles escaped his captors in 1647, the fighting was briefly renewed. It was ended once and for all in 1648. When the king was tried the following year, Cromwell was among those leading the charge to have Charles executed. The king's opponents had their way and, having abolished the monarchy, soon established the Commonwealth of England, installing Oliver Cromwell as "Lord Protector."

Though Cromwell endeavored to bring religious tolerance to England and was somewhat successful in setting up a quasi-democratic government (he declined to take the title of king in 1657), his leadership was constantly challenged by those who wished to restore the Stuart monarchy. When Cromwell died in 1658, he was succeeded by his son, Richard (1626–1712), whose talents were not up to the challenges put to the Lord Protector. The movement to restore the monarchy—particularly the Stuart line—gained impetus, and Richard Cromwell was soon dismissed and went to live outside of England for the next 20 years.

Charles II (1630–1685), the son of Charles I, ascended the throne in 1660, beginning the eight-year period known as the Restoration. Both Charles and his brother, James II (1633–1701), who succeeded him in 1685, worked to reassert the absolutism of the Stuart monarchy. But both kings butted heads with Parliament, particularly when it came to financial matters. Finally in 1688, James II was deposed in the so-called Glorious Revolution. William III (1650–1702), grandson of Charles I, and his wife, Queen Mary II (1662–1694), daughter of James II, were placed on the throne the following year. Though the House of Stuart remained in power, there was an important hitch: Parliament compelled William and Mary to accept the Bill of Rights (of 1689), which asserted that the Crown no longer had absolute power in England and that it must rule responsibly through the nation's representatives sitting in Parliament. Thus, the English civil war (also called the Protestant Revolution) and the influence of Oliver Cromwell and other parliamentarians laid the foundation for England's constitutional monarchy.

What was the **Bill of Rights of 1689**?

The English bill, accepted by King William III (1650–1702), who ruled jointly with his Protestant wife Queen Mary II (1662–1694), seriously limited royal power. After a struggle between the Stuart kings and Parliament and the subsequent ousting of King James II (1633–1701), Parliament presented the Bill of Rights to King William and Queen Mary as a condition of their ascension: The document not only described certain civil and political rights and liberties as "true, ancient, and indubitable," but it also ironed out how the throne would be succeeded. This point was of critical importance to the future of England—the article stipulated that no Roman Catholic would rule the country. Since the Bill of Rights served to assert the role of Parliament in the government of England, it is considered one of the seminal documents of British constitutional law.

What was the **Act of Settlement**?

The 1701 decree reiterated the Bill of Rights of 1689 by stating that the king or queen of England must not be a Roman Catholic and must also not be married to a Roman Catholic. The Act of Settlement further stipulated that the sovereign must be a member of the Protestant Church of England. The measure was passed by Parliament, in part due to fear of the Jacobites, supporters of the Stuart king, James II (1633–1701), who had been exiled in 1688 in the Glorious Revolution. The Jacobites (not to be confused with the French Jacobins, the political party that came to power during the French Revolution), sought to keep the House of Stuart in power. Due to the line of succession, however, the Act of Settlement did not take effect until 1714, when Queen Anne (1665–1714) died and left no successors to the throne. She was the last Stuart ruler of England.

The phrase "Let them eat cake" was incorrectly attributed to Marie Antoinette, the daughter of Holy Roman Emperor Francis I.

Did **Marie Antoinette** really say, "Let them eat cake"?

No, the widely quoted phrase was incorrectly attributed to her, and the entire story was probably made up. Nevertheless, the legend is not far from fact: As the daughter of a Holy Roman Emperor (Francis I), the beautiful Marie Antoinette (1755–1793) was accustomed to a life of luxury. Unhappy in her marriage to Louis XVI (1754–1793), king of France, she pursued her own pleasurable interests with abandon. Despite the economic problems that plagued France at the time, she lived an extravagant lifestyle, which included grand balls, a "small" palace at Versailles, theater, gambling, and other frivolities. She was completely disinterested in the affairs of the nation. Many French people blamed her for corruption in the court. In short, she did much to earn herself the terrible reputation that has followed her through history.

Unpopular in her own day, one of the stories that circulated about her had Marie Antoinette asking an official why the Parisians were angry. When he explained to the queen that it was because the people had no bread, she replied, "Then let them eat cake." The French Revolution, which began in 1789, soon put an end to Marie Antoinette's excesses. Along with her husband, she was put to death by guillotine in 1793.

What were the **Rights of Man**?

In 1789 the French assembly made the Declaration of the Rights of Man and of the Citizen, meant to flesh out the revolutionary cry of "liberty, equality, and fraternity." Influenced by the U.S. Declaration of Independence (1776) as well as the ideas of the Enlightenment, the document guaranteed religious freedom, the freedom of speech and the press, and personal security. It proclaimed that man has natural and inalienable rights, which include "liberty, property, personal security, and resistance to oppression.... " The declaration further stipulated that "No one may be accused, imprisoned, or held under arrest except in such cases and in such a way as is prescribed by law" and that "Every man is presumed innocent until he is proved guilty.... "

The declaration was subsequently written into the preamble of the French constitution (1791). However, the Code Napoleon superseded many of the ideas it set forth.

What were **Napoleon's Hundred Days**?

The term refers to Napoleon Bonaparte's last 100 days as ruler of France. Having been defeated by his enemies—a coalition of European powers Britain, Sweden, Austria, and Prussia, who aligned themselves against Napoleon's domination—the emperor abdicated the throne in the spring of 1814, and was exiled to the island of Elba in the Mediterranean Sea. There he heard of the confusion and discontent that came after he had descended the throne. He left Elba, and with more than 1,000 men, arrived on the French coast at Cannes and marched inland to Paris. Hearing of his arrival, the new Bourbon king, Louis XVIII (1755–1824), fled. On March 20 Napoleon began a new reign, but it was only to last until the European allies defeated him again, at the Battle of Waterloo, June 12 through 18. After that battle, the so-called Little Corporal (for his diminutive stature) was permanently exiled to the British island of St. Helena, where he remained until his death in 1821.

Why is **Napoleon still controversial**?

Even history has not been able to sort out the widely disparate opinions of the diminutive French ruler. And both the detractors and the champions (or some would say, the apologists) continue to publish their arguments and supporting research. The most obvious point on which scholars differ centers on the fact that first and foremost Napoleon Bonaparte (1769–1821) was a military man. Here opinion divides quite naturally. Not long after Napoleon assumed power (in the coup of 1799), he proceeded to keep France—and the rest of Europe—at war for more than 10 years. From the French perspective, Napoleon was a great man, a brilliant strategist who could not only muster his troops but could keep them motivated to fight one campaign after another. The targets of these campaigns—England, Russia, Austria, Germany, Spain, and Portugal among them—view Napoleon in quite a different light, as would be expected. Researchers from these countries have seen and rendered Napoleon's dark side, calling him a megalomaniac and a psychopath, and even seeing him as a forerunner of Adolf Hitler (1889–1945).

To further complicate the matter of how history views Napoleon, before he declared himself emperor for life (in 1804) and launched his military conquests throughout the continent and beyond (1805), Napoleon enjoyed a brief period in which many Europeans—not just the French —believed him to be a hero. After all, he assumed leadership of France after the hideous period of Robespierre's Terror and the ineffectual government of the Directory, and then he proceeded to make peace with the Americans, the Russians, *and* the British. Many believed Napoleon was just the man to bring order to the chaos France had known since the storming of the Bastille, and he extended an olive branch to France's longtime enemies. It looked like he would restore order at home and abroad. Of course this honeymoon did not last: By 1805 the leader many had looked to to end the turmoil only became the cause of more turmoil. His "compulsive war-making," as one writer put it, soon swept over the continent, ultimately uniting various countries in an effort to rid Europe of the scourge that was Napoleon and his Grand Army.

Thus, students and readers are left to sort through the diverging accounts of this controversial figure. Napoleon has been called the "Emperor of Kings," credited for his vision, insight, courage, and even with the development of the modern liberal democracy; he has also been described as a compulsive tyrant who had an insatiable appetite for battle, a man whose own ambitions left millions dead. But on a few descriptions both sides can agree: He was a brave soldier, an inspired military leader, and, at least for a time, a charismatic ruler.

American Democracy

What were the **Intolerable Acts**?

The so-called Intolerable Acts, also known as the Coercive Acts, were five laws passed by the British parliament early in 1774. Intended to assert British authority in the Massachusetts colony, the measures were seen as punishment for the Boston Tea Party (December 1773). In brief, the laws enacted the following: closure of the port of Boston; an English trial for any British officer or soldier who was charged with murder in the colonies; the change of the charter of Massachusetts such that the council had to be appointed by the British and that town meetings could not be held without the (British-appointed) governor's permission; the requirement that the colonists house and feed British soldiers; and the extension of the province of Quebec southward to the Ohio River.

While the British intention was to bring the Massachusetts colony under control (and actually the fifth act was not intended to have any punitive effect on the colony), the result was instead to unite all the colonies in opposition to British rule. In this regard, the acts are seen as a precursor to the American Revolution (1775–83).

Who started the **Boston Tea Party**?

Many believe that on December 13, 1773, it was patriot Samuel Adams (1722–1803) who gave the signal to the men, who may have numbered more than 100 and were dressed as Indians, to board the ships in Boston Harbor and dump the tea overboard. Whether or not it was Adams who started the Tea Party, about this there can be no doubt: He was most certainly a leader in the agitation that led up to the event. The show of resistance was in response to the recent passage by the British parliament of the Tea Act, which allowed the British-owned East India company to "dump" tea on the American colonies at a low price and also required that the colonists pay a duty for said tea. Colonists feared the act would put local merchants out of business and that if they conceded to pay the duty to the British, they would soon be required to pay other taxes as well.

Once the ships carrying the tea had arrived in Boston Harbor, the colonists tried to have them sent back to England. But when Governor Thomas Hutchinson (1711–1780) of Massachusetts refused to order the return of the ships, patriots organized their show of resistance, which came to be known as the Boston Tea Party.

Why were there two **Continental Congresses**?

Both meetings were called in reaction to British Parliament's attempts to assert its control in the American colonies. When colonial delegates to the First Continental Congress met, they developed a plan but were obviously prepared for it not to work, since even before dismissal they agreed to reconvene if it were necessary to do so. In short, the first Congress developed Plan A; the second resorted to Plan B (which was one last appeal to the king) and then to Plan C (finally declaring independence from Britain).

The First Continental Congress convened on September 5, 1774, in Philadelphia, Pennsylvania. The meeting was largely a reaction to the so-called Intolerable Acts (or the Coercive Acts), which British Parliament had passed in an effort to control Massachusetts after the rebellion of the Boston Tea Party (December 1773). Sentiment grew among the colonists that they would need to band together in order to challenge British authority. Soon 12 colonies dispatched 56 delegates to a meeting in Philadelphia. (The thirteenth colony, Georgia, declined to send representatives but agreed to go along with whatever plan was developed.) Delegates included Samuel Adams (1722–1803), George Washington (1732–1799), Patrick Henry (1736–1799), John Adams (1735–1826), and John Jay (1745–1829). Each colony had one vote, and when the meeting ended on October 26, the outcome was this: The Congress petitioned the king, declaring that the British parliament had no authority over the American colonies, that each colony could regulate its own affairs, and that the colonies would not trade with Britain until Parliament rescinded its trade and taxation policies. The petition stopped short of proclaiming independence from Great Britain, but the delegates agreed to meet again the following May—if necessary.

But King George III (1738–1820) was determined that the British Empire be preserved at all costs; he believed that if the empire lost the American colonies, then there may be a domino effect, with other British possessions encouraged to also demand independence. He feared these losses would render Great Britain a minor state, rather than the power it was. Britain unwilling to lose control in America, in April 1775 fighting broke out between the redcoats and patriots at Lexington and Concord, Massachusetts. So, as agreed, the colonies again sent representatives to Philadelphia, convening the

Illustration of the 1776 signing of the Declaration of Independence, which has long been regarded as history's most eloquent statement of the rights of the people. (Original painting by Asher Durand.)

Second Continental Congress on May 10. Delegates—including George Washington, John Hancock (1737–1793), Thomas Jefferson (1743–1826), and Benjamin Franklin (1706–1790)—organized and prepared for the fight, creating the Continental army and naming Washington as its commander in chief. With armed conflict already under way, the Congress nevertheless moved slowly toward proclaiming independence from Britain: On July 10, two days after issuing a declaration to take up arms, Congress made another appeal to King George III, hoping to settle the matter without further conflict. The attempt failed, and the following summer the Second Continental Congress approved the Declaration of Independence, breaking off all ties with the mother country.

What does the **Declaration of Independence** say?

The Declaration of Independence, adopted July 4, 1776, has long been regarded as history's most eloquent statement of the rights of the people. In it, not only did the 13 American colonies declare their freedom from Britain, they also addressed the reasons for the proclamation (naming the "causes which impel them to the separation") and cited the British government's violations of individual rights, saying "the history of the present King 'George III' of Great Britain is a history of repeated injuries and usurpations," which aimed to establish "an absolute Tyranny over these States."

The opening paragraphs go on to state the American ideal of government, an ideal that is based on the theory of natural rights. The Declaration of Independence puts forth the fundamental principles that a government exists for the benefit of the people and that "all men are created equal." As the chairman of the Second Continental Congress committee that prepared the Declaration of Independence, it was Thomas Jefferson (1743–1826) who wrote and presented the first draft to the Second Continental Congress on July 2, 1776.

The passage that is most frequently quoted is this:

> We hold these truths to be self-evident, that all men are created equal, that they are endowed by their Creator with certain unalienable Rights, that among these are Life, Liberty and the pursuit of Happiness. That to secure these rights, Governments are instituted among Men, deriving their just powers from the consent of the governed, That whenever any Form of Government becomes destructive of these ends, it is the Right of the People to alter or to abolish it, and to institute new Government, laying its foundation on such principles and organizing its powers in such form, as to them shall seem most likely to effect their Safety and Happiness.

Who are considered the **Founding Fathers** of the United States?

The term is used to refer to a number of American statesmen who were influential during the revolutionary period of the late 1700s. Though definitions vary, most include the authors of the Declaration of Independence and the signers of the U.S. Constitution among the nation's Founding Fathers.

Of the 56 members of the Continental Congress who signed the Declaration of Independence (July 4, 1776), the most well-known are John Adams (1735–1826) and Samuel Adams (1722–1803) of Massachusetts, Benjamin Franklin (1706–1790) of Pennsylvania, John Hancock (1737–1793) of Massachusetts, and Thomas Jefferson (1743–1826) of Virginia.

The 39 men who signed the U.S. Constitution on September 17, 1787, include notable figures such as George Washington (1732–1799), who would go on, of course, to become the first president of the United States; Alexander Hamilton (1755–1804), who, as a former military aid to George Washington, went on to become the first U.S. secretary of the treasury; and James Madison (1751–1836), who is called the "father of the constitution" for his role as negotiator and recorder of debates between the delegates. At 81 years of age, Benjamin Franklin was the oldest signer of the Constitution and was among the six statesmen who could claim the distinction of signing both it and the Declaration of Independence; the others were George Clymer (1739–1813), Robert Morris (1734–1806), George Read (1733–1798), Roger Sherman (1721–1793), and James Wilson (1742–1798).

Patriots and politicians conspicuous by their absence from the Constitutional Convention of 1787 were John Adams and Thomas Jefferson, who were performing other government duties at the time and would each go on to become U.S. president; Samuel Adams and John Jay (1745–1829), who were not appointed as state delegates but who continued in public life, holding various federal and state government offices (including governor of their states); and Patrick Henry (1736–1799) of Virginia, who saw no need to go beyond the Articles of Confederation (1777) to grant more power to the central government. Henry's view on this issue foreshadows the discontent that crested nearly 100 years later when 12 southern states (including Virginia) seceded from the Union, causing the Civil War (1861–65) to break out.

Adams, Franklin, Hancock, Jefferson, Washington, Hamilton, Madison, Jay, and Henry: These are the names that come to mind when the words "Founding Fathers" are uttered. Each of them had a profound impact in the political life of the United States—even beyond their starring roles as patriots and leaders during the American revolutionary era. However, it's important to note that in many texts and to many Americans, the term *Founding Fathers* refers *only* to the men who drafted the U.S. Constitution since it is that document that continues—more than 200 years after its signing—to provide the solid foundation for American democratic government.

Why did **John Hancock** go down in history as *the* notable signer of the Declaration of Independence?

Most Americans know that when they're putting their "John Hancock" on something, it means they're signing a document. It's because of the 56 men who signed their names to the historic document, it was Hancock (1737–1793) who, as president of the Second Continental Congress, signed the declaration first.

The events were as follows: On July 2, 1776, Thomas Jefferson (1743–1826) presented the draft of the declaration to the Second Continental Congress, which had convened in Philadelphia, Pennsylvania, more than a year earlier (on May 10, 1775). The congressional delegates of the 13 colonies then deliberated and debated the draft, making some changes: A section was deleted that condemned England's King George III (1738–1820) for encouraging slave trade. Other changes were cosmetic in nature. On July 4 the final draft of the declaration was adopted by Congress, and it was then that it was signed by Hancock. The document was then printed. A few days later, on July 8, the declaration was read to a crowd who assembled in the yard of the state house. On July 19 the Congress ordered that the Declaration of Independence be written in script on parchment. It is that copy that in early August was signed by all 56 members of the Second Continental Congress. The Declaration of Independence is housed, along with the U.S. Constitution (1788) and the Bill of Rights (1791), in the National Archives Building in Washington, D.C., where it is on display to the public. John Hancock went on to become governor of Massachusetts, from 1780 to 1785 and from 1787 to 1793.

What were the **Articles of Confederation**?

This American document was the forerunner to the U.S. Constitution (1788). Drafted by the Continental Congress at York, Pennsylvania, on November 15, 1777, the Articles of Confederation went into effect on March 1, 1781, when the last state (Maryland) ratified them. The articles had shortcomings that were later corrected by the Constitution: They provided the states with more power than the central government, stipulating that Congress rely on the states both to collect taxes and to carry out the acts of Congress.

It is largely thanks to Alexander Hamilton (1755–1804) that the articles were thrown out: realizing they made for a weak national government, Hamilton led the charge to strengthen the central government—even at the expense of the states. Eventually, he won the backing of George Washington (1732–1799), James Madison (1751–1836), John Jay (1745–1829), and others, which led to the convening of the Philadelphia Constitutional Convention, where the ineffectual Articles of Confederation were thrown out and the Constitution was drafted.

One lasting provision of the Articles of Confederation was the ordinance of 1787. Signed in an era of westward expansion, the ordinance set the guidelines for how a territory could become a state: A legislature would be elected as soon as the population had reached 5,000 voting citizens (which were men only), and the territory would be eligible for statehood once its population had reached 60,000.

Which states were the **original 13**?

In order of admission they are Delaware, Pennsylvania, New Jersey, Georgia, Connecticut, Massachusetts, Maryland, South Carolina, New Hampshire, Virginia, New York, North Carolina, and Rhode Island. Vermont was fourteenth and the first free state (the first state without slavery).

THE U.S. CONSTITUTION

What was the **Virginia Plan**?

It was the famous plan, drafted by James Madison (1751–1836), put forth by the Virginia delegates to the Constitutional Convention, which convened on May 25, 1787. After taking a few days to set the ground rules and elect officers, on May 29 the delegation from Virginia, led by Edmund Jennings Randolph (1753–1813), proposed a plan to write an all-new constitution rather than attempting to revise and correct the weak Articles of Confederation. There was opposition (sometimes called the New Jersey Plan) and the issue was debated for weeks. Eventually, a majority vote approved the Virginia Plan and the delegates began work drafting a document that would provide a strong national government for the United States.

Who wrote the **U.S. Constitution**?

In spirit the U.S. Constitution was created by all of the 55 delegates to the meeting that convened on May 25, 1787, in Philadelphia's Independence Hall. Thomas Jefferson (1743–1826) called the Constitutional Convention "an assembly of demi-gods," and with good cause: The delegates were the young nation's brightest and best. When the states had been called upon to send representatives to the meeting, 12 states answered by sending their most experienced, most talented, and smartest men; Rhode Island, which feared the interference of a strengthened national government in state affairs, sent no one to Philadelphia.

Even in such stellar company, the document did have to be written. While many had a hand in this process, it was New York lawyer and future American politician and diplomat Gouverneur Morris (1752–1816) who actually took on the task of penning the Constitution, putting into prose the resolutions reached by the convention. Morris had the considerable help of the records that James Madison (1751–1836) of Virginia had kept as he managed the debates among the delegates and suggested compromises. In that capacity and in that he designed the system of checks and balances among the legislative (Congress), the executive (the president of the United States), and the judicial (Supreme Court), Madison had considerable influence on the document's language, quite rightfully earning him the designation "father of the constitution."

The original document, drafted by Morris, is preserved in the National Archives Building in Washington, D.C. While the Constitution has been amended by Congress, the tenets set forth therein have remained with Americans for more than two centuries, and they have provided proof to the countries of the world that a constitution outlining the principles and purposes of its government is necessary to good government.

When was the **U.S. Constitution ratified**?

The Constitution was ratified by the required nine states by June 21, 1788. It went into effect the following year, superseding the Articles of Confederation (1781).

What rights are protected in the **Bill of Rights**?

The first 10 amendments to the U.S. Constitution are collectively called the Bill of Rights, which became law on December 15, 1791, and are meant to guarantee individual liberties.

The First Amendment, which is perhaps most often cited by Americans, guarantees freedom of religion, speech, and the press, as well as the right to assemble peaceably and the right to petition the government for a redress of grievances. The Second Amendment guarantees the right to keep and bear arms (stating that "a well regulated militia being necessary to the security of a free State"). The Third Amendment forbids

Why was the Bill of Rights added to the U.S. Constitution?

The United States Constitution, ratified in 1788, contained few personal guarantees. In fact, initially there was some opposition to the new Constitution—much of it based on the lack of specific guarantees of individual rights. It was the "father of the constitution" and future president of the United States James Madison (1751–1836), then a member of the U.S. House of Representatives, who in December 1791 led Congress to adopt the 10 constitutional amendments that became known as the Bill of Rights. Most of the rights focus on individual liberties that had been cited in the Declaration of Independence as having been violated by the British. Most of these specific grievances had not been addressed by the Constitution; therefore, the Bill of Rights was added to cover this ground.

peacetime quartering of soldiers in private dwellings without consent of the owner. The Fourth Amendment forbids unreasonable searches and seizures.

The Fifth through the Seventh amendments establish basic standards of jurisprudence. The Fifth (which long ago fell into common usage with the phrase, "s/he's pleading the Fifth") guarantees that a person will not be compelled to testify against himself. The amendment also ensures that a criminal indictment can only be handed down by a grand jury (12 to 23 people who determine if a trial is necessary) and prohibits double jeopardy (being prosecuted twice for the same criminal offense). The Sixth Amendment protects the rights of accused persons in criminal cases by guaranteeing a speedy and fair trial, an impartial jury, and the right to counsel. The Seventh Amendment guarantees trial by jury. The Eighth Amendment prohibits excessive bail, excessive fines, and cruel and unusual punishment.

The Ninth Amendment, which is one that many Americans are probably unable to cite, is an important one nevertheless: It states that simply because a right is not enumerated in the Constitution, it does not mean that the people do not retain that right.

The Tenth Amendment relinquishes to the state governments those powers the Constitution did not expressly grant the federal government or deny the states. In other words, it limits the power of the federal government to that which is granted in the Constitution.

Who **determines whether a law violates the liberties** guaranteed by the Bill of Rights?

It is the job of the U.S. Supreme Court to decide whether or not a law impinges upon the liberties listed in—or implied by—the Bill of Rights (1791). The difficult task before the Supreme Court justices is in determining what rights are implied. Such questions prompt months of hearings and deliberations before a decision can be reached as to the constitutionality of a contested law. The judicial body makes its determinations based on a majority vote of the nine justices (one chief justice and eight associates). Established as the highest court in the country by Article 3 of the Constitution, the Court has ultimate authority in all legal questions that arise pertaining to the Constitution. Called the "court of last resorts," the Supreme Court both interprets the acts of Congress (including laws and treaties) and determines the constitutionality of federal and state laws (under the Fourteenth Amendment, the Court has upheld that most of the Bill of Rights also applies to state governments).

How are **amendments** made to the U.S. **Constitution**?

There are two paths that proposed amendments can take to become law. The first path is this: An amendment is proposed in Congress; two-thirds of both houses must then approve it (if they do not, then the proposal ends here); if approved in both houses of the U.S. Congress, the proposed amendment is sent to the legislatures (or conventions) of each state of the union; three-fourths of all the state legislatures must then approve it (by whatever rules each state legislature uses); once three-fourths of the states approve it, the amendment is made (if three-fourths of the states do not approve it, the amendment fails to become law).

The second path is this: The legislatures of two-thirds of the states ask for an amendment to be made to the U.S. Constitution; Congress then calls a convention to propose it; then the proposed amendment becomes a law when it is ratified by the legislatures in three-fourths of the states. While this path has never been taken, it's an important provision nonetheless since it allows for a popular, state-based proposal to be considered.

How many amendments have been made to the U.S. Constitution?

There have been 27 amendments to date. The following list gives brief summaries and the year each became part of the U.S. Constitution.

First Amendment through the Tenth Amendment (1791): Comprise the Bill of Rights.

Eleventh Amendment (1798): Declares that U.S. federal courts cannot try any case brought against a state by a citizen of another state or country.

Twelfth Amendment (1804): Revised the presidential and vice presidential election rules such that members of the electoral college, called electors, vote for one person as president and for another as vice president. Prior to the passage of this amendment, the electors simply voted for two men—the one receiving more votes became president and the other became vice president.

Thirteenth Amendment (1865): Prohibits slavery. Along with Amendments 14 and 15, these are sometimes called the Civil War amendments.

Fourteenth Amendment (1868): Defines U.S. citizenship and gives all citizens equal protection under the law. (This amendment made former slaves citizens of both the United States and the state where they lived. It further forbade states to deny equal rights to any person.)

Fifteenth Amendment (1870): States that the right of U.S. citizens to vote shall not be denied or abridged by the United States or by any state on account of race, color, or previous condition of servitude. (This amendment was meant to extend suffrage to black men.)

Sixteenth Amendment (1913): Authorizes a federal income tax.

Seventeenth Amendment (1913): Provides for the direct election of senators. Before this passed, state legislatures elected senators to represent them; this amendment gave that power to the people of each state.

Eighteenth Amendment (1919): Made prohibition legal. (In other words, the manufacture and distribution of alcohol became illegal.)

Nineteenth Amendment (1920): Grants women the right to vote.

Twentieth Amendment (1933): Also called the "lame duck amendment," it changed congressional terms of office and the dates of the presidential inauguration so that newly elected officials take office closer to election time.

Twenty-first Amendment (1933): Repealed Amendment 18 to end prohibition.

Twenty-second Amendment (1951): Limits presidential tenure to two terms in office. (A president can hold office for no more than 10 years—two years as an unelected president, and two terms as an elected president.)

Twenty-third Amendment (1961): Grants residents of Washington, D.C., the right to vote in presidential elections.

Twenty-fourth Amendment (1964): Outlaws the poll tax in all federal elections and primaries. (Some states had used poll taxes as a way of keeping certain populations of voters from casting their ballots; the practice had served to disenfranchise blacks and poor people.)

Twenty-fifth Amendment (1967): Provides for procedures to fill the vice presidency and further clarifies presidential succession rules. (Upon removal, resignation, or death of the president, the vice president assumes the presidency; if a vice president is removed, resigns, or dies while in office, the president nominates a vice president who takes office upon confirmation by a majority vote of both houses of Congress.)

Twenty-sixth Amendment (1971): Lowers the voting age for federal and state elections to 18.

Twenty-seventh Amendment (1992): Prevents Congress from passing immediate salary increases for itself; it requires that salary changes passed by Congress cannot take effect until after the next congressional election. (This amendment had been passed by Congress in 1788 and was then sent to the states for ratification. Since the amendment had no time limit for ratification, it became part of the Constitution in 1992, after Michigan became the thirty-eighth state to ratify it.)

THE AMERICAN PARTY SYSTEM

Who were the **Whigs**?

They were members of political parties in Scotland, England, and the United States. The name is derived from *whiggamor* (meaning "cattle driver"), which was a derogatory term used in the seventeenth century to refer to Scottish Presbyterians who opposed King Charles I of England (1600–1649). Charles, who ruled from 1625 to 1649, was deposed in a civil war and subsequently tried in court, convicted of treason, and beheaded. The British Whigs, who were mostly merchants and landed gentry, supported a strong Parliament. They were opposed by the aristocratic Tories who upheld the power of the king. For a short period during the eighteenth century, the Whigs dominated political life in England. After 1832 they became part of the Liberal Party.

At about the same time, the Whig Party in the United States emerged as one of the two major American political parties. The other was the Democratic Party that Americans still know today, which supported President Andrew Jackson (1767–1845) for reelection in 1832. Though Jackson's first term of office had proved to be somewhat controversial, the Whigs were unable to elect their candidate (Henry Clay, 1777–1852, of the so-called Southern "Cotton" Whigs), and Jackson, called "Old Hickory," went on to a second term. In the election of 1840 the Whigs, whose leadership had succeeded in uniting the party, were finally successful in putting their candidate in the White House. But William Henry Harrison (1773–1841) died after only 32 days in office, and his successor, John Tyler (1790–1862), alienated the Whig leaders in Congress, and they ousted him from the party. In 1848 the Whigs put Zachary Taylor (nicknamed "Old Rough and Ready," 1784–1850) in the White House, but two years later he, too, died in office. His successor, Millard Fillmore (1800–1874), remained loyal to the Whigs, but there were problems within the party. The last Whig presidential candidate was General Winfield Scott ("Old Fuss and Feathers," 1786–1866) in 1852, but he was defeated by Franklin Pierce (1804–1869). Shortly thereafter the Whig party broke up

Who were the Know-Nothings?

The Know-Nothings were members of a U.S. political movement during the mid-1800s: Americans who feared the foreign influence of immigrants (there was an influx of new arrivals in the 1840s) banded together, sometimes in secret societies, in order to uphold what they believed to be the American view. When people who were thought to be members of these groups were asked about their views and activities, the typical response was, "I don't know," which gave the movement its name.

The Know-Nothings worked to elect only "native" Americans (U.S.-born citizens) to political office, and they advocated the requirement for citizenship be 25 years of residence in the United States. Since many immigrants came from European countries and were Roman Catholics, the Know-Nothings also opposed the Catholic Church.

In 1843 Know-Nothings formed the American Republican Party. By 1854 they had allied themselves with factions within the Whig Party and in the state elections held that year, Know-Nothings swept the vote in Massachusetts and Delaware, nearly carried New York and Pennsylvania, and pulled substantial votes in the South. The following year, the Know-Nothings dropped much of their secrecy and became known simply as the American Party. It was the issue of slavery that finally split the party in the national election of 1856, and the group dissolved after that. Antislavery members of the American Party joined the newly formed Republican Party.

over the slavery issue; most of the Northern Whigs joined the Republican Party, while most of the Southern "Cotton" Whigs joined the Democratic Party.

How did the **Republican Party** begin?

The Republican Party, one of the two principal political parties of the United States, was founded in 1854 by those opposing the extension of slavery into new territories. The party mustered enough support to elect their candidate in 1860, Abraham Lincoln (1809–1865). During the 1880s party members nicknamed themselves the Grand Old Party; the vestige of this nickname is still around today, as the GOP. There have been 17 Republican presidents.

How did the **Democratic Party** begin?

The other—and older—principal party in the United States today, the Democratic Party was founded around electing Thomas Jefferson (1743–1826) to office in 1800 (against

Alexander Hamilton's Federalist Party). The party's platform favored personal liberty and the limitation of federal government. Installing Jefferson in office, the party—then called the Democratic-Republicans—went on to get its candidates into the White House for the next 25 years. In 1828 they became known simply as Democrats, dropping the suffix. Depending on how one counts, there have been either 18 or 19 Democratic presidents; Andrew Johnson (1808–1875) is problematic since he was a Democrat before joining the National Union Party ticket as the vice presidential candidate in 1864. (Some sources list both party affiliations, as in Democratic/National Union.)

Did all **Southern lawmakers leave Washington** once the South seceded?

All but one: Even after the South seceded and the first shot of the war was fired at Fort Sumter, South Carolina, Senator Andrew Johnson (1808–1875) of Tennessee opted not to leave the Union. The fact that Johnson did not stick with the state he represented may seem a surprising move, but it reveals one of his most fundamental and fiercely held beliefs: He maintained an unswerving trust in the Constitution. Consequently, he viewed secession as not only treasonous but illegal.

His decision to remain with the Union proved politically advantageous to Johnson, a Democrat: In 1862 President Abraham Lincoln (1809–1865) appointed him military governor of Tennessee. When Abraham Lincoln went on to a second term in office (in March 1865), Johnson was his vice president. He had held this job for a scant six weeks when Lincoln was assassinated (April 14) and Johnson assumed the presidency.

How were the **Southern states brought back into the Union**?

Even before the Civil War had ended, Washington, D.C., considered the difficult problem of how to rejoin the seceding states with the North. Some lawmakers felt the Southern states should be treated as if they were territories that were gained through war. Others, including both Abraham Lincoln (1809–1865) and Andrew Johnson (1808–1875), reasoned that since secession was illegal, the South belonged—and always had—to the Union, and therefore the states ought to be brought back into their "proper relationship" with the federal government. They favored punishing the Southern leaders—but not the states themselves.

President Abraham Lincoln developed his 10 percent plan: As soon as 10 percent of a state's population had taken an oath of loyalty to the United States, the state would be allowed to set up a new government. But Congress opposed it, proclaiming the policy too mild, and responded by passing the Wade-Davis Bill (June 1864), making the requirements for statehood more rigid. Instead of Lincoln's 10 percent, Congress required that a majority of voters in each state would need to swear their loyalty, in an "ironclad oath," before statehood could be restored. Further, the bill stipulated that the constitution of each state had to abolish slavery and that Confederate military leaders were to be prohibited from holding political office and otherwise disenfranchised. Lincoln opposed the bill and neither signed nor returned it before Congress was dismissed, and so the Wade-Davis measure failed to become law.

When Lincoln was assassinated the following April, the matter remained unsettled. His successor, President Andrew Johnson, soon put forth a plan to readmit the states. He called for each state constitution to abolish slavery and repudiate the Confederate war debt; further, a majority of voters in each state needed to vow allegiance to the Union. Once a state had reorganized itself under this plan, Johnson required the state legislature to approve the Thirteenth Amendment (abolishing slavery in the United States). When Congress reconvened in December 1865 for the first time since Lincoln's assassination, all former Confederate states except Texas had complied with the president's specifications for statehood. But these new states had also set up Black Codes, severely restricting the rights of blacks. Further, there was violence against blacks who were the victims of attacks by white Southerners—including members of the newly formed Ku Klux Klan, a secret white organization that spread terror across the South.

Congress became determined to fight the readmission of the Southern states by Johnson's lenient standards, and it refused to seat any representatives from the South. The move angered President Johnson, and political volleying between the legislature and the executive office began. Ultimately it was Congress that determined the process by which the Southern states were readmitted.

By the summer of 1868 the legislatures of 7 (of 11) Southern states had approved the Fourteenth Amendment. The remaining four states—Georgia, Mississippi, Texas, and Virginia—complied with the requirements for statehood by 1870, at which time the Union was restored and Congressional representatives from the South were again welcomed in Washington.

In the intervening period (between Congress's rejection of President Johnson's plan for statehood and the ratification of the Fourteenth and Fifteenth Amendments), the South was governed by military administrators who protected people and property and oversaw the reorganization of government in each state.

RECONSTRUCTION

What was the **Reconstruction**?

The Reconstruction was the 12-year period (1865–77) of rebuilding that followed the Civil War. The last battle over, the South lay in ruins: Food and other supplies were scarce, people were homeless, city centers had been destroyed, schools were demolished, railways torn up, and government was nonexistent. Further, the nation had new citizens to enfranchise—and protect—the freed slaves. There was also the question of how to readmit each Southern state to the Union.

In short, the nation's wounds needed to heal. But the long years of the Reconstruction brought only more divisiveness and quarrels. This time the battlefield was not Gettysburg or Chattanooga, but Washington, D.C. President Andrew Johnson (1808–1875), a Southern Democrat and former slave owner, squared off with Congress, led by a radical Republican faction. The two branches of government fought over who should guide Reconstruction policy. Johnson favored a more tolerant and swifter approach to reuniting the nation, but his measures failed to protect the country's black citizens. Congress proceeded more cautiously, setting up military administrators in the South as an interim form of government until readmission of the states could be effected. In the end, Congress won out by overriding President Johnson's vetoes again and again.

Congress passed the Civil Rights Act of 1866, which took a first step toward enfranchising the black population by guaranteeing the legal rights of former slaves; the Reconstruction acts (1867), outlining how each Southern state would be readmitted to the Union; the Freedmen's Bureau Bill, which extended the life of the wartime agency in order to help Southern blacks and whites get back on their feet; and adopted the Thirteenth, Fourteenth, and Fifteenth Amendments (the so-called Civil War amendments). Since the South was based on agriculture, the economy slowly recovered, eventually becoming more industrial. Public schools were established in each state. And the state governments became more open than they had been, with more offices up for election rather than appointment. In addition, blacks were guaranteed the vote—and the right to run for office.

But there was resistance to all these measures. And the post–Civil War recovery was not smooth. Many historians believe that the controversy that ensued in the years that followed the Confederate surrender at Appomattox (April 1865) laid the groundwork for segregation and other injustices that brought on the civil rights movement. Many also believe that the problems are still with the country today.

What does **"40 acres and a mule"** mean?

The term originated with Union general William T. Sherman (1820–1891) toward the end of the Civil War. On January 16, 1865, in his "March to the Sea," the military advancement he led from Atlanta to Savannah, Sherman issued Special Field Order No.15, which set aside a tract of land along the South Carolina, Georgia, and Florida coasts for the exclusive settlement of 40,000 freed slaves. According to the order, each black family was to be given "a plot of not more than (40) forty acres of tillable ground…in the possession of which land the military authorities will afford them protection, until such time as they can protect themselves, or until Congress shall regulate their title." All of the lands were former plantations that had been confiscated during the war; they became the jurisdiction of the Freedmen's Bureau. Though there was a provision for "one or more of the captured steamers to ply between the settlements and one or more of the commercial points…to sell the products of their land and labor," there was no mention of providing a mule, or any other animal, to each freedman. The mules, tired army work animals, were distributed later to the landholders. News of "40 acres and a mule" spread among the freed slaves.

But soon after the war ended in April 1865, this promise to materially assist the freedmen was abandoned when President Andrew Johnson (1808–1875) issued pardons for the ex-Confederates and ordered their lands returned to them. The somewhat obscure term "40 acres and a mule" resurfaced in the 1990s and early 2000s as the issue of reparations came to the fore in the United States.

What were **carpetbaggers**?

Carpetbaggers was a derisive term that referred to Northerners who arrived in the South in the early days of Reconstruction (1865–77), the 12-year period of rebuilding that followed the American Civil War (1861–65). Even though many of these Northern businessmen intended to settle in the South, Southerners viewed them as outsiders and, worse, as opportunists who only intended to make a quick profit before returning north. They were called carpetbaggers because many carried carpetbags as luggage; some Southerners even quipped that these Northerners could carry all their belongings in a carpetbag, implying that they were nothing more than transients. Nevertheless, Northerners who relocated to the South following the Civil War played an important role during Reconstruction. Some, aided by the black vote, gained public office and impacted state and local policy. But others proved to be corrupt. Because of the latter, the term "carpetbagger" became synonymous with a meddling, opportunistic outsider.

What were **scalawags**?

Scalawags was a derogatory term used by white Southerners to refer to those white Southerners who supported or were sympathetic to the goals of the Republican lawmakers during Reconstruction (1865–77), the years of rebuilding that followed the American Civil War (1861–65). The Republican Party was the object of much contempt in the American South during the mid- to late 1860s: Congress was dominated by radical Republicans who were able to overturn presidential vetoes to achieve their goals. While President Andrew Johnson (1808–1875) supported a lenient policy toward the Southern "rebel" states, Congress favored a more gradual approach to reunification. In order to prevent another resurgence in the South, Congress delayed the readmission of former Confederate states until Republican governments could be established in them. Congress was also determined to enact legislation and sponsor constitutional amendments that would extend the rights of citizenship to freed slaves. After four years of brutal combat during the war, many Southerners saw fit to "side" with the Union and with the Republican lawmakers who were determining Reconstruction policy. President Andrew Johnson had proved himself an ineffectual leader, and Congress had mustered enough votes to repeatedly override him; in fact, they impeached him in 1868. Some whites in the South, contemptuously called scalawags, joined with Northern businessmen and politicians (derisively called carpetbaggers) as well as the freed slaves to forge a new South.

Why was **President Andrew Johnson impeached**?

In late February 1868 nine articles of impeachment were brought against Andrew Johnson over political and ideological differences between the president and Congress.

Johnson—self-educated, self-made, and outspoken—inspired people to either love or hate him. A Southern Democrat in the U.S. Senate, he broke bonds of home and party when he swore allegiance to the Union after the outbreak of the Civil War (1861–65). This he did because of his strong personal belief that the Southern states had violated the U.S. Constitution when they seceded from the Union. Soon this Tennessean and former Democrat shared the Union Party ticket with Republican Abraham Lincoln (1809–1865) as he ran for re-election to the presidency in the fall of 1864. Inaugurated in March, Vice President Johnson became President Johnson that fateful mid-April day when Lincoln was shot as he sat watching a play at Washington, D.C.'s Ford's Theater. But Johnson's troubles had already begun: As he and Lincoln took the oath of office in March, Johnson appeared to be drunk. Some attributed this to the fact that he was recovering from typhoid fever, but one journalist labeled him a "drunken clown," and a group of senators began calling for his resignation. Lincoln met with his vice president for the first—and what would turn out to be the last—time on April 14, just hours before Lincoln's life was claimed by assassin John Wilkes Booth (1838–1865).

As president, Johnson's true colors shined through. Again allegiant to his homeland, his policies toward Southern states were lenient; ever class-conscious, he used the power of his office to demonstrate to the Southern aristocrats, whom he openly despised, just how far a poor man from North Carolina had come; as a states' rights advocate, he was ever-watchful of any congressional bills that might impinge upon the freedoms of the individual states; and as a racist, he proved reticent to grant rights or protection to blacks.

All of these traits combined to create sticking points between Johnson and Congress: In February 1866 Congress voted to extend the life of the Freedmen's Bureau, a War Department agency that assisted blacks and whites. But Johnson vetoed the measure, and Congress was unable to overturn his veto. Later that year Congress passed the Civil Rights Act of 1866, a bill that extended citizenship to freed slaves and guaranteed them "equal protection of the laws." Believing this piece of legislation overstepped the boundaries of central government (he felt this sort of lawmaking was up to each state), he again vetoed it. But this time Congress mustered the votes it needed to overturn a presidential veto. It was the first of many veto overrides during Johnson's administration. Feeling Johnson was ill-equipped to run the nation, Congress moved its meeting time so that it could keep an eye on the executive branch. Meantime, Congress was guiding Reconstruction policy. The Southern states were being run by their military administrators, reporting to General Ulysses S. Grant (1822–1885).

In 1867 Congress passed a law, the Tenure of Office Act, preventing the president from removing any cabinet member without Congress's permission. By this time, Congress has already begun to consider whether Johnson ought to be impeached. That fall, President Johnson pardoned many Confederate generals and officials, further raising the ire of Congress—and the nation. Johnson's popularity was waning. The following February, Grant attempted to replace Edwin Stanton (1814–1869) as secretary of war. Stanton, who was favored by Congress, refused to leave his office, physically chaining himself to his desk. Congress viewed Johnson's move as a violation of the Tenure of Office Act and proceeded to hold impeachment hearings in the House of Representatives. Within a few days, the House approved a resolution of impeachment. On March 13, the trial began in the Senate. On May 19 the Senate voted on one of the articles of impeachment—it was considered to be the one most likely to receive the two-thirds majority vote required to convict the president. The measure failed—by one vote. Subsequent votes resulted in the same tally.

While many believe Johnson was an inadequate and unpopular president who made numerous mistakes while in office, many others believe he was not guilty of the high crimes and misdemeanors called for in Article 2 (Section 4) of the Constitution. In fact, the law that he was accused of breaking, the Tenure of Office Act, was later overturned as unconstitutional.

THE AMERICAN PRESIDENCY

Why does the president of the United States give a **State of the Union Address**?

The U.S. Constitution requires the president to annually present a joint session of Congress (attended by representatives and senators) with a status report on the nation. Presidents George Washington (1732–1799) and John Adams (1735–1826), the first and second presidents, delivered their messages in person. Thereafter the State of the Union was sent as a written message, which was read in Congress. But President Woodrow Wilson (1856–1924) delivered his messages in person, including that of January 1918, when he delivered the Fourteen Points—his formulation of a peace program for Europe once World War I (1914–18) had ended. Since Franklin D. Roosevelt (1882–1945) held office (beginning in 1933), all U.S. presidents have made formal addresses to Congress.

Which U.S. president held the **first press conference**?

President Woodrow Wilson (1856–1924) was the first president to routinely assemble the press to answer questions for the public. On March 15, 1913, shortly after his inauguration, he called the first presidential press conference. More than 100 news reporters attended the event. Decades later, President John F. Kennedy (1917–1963) became known for his frequent use of the televised press conference to directly communicate with Americans.

What was the **Kitchen Cabinet**?

It was the name given to President Andrew Jackson's unofficial group of advisers, who reportedly met with him in the White House kitchen. The group included the then secretary of state Martin Van Buren (1782–1862), who went on to become vice president (during Jackson's second term) and president from 1837 to 1841; F. P. Blair (1791–1876), editor of the *Washington Post,* who was active in American politics and later helped get Abraham Lincoln elected to office (1860); and Amos Kendall (1789–1869), a journalist who was also a speech writer for Jackson and went on to become U.S. postmaster general. The Kitchen Cabinet was influential in formulating policy during Jackson's first term (1829–33), many believe because the president's real cabinet, which he convened infrequently, had proved ineffective. But Jackson, the seventh president of the United States, drew harsh criticism for relying on his cronies in this way. When he reorganized the cabinet in 1831, the Kitchen Cabinet disbanded.

Jackson's favoritism to his circle of friends did not end with the Kitchen Cabinet, however. During his presidency the "spoils system" was in full force: Jackson gave public offices as rewards to many of his loyal supporters. Though the term *spoils system* was popularized during Jackson's terms in office (it was his friend, Senator William Marcy, who coined the phrase when he stated, "to the victor belong the spoils of the enemy"), Jackson was not the first president to grant political powers to his party's members. And the practice continued through the nineteenth century. However, beginning in 1883 laws were passed that gradually put an end to, or at least limited, the spoils system.

What was **Teapot Dome**?

Teapot Dome was a notorious political scandal that was on a level with Watergate (1972). While the early 1920s abuses of power affected President Warren G. Harding

Who Says a Watched Pot Never Boils?

A 1924 illustration of Teapot Dome, a notorious scandal involving the Warren Harding administration.

(1865–1923), it was not Harding who was implicated in the crimes. Albert Bacon Fall (1861–1944), Harding's secretary of the interior, secretly transferred government oil lands at Elk Hills, California, and Teapot Dome, Wyoming, to private use and he did so without a formal bidding process. Fall leased the Elk Hills naval oil reserves to American businessman Edward L. Doheny (1856–1935) in exchange for an interest-free "loan" of $100,000. Fall made a similar arrangement with another businessman, Harry F. Sinclair (1876–1956) of Sinclair Oil Corporation—leasing the Teapot Dome reserves in exchange for $300,000 in cash, bonds, and livestock.

The scandal was revealed in 1922, and committees of the U.S. Senate and a special commission spent the next six years sorting it all out. By the time the hearings and investigations were concluded in 1928, Harding had died; Fall had resigned from office and taken a job working for Sinclair; all three players—Doheny, Sinclair, and Fall—had faced charges; and the government had successfully sued the oil companies for the return of the lands. The punishments were light considering the serious nature of the charges: Fall was convicted of accepting a bribe, fined $100,000, and sentenced to a year in prison, while Doheny and Sinclair were both indicted but later acquitted of the charges against them, which included conspiracy and bribery.

How were **Theodore and Franklin Roosevelt** related?

The two men, among America's most well-known presidents, were distant cousins. Theodore Roosevelt (1858–1919) was born in New York City, and after a career in public service that included organizing the first volunteer cavalry regiment that was known as the Rough Riders, the ardent outdoors enthusiast became vice president in 1901. When President William McKinley (1843–1901) died in office later that year (on September 14), "Teddy" Roosevelt succeeded him as president. He was elected in his own right in 1904 and went on to serve until 1909, spending nearly two full terms in the White House.

Teddy Roosevelt was president of the United States when he walked his niece, Eleanor Roosevelt (1884–1962), down the aisle on March 17, 1905. The young woman was marrying her distant cousin Franklin Delano Roosevelt (1882–1945), who had been courting her since he entered college at Harvard in 1900.

Franklin D. Roosevelt was born in Hyde Park, New York. Like his fifth cousin Theodore, Franklin went on to a life of public service, which bore some remarkable similarities to that of his cousin: Both Theodore and Franklin served as assistant secretary of the U.S. Navy (1897–98 and 1913–20, respectively) and both were governors of New York (1899 to 1900 and 1929 to 1933, respectively). As presidents, both served the nation for more than one term—but Franklin Roosevelt made history for being the only president to be elected for third and fourth terms. (In 1951 the U.S. Congress voted in favor of the Twenty-second Amendment, limiting presidential tenure to just two terms.) Both served the country in times of conflict: For Theodore it was the

Russo-Japanese War (1904–05)—which he was instrumental in ending with the Treaty of Portsmouth (New Hampshire) on September 5, 1905, and for which he was awarded the Nobel peace prize the following year. Franklin Roosevelt was one of the so-called Big Three leaders: Along with Britain's Sir Winston Churchill (1874–1965) and the Soviet Union's Joseph Stalin (1879–1953), he coordinated the Allied nations' effort against Nazi Germany and Japan during World War II (1939–45). He, too, was a champion of peace, having been central in laying plans for the United Nations.

President Franklin D. Roosevelt, pictured in 1937, began broadcasting his "fireside chats" to reassure the American public during the Great Depression.

It's an interesting note, however, that when Teddy Roosevelt ran for president in 1912, he was opposed by his young Democratic cousin Franklin, then a state senator in New York, who supported Woodrow Wilson (1856–1924) in the presidential race. After Wilson was elected, he appointed Franklin Roosevelt assistant secretary of the navy—a post that delighted him for combining his vocation (politics) with his avocation (ships), and one that certainly furthered his political career. By the end of World War I (1914–18), Franklin Roosevelt was a well-known national figure.

Theodore and Franklin also shared an interest in outdoor activities. But Franklin's participation in sports was curtailed when he was stricken with polio in August 1921. The 39-year-old Roosevelt was paralyzed for a time, and though he later regained movement and was able to walk with braces, he never fully recovered. Through fierce determination he continued his life of public service, becoming president in 1933. He saw the country through two of its most trying periods: the Great Depression (1929–39) and World War II (1939–45). He died suddenly of a brain hemorrhage in April 1945.

What were the **"fireside chats"**?

They were radio broadcasts of President Franklin D. Roosevelt's (1882–1945) messages to the American people. FDR began making the informal addresses on March 12, 1933, during the long and dark days of the Great Depression. In his efforts to reassure the nation, FDR urged listeners to have faith in the banks and to support his New Deal measures. Sometimes beginning his talks with "My friends," the radio broadcasts were enormously successful and attracted more listeners than even the most popular broadcasts during this "golden age" of radio. FDR continued his fireside chats into the 1940s, as Americans turned their attention to the war effort.

Why was **Eleanor Roosevelt** called "the people's First Lady"?

While several First Ladies before her had also been active in the nation's life, Eleanor Roosevelt (1884–1962), wife of thirty-second president of the United States Franklin D. Roosevelt (1882–1945), stands out as one of the country's most active first ladies and as a woman of enormous accomplishment in her own right. During her husband's administration, which began in the dark days of the Great Depression (1929–39) and continued as the world again went to war, Eleanor Roosevelt acted not only as an adviser to the president, but as the president's eyes and ears on the nation—traveling in a way that his physical condition prevented him from doing.

From the start, Eleanor Roosevelt remained in constant communication with the American people: She was known for her weekly press conferences, numbering some 350 by the end of the Roosevelt presidency, that were open only to women reporters. In 1934 she began a radio program, which became so popular that she was soon dubbed "the First Lady of Radio." Beginning in 1936 she authored a daily column called "My Day," which was syndicated to newspapers around the country. These forums gave the First Lady an unprecedented voice in American life and gave Americans a clear understanding of their First Lady and her concerns.

Concerned about the effects of the Great Depression on American children, she was instrumental in creating the National Youth Administration, which helped high school and university students complete their studies before joining the workforce. She was a champion of minority groups, declaring that the right to work "should know no color lines" and resigning from the Daughters of the American Revolution when the group refused to allow black singer Marian Anderson (1897–1993) to perform at Constitution Hall.

Eleanor Roosevelt was known for getting out among the people; she lectured frequently and made other public appearances in which she met and spoke face to face with the American people. A famous cartoon depicted a coal miner pausing in his work to exclaim, "For gosh sakes, here comes Mrs. Roosevelt." During World War II (1939–45) she made a remarkable 23,000-mile trip across the South Pacific, where she untiringly visited American GIs in field hospitals and on the lines.

Mrs. Roosevelt was an advocate for the people, and it just so happened that she lived in the White House. A beloved First Lady who actively supported liberal causes and humanitarian concerns, she has been a model to subsequent first ladies and women politicians and activists.

Why is the **Kennedy presidency** called "Camelot"?

The term was actually assigned by Jacqueline Bouvier Kennedy (1929–1994). Shortly after President John F. Kennedy was assassinated (November 22, 1963), the former First Lady was talking with a journalist when she described her husband's presidency as an American Camelot, and she asked that his memory be preserved. Camelot refers, of course, to the time of King Arthur and the Knights of the Round Table, and has come to refer to a place or time of idyllic happiness. John Fitzgerald Kennedy's widow, who had with fortitude and grace guided her family and the country through the sorrow and anguish of the president's funeral, quite naturally held sway over the American public. So when she suggested that the shining moments of her husband's presidency were reminiscent of the legends of Camelot, journalists picked up on it. Despite subsequent revelations that there were difficulties in the Kennedy marriage, public opinion polls indicate that the image of Camelot—albeit somewhat tarnished—has prevailed.

How many **U.S. presidents** have been **assassinated**?

Four American presidents were assassinated in office: Abraham Lincoln, James Garfield, William McKinley, and John F. Kennedy.

Abraham Lincoln (1809–1865) was shot on the evening of April 14, 1865, as he sat in the presidential box of Ford's Theatre in Washington, D.C., watching a performance of *Our American Cousin*. The man who fired the shot was actor John Wilkes Booth (1838–1865), who then jumped onto the stage, fell (breaking a leg), and limped away, calling out, "*Sic semper tyrannis*" (a Latin phrase meaning "Thus always to tyrants"). The president lived through the night, attended by family. He died just after 7:00 A.M. on April 15. He was succeeded in office by Vice President Andrew Johnson (1808–1875). On April 26 a search party found Booth in a Virginia barn, where he was fatally shot.

James Garfield (1831–1881) was en route to a class reunion at Williams College (Williamstown, Massachusetts), on July 2, 1881, when his assailant fired two shots at him in a Washington, D.C., train station. The shooter was Charles J. Guiteau (1841–1882), who held a grudge against the president. One of Guiteau's bullets had only grazed the president; the other was fixed in his back, and doctors were unable to locate it. Today the president's life might well have been spared, but the medical treatment of the late 1800s, which lacked both the X-ray machine and antiseptics, could not save him. He lived 80 days more, dying at a cottage on the New Jersey shore on September 19. He was succeeded in office by Vice President Chester Arthur (1830–1886). Guiteau's trial lawyer would later claim that Garfield's assassin was insane, but it was an unsuccessful plea for his life: In 1882 he was convicted and hung.

On September 6, 1901, President William McKinley (1843–1901) was attending a reception in Buffalo, New York, where the previous day he had delivered a speech. As he approached a man to shake his hand, the fellow fired two shots at McKinley. One bullet delivered only a minor flesh wound, but the other lodged in his stomach. Sur-

On September 6, 1901, anarchist Leon Czolgosz shot President William McKinley with a concealed revolver; McKinley died eight days later. (Original drawing by T. Dart Walker.)

geons operated, but gangrene and infection set in, claiming the president's life the morning of September 14. He was succeeded in office by Vice President Theodore Roosevelt (1858–1919). The shooter was identified as avowed anarchist Leon F. Czolgosz (1873–1901); he was tried, convicted, and put to death in 1901.

President John F. Kennedy (1917–1963), accompanied by his wife, Jacqueline Bouvier Kennedy (1929–1994), traveled in a motorcade through the streets of Dallas, Texas, on November 22, 1963. They were en route to the Dallas Trade Mart, where the president was scheduled to make a lunchtime speech. At 12:30 P.M., shots rang out; the president, who was riding in the back seat of a convertible, was hit in the neck and head. He was rushed to a nearby hospital, where he died at 1:00 P.M. The nation's loss was immediately felt, as television and radio stations broadcast the message live that Kennedy had been shot and killed. He was succeeded by Vice President Lyndon Baines Johnson (1908–1973), who took the oath of office aboard an airplane just after 2:30 P.M.

Additionally there were assassination attempts on the lives of Presidents Andrew Jackson (April 14, 1835), Theodore Roosevelt (October 14, 1912), Franklin D. Roosevelt (February 15, 1933), Harry S. Truman (November 1, 1950), Gerald R. Ford (two attempts, both in September 1975), and Ronald Reagan (March 30, 1981). Theodore

Roosevelt and Ronald Reagan recovered from their injuries; the others were not injured in the attempts.

What happened at **Watergate**?

Watergate is a complex of upscale apartment and office buildings in Washington, D.C. In July 1972 five men were caught breaking into the Democratic Party's national headquarters there. Among these men was James McCord Jr. (1924–), the security coordinator of the Committee for the Re-election of the President (CRP). McCord was among those working to get President Richard Nixon (1913–1994), a Republican, elected to a second term in office.

All five men who were caught in the break-in were indicted on charges of burglary and wiretapping, as were CRP aide G. Gordon Liddy (1930–) and White House consultant E. Howard Hunt (1918–). Five of the men pleaded guilty to the charges. McCord and Liddy were tried and found guilty.

In February 1972—five months before the break-in at Watergate—President Nixon had traveled to China, becoming the first U.S. president to visit that country. In May he traveled to Moscow, where he signed the Strategic Arms Limitation Treaty (SALT-1 treaty), the first such treaty between the United States and the U.S.S.R. When the election was held in November, Nixon won in a landslide victory over the Democratic candidate George McGovern (1922–1998).

But early in Nixon's second term, which began in 1973, the Watergate affair became a full-blown political scandal when convicted burglar James McCord wrote a letter to District Court judge John Sirica (1904–1992), charging a massive cover-up in the Watergate break-in. A special Senate committee began televised investigations into the affair. Before it was all over, about 40 people, including high-level government officials, had been charged with crimes including burglary, wiretapping of citizens, violating campaign finance laws by accepting contributions in exchange for political favors, the use of government agencies to harm political opponents, and sabotage.

Among those prosecuted were John Dean (1938–), former White House counsel, and Attorney General John Mitchell (1913–1988). It was revealed that members of the Nixon administration had known about the Watergate burglary. It was also discovered that the president had taped conversations in the Oval Office. When Dean and Mitchell were convicted, public confidence in President Nixon plummeted. In July 1974 the Judiciary Committee of the House of Representatives was preparing articles of impeachment (including one that charged the president with obstruction of justice) against the president. The impeachment proceedings would not make it as far as the Senate: Nixon chose to resign on August 9, 1974. He was the first and so far only U.S. president to resign from office.

Former White House aide John Dean is sworn in by Senate Watergate Committee chairman Sam Ervin, June 25, 1973.

Shortly after taking office, Nixon's successor, Gerald R. Ford (1913–), pardoned Nixon. But Watergate remains a dark chapter in the nation's history.

What was the **Iran-Contra** affair?

It was a series of actions on the part of U.S. federal government officials, which came to light in November 1986. The discoveries had the immediate effect of hurting President Ronald Reagan (1911–2004), whose policy of antiterrorism had been undermined by activities initiated from his own executive office. Following in-depth hearings and investigations into "who knew what, when," special prosecutor Lawrence Walsh (1912–) submitted his report on January 18, 1994, stating that the dealings with Iran and with the contra rebels in Nicaragua had "violated United States policy and law."

The tangled string of events involved Reagan's national security advisers Robert McFarlane (1937–) and Admiral John Poindexter (1936–), Lieutenant Colonel Oliver North (1943–), Poindexter's military aide, the Iranian government, and Nicaraguan rebels.

The U.S. officials evidently had begun their dealings with both the Iranian government and the Nicaraguan rebels with the goal of freeing seven Americans who were held hostage by Iranian-backed rebels in Lebanon. President Reagan had met with the families of the captives and was naturally concerned about the hostage situation. Under pressure to work to free the hostages, McFarlane, Poindexter, and North arranged to sell an estimated $30 million in spare parts and antiaircraft missiles to Iran (then at war with neighboring Iraq). In return, the Iranian government would put pressure on the terrorist groups to release the Americans.

Profits from the arms sale to Iran were then diverted by Lieutenant North to the contras in Central America who were fighting the dictatorial Nicaraguan government. Congress had already passed laws that prohibited U.S. government aid to the Nicaraguan rebels; the diversion of funds certainly appeared to violate those laws.

The Iran-Contra affair led to North's dismissal and to Poindexter's resignation. Both men were prosecuted. Though the hostages were freed, Reagan's public image was seriously damaged by how the release had been achieved.

During the Iran-Contra hearings in 1987, National Security Commission officials revealed that they had been willing to take the risk of providing arms to Iran in exchange for the safe release of the hostages because they all remembered the U.S. government's failed attempt in 1980 to rescue hostages held at the American Embassy in Tehran, Iran.

Nevertheless, the deal with Iran had supplied a hostile country with American arms that could then be used against the United States. In 1987 Iran did launch an offensive when it attacked Kuwaiti oil-tankers that were registered as American and laid mines in the Persian Gulf. The United States responded by sending in the navy, which attacked Iranian patrol boats. During this military initiative, the U.S. Navy accidentally shot down a civilian passenger jet, killing everyone on board.

Why was **President Clinton** impeached?

Some believe the proceedings were nothing more than a "vast right-wing conspiracy," a term coined by First Lady Hillary Rodham Clinton (1947–) early in 1998. Still, others—including enough members of the U.S. House of Representatives to bring 11 counts of impeachment against President Clinton in December 1998—felt the nation's chief had perjured himself and obstructed justice. Many also believed he had jeopardized the authority of the U.S. presidency. Accused of having an affair with White House intern Monica Lewinsky, President Clinton vehemently denied it. Upon continued investigation, conducted by Special Prosecutor Kenneth Starr's office, the allegations proved to be true. Since the president had been so adamant in his statements to the contrary, evidence began to accumulate that he had lied about his relationship with the young woman and that he had tried to cover up the matter.

Many believed the charges against Clinton did not constitute the high crimes and misdemeanors called for by the U.S. Constitution to remove a president from office. Nevertheless, in January 1999 the U.S. Senate organized itself to hear the charges against the president. When the trial concluded in February, Clinton was acquitted of

On July 31, 1998, President Bill Clinton gestures to reporters that he will take no questions about the Monica Lewinsky investigation.

both perjury and obstruction of justice. He served out his second term and left office with high approval ratings, despite being the subject of the longest criminal investigation of a president in history.

The Clinton impeachment was the first time the federal legislature had convened itself as a court in more than 130 years—since the impeachment hearings of President Andrew Johnson (1808–1875) in 1868.

What is a **hanging chad**?

A hanging chad, also called a bulging chad or a pregnant chad, is the scored piece of paper that remains on a punch-card ballot if the voter fails to fully punch the card to dislodge the paper. The chad gained national and international attention in the presidential election of 2000 when the race between Republican candidate George W. Bush (1946–) and Democrat Al Gore (1948–) hinged on the ballot count in Florida. The vote tally in the Sunshine State was so close that manual counting took place in the days following the November 7 election. News photos showed counters holding punch-card ballots up to the light to try to determine if a voter had meant to poke a chad through, thereby indicating a vote.

But Florida was not the first state where a disputed election came down to ballot inspection to try to determine voter intent. In 1996 a Democratic primary for a House seat in Massachusetts was settled by carefully inspecting ballots and counting those that were merely dimpled by the voter. In support of its decision to scrutinize ballots to determine voter intent, the Massachusetts Supreme Court stated, "If the intent of the voter can be determined with reasonable certainty from an inspection of the ballot...effect must be given to that intent and the vote counted." The conclusion was that a voter should not be disqualified for failing to completely express his or her intent. A 1990 case in Illinois and a 1981 case in Indiana also ruled in favor of examining ballots to try to determine what the voter meant to do. In all cases, ballots where no clear intent could be determined were set aside. In 2000 the Florida recount ended with a tally of 2,909,176 votes for Bush and 2,907,451 for Gore. In winning Florida, the national electoral vote swung in Bush's favor. Though Gore won the popular vote with a national total of 51,003,894 votes to Bush's 50,459,211, Bush managed to carry the electoral college.

Are the **Bushes** the **first father-son presidents**?

No, the nation's 41st (George H. W. Bush) and 43rd (George W. Bush) presidents were preceded as father-son presidents by John Adams (1735–1826), the 2nd president of the United States, and John Quincy Adams (1767–1848), the 6th.

There have been other presidents whose relatives held the office before them: Benjamin Harrison, the 23rd president, was the grandson of the nation's 9th, William Henry Harrison. Zachary Taylor, the 12th president, and James Madison, the 4th, were second cousins. Franklin Delano Roosevelt was preceded in the office by his distant cousin, Teddy Roosevelt. Genealogists determined that FDR had ties to 10 other presidents as well, four of them were blood relatives and six were relatives by marriage: John Adams, John Quincy Adams, Ulysses S. Grant, William Henry Harrison, Benjamin Harrison, James Madison, William Howard Taft, Zachary Taylor, Martin Van Buren, and George Washington.

ABSOLUTISM

Was the **holocaust** directed by the German government?

Yes. When fervent nationalist Adolf Hitler (1889–1945) rose to power in Germany in 1933 he quickly established a reign of terror based on his philosophy that the German (Aryan) race is superior to all others. He established a violent policy against Jews: Those who did not flee the country were rounded up and sent to concentration camps, where they were kept without cause. This was before Hitler's acts of military aggression in Europe. But after German troops invaded Poland and World War II began, the führer's anti-Semitic campaign was accelerated. Jews in Germany and in Nazi-occupied countries of Europe were severely persecuted. Those who were put into concentration camps—including Auschwitz, Treblinka, Buchenwald, and Dachau—were exterminated, many of them in gas chambers. By the end of the war, in 1945, Hitler's "final solution to the Jewish question" had been under way for some 12 years, and 6 million Jews had been systematically murdered by the Nazis during the Holocaust, or Shoah (Hebrew). As his defeat was imminent, the despotic ruler took his own life in 1945. By then he had destroyed Europe's Jewish community. Many of Hitler's leaders were later tried by an international court at Nuremberg.

Why do historians draw comparisons among **Stalin, Hitler, Napoleon, and Alexander the Great?**

All four men were fiercely powerful—and seemingly indomitable—leaders. And their early lives and careers shared some interesting characteristics as well.

Inmates of the Nazi government's Buchenwald concentration camp, April 1945. By the end of World War II, about 6 million Jews had died in the holocaust.

First, none of them were native sons of the countries they ended up ruling (and some would say, defining). Soviet leader Joseph Stalin (1879–1953) was born in Georgia, German chancellor and führer Adolf Hitler (1889–1945) in Austria, French emperor Napoleon Bonaparte (1769–1821) in Corsica (which, though technically a new possession of France, was decidedly more Italian in nature), and king of the Macedonians Alexander the Great (356–323 B.C.) was from the fringe city-state of Macedonia. Yet it was, respectively, Russian, German, French, and Greek culture that each leader spread. All four men rose to power through the military, or, in Alexander's case (who succeeded his father), after having carried out military strategy. Once installed as rulers, all led fantastic military campaigns to expand their spheres of influence. The wars brought on by their nationalism left millions dead. Finally, all four men inspired, and some would say continue to inspire, terror in the human soul.

Still, history has been more forgiving of both Alexander and Napoleon, who today seem to be portrayed as often as heroes as they are as megalomaniacs. With Hitler's and Stalin's horrible legacies a more recent memory, the view of these men is decidedly and far more consistently dark and loathsome.

What were the **gulags**?

Gulags were prisons for political dissidents in the Soviet Union; the prisons existed from 1919 into the 1950s. Gulag is an abbreviation of the Russian name of the system, *Glavnoye Upravleniye Ispravitelno-trudovykh Lagerey,* which translates to Chief Administration of Corrective Labor Camps. The camps were first used during the collectivization of agriculture in the late 1920s and early 1930s. Under Soviet leader Joseph Stalin's (1879–1953) purges of the 1930s, anyone who posed, or seemed to pose, a threat to his hard-line Communist regime was rounded up and sent to a gulag. During World War II (1939–45), prisoners of war were held in the gulags. And after the war, Stalin continued to use the camps to punish those who opposed him. Though exact figures are unknown, it is believed that as many as 30 million people were imprisoned in gulags, where they faced forced labor, grueling conditions, and maltreatment including starvation. (Official Soviet figures place the number around 10 million.) Millions are believed to have died in the gulags. After Stalin died in 1953, the system was dismantled, with some of the prisoners receiving amnesty.

What was the **Long March**?

The Long March began in October 1934 when Mao Tse-tung (1893–1976) led Chinese Communist forces (the Red Army of China), numbering 100,000 men and women, on an epic walk across China. With the nationalist army in pursuit, the Communist marchers crossed 18 mountain chains and 24 rivers to cover 6,000 miles. Almost all women and children died along the way. In 1935, 20,000 to 30,000 people finally reached Shaanxi (Shensi) Province in the north, where the Red Army established a stronghold. It was there that Mao, one of the earliest members of the Chinese Communist Party, formulated his own philosophy that came to be known as Maoism. He had adapted Marxism to the Chinese conditions—replacing German politician and socialist Karl Marx's (1818–1883) urban working class with the peasant farmers as the force behind revolution. The Red Army went on to defeat the nationalists in 1949; Mao was named chairman of the People's Republic of China, a Communist state, that same year.

THE POSTWAR ERA

How was the **United Nations** formed?

Officially, the United Nations (UN) was not formed until October 1945. However, events during World War II (1939–45) had paved the way for the founding of the international peacekeeping organization that today is so familiar to people around the globe.

Joseph Stalin, Franklin D. Roosevelt, and Winston Churchill during the Tehran Peace Conference, 1943.

Fervent German nationalist Adolf Hitler (1889–1945) and his troops invaded Poland in 1939, and soon Nazi Germany had conquered much of Europe. Leaders of nine nations—Belgium, Czechoslovakia, France, Greece, Luxembourg, the Netherlands, Norway, Poland, and Yugoslavia—met with Britain and its Commonwealth states in London. There, on June 12, 1941, the countries signed the Inter-Allied Declaration, vowing to work together for a free world. Two months later, on August 14, U.S. president Franklin Delano Roosevelt (1882–1945) and British prime minister Winston Churchill (1874–1965) signed the Atlantic Charter. In it the two leaders outlined their aims for peace.

On January 1, 1942, the "declaration by the United Nations" was signed by 26 countries who pledged to work together to fight the Axis powers (Germany, Italy, and Japan), and they agreed not to make peace separately. The term *United Nations,* otherwise known as the UN, is believed to have originated with Roosevelt.

In late November and early December 1943, the Big Three—Roosevelt, Churchill, and Soviet premier Joseph Stalin (1879–1953)—met in Tehran, Iran, for the first time during the war. There these Allied nations leaders cited the responsibility of a United Nations organization in keeping the peace once the war was over. Though ending the war was first and foremost in the minds of these leaders, all had seen two world wars fought in close succession and were determined that the nations of the world could work together to prevent such an event from happening again. In August 1944, at the Dumbarton Oaks Conference in Washington, D.C., representatives of Britain, the United States, the Soviet Union, and China met to make plans for the peacekeeping organization that had been envisioned. The outcome of that meeting, which lasted into October, was the basic concept for the UN Security Council as we know it—with the world's (five) major powers having permanent seats on the council and a limited and rotating membership beyond that.

When the Big Three met again at the Yalta Conference (in the Soviet Union) in February 1945, they discussed matters that were central to ending the fighting with Germany and with Japan. But Roosevelt, Churchill, and Stalin also announced that a conference of the United Nations would open in San Francisco on April 25 of that year.

Having directed the United States' massive war effort, Roosevelt did not live to see the end of World War II or the creation of the international peacekeeping body. Roosevelt died suddenly on April 12, 1945. The war in Europe ended May 7. And during the closing days of World War II, the UN was chartered—the representatives of the Allied nations met as promised in San Francisco. On June 26, 1945, the governing treaty was signed by the delegates. On October 24, 1945, shortly after the war ended with Japan, the United Nations officially came into existence when the required number of nations approved the charter. Fifty nations were members of the UN at the outset; today the membership numbers 191.

When did the **United Nations begin** its work?

The United Nations (UN) began fulfilling its mission very soon after it was chartered: The first meeting was held January 10, 1946, when the inaugural session of the General Assembly opened in London. On February 14, the UN voted to make its headquarters in the United States. Its complex was built in Manhattan, overlooking New York's East River, in 1950.

What does the **UN** do?

Representatives of member countries work to keep peace and ensure security for people around the globe. Over the decades, the UN has expanded its role as a provider of humanitarian aid and stepped up its efforts in the area of human rights. When disputes arise, the UN works toward diplomatic resolution. Though not always successful in its role as peacekeeper or peace negotiator, the organization has provided a forum for debate, which has prevented some disputes from developing into major wars. And through its various agencies, the UN provides assistance to developing nations, promotes humanitarian causes, and sends relief to war-torn areas.

However, critics have charged the international organization with arbitrarily defining borders between countries, saying that these drawn boundaries divide ethnic groups and result in conflicts (such as in the Middle East and Africa).

How many nations are **members of the UN** today?

In 2005 the United Nations (UN) membership stood at 191 nations, with the most recent additions having been Switzerland and Timor-Leste (East Timor) in 2002. The membership of the UN grew steadily after it was founded in 1945. Originally there were 51 member states, including Canada, France, the United Kingdom, the USSR (today the Russian Federation), and the United States. Within a decade, 25 more countries had been added to the UN's membership roster. By 1965 there were 117 members; by 1975, 144 members; by 1985, 159 members; and by 1995, 185 members.

What are the **bodies and agencies of the United Nations**?

The UN's charter establishes six main bodies and explains their duties and operating methods. The General Assembly is the major forum: all member nations are represented there, and the assembly can discuss any issue that is deemed relevant and important to the UN; the Security Council has the major responsibility for preserving peace; the Economic and Social Council investigates economic questions and works to improve living standards; the Secretariat is the UN's administrative body, helping all organs do their work; the Trusteeship Council assists non-self-governing territories; and the International Court of Justice hears disputes between member nations. Except for the last one, all bodies convene in the UN headquarters in New York City. The International Court of Justice meets in The Hague.

Since the charter was written in 1945, the United Nations has established numerous agencies, committees, and commissions to help carry out its work around the world. Among those that are most well known to the public are the United Nations Educational, Scientific, and Cultural Organization (UNESCO), which encourages the exchange of ideas among nations; the United Nations International Children's Emergency Fund (UNICEF), which assists children and adolescents worldwide, particularly those in devastated areas or developing countries; the World Health Organization (WHO), which promotes high health standards around the globe; the International Labor Organization (ILO), which works to improve labor conditions and protect workers; and the International Monetary Fund (IMF), which addresses currency and trade issues.

Who are the **members of the United Nations Security Council**?

According to the UN charter, the Security Council has 15 members: 5 permanent and 10 that are elected by the General Assembly for two-year terms. The permanent members are China, France, the Russian Federation, the United Kingdom, and the United States. The 10 temporary slots are elected annually, 5 nations at a time, so that on January 1, 2004, the following nations began two-year memberships: Algeria, Benin, Brazil, the Philippines, and Romania. On January 1, 2005, Argentina, Denmark, Greece, Japan, and the United Republic of Tanzania began two-year memberships. In 2005 Japan, Germany, India, and Brazil made bids for permanent seats on the council.

What does the **Security Council** do?

According to the UN charter, the functions and powers of the Security Council are:

> to maintain international peace and security in accordance with the principles and purposes of the United Nations; to investigate any dispute or situation which might lead to international friction; to recommend methods of adjusting such disputes or the terms of settlement; to formulate plans for the establishment of a system to regulate armaments; to determine the existence of a threat to the peace or act of aggression and to recommend what action should be taken; to call on Members to apply economic sanctions and other measures not involving the use of force to prevent or stop aggression; to take military action against an aggressor; to recommend the admission of new Members; to exercise the trusteeship functions of the United Nations in 'strategic areas'; and to recommend to the General Assembly the appointment of the Secretary-General and, together with the Assembly, to elect the Judges of the International Court of Justice.

What is the **G-8**?

It is the Group of Eight, an annual meeting of the world's leading industrial democracies: Canada, France, Germany, Italy, Japan, Russia, the United Kingdom, and the United States. It began as the Group of Six, with a 1975 conference in France, where representatives of six nations (France, Germany, Italy, Japan, the United Kingdom, and the United States) met to discuss international economic and political issues. The goal of the meeting was to shore up cooperation on matters of concern to the member nations. Canada joined the group in 1976 and Russia in 1994 (though the nation did not participate fully in the sessions until 1998). Hosting responsibilities rotate among the eight member countries. In 2004 the United States hosted the 30th meeting, on an island off the coast of Georgia. Among the agenda items were the training of international peacekeepers, setting up a global initiative to develop a human immunodeficiency virus (HIV) vaccine, and developing a plan to end famine in the Horn of Africa (easternmost Africa, including Somalia, Ethiopia, and Djibouti) by 2009. The leaders and representatives of non-G-8 countries were invited to participate in discussions relevant to them.

Why was **Mohandas Gandhi** called "Mahatma"?

Mohandas Gandhi was called "Mahatma" (meaning "great-souled") by the common people, who viewed him as India's national and spiritual leader. He is considered the father of his country. He was born in India on October 2, 1869. As a young man, Gandhi studied law in Britain. Practicing briefly in India, he then traveled to British-controlled South Africa on business. Observing oppressive treatment of Indian immigrants there, he held his first campaign of passive resistance. Gandhi would later become very well known for this method of protest, called *satyagraha* (meaning "firmness in truth").

Back in India as of 1915, Gandhi organized a movement of the people against the British government there: Britain had taken control of India during the 1700s and remained in power. After World War I (1914–18), Indian nationalists fought what would be a long and sometimes bitter struggle for political independence. While Gandhi's protests took the form of nonviolent campaigns of civil disobedience, such as boycotts and fasts (hunger strikes), he was more than once arrested by the authorities for causing disorder, as his actions inspired more extreme measures on the part of his followers, whose protests took the form of rioting.

As a member and, later, the president of India's chief political party, the Indian National Congress, Gandhi led a fight to rid the country of its rigid caste system, which organizes Indian society into distinct classes and groups. In Gandhi's time, not only were there four *varna,* or social classes, but there was a fifth group of "untouchables" who ranked even below the lowest class of peasants and laborers. Improving the lot of the untouchables was of tantamount importance to the leader, who by this time had abandoned Western ways in favor of a life of simplicity.

Beginning in 1937 Gandhi became less active in government, giving up his official roles, but he continued to be regarded as a leader of the independence movement. During World War II (1939–45), he was arrested for demanding British withdrawal from the conflict. Released from prison in 1944, Gandhi was central to the postwar negotiations that in 1947 resulted in an independent India. A believer in the unity of humankind under one God, he remained tolerant to Christian and Muslim beliefs. Amidst an outbreak of violence between Hindus and Muslims, Gandhi was on a prayer vigil in New Delhi when a Hindu fanatic fatally shot him in 1948.

Was **Indira Gandhi** related to Mohandas Gandhi?

No, the two were not related, except by events. After India achieved independence in 1947, the country's first prime minister was Jawaharlal Nehru (1889–1964), who had been a follower of Mohandas Gandhi (1869–1948), the great leader of India's long struggle for autonomy from Great Britain. During his entire tenure (1947–64) as leader of

India, Nehru was assisted by his only child, Indira (1917–1984), who in 1942 married a man named Feroze Gandhi—of no relation to Mohandas Gandhi. Indira Gandhi took an active role in India's national affairs. After her father died, she went on to become prime minister in 1966. However, hers was a troubled tenure. Found guilty of employing illegal election practices, Indira Gandhi was ousted by her political opponents in 1977. Determined to return to power, she was re-elected to parliament in 1980 and again served as prime minister until her death in 1984. She was assassinated by two of her own security guards, Sikhs who were motivated by religious reasons. Her son and successor, Rajiv Gandhi (1944–1991), was also assassinated, in 1991.

Mohandas Gandhi (pictured c. 1945) was called "Mahatma," meaning "great-souled," by the common people, who viewed him as India's national and spiritual leader.

The Middle East

When was **modern Israel** established?

As a modern state, Israel was formed by decree in 1948. In the wake of World War II (1939–45), the United Nations (UN) formed a special committee to address the British control of Palestine, the region in the Middle East (southwest Asia) that borders the Mediterranean Sea to the west, Lebanon to the north, Syria and Jordan to the east, and Egypt (the Sinai Peninsula) to the southwest; the narrow piece of land comes to a point in the south, where it fronts the Gulf of Aqaba. In November 1947 the United Nations (UN) carved Israel out of the Palestine region; areas of Palestine that were not designated as Israel were divided between neighboring Arab countries.

Modern Israel's first leader, David Ben-Gurion (1886–1973), proclaimed an independent Israel on May 14, 1948. Born in Poland, Ben-Gurion had arrived in Palestine as a young man of about 20 and became extremely active in efforts to assert Jewish autonomy in the region. He served as prime minister from 1949 to 1953 and again between 1955 and 1963. But Israel's history goes back much farther than these twentieth-century events. And, having such a long history, it is also a complicated one.

Israel was an ancient kingdom in Palestine, formed under King Saul in 1020 B.C. Israel included the lands in Canaan, the Promised Land of the Hebrew tribes who descended from the people that Moses (fourteenth–thirteenth century B.C.) led out of Egypt. But the kingdom was subsequently divided and by the eighth century B.C. it had ceased to exist. Nevertheless, the area remained home—and holy land—to the Hebrews (Israelites) who had settled there.

The entire region of Palestine, including the kingdom of Israel, subsequently came under the control of various empires. Palestine saw the rule of the Assyrians, the Chaldeans, the Persians, the Macedonians under Macedonian king Alexander the Great (356–323 B.C.), and the Romans (the area was the Roman province of Syria in the time of Jesus). After Roman rule, Palestine was, with only one exception, ruled by various Muslim (Islamic) dynasties, including the Ottoman Empire (1516–1917). It was in 1917 that Palestine came under control of the British, who proclaimed in the Declaration of Balfour to support the establishment of a national home for the Jews living there. However, Britain reversed this policy in 1939 at the same time the area was seeing an influx of Jewish people who were escaping persecution in Europe. Jews in Palestine opposed British control, and at the same time fighting intensified between Jews and Arabs.

The 1947 decision by the UN to establish a Jewish homeland resulted in nearly two years of fighting between Israelis and Arabs in the region. And though boundaries among the various states were determined anew in 1949, fighting in the region continued with the Arab-Israeli wars of 1956, 1967, 1973–74, and 1982. Unrest prevailed throughout the 1980s and into the 1990s, when the two sides began discussions to resolve the long conflict.

What is the **Palestine Liberation Organization**?

It is a group formed in 1964 by Arabs in Palestine, a region coextensive with the nation of Israel. Known as the PLO, the organization seeks to establish an area of self-rule for Muslims. Dominated by then guerrilla leader Yasser Arafat (1929–2004), the PLO regarded Israel as an illegitimate state and became determined to establish a Palestinian state in the region. In 1974 the PLO was recognized by the United Nations and by Arab countries as the governing body of the Palestinian people; however, the Palestinians remained without a homeland and continued to fight for one, often resorting to terrorist tactics.

In 1993 Israel and the PLO, still under Arafat's leadership, officially recognized each other: In an internationally brokered agreement, Israel agreed that by early 1996 it would withdraw its troops from the Gaza Strip, a tiny ribbon of land along the Mediterranean and bordering Egypt, and most cities and towns of the West Bank, a larger region lying west of the Jordan River and Dead Sea. (The city of Jerusalem, recognized by both Israelis and Palestinians as their capitals, straddles the western border of the West Bank and is divided into an Arab East Jerusalem and a Jewish-Israeli West Jerusalem.) The 1993 Arab-Israeli agreement, called the Oslo Accords, effectively carved out an autonomous Palestinian homeland, or at least an autonomous region. To govern this region, the Palestinian Authority (PA) was set up.

In 1994 Arafat and Israeli leaders Yitzhak Rabin (1922–1995) and Shimon Peres (1923–) were awarded the Nobel peace prize for their efforts. In January 1996 Palestinians in the Gaza Strip and the Palestinian-controlled parts of the West Bank elected a legislature and a president (Arafat) to govern these areas.

But the accords foundered: both sides contributed to an escalation of violence in the long-troubled region. In 2003 the Roadmap for Peace, developed by the United States, in cooperation with Russia, the European Union, and the United Nations, was presented to Israel and the Palestinian Authority (PA). The plan outlined clear goals and timelines for their achievement for both sides. Still the violence continued, with both sides failing to take the reciprocal steps necessary to peaceably coexist in the region. Cautious hopes for peace were again raised in February 2005 when Palestinian leader Mahmoud Abbas (who in January succeeded Arafat) and Israeli Prime Minister Ariel Sharon met at an Egyptian resort for summit talks. The two announced a verbal cease-fire pledge: Sharon promised that the Israeli military would end assaults on Palestinians, and Abbas vowed to bring an end to militant attacks on Israelis. But there were obstacles to the agreement: Israeli settlements in the Gaza Strip and West Bank needed to be dismantled; the Palestinians needed to gain control of the militant group Hamas, responsible for hundreds of acts of terrorism against Israelis; and the Israeli military needed to step down its responses. The Palestinian government remained tied to the PLO.

The Soviet Bloc

Who coined the term **"iron curtain"**?

It was former British prime minister Winston Churchill (1874–1965). In a March 1946 speech in Fulton, Missouri, he remarked that "an iron curtain has descended across the Continent." The statesman, who had been instrumental in coordinating the Allied victory in World War II (1939–45), was commenting on Soviet leader Joseph Stalin's (1879– 1953) tactics in Eastern Europe, which indicated the Soviets were putting up barriers against the West—and building up Soviet domination behind those barriers.

Just as he had issued warnings of the threat posed by Nazi Germany prior to World War II, Churchill astutely observed the rapidly emerging situation in Eastern Europe: In 1946 the Soviets installed Communist governments in neighboring Romania and in nearby Bulgaria; in 1947 Hungary and Poland came under Communist control as well; and the following year, Communists took control of Czechoslovakia. These countries, along with Albania, Yugoslavia, and East Germany, soon formed a coalition of Communist allies—and the Eastern bloc was formed. The United States and its democratic allies formed the Western bloc. The stage was set for the Cold War (1947–89).

How was the **Soviet Union formed**?

The Soviet Union was officially created in 1922 when Russia joined with Ukraine, Belorussia, and the Transcaucasian Federation (Armenia, Azerbaijan, and Georgia) to form the Union of Soviet Socialist Republics (U.S.S.R.). These republics were later joined by nine others, and territories were redrawn so that by 1940 the union consisted of 15 Soviet Socialist republics: Armenia, Azerbaijan, Belorussia (now Belarus), Estonia, Georgia, Kazakhstan, Kirghiz (now Kyrgyzstan), Latvia, Lithuania, Moldavia (now Moldova), Russia, Tadzhikistan (also spelled Tajikistan), Turkmenistan, Ukraine, and Uzbekistan.

How many **leaders** did the **Soviet Union** have?

From its formation in 1922 (just five years after tsarist Russia had fallen in the revolution of 1917), the Union of Soviet Socialist Republics (U.S.S.R.) had only 10 leaders. But just five of these had meaningful tenure, either due to length of time served or true authority: Lenin, Stalin, Khrushchev, Brezhnev, and Gorbachev.

After tsarist Russia ended with the revolution of 1917, Bolshevik leader Vladimir Lenin (1870–1924) became head of the Soviet Russian government as chairman of the Council of People's Commissars (the Communists), dissolving the elected assembly and establishing a dictatorship. This lasted six years: When Lenin died of a stroke in 1924, Joseph Stalin (1879–1953)—who had been an associate of Lenin—promptly eliminated his opposition and in 1929 established himself as a virtual dictator. Stalin ruled the U.S.S.R. during World War II (1939–45), and though he was aligned with the United States, Britain, and the other Allied nations during that conflict, soon after the war, he began a buildup of power in Eastern Europe, leading to the Cold War (1947–89). Even though Stalin's domestic policies were extremely repressive and he ruled largely by terror, he remained in power until his death in 1953.

After Stalin died, the Soviet Union entered a brief period of struggle among its top leaders: Deputy Premier Georgy Malenkov (1902–1988), a longtime Stalin aide, came to power. In 1955 Malenkov was forced to resign, and he was succeeded by his (and Stalin's) former defense secretary, Nikolai Bulganin (1895–1975). However, Bulganin was a premier in name only; the true power rested with Communist Party secretary Nikita Khrushchev (1894–1971), who expelled Bulganin and officially took power as premier in 1958.

Khrushchev denounced the oppression of the long Stalin years, which had ended only five years earlier, and worked to improve living standards. On the international front, he pursued a policy of "peaceful coexistence" with the West and even toured the United States in 1959, meeting with President Dwight D. Eisenhower (1890–1969). In 1960 a U.S. reconnaissance plane was shot down over the U.S.S.R., raising doubts among the Soviets about Khrushchev's policy toward the West. Further troubles at home resulted from widespread hunger due to crop failures. Meantime, Khrushchev advanced the cause of Soviet space exploration, beginning the so-called space race with the United States. Eventually, his stance on international issues, which included a rift with Communist China, led to his downfall. He was removed from power in October 1964.

Vladimir Lenin and Josef Stalin in 1922. When Lenin died of a stroke in 1924, his associate Stalin established himself as a virtual dictator of the Soviet Union.

Khrushchev's ouster (which was a forced retirement) had been engineered by his former ally and political adviser Leonid Brezhnev (1906–1982). With Khrushchev out of the way, technically Brezhnev was to lead the country along with premier Alexei Kosygin (1904–1981). But as head of the Communist Party, it was Brezhnev who truly held the power. By the early 1970s, Brezhnev emerged as the Soviet chief—even though Kosygin remained in office until 1980. During his administration, Brezhnev kept tight control over the Eastern bloc (Communist countries), built up the Soviet Union's military (in what became an arms race with the United States), and did nothing to try to reverse the downward trend of the Soviet Union's economy.

When Brezhnev died in 1982, he was succeeded by Yuri Andropov (1914–1984). However, Andropov died two years later and Konstantin Chernenko (1911–1985) replaced him as premier. When Chernenko, too, died an untimely death in March 1985, Mikhail S. Gorbachev (1931–) became head of the Communist Party and leader of the Soviet Union. With Gorbachev, the reign of the old guard of Stalin-trained leaders had come to an end. Gorbachev's policies of openness to the West and economic development led to the disintegration of the Soviet Union, with Communist rule ending in 1991 and each Soviet republic setting up its own government.

What were the **five-year plans**?

These were the plans initiated by premier Joseph Stalin (1879–1953) of the Soviet Union to speed industrialization of the U.S.S.R. and organize agriculture under the

What was the détente?

A détente is a relaxation of strained relations, particularly between nations. The détente of the Cold War era began after Premier Nikita Khrushchev (1894–1971) rose to power in the Soviet Union in 1958 and initiated a plan of peaceful coexistence with the West. During the 1960s the United States and the Union of Soviet Socialist Republics (U.S.S.R.) entered a phase of improved relations, which saw the signing of the Nuclear Nonproliferation Treaty (1968), the Strategic Arms Limitation Treaty (known as the SALT-1 treaty; 1972), and the Helsinki Accords (1975)—which pledged increased cooperation between the nations of Eastern and Western Europe.

Some historians refer to the détente as the end of the Cold War, while others view it as an intermission: When the Soviet Union under Premier Leonid Brezhnev (1906–1982) invaded neighboring Afghanistan in 1979 to put down an anti-Communist movement there, tensions between the two superpowers (the U.S. and the U.S.S.R.) heightened dramatically. Further, Brezhnev had been steadily building up Soviet arms during his tenure. These events brought an end to the détente.

collective control of the Communist government. The first five-year plan began in 1928, and subsequent plans were carried out until 1958, at which time the new Soviet leadership developed a seven-year plan (1959–65) aimed at matching—and surpassing—American industry. Under Premier Leonid Brezhnev (1906–1982), the five-year plans were reinstated in 1966 and continued until the dissolution of the Soviet Union during 1990 and 1991. Other Communist countries also instituted five-year plans, all with the goal of bringing industry, agriculture, and the distribution of goods and services under government control.

MEXICO

Why was the election of **Vicente Fox** a landmark?

The July 2000 election of businessman-turned-politician Vicente Fox (1942–) was monumental in Mexico's political history because it ousted the Institutional Revolutionary Party (PRI) that had ruled the nation for 71 years. It was a victory not only for Fox but for his National Action Party (PAN). The former Coca-Cola Company executive had turned to politics in the 1980s, following a highly successful career in business. In 1987 Fox was elected to the national Chamber of Deputies, and in 1995 he was voted in as governor of his home state of Guanajuato. He promised voters economic and political reform, with a particular emphasis on ridding the nation's government of seemingly endemic corruption. Fox was sworn into office on December 1, 2000, for a six-year term. In 2003 mid-term elections, Fox's PAN party lost ground to the PRI.

THE POST-9/11 ERA IN THE UNITED STATES

When was the **Office of Homeland Security** formed?

The Office of Homeland Security was organized in the days following the September 11, 2001, terrorist attacks. President George W. Bush (1946–) chose Pennsylvania governor Tom Ridge (1945–) as the first Office of Homeland Security advisor. Ridge was sworn in on October 8, 2001. The office was elevated to the department level on November 25, 2002, when President Bush signed into law the Homeland Security Act, creating the Department of Homeland Security (DHS) and making Ridge a cabinet-level administrator.

The DHS consolidated several existing agencies and pledged to carry out new initiatives to, the extent possible, protect the nation from further attacks. Agencies and sub-departments within the DHS's purview eventually included the Transportation Security Administration, Customs and Border Protection, Immigration and Customs Enforcement, the Federal Emergency Management Agency (FEMA), Information Analysis and Infrastructure Protection offices, U.S. Citizenship and Information Services (formerly the Immigration and Naturalization Service, or INS), an office for Civil Rights and Civil Liberties, the U.S. Coast Guard, and the U.S. Secret Service.

On February 15, 2005, Ridge was succeeded by Michael Chertoff (1953–), a former U.S. Circuit Court judge. Chertoff had also worked as an assistant attorney general; in that position, he helped trace the 9/11 terrorist attacks to the al Qaeda network and worked to increase information sharing within the Federal Bureau of Investigation (FBI) and with state and local officials.

What is the **Patriot Act**?

The Patriot Act is a controversial law passed by a wide majority of Congress and signed by President George W. Bush (1946–) in October 2001; it was designed to strengthen national security following the 9/11 terrorist attacks. The legislation relaxes federal surveillance laws, granting authorities broad leeway to gather information on U.S. citizens and resident foreigners. It also expands the government's prosecutorial powers against suspected terrorists and their associates. The complex act, which contains 168 sections, allows the nation's intelligence and law enforcement agencies to, among other things, monitor email and financial transactions without securing a subpoena, use wiretapping without a court order, and require Internet service providers (ISPs) to hand over usage data on customers.

One of the most controversial sections of the Patriot Act is the so-called "library provision," which allows government officials to secretly subpoena books, records, papers, documents, and other items from businesses, hospitals, and other organizations. Critics feared that the government could use the provision to snoop into the reading habits of innocent Americans. The reaction to the provision was so strong that, according to the American Civil Liberties Union (ACLU), 5 states and 375 communities in 43 states had passed anti-Patriot Act resolutions by spring 2005. Another contentious section of the Patriot Act allows the delayed notification of search warrants; this is called the "sneak and peek" provision because it lets federal officials search a suspect's home without telling the individual until later.

While many legislators and security experts hailed the Patriot Act provisions as necessary in combating terrorism and securing the homeland, others immediately saw the legislation as a serious infringement of civil rights. Supporters pointed to the hundreds of charges brought against suspected terrorists, as well as hundreds of convictions, as a result of the Patriot Act. But critics, including legislators, the ACLU, conservative groups, and many citizens, called the act unconstitutional—and unpatriotic. A top ACLU representative said, "Cooler heads can now see that the Patriot Act went too far, too fast and that it must be brought back in line with the Constitution." The fallout included charges of abuses by law enforcement, the introduction of alternate legislation in Congress to revise or repeal sections of the act, as well as challenges in court (in 2004 at least two sections were found to be unconstitutional in district court).

As the debate continued over the constitutionality of the Patriot Act, some provisions were set to expire the end of 2005. In April two of the Bush administration's top law enforcement officials urged Congress to renew every provision of the antiterror act: Attorney General Alberto Gonzales (1955–) said that some of the most controversial provisions of the Patriot Act had proven invaluable in fighting terrorism; FBI director Robert Mueller (1944–) said sections of the law that allow intelligence and law enforcement agencies to share information were especially important.

What was the controversy about the **Guantanamo detainees**?

After the 9/11 attacks, the U.S. military began holding terror suspects at a detention center at the naval base at Guantanamo Bay, Cuba. (The U.S. Navy has occupied Guantanamo since the Spanish-American War, in 1898, paying an annual lease to Cuba.) The White House labeled the detainees "enemy combatants"; the controversy came when they were not charged with any crimes, yet they continued to be held.

Taliban and al Qaeda detainees kneel under the watchful eyes of U.S. military police at the naval base at Guantanamo Bay, Cuba, January 2002.

The first detainees were transported to Guantanamo, or "Camp Gitmo," in January 2002, after being captured in Afghanistan. But no charges were made against any detainees until more than two years later, in February 2004. American lawyers challenged the Bush administration policy at Guantanamo, saying that it was a violation of the due process clause of the U.S. Constitution. In January 2005 one district court judge agreed with the prosecution, saying that the Constitution applied to the prisoners: they could not be deprived of their liberty without due process of the law. The Bush administration immediately moved to appeal the ruling. Some in the international community also strongly criticized the U.S. government for holding the suspects. One of Britain's most senior judges called the policy a "monstrous failure of justice."

In response to the widespread criticism, the U.S. Defense Department considered making major changes to the tribunals set up to prosecute terror suspects at Guantanamo. The changes were to bring the tribunals in line with the judicial standards of U.S. court-martials.

But questions also arose about the treatment of the detainees and the methods used in interrogating them. Human rights groups made charges of abuses. In spring 2005 United Nations officials working in the area of war crimes were awaiting a visit to

Guantanamo; the hold up was that they requested full access to the facilities and the prison population, conditions Washington was reluctant to allow. According to a February 2005 American Forces Information Service report, there were about 545 people from some 40 countries being held at the Guantanamo detention center at that time. The government also held terror suspects, some of them senior members of the terrorist network al Qaeda, in navy brigs in South Carolina, Iraq, and Afghanistan, and on navy vessels at sea.

LAW AND FAMOUS TRIALS

What was **Roman law**?

It was the system of law used by the Romans from the eighth century B.C. until the fall of the empire (Rome, in the West Roman Empire, was toppled in A.D. 476; the East Empire fell in A.D. 1453). Justinian the Great (483–565), the emperor of the East Roman Empire, is credited with codifying (writing and organizing) Roman law by ordering the collection of all imperial statutes and of all the writings of the Roman jurists (judges and other legal experts). Justinian appointed the best legal minds in the empire to assemble, write, publish, and update the code; work began in and continued until the time of Justinian's death, in 565. The result was the *Corpus juris civilis* (Body of Civil Law), also called the Justinian Code. It consists of four parts: the Codex (a collection of imperial statutes), the Digest (the writings and interpretations of Roman jurists), the Institutes (a textbook for students), and the Novels (the laws enacted after the publication of the Codex).

Though largely suspended during the Middle Ages (500–1350), it was kept alive in the canon law of the medieval church and was handed down through the centuries. It forms the basis of modern civil law in most of continental Europe and in other non-English-speaking countries, as well as in the state of Louisiana. Nations or states whose systems of justice are based on civil law rely not on precedents set by the courts (which is the common law system of the United States and Great Britain), but rather on the letter of the law—the statutes themselves.

What is **common law**?

It is the system of justice that prevails in Great Britain and the United States (except Louisiana), where the precedents (past decisions) of the courts are used as the basis of the legal system. It is sometimes referred to as customary law since justices consider

Witchcraft trial of George Jacobs of Salem, Massachusetts. Many of those accused of witchcraft were put through trial by ordeal; those found guilty were put to death. (Original painting by Tompkins Harrison Matteson.)

prevailing practices (customs) in order to arrive at their decisions. In the U.S., Louisiana is the only state where judges do *not* rely on precedent or custom to decide private cases—they rely on civil law, the letter of the code, and are free to disregard the decisions of similar cases. (In all other states, the precedents must be considered.) The exception of Louisiana to the prevailing system of common law is explained by its unique history—it was long a French holding and that influence is still felt.

In many countries, the justice system is a combination of the civil law handed down by the Romans under Justinian and the common law formulated in England. Private cases (often and confusingly called civil cases) are largely the realm of civil law (in other words, the statutes prevail); whereas criminal cases (in which crimes have been committed against society) are the realm of common law (i.e., decisions are based on precedent).

What was **trial by ordeal**?

It was an irrational way of determining someone's guilt or innocence. After the fall of Rome (476), Roman law gave way to the laws of the various Germanic (also called barbarian) tribes in Europe. If someone was charged with a crime, he or she was deliberately injured in some way: If the injury (from a heated iron bar or immersion into hot water, for example) healed within a prescribed number of days (usually three), the person was declared innocent. If the wound failed to heal, the verdict was guilty. This method for determining innocence or guilt was also called divination, since the court was trying, through the ordeal, to *divine* (discover intuitively) whether the accused person was guilty.

Trial by ordeal gave way to a far more practical, and certainly more rational, form of trial, in which judge and jury presided over the presentation of a case and employed written code or precedent or both to arrive at a verdict. But divination (literally, to predict by supernatural means) was used as recently as the 1600s, when women in Puritan New England were charged with witchcraft. A suspect was bound up with rope and immersed in water. If she sank, she was innocent; if she floated, she was declared guilty (the "reasoning" being that only someone with supernatural power could float under the circumstances). Those found guilty by this form of trial were put to death.

What was **trial by battle**?

Like trial by ordeal, trial by battle was a method of "justice" used predominately during the Middle Ages (500–1350). When noblemen had disputes, they would engage in a duel with one other: The assumption was that the person who was in the right would have God on his side, and he would emerge the victor in combat. No questions asked. This form of trial was gradually replaced by trial by jury.

What was the **Code Napoleon**?

In 1800, just after Napoleon Bonaparte (1769–1821) had come to power in France, he appointed a commission of legal experts to consolidate all French civil law into one code. The process took four years; the so-called Code Civil went into effect on March 21, 1804, the same year that Napoleon named himself emperor of France (which he did in December), The laws thus took on the alternate name of the Code Napoleon or Napoleonic Code. It went into force throughout France, Belgium, Luxembourg, and in other French territories and duchies in Europe.

The code represented a compromise between Roman law and common (or customary) law. Further, it accommodated some of the radical reforms of the French Revolution (1789–99). The Code Civil set forth laws regarding individual liberty, tenure of property, order of inheritance, mortgages, and contracts. It had broad influence in Europe as well as in Latin America, where civil law is prevalent. As opposed to the common law of most English-speaking countries, civil law judgments are based on codified principles, rather than on legal precedent. For example, under the Code Civil an accused person is guilty until proven innocent (as opposed to common law, which holds that a person is innocent until proven guilty).

What is **habeas corpus**?

The writ of habeas corpus (which is roughly translated from the Latin as "you should have the body") is considered a cornerstone of due process of law. It means that a person cannot be detained unless he or she is brought in person before the court so the court can determine whether or not the person is being lawfully held.

The notion dates back to medieval England; many historians believe that habeas corpus was implied by the Magna Carta (1215)—Article 39 states, "No freemen shall be taken or imprisoned...or exiled or in any way destroyed...except by the lawful judgment of his peers or by the law of the land." The writ was reinforced by Britain's Habeas Corpus Amendment Act of 1679, which stated that the Crown (king or queen) cannot detain a prisoner against the wishes of Parliament and the courts. The English introduced the concept in the American colonies. And when the U.S. Constitution (1788) was written, it declared (in Article I, Section 9) that habeas corpus "shall not be suspended, unless when in cases of rebellion or invasion the public safety may require it." For example, during the Civil War (1861–65), President Abraham Lincoln (1809–1865) suspended habeas corpus; and during Reconstruction (1865–77), it was again suspended in an effort to combat the activities of the Ku Klux Klan.

International Law

What is **international law**?

As interpreted by Dutch jurist and humanist Hugo Grotius (1583–1645), natural law prescribes the rules of conduct among nations, resulting in international laws. His 1625 work, titled *Concerning the Law of War and Peace,* is considered the definitive text on international law, asserting the sovereignty and legal equality of all states of the world. But the notion also had its detractors, English philosopher Thomas Hobbes (1588–1679) among them. Hobbes insisted that since international law is not enforced by any legal body above the nations themselves, it is not legitimate.

Since the seventeenth century, however, international law has evolved to become more than just theory. During the 1800s and early 1900s, the Geneva Conventions (1864, 1906, 1929, 1949) and the Hague Conferences (1899, 1907) set forth the rules of war. Today, treaties (between two or among many countries), customary laws, legal writings, and conventions all influence international law, which is also referred to as "the law of nations." Further, it is enforced by the International Court of Justice (a United Nations body) as well as by world opinion, international sanctions, and the intervention of the United Nations (apart from the International Court of Justice).

What are the **Geneva Conventions**?

The Geneva Conventions are humanitarian treaties signed by almost all of the approximately 200 nations in the world today (there were 189 signatories as of 2003). The treaties were forged in Geneva, Switzerland, in 1864, 1906, 1929, and 1949. (The initial protocols, of 1864, gave rise to the Red Cross.) There were two amendments, called protocols, in 1977. In their entirety, the Geneva Conventions set standards for how signatory nations are to treat the enemy during war; they cover access to and treatment of battlefield casualties, treatment of prisoners of war (POWs), and the treatment of civilians.

The summaries of each convention and protocol are as follows:

Convention I: For the Amelioration of the Condition of the Wounded and Sick in Armed Forces in the Field. Sets forth the protections for members of the armed forces who become wounded or sick.

Convention II: For the Amelioration of the Condition of Wounded, Sick, and Shipwrecked Members of Armed Forces at Sea. Extends protections to wounded, sick, and shipwrecked members of the naval forces.

Convention III: Relative to the Treatment of Prisoners of War, Geneva. Lists the rights of prisoners of war.

Convention IV: Relative to the Protection of Civilian Persons in Time of War, Geneva. Deals with the protection of the civilian population in times of war. [All four Conventions were ratified as a whole in 1949.]

Protocol I: Additional to the Geneva Conventions of 12 August 1949, and relating to the Protection of Victims of International Armed Conflicts. Extends protections to victims of wars against racist regimes and wars of self determination.

Protocol II: Additional to the Geneva Conventions of 12 August 1949, and relating to the Protection of Victims of Non-International Armed Conflicts. Extends protections to victims of internal conflicts in which an armed opposition controls enough territory to enable them to carry out sustained military operations.

The Geneva Conventions, along with the Hague Conventions (1899, 1907), comprise much of what is called International Humanitarian Law (IHL). Because so many nations of the world have ratified both the Geneva Conventions and the Hague Conventions, they are considered customary international law, which means they are binding on all nations.

What are the **Hague Conventions?**

They are international treaties (1899, 1907) covering the laws and customs of war. The first Hague Convention developed out of the Peace Conference of 1899, held in The Hague (The Netherlands) and convened by Russian tsar Nicholas II (1868–1918). Among the original goals was limiting the expansion of armed forces. Though the representatives there, from 26 nations (including the United States), failed to agree on a resolution to limit such expansion, they did agree on certain rules of engagement for war on land and at sea. Also, very importantly, they adopted the Convention for the Pacific Settlement of International Disputes. This convention set up the permanent international court of arbitration and justice, still in existence today. The court is in The Hague, where it is housed in the Peace Palace, a gift of American industrialist and philanthropist Andrew Carnegie (1835–1919). A later convention, the Second Hague Peace Conference, was held in 1907; representatives of 44 nations met for a period of four months. The convention of 1907 modified and added to the first. Delegates resolved to meet again in 1915, but that conference was not held due to the outbreak of the first World War. The Hague Conventions were the forerunners of the League of Nations and the United Nations.

U.S. Law and Justice

How was the **makeup of the U.S. Supreme Court** decided?

Article 3 of the U.S. Constitution (1788) states that the "judicial Power of the United States, shall be vested in one supreme Court." It goes on to describe the high court's jurisdiction, but it does not specify how the court was to be formed or how many justices it would consist of. These matters were left to Congress, which in 1788 (the year the Constitution was ratified) passed the first Judiciary Act, formulating the court's original organization; it had just six members: the chief justice, plus five associates. Subsequent congressional acts have modified the Supreme Court's organization and jurisdiction.

Since 1869 the Court has consisted of nine members: the chief justice and eight associates who, once named, serve for life. Justices are appointed by the president but must be approved by the Senate (according to Article 2 of the Constitution). To avoid partisanship, Congress is prevented from lowering the salaries of any of the justices, and justices can only be removed from the bench by impeachment (a formal document that charges a public official with misconduct). Cases reach the high court through appeal (lower-court decisions that are formally challenged), and the justices make their rulings based on a majority vote.

Who was the **first chief justice** of the U.S. Supreme Court?

President George Washington appointed John Jay (1745–1829) to the post, along with five associate justices: James Wilson, John Rutledge, William Cushing, John Blair, and

What are blue laws?

The term "blue laws" refers to laws that are intended to enforce moral conduct. They originated in colonial times in Puritan New England and got the name because they were printed on blue paper. Some blue laws prescribed proper conduct for the Sabbath, which included no working, sports, or drinking. The early blue laws of New Haven, Connecticut, were widely publicized in a 1781 book, *A General History of Connecticut.* The author, Samuel Peters (1735–1826), took some freedoms with the text, however, and even invented a few laws of his own. Since blue laws violate individual freedoms, most of them have been repealed through the years. If a community still has blue laws on its books today, they are not likely to be enforced.

James Iredell. Jay's was an impressive resume by the time of his appointment: He had been a member of the Continental Congress, over which he presided as president in 1778 and 1779; served as U.S. minister to Spain (1779, during the American Revolution); and joined American statesman Benjamin Franklin (1706–1790) and the rest of the American peace commission in Paris (in 1782) to draw up the treaty ending the war with Britain. Once the new republic was established, Jay remained at the fore: he became President Washington's secretary of foreign affairs (1784–89) and, along with Alexander Hamilton (1755–1804) and James Madison (1751–1836), authored the *Federalist* (1787–88), which explained the Constitution for the benefit of the states as they considered ratification.

What were the **Jim Crow laws**?

They were laws or practices that segregated blacks from whites. They prevailed in the American South during the late 1800s and into the first half of the 1900s. Jim Crow was a stereotype of a black man described in a nineteenth-century song-and-dance act. The first written appearance of the term is dated 1838, and by the 1880s it had fallen into common usage in the United States. Even though in 1868 Congress passed the Fourteenth Amendment, prohibiting states from violating equal protection of all citizens, southern states passed many laws segregating blacks from whites in public places. In short, the laws were both manifestation and enforcement of discrimination. Thanks to the civil rights movement, the laws were finally found to be unconstitutional during the 1950s and 1960s.

What was the **Supreme Court's role in racial segregation**?

Though most segregation laws (or "Jim Crow laws") were overturned by decisions of the Supreme Court during the 1950s and 1960s, the Court was righting its own wrong:

A view of school segregation: White students in class and a black student in an anteroom at the University of Oklahoma, 1948.

In the late 1800s, during the years following the Civil War (1861–65) and the abolition of slavery, the Supreme Court made rulings that actually supported segregation laws at the state level. The most famous of these was the 1896 case of *Plessy* v. *Ferguson,* in which the highest court in the land upheld the constitutionality of Louisiana's law requiring "separate-but-equal" facilities for whites and blacks in railroad cars. (One strong dissenting opinion came from Associate Justice John Marshall Harlan, who declared "the Constitution is color blind.") Following the *Plessy* v. *Ferguson* decision, states went on to use the separate-but-equal principle for 50 years, passing Jim Crow laws that set up racial segregation in public schools, transportation, and in recreation, sleeping, and eating facilities. This meant there were drinking fountains, benches, restrooms, bus seats, hospital beds, and theater sections designated as "Whites Only" or "Colored." One Arkansas law even provided that witnesses being sworn in to testify in a courtroom be given different Bibles depending on the color of their skin.

Two landmark Supreme Court decisions came in 1954 and 1960 in the cases of *Brown* v. *the Board of Education* and *Boynton* v. *Virginia.* In the first case, parents of black children in Topeka, Kansas, elementary schools charged that the segregation of white and black students in the public schools denied black children the equal protection cited in the Fourteenth Amendment (1868). The parents were supported in their fight by the NAACP (National Association for the Advancement of Colored People), whose legal counsel included Thurgood Marshall (1908–1993). On May 17, 1954, the Supreme Court ruled that segregated schools do violate the equal protection clause, overturning the separate-but-equal doctrine previously upheld by *Plessy* v. *Ferguson.* The federal government again made a stand against state segregation laws when, in December 1960, chief appellant lawyer Thurgood Marshall argued the case of Howard University law student Bruce Boynton before the Supreme Court. Again ruling in favor of the plaintiff, who had charged that the segregation laws at the Richmond, Virginia, bus station violated the federal antisegregation laws, Washington sent a clear message to the states that public facilities are for the use of all citizens, regardless of color. These decisions combined with the activism of the civil rights movement to outlaw racial segregation.

What was **Operation Falcon**?

It was the code name for the mid-April 2005 roundup of more than 10,000 fugitives in one week; the coordinated nationwide effort was led by the U.S. Marshals Service. Together with officers from 960 federal, state, and local law enforcement agencies, the marshals arrested 10,340 people who were wanted for various crimes, many of them violent. The operation took place during Crime Victims Rights Week. More than 150 of the fugitives were wanted for murder, 550 for sexual assault charges, and more than 600 for armed robberies. There were also escaped prisoners and criminal suspects among those arrested. Operation Falcon was a landmark in law enforcement because of the sheer number of arrests; previous coordinated efforts had nabbed only hundreds of fugitives.

FAMOUS TRIALS

What were the **Salem witch trials**?

A series of trials in Salem, Massachusetts, in 1692, the proceedings against 200 people accused of witchcraft became allegory for searching out or harassing anyone who holds unpopular views. Indeed, the 19 hangings that resulted from the witch hunt provide students of history with a cautionary tale about the hazards of mass hysteria.

In the seventeenth century, people widely believed in witchcraft and that those who wielded its supernatural power could perform acts of ill will against their neighbors. Courts somewhat regularly heard cases involving the malice of witches: Before the notorious trials of 1692, the records of colonial Massachusetts and Connecticut show that 70 witch cases had been tried, and 18 of the accused were convicted. But nothing had reached the scale of the 1692 witch hunt. In January of that year, the daughter and niece of a Reverend Samuel Parris began exhibiting strange behavior. Upon examination by a doctor, the conclusion was that the young girls (ages nine and eleven) were bewitched. Compelled to name those who had bewitched them, the girls named a Carib Indian slave who worked in the minister's home, and two other women—one a derelict and the other an outcast. They were arrested. Hearings were held and others were accused, including upstanding members of the community whose only "crime" seemed to be their opposition to Reverend Parris. Members of his congregation became corroborative witnesses. By May jails in Salem and Boston were filled with suspected witches awaiting trial. The court, now with a docket of some 70 cases, convened on June 28.

Through the summer months and into September, 50 of the accused confessed to practicing witchcraft, 26 were convicted, and 19 were executed. The colonial governor of Massachusetts became alarmed by the number of convictions; he ordered the Salem court to disband and commenced hearings of the remaining cases in a superior court. Of 50 still accused, the court indicted only 23; of these, there were only three convictions—all of which were overturned. In 1693 the colonial governor pardoned those whose cases were still pending and declared that witchcraft was no longer an actionable offense.

Who was **John Peter Zenger**?

John Peter Zenger (1697–1746) was a New York City printer who was accused of seditious libel in 1735. His case changed the definition of libel in American courtrooms and laid the foundation for freedom of the press.

The German-born Zenger immigrated to the American colonies in 1710, when he was 13 years old. He found a job as a printer's apprentice, working on the colony's official newspaper, the *New York Gazette*. Fifteen years later he began his own operation, which was mostly concerned with printing religious pamphlets. In 1733 New York received a new colonial governor from England: William Cosby quickly earned the contempt of the colonists, both rich and poor. Prosperous businessmen who opposed Cosby and his grievous tactics approached Zenger, offering to back a newspaper that he would both edit and publish. Zenger agreed and on November 5, 1733, the first issue of the *Weekly Journal* was released. It included scathing criticisms of the royal governor, raising Cosby's ire. After burning several issues of the papers, Cosby had Zenger arrested in November 1734. The editor-publisher continued to operate the journal from inside his jail cell, dictating editorials to his wife through the door.

Zenger's case went to trial in August 1735. Prominent Philadelphia attorney Andrew Hamilton (1676–1741), considered the best lawyer in the colonies, came to Zenger's defense. Hamilton admitted his client was guilty of publishing the papers, but, he argued, that in order for libel to be proved, Zenger's statements had to be both false and malicious. The prosecution contested the definition of libel, asserting that libelous statements are any words that are "scandalous, seditious, and tend to disquiet the people." The court agreed with the prosecution, and Hamilton was therefore unable to bring forth any evidence to support the truth of the material Zenger printed in the *Weekly Journal.* The defense argument was not heard until the closing statement was made by Hamilton; his summation stands as one of the most famous in legal history. He accused the court of suppressing evidence, urging the jury to consider the court's actions "as the strongest evidence," and went on to declare that liberty is the people's "only bulwark against lawless power... Men who injure and oppress the people under their administration provoke them to cry out and complain." The brilliant attorney closed by urging the gentlemen of the jury to take up the cause of liberty, telling them that by so doing, they will have "baffled the attempt of tyranny." The seven jury members were convinced by Hamilton's impassioned speech and found Zenger not guilty.

Discharged from prison the next day, Zenger returned to his printing business, publishing the transcripts of his own trial. While colonial officials were reluctant to accept the case's ruling on the definition of libel, the case became famous throughout the American colonies. And once the colonists had thrown off England's royal rule and established a new republic, the nation's founding fathers codified the Zenger trial's ruling in the Bill of Rights: The First Amendment to the U.S. Constitution guarantees freedom of press.

Why was the **Dred Scott decision** important?

The decision in the case of Dred Scott pronounced the Missouri Compromise (1820) unconstitutional and served to deepen the divide between North and South, helping pave the way for the Civil War (1861–65).

In the mid-1800s Dred Scott (c. 1795–1858), who had been born into slavery in Virginia, tried to claim his freedom on the basis that he had traveled with his owner, a doctor, in Wisconsin and Illinois, where slavery had been prohibited by the Missouri Compromise. By the compromise, Congress decided to admit Missouri as a slavery state and Maine as a free state, and declared that the territories north of the 36th parallel (present-day Missouri's southern border) were free, with the exception of the state of Missouri.

After a lifetime of slavery, Dred Scott sued Missouri for his freedom in April 1846. The case, which hinged on Scott's travels in free territories in the North, went through two trials; the second was granted due to a procedural error in the first. In 1850, at the conclusion of the second trial, a Missouri jury ruled Scott a free man based on precedents that indicated residence in a free territory or state resulted in emancipation, regardless of the fact that Missouri itself was a slave state. John F. A.

What was Dr. Mudd tried for?

A doctor who treated the broken ankle of John Wilkes Booth (1838–1865) after Booth shot President Abraham Lincoln (on April 14, 1865; he died the next day), Samuel Mudd was later charged as an accomplice in the president's assassination and was charged with treason and conspiracy. He was tried before a military commission and on June 30, 1865, was found guilty and sentenced to life in prison. Three years later Mudd was pardoned by President Andrew Johnson (1808–1875). The official reason for the pardon was Mudd's humanitarian efforts to save lives during a prison epidemic. But the case against him had been flimsy at best, and history has credited the guilty verdict to overly ambitious politics and a commission bent on retribution. Nevertheless, Mudd's name remained tainted, giving the popular culture the phrase, "His name is Mudd."

Sanford, the lawyer for Scott's owner, immediately appealed the decision before the Missouri Supreme Court, where a pro-slavery judge reversed the ruling, rescinding Scott's freedom.

But the case was not over yet: Because Sanford had filed the court papers under his own name rather than that of Scott's former owner, the case of *Scott* v. *Sanford* (because of a filing error, the case is also rendered as *Scott* v. *Sandford*) took an interesting twist. Scott hired a new lawyer who was able to get the case before the federal court: Sanford had moved to New York, and since the appellant (Scott) and the registered defendant (Sanford) were now residents of different states, the case came under federal purview. In 1854 a circuit court in St. Louis again heard Dred Scott's case, but his freedom was again denied. This decision was appealed to the U.S. Supreme Court, which began hearing the case in 1856.

In March 1857 the Supreme Court, which had a southern majority, ruled that Scott's residence in Wisconsin and Illinois did not make him free, that a black (a "Negro descended from slaves") had no rights as an American citizen and therefore could not bring suit in a federal court, and that Congress never had the authority to ban slavery in the territories. Dred Scott died the following year.

Why was **Susan B. Anthony** tried?

Susan B. Anthony (1820–1896) was tried for violating federal voting laws. The suffrage movement was in full swing in 1872 when Anthony and 14 female companions went to the Rochester, New York, voter registration office and demanded to be registered. When the officials refused, Anthony argued with them, showing them the written opinion of a Judge Henry R. Selden, who agreed with her (and others) that the Fourteenth Amendment (1868) also protects women's rights, including the right to vote. She threatened the registrars that she would sue them if they did not allow her to participate in elections. They gave in and the women signed up to vote. On election day, November 5, they did just that. Twenty-three days later, all 15 women were arrested for having done so. Bail was set, and eventually all the women were released. The following June, Anthony's trial got under way. The U.S. district attorney presented the government's case against her: she had "upon the 5th day of November, 1872, ...voted...At which time she was a woman." She was found guilty and ordered to pay a fine of $100. In another act of civil disobedience, Anthony refused to pay it, saying, "Resistance to tyranny is obedience to God."

In the coming years, the nation's courts continued to narrowly interpret the Fourteenth Amendment to the U.S. Constitution, to the exclusion of women. Anthony died 24 years before American women were granted suffrage (after the Nineteenth Amendment was made in 1920).

Was **Mata Hari** really a spy?

When Dutch-born Margaretha Zelle MacLeod (1876–1917) was arrested in Paris on February 13, 1917, there was scant hard evidence that this woman, known throughout Europe as Mata Hari, was actually a spy for the Germans during World War I (1914–18); but there was plenty of evidence that she had long consorted with the enemy and had been paid by them, but for exactly what was never discovered. Nevertheless, the testimony heard by the jury over two days in July in a closed Parisian courtroom was enough to convince them that this former exotic dancer, who could count as her lovers a "who's who" of European men, was, in fact, a spy. She was sentenced to death.

At age 18, the result of a matrimonial advertisement, she married a middle-aged colonial captain in the Dutch army, John Rudolph Campbell MacLeod. He was posted to duty on the island of Java, where his young and beautiful wife, now 21 years old, learned not only the Malay language, but native dances as well. Her Javanese friends named her Mata Hari, meaning "the eye of the dawn." Upon returning to Holland, Mata Hari secured a separation from her husband and moved to Paris where she enjoyed a life of excess and soon became known as an exotic Hindu dancer. She performed throughout Europe, all the while engaging in liaisons with powerful and wealthy men. In 1914 she moved to Germany, where she is believed to have been trained as a spy in Antwerp.

With World War I on, Mata Hari returned to Paris; she was permitted to enter France since she owned property there and was a citizen of neutral Holland. She renewed her ties with men of influence and in that capacity collected information for the Germans. The Allied nations kept a close eye on her, and, suspecting her of espionage, set a trap for her. She became a double agent. The French sent her to Spain to work, but there she reportedly met regularly with German intelligence agents. When the Germans ordered her back to Paris, Allied officials—having intercepted a German cable for her—awaited her return. They arrested Mata Hari, who was found in possession of a check from the Germans. At her trial, a report compiled by the French and holding Mata Hari responsible for the deaths of some 50,000 Allied soldiers, was brought into evidence. She was killed by firing squad on October 15, 1917, the war still more than a year from ending.

Who were **Leopold and Loeb**?

Nathan "Babe" Leopold (1904–1971) and Richard "Dickie" Loeb (1905?–1936) were privileged, well-educated, even brilliant young men who committed what they believed to be the perfect murder. Both were from well-to-do Chicago families. In May 1924 Loeb, then 18 years old, became the youngest graduate of the University of Michigan; he was to go on to postgraduate studies at the University of Chicago. Nineteen-year-old Leopold was a member of Phi Beta Kappa and a law student there. The two became friends, and, as testimony would later reveal, in the fall of 1923 became convinced that they could literally get away with murder—that they could plan it, carry it out, and never get caught.

On May 21, 1924, the pair carried out their dastardly plan. Their victim was 14-year-old Bobby Franks, son of a millionaire and cousin to Loeb. Franks's body was found, as were a pair of eyeglasses belonging to Leopold. The spectacles were traced to him, and he and Loeb (who was part of Leopold's alibi) were grilled by the police. They stuck to their story for exactly one day. Then Loeb, believing Leopold had betrayed him, confessed. They were charged with murder and kidnapping. Under the counsel of noted defense attorney Clarence Darrow (1857–1938), who had been hired by their families, the pair pled guilty, reducing what would have otherwise been death sentences to life in prison plus 99 years. In 1936 Loeb was killed by a fellow prison inmate. In 1958 Leopold was freed—his sentence had been reduced by Illinois governor Adlai Stevenson in exchange for the inmate's contribution to testing for malaria during World War II (1939–45). He lived out his life in Puerto Rico, where he married, earned a master's degree, performed charitable works, and taught.

What was the importance of **Hitler's beer hall *putsch* trial?**

The 1924 trial of German chancellor and führer Adolf Hitler (1889–1945) and nine other men, charged with treason for their attempted coup (in German, *putsch*) of late 1923, marked the beginning of Hitler's seemingly unstoppable rise to power.

As the leader of the Nazi Party (National Socialists German Workers' Party), Hitler had gained enough of a following to believe that on the night of November 8, 1923, as Bavarian leader Gustav von Kahr spoke in a Munich beer hall, Hitler and his followers—all of them determined to recreate a powerful German empire and rid it of its "mongrel-like" quality—could topple the weak German government, merely by demonstrating that the Nazis, and not the official government, had gained the support of the people. But in a march through Munich the following day, the still-loyal Germany regular army and the Bavarian state police opened fire on the Nazi demonstrators and their sympathizers, killing 16 and arresting Hitler and his nine co-conspirators. Their trial began on February 26, 1924: Over the course of 25 days, aided by radio and newspaper coverage, Hitler held forth (in one case taking four hours to respond to a single question), earning him the overwhelming support of the German people. His impassioned appeals turned what ought to have been a open-and-closed case of treason against him into an indictment of the German government. His basic argument was this: "I cannot declare myself guilty. True, I confess to the deed, but I do not confess to the crime of high treason. There can be no question in an action which aims to undo the betrayal of this country in 1918." Hitler was referring to the German surrender in World War I (1914–18).

Nevertheless, he and nine others were convicted of treason. Hitler was sentenced to five years in prison, where he wrote the first volume of his infamous work *Mein Kampf* (My Struggle), which revealed his frightening theories of racial supremacy and his belief in the Third Reich. Released after only nine months, Hitler walked out of prison more popular than he had been before his highly publicized trial.

What was the **"monkey trial"**?

The July 1925 trial of Dayton, Tennessee, public schoolteacher John T. Scopes (1900–1970) was dubbed the "monkey trial" because at issue was Scopes's teaching of evolution in his classroom. Having yielded to religious beliefs in creationism (the story of human origins told in the Bible's book of Genesis), Tennessee state law prohibited teaching public school students about the theories of English naturalist Charles Darwin (1809–1882). Darwin's scientifically credible work *The Origin of Species* argued that humans had descended from apelike creatures. Celebrated attorney Clarence Darrow (1857–1938) defended Scopes; lawyer and former presidential candidate William Jennings Bryan (1860–1925), known as the "Great Commoner," argued the prosecution. For 12 days in the summer of 1925, the small town in eastern Tennessee became the site of a showdown between modern scientific thought and traditional fundamentalism, or as some observed, between cosmopolitan and rural America. Spectators crowded the courtroom, eventually forcing the proceedings to be moved to the courthouse lawn. Journalists issued daily reports, which were published in newspapers across the country. It was headline writers who dubbed the case the "monkey trial."

Darrow made history when he called Bryan himself to the stand; it was a daring move on the defense attorney's part, but since Bryan eagerly accepted the summons, the judge allowed the questioning. Darrow first got Bryan to agree that every word in

The July 1925 trial of Tennessee school teacher John T. Scopes was dubbed the "monkey trial" because at issue was Scopes's teaching of evolution in his classroom.

the Bible is true; then he set in to reveal the hazards of such a literal interpretation, asking, for example, how Cain had found himself a wife if he, Adam, Eve, and Abel were the only four people on Earth at the time. Darrow succeeded in shaking the prosecutor, who finally admitted that he did not believe Earth was made in six days. Bryan retaliated by accusing Darrow of insulting the Bible, to which Darrow replied, "I am examining you on your fool ideas that no Christian on earth believes." It was drama better than any novelist could write. Darrow lost the case, which was later overturned on a technicality. Scopes had only been charged a $100 fine for violating the state law, which was repealed in 1967. But the trial, preserved in the play and film *Inherit the Wind,* is still remembered today: Scopes's crime was not sensational, his trial did not break any legal ground, and the defense had not won a brilliant victory, but the proceedings, carried out in the midsummer heat of the American South, epitomized the era and, ultimately, made for a great story.

Why is the court-martial of **Billy Mitchell** famous?

The 1925 military trial of William "Billy" Mitchell (1879–1936) made headlines because of the defendant's open and controversial criticism of the U.S. military.

A U.S. general in World War I (1914–18), Mitchell returned from the experience convinced that the future military strength of the country depended on air power. In fact, he had commanded the American expeditionary air force during the war in Europe and had even proposed to General John Pershing that troops be dropped by parachute behind German lines; Pershing dismissed the idea. The war over, in 1921 Mitchell declared that "the first battles of any future war will be air battles." But when the navy and war departments failed to develop an air service, Mitchell was outspoken about it, charging the military with incompetence, criminal negligence, and describing the administration as treasonable. Those were fighting words. Charged with insubordination and "conduct of a nature to bring discredit upon the military service," Mitchell's trial began on October 28, 1925. After lengthy hearings, on December 17 of that year Mitchell was found guilty and was suspended without pay from the military for a period of five years. Congress entered the fray, proposing a joint resolution to restore Mitchell's rank, but President Calvin Coolidge (1872–1933) upheld the court's decision. Mitchell responded by resigning. He returned to civilian life but continued to write and speak about his belief in an air force. He died in 1936, about five years too soon to see his predictions come true: In surprise air raids on December 7, 1941, the Japanese attacked U.S. military installations in the Philippines and Hawaii. Though the U.S. military rose to the occasion, entering World War II and building an impressive and mighty air fleet, many observers felt the military could have been better prepared to stage that monumental effort had Mitchell's advice been heeded years earlier.

What was **Al Capone** tried for?

Notorious American gangster Alphonse "Scarface Al" Capone (1899–1947), whose crime syndicate terrorized Chicago in the 1920s, was brought to trial for income tax evasion. After Chicago police had been unable to bring Capone to justice for his criminal activities, which included trafficking bootleg liquor, gambling, prostitution, and murder, the Federal Bureau of Investigation determined that the only way to prosecute the crime boss would be through violation of the tax laws. For two and a half weeks in October 1931, the case against Capone was heard in a Chicago courtroom. He was found guilty on five counts of tax evasion, sentenced to 11 years in prison, and charged $50,000 in fines and $30,000 in court costs. While his first jail cell, in Illinois's Cook County Jail, allowed him the luxuries of a private shower, phone conversations, telegrams, and even visits by other gangsters, including "Lucky" Luciano and "Dutch" Schultz, Capone was eventually moved to Alcatraz Island in San Francisco Bay, where he received no privileges. Released in 1939, Capone lived out his remaining years with his wife and son in Miami Beach, where he was reportedly haunted by imaginary killers.

Who was tried at **Nuremberg**?

Following World War II (1939–45), 22 leaders of Nazi Germany were put on trial at Nuremberg's Palace of Justice. The International Military Tribunal began the proceedings on November 25, 1945, and they were not concluded until September 30 of the following year; the verdicts were announced on October 1. The site was deliberately selected by the Allies; the now bombed-out city of Nuremberg was considered a seat of Nazi power.

Though many, including Soviet leader Joseph Stalin (1979–1953) thought that Hitler's henchmen ought only to be tried as a show of justice before they were executed, others, notably U.S. chief prosecutor Robert Jackson (1892–1954), believed due process of law must be observed. The American view prevailed.

The tribunal indicted 23 Nazi leaders on four counts: conspiracy, crimes against peace, war crimes, and crimes against humanity. One of the defendants, Robert Ley (1890–1945), committed suicide in prison before the trial began. The case against the Nazis was based on a mountain of written evidence, such as orders, reports, manifests, logs, letters, and diaries; the Germans had scrupulously recorded their evil deeds. The presentation of the documents was punctuated with live testimony of a German civilian contractor who, out of curiosity, had followed a Nazi detachment to an embankment where several thousand Jewish men, women, and children were shot and buried in a pit; and of a French woman, a survivor of the horrors of Auschwitz, recollecting a night when "children had been hurled into furnaces alive," since the Nazis had run out of fuel. The atrocities were rendered unimaginably horrific by the sheer number of Nazi victims, which included 3.7 million (of the 5.7 million captured) Soviet troops who died in prison, 4 million Jews who died in extermination camps, and the murder of at least 2 million more Jews elsewhere. The defense was prohibited from employing a "you did it, too" argument, which would have been an attempt to justify their actions by claiming it was all part of war. The Allies were determined to bring the Nazis to justice for their appalling and diabolical acts.

Among those tried at Nuremberg were Hitler's chief deputy Hermann Goering (1893–1946, whom a *New Yorker* correspondent covering the trials described as "a brain without a conscience"), foreign minister Joachim von Ribbentrop (1893–1946), and armaments minister Albert Speer (1905–1981). Goering and Ribbentrop were among the five men found guilty on all four counts against them; they were sentenced to hang. Six others were found guilty of crimes against humanity and were all sentenced to hang. (A seventh man, Martin Bormann, 1900–45, who had been tried in absentia, was also sentenced to hang—if he were found to be alive.) Seven others were also found guilty on one or two counts and were sentenced to prison terms, ranging from 10 years to life. Three were acquitted on all four counts. Goering escaped his hanging: Though he was to be closely monitored by his jailers, he managed to secure a vial of cyanide, which he swallowed a few hours before his scheduled execution. Since Bormann was at large, 10 Nazis died in the three gallows that had been constructed in the prison gym of the Palace of Justice.

The trials at Nuremberg cemented the principle that wartime leaders are accountable under international law for their crimes and immoralities.

Why was **Alger Hiss** tried?

U.S. public official Alger Hiss (1904–1996) was tried for perjury during 1949 and 1950. His first trial ended in a hung jury, and the second trial concluded with a guilty verdict and a sentence of five years in prison. Hiss served four years and eight months before he was released and returned to private life. To this day many believe Hiss was framed by Republican politicians who charged President Harry Truman's (1884–1972) administration with employing Communists who acted as secret agents for the Soviet Union. The politically charged case was packed with intrigue, including the testimony of a *Time* magazine senior editor who was later revealed to be a perjurer and who used at least seven different aliases in a 14-year period; microfilm evidence stored in a hollowed-out pumpkin in the middle of a farm field; and an old typewriter, which later evidence and testimony revealed was probably a fake.

The case against Hiss was made amidst the Investigation of Un-American Activities of the House of Representatives. It was 1948 and the Cold War was on; distrust was running high. And when a man named Whittaker Chambers (an editor at *Time*) appeared before the House committee and claimed that Hiss had been a courier who had transported confidential government documents to the Soviets, Hiss, then president of the Carnegie Endowment for International Peace, became the subject of investigation. He was indicted and stood trial. In spite of his distinguished career as a public servant (he had served in the State Department for 11 years); a parade of character witnesses who testified of his integrity, loyalty, and veracity; and his own vehement denial of the charges, the prosecution managed to bring enough evidence against him to convince a second jury that Hiss lied when he said the charges that he was a secret agent were "a complete fabrication." (The jury in Hiss's first trial deadlocked following more than 14 hours of deliberation.)

Even after his conviction, Hiss's lawyers worked tirelessly to appeal the case; all attempts were denied. In 1957 Hiss published his own account of the case, *In the Court of Public Opinion,* in which he reasserted his innocence. Then in 1973, during the Watergate hearings, former White House counsel John Dean's (1938–) explosive testimony included the statement that he heard President Richard Nixon (1913–1994) say, "The typewriters are always the key...We built one in the Hiss case." In 1988 Hiss published again; the book was titled *Recollections of a Life.* Four years later, at the age of 87, Hiss appealed to the Russian government to examine their intelligence archives to see what they revealed about him; the response came back that there was "not a single document" substantiating the allegations that Hiss had collaborated with the Soviet Union's intelligence service. That same year, 1992, Hiss's son, Tony, wrote an article for the *New Yorker* magazine; it was titled "My Father's Honor."

Why were the **Rosenbergs** tried?

Husband and wife Julius (1918–1953) and Ethel Rosenberg (1915–1953) were tried for conspiracy to commit wartime espionage. Arrested in 1950, the Rosenbergs were

Ethel and Julius Rosenberg en route to Sing Sing Prison after they were found guilty of conspiracy to commit espionage in 1953.

charged with passing nuclear weapons data to the Soviets, enabling the Communists to develop and explode their own atomic bomb—an event that had been announced to the American public by President Truman on September 23, 1949. As the realization set in that the United States could now be the victim of an atomic attack, the anxieties of the Cold War heightened. Citizens were encouraged to build bomb shelters, school children participated in air-raid drills, civil-defense films (such as *How Can I Stay Alive in an Atom Bomb Blast?*) were screened, and entire towns conducted tests of how residents would respond in the event of an "A-bomb."

Meantime, the leak of top-secret information from the Manhattan Project at Los Alamos, New Mexico, was traced to New York City machine-shop owner Julius Rosenberg, his wife, and her brother, David Greenglass. Historian Doris Kearns Goodwin writes that the "short, plump Mrs. Rosenberg looked more like one of my friends' mothers than an international spy." Indeed, the case marked the first time American civilians were charged with espionage; and the trial made international headlines. Though the Rosenbergs were only two of many involved in the conspiracy, theirs was the heaviest of the punishments handed down in the cases against the spy ring. For their betrayal and their refusal to talk, the Rosenbergs were sentenced to death; in issuing the sentence, Judge Irving Kaufman accused the couple of having "altered the course of history." The penalty rocked the world: As Supreme Court Justice Felix Frankfurter put it, they "were tried for conspiracy and sentenced for treason." They were electrocuted the evening of June 19, 1953, as New York's Union Square filled with an estimated 10,000 protesters.

What was the lasting effect of the **Clarence Earl Gideon trials**?

A 51-year-old drifter charged with burglary in Panama City, Florida, Clarence Earl Gideon had two trials, in 1961 and 1963. But it's what happened between the two trials that is important to every American today. What might have been pretty standard fare in the day-to-day business of the American justice system (Gideon was charged with robbing a cigarette machine and a jukebox), the Gideon case instead made history when the defendant successfully argued that his constitutional rights had been denied when he was refused an attorney. Though he had a limited education, after a guilty verdict was handed down in his 1961 trial, Gideon knew enough about his rights to petition the Supreme Court, saying that his right to a fair trial (guaranteed by the Sixth Amendment) had been violated: Since he was not able to hire a lawyer to defend himself, the trial had not been fair. The petition, one of thousands the Supreme Court receives each year, somehow rose to the top. The high court heard Gideon's case and agreed with his conclusion, calling it "an obvious truth," and clearly stating that "any person hailed into court, who is too poor to hire a lawyer, cannot be assured a fair trial unless counsel in provided for him." For Gideon, the opinion served to throw out the first trial; for the rest of America, it was assurance that regardless of the crime, a defendant would be guaranteed legal counsel. With the benefit of that counsel, Gideon's case was retried in 1963. He was acquitted on all charges.

What is the **Miranda warning**?

Familiar to many Americans from TV police dramas, the Miranda warning is a reading of the arrested person's rights: "You have a right to remain silent…anything you say can and will be used against you in a court. You have a right to consult with a lawyer…if you cannot afford a lawyer, one will be appointed for you.… "

Reading the defendant his rights became a requirement after the 1963 trial of Ernesto Miranda, a Mexican, who was accused of rape. He was found guilty and sentenced to 20 to 30 years imprisonment. But Alvin Moore, Miranda's court-appointed lawyer, had revealed through his questioning of a police officer that the defendant had not been notified of his right to the services of an attorney. The same police officer had taken Miranda's written confession following two hours of interrogation. Moore, convinced that the confession should not have been admissible in court because of the procedural error of not informing the defendant of his rights, appealed the Miranda case all the way to the Supreme Court. On June 13, 1966, the high court ruled, in a five-to-four decision, that Moore was right. Chief Justice Earl Warren (1891–1974) reasserted that "prior to any questioning a person must be warned that he has a right to remain silent, that any statement he does make may be used as evidence against him, and that he has the right to…an attorney."

Miranda's first trial was thrown out, and in 1967 he again stood trial in Arizona. But the prosecution secured new evidence; the testimonial of his estranged girlfriend

Thousands of men and women protest (on January 22, 1973) the U.S. Supreme Court's *Roe* v. *Wade* decision, which ruled against state laws that criminalize abortion.

that Miranda had confessed to her the rape he was charged with. He was convicted and again sentenced to 20 to 30 years in prison. Released on parole, Miranda died in a bar fight in January 1976. But police officers, the courts, and defendants still remember the importance of the case—even if they can't recall Miranda's name or crime.

Why is the ruling in *Roe* v. *Wade* controversial?

The 1973 Supreme Court decision in the case of *Roe* v. *Wade* legalized abortion in the United States and has probably engendered more public controversy than any other legal decision of the late twentieth century. Women's access to safe abortion continues to be the subject of debate, at issue in legal cases, and has inspired overzealous antiabortion activists to violence against doctors who perform abortions and office workers in women's health clinics. The seven Supreme Court justices who issued the majority decision became the recipients of thousands of letters of hatred, some of them threatening.

The case was brought as a class-action suit (representing all pregnant women) by 21-year-old Norma McCorvey (1947–), who has since the ruling reversed her feeling and joined the antiabortion camp as a "Right to Life" advocate. But in 1969, under the alias Jane Roe, McCorvey claimed that Texas's abortion law (on the books since 1859) violated her constitutional rights and those of other women. The other party named in the case was Texas district attorney Henry B. Wade (1914–2001), who argued to uphold Texas state law that punished anyone who gave an abortion. Despite the fact that the ruling in the case would do nothing to help McCorvey, for whom even a favorable decision would come too late to end her unwanted pregnancy, her lawyers, Linda Coffee and Sarah Weddington, agreed to pursue the case as a test. The crux of the plaintiff's case is best summed up by arguments made before the Supreme Court, when in December 1971, Weddington argued that Texas's ability to compel women to bear children infringed on a woman's right to control her own life. It was therefore a violation of the Constitution (the Fourteenth Amendment), which forbids states to "make or enforce any law which shall abridge the privileges or immunities of citizens." In response to defense claims that the fetus is entitled to protection, Weddington averred, "the Constitution as I read it...attaches protection to the person at the time of birth." These arguments, and those of the defense, were presented twice to the Supreme Court; after the first presentation the seven justices then seated concluded that such an important decision should not be made until the two newly appointed justices could participate. In October 1972 the case was heard again. As it turned out, the two new judges represented one majority vote and one dissenting vote. The majority decision was read by Justice Harry Blackmun on January 22, 1973: The high court overturned all state laws restricting women's access to abortions.

The decision was based on the Court's opinion that existing laws banning abortions had been enacted to protect the health of American women (since abortion had previously been a risky medical procedure) and that with advances in medicine this protection was no longer necessary or valid. The court also agreed that the Constitution's implied right to privacy, as found in the "Fourteenth Amendment's concept of personal liberty...or in the Ninth Amendment's reservation of rights to the people, is broad enough to encompass a woman's decision to terminate her pregnancy." Two justices dissented in the opinion, with Justice Byron White writing that the court had sustained a position that "values the convenience, whim or caprice of the putative mother more than life or the potential life of the fetus." Nearly three decades after the landmark decision, opinion continues to divide along such lines.

What was **ABSCAM**?

ABSCAM was an undercover operation conducted by the Federal Bureau of Investigation (FBI) to ferret out corrupt government officials and prosecute them. In 1978 agents began posing as American representatives of Arab businessmen whose company, the fictitious Abdul Enterprises Limited, was willing to buy political influence in the United States. (ABSCAM comes from the first two letters of the business name, with the word "scam" added to the end.) The trap was set. The first to be caught was Congressman Michael "Ozzie" Myers (1943–) of Pennsylvania, who was videotaped accepting a $50,000 bribe and saying, "I'm going to tell you something real simple and short—money talks in this business." Other officials also fell prey to the sting operation. All of them were arrested on charges of bribery and conspiracy. The first of seven trials got under way in 1980. None of the officials were acquitted; most faced fines and/or imprisonment; and all of them lost their offices. The wide net cast by ABSCAM had caught one U.S. senator, six representatives in Congress, one mayor, three members of the Philadelphia city council, one INS (Immigration and Naturalization Service) inspector, one lawyer, one accountant, and many of their associates. The FBI operation and the resulting trials and punishments sent a loud warning to any public official subject to influence.

What was the **Gang of Four**?

The Gang of Four was a group within China's Communist Party that, under the leadership of Mao Tse-tung's wife, Jiang Qing (1914–1991), 21 years his junior, carried out its own power-hungry agenda and plotted the takeover of the government from Chairman Mao (1893–1976).

A former stage and movie actress, Jiang was also an astute student of politics. In the late 1960s (at which time she had been married to Mao for some 30 years), she became associated with former army commander Lin Bao, and the pair conspired to stage a coup. In 1970, at a Communist Party conference, they announced that Lin had surpassed Mao as the leader of the people; one year later Lin and Jiang tried to overthrow Mao's government. Failing, Lin fled the country (his plane was later shot down), and Jiang succeeded in covering up her involvement in the affair. But she continued her subversive activities, associating with three other members of the politburo (the chief executive and political committee of the Communist Party). In 1974 Mao publicly admonished his wife and her cohorts, Wang Hongwen, Yao Wenyuan, and Zhang Chunquiao, to cease their power-seeking activities. In-fighting in the party had already resulted in Mao's loss of influence. Two years later, on September 9, 1976, Mao died. The Gang of Four were arrested and thrown into prison. There they remained for years while the case against them was formulated, resulting in an indictment that consisted of 20,000 words.

Finally, on November 20, 1980, the Gang of Four, expanded to include six other conspirators, were put on trial—charged with counterrevolutionary acts, including sedition and conspiracy to overthrow the government, persecution of party and state leaders, suppression of the people, and plotting to assassinate Mao. During nearly six weeks of testimony, Jiang's machinations were revealed to the 600 representatives who attended the trial, held in an air force auditorium in western Beijing, as well as to the Chinese press (foreign press was prohibited from attending). Her laundry list of malicious acts as ringleader of the Gang of Four included public humiliation and even torture of Communist Party rivals, execution of her personal enemies, inspiring the fear of the masses, and purging the arts of anything that did not carry a revolutionary theme. Jiang, while not denying many of these acts, insisted that she had all along acted at the behest of her husband, Mao. During the explosive testimony and presentation of evidence, which included tapes and documents substantiating the state's case against Jiang, she made outbursts, was temporarily expelled from the courtroom, was dragged screaming from the courtroom twice, and even taunted her accusers into executing her, saying it would be "more glorious to have my head chopped off."

In the end Jiang and one other conspirator were found guilty and sentenced to death (later commuted to life in prison), and the eight others were also found guilty and charged with sentences ranging from 16 years to life in prison. Jiang died on May 14, 1991, in what appeared to be a suicide.

Was despotic Romanian leader **Nicolae Ceausescu** brought to justice?

In the 1989 "trial" of Nicolae Ceausescu (1918–1989) and his wife Elena (1919–1989), justice may not have been served, but many believed the tyrannical Communist leader of Romania had indeed met with just desserts. The December 25 trial of the Ceausescus lasted all of 60 minutes: 55 minutes of questioning, to which the president's response was, "I do not recognize you…I do not recognize this court," followed by five minutes of deliberation. The court and judge were made up of the leaders in the popular rebellion that had begun December 16 when a pro-democracy rally attended by some 350,000 people ended in the Romanian army's and Ceausescu's secret police attacking unarmed demonstrators, killing several hundred men, women, and children. In demonstrations that followed, the Romanian army, long resentful of the privileged status enjoyed by the president's secret police, turned on Ceausescu's government, handing over automatic weapons to insurgents, whom they now joined in a popular uprising.

On December 21 state television and radio came under the people's control, as did the Communist Party's central building and the royal palace, which were later found to be replete with luxuries and were also connected by a maze of tunnels. The Ceausescus and a few of their close associates tried to flee, but were captured on December 22—the same day that mass graves were found, revealing the secret police's torture and destruction of several hundred men, women, and children. The rebels drove the Ceausescus around for three days, averting the still-loyal secret police. Realizing that time was not on their side, the captors assembled an "extraordinary military tribunal" in a small schoolroom at an army barracks. A defense lawyer was provided for Ceausescu; counsel urged the former president to plead guilty by reason of insanity. He refused. The charges against Ceausescu included genocide, the massacre of demonstrators, and subversion of the economy for his own benefit. One hour later, the guilty verdict was delivered. Asked if they wished to appeal the decision, the Ceausescus remained silent. They were promptly taken outside, where a squad opened fire on the former president and his wife. Videotape of the brutal killings (the squad had fired as many as 30 rounds) was shown on Romanian television. By December 30 the country was controlled by rebel forces.

What was *the* trial of the twentieth century?

As the century drew to a close, American historians, legal experts, and the public considered which of the many trials hailed as *the* trial of the twentieth century actually was. But the criteria used by each person varied: some believed the most important trial was the most highly publicized; others believed it was a trial in which the verdict affected everyone; some thought it was a trial that most epitomized an era; and some believed the most important trial was the one that inspired the most public debate. Still others looked for a single trial that seemed to "have it all": notoriety, impact, reflections of society at large, and a controversial outcome.

Among the courtroom dramas that were mentioned were:

The 1907 to 1908 trial of Harry Thaw (1871–1947), whose lawyers went through two trials (the first ended in a deadlocked jury) to convince jurors that Mr. Thaw suffered from "dementia Americans," a condition supposedly unique to American men that had caused Thaw to experience an uncontrollable desire to kill a man who had had an affair with his wife; the case took "innocent by reason of insanity" to new heights. The well-to-do, Harvard-educated Thaw was declared not guilty.

The 1921 case of Nicola Sacco (1891–1927) and Bartolomeo Vanzetti (1888–1927), Italian-born anarchists charged with and, amidst international uproar, found guilty of murder and robbery. So many people were convinced of the pair's innocence that demonstrations were mounted in cities around the world. They were executed in August 1927, but 50 years later Governor Michael Dukakis of Massachusetts signed a special proclamation clearing their names.

The Bruno Richard Hauptmann (1899–1936) trial of 1935: The German-born defendant was convicted of murdering the 20-month old son of celebrated aviator Charles A. Lindbergh (1902–1974) and his noted wife Anne Morrow Lindbergh (1906–2001), after the child was kidnapped from the family's Hopewell, New Jersey, home on March 1, 1932. For two and a half months, the world had prayed for the safe return of Charles Jr. But the toddler's body was found on May 12, two miles from the Lindbergh home. Public outrage demanded justice. Evidence surfaced that implicated Hauptmann, who was tried January 2 to February 13, 1935. Found guilty, he died by electrocution. Influential journalist H. L. Mencken noted that the trial, in which the conviction seemed to hinge on circumstantial evidence and which was attended by a "circuslike" atmosphere, was the "biggest story since the Resurrection." Though many remained convinced that officials had acted hastily to bring a case against

Hauptmann and maintained that he'd been framed, efforts to clear his name continued to be denied into the 1990s.

The 1931 to 1937 trials of the so-called "Scottsboro boys," nine men, ranging in age from 12 to 20, who had been seized from several points along a 42-car train in northeastern Alabama and were promptly charged with raping two white women. Upon medical examination, the women showed no signs of having been raped—or even of having had intercourse in the time frame in question. Nevertheless, the court of public opinion in the segregated South saw to it that eight of the nine were convicted of the crime, in spite of overwhelming evidence and testimony supporting their innocence.

The 1995 case of former football player O. J. Simpson (1947–), who was tried and acquitted in the murders of his former wife, Nicole, and her friend, Ronald Goldman. One observer said this trial had it all: "women, minorities, public interest, domestic violence, fallen hero," and through its live media coverage had "exposed the legal system to the public."

Other trials routinely mentioned in considering the question included the cases of convicted murderers Leopold and Loeb; the infamous Scopes "monkey trial," which pitted faith against reason, religion against science, and tradition against modernity; the Nuremberg trials, which established a process that brought war criminals to justice; the case of Alger Hiss, who was "either a traitor or the victim of a framing for political advantages at the highest levels"; and the Rosenberg espionage case. Undoubtedly there are trials missing from even this long list; there can be no definitive answer to the question.

What were the sentences in the **Beltway sniper cases**?

The October 2002 shootings that left ten people dead and three critically injured in the Washington, D.C., area set off a flurry of legal activity, which was still under way three years later. Two men had been found guilty by juries in separate trials in Virginia in late 2003: John Allen Muhammad (born John Allen Williams [1960–]) was sentenced to death, and accomplice Lee Boyd Malvo (a.k.a. John Lee Malvo [1985–]) received life imprisonment without parole (since he was a minor at the time of the shooting spree, the death penalty could not be applied in his case). In October 2004, under a plea deal in two other Beltway sniper cases, Malvo received other sentences of life imprisonment without parole and eight years on gun charges.

Muhammad and Malvo were arrested on October 24, 2002, in connection with the Beltway sniper attacks. The random shootings, which terrified suburban Washington and gripped the nation, began with a killing in a grocery store parking lot in Wheaton, Maryland, on the evening of October 2; within the next 16 hours there were four more shootings and four more dead. By October 22 there were eight more sniper shootings; three of the victims survived. People were targeted as they went about their daily business: pumping fuel outside gas stations, loading parcels into their cars in parking lots, arriving at school, and crossing streets. Tips led investigators to issue federal warrants for Muhammad's and Malvo's arrests on October 23, 2002; they were taken into custody early the following morning at a rest stop in Maryland.

The two men faced charges in several states as well as the District of Columbia. In addition to Virginia and Maryland, the pair had been tied to crimes in the state of Washington, where their journey had begun, as well as in Louisiana, Alabama, and Georgia. In 2005, with Malvo behind bars for life and Malvo awaiting a death sentence, prosecutors still wanted the two tried in other cases, as insurance against reversals on appeals and as closure for the families of victims. But legal analysts questioned the high price of further trials; defense costs alone for Muhammad and Malvo, who received court-appointed (and publicly paid) defenders, were expected to near $1 million.

The highly publicized sniper shootings added strength to the gun-control lobby. According to federal law, neither Muhammad nor Malvo could buy firearms; the weapon they used in their deadly spree was shoplifted from a Tacoma, Washington, gun store. The store owner and the gun manufacturer were named in a civil suit by the Legal Action Project of the Brady Center to Prevent Gun Violence on behalf of the victims and their families. In 2004 a $2.5 million settlement was reached in the case.

Were any **9/11 conspirators** convicted?

Yes, but by 2005 there had been only one: French citizen Zacarias Moussaoui (1968–) was convicted in a U.S. court in Alexandria, Virginia, in connection with the September 11, 2001, attacks that claimed nearly 3,000 lives. Moussaoui was taken into custody by the Federal Bureau of Investigation (FBI) in August 2001; a flight instructor in Minnesota, where he was training, had reported him as suspicious. After the September terrorist attacks, Moussaoui continued to be held as the possible 20th hijacker (one of the flights on 9/11 had four hijackers; the other three flights each had five). For the next three years, the suspected terrorist was the subject of a sometimes dramatic legal battle: Moussaoui insulted the U.S. District judge hearing his case, attempted to fire his lawyers, and pleaded guilty only to later change his mind. On April 22, 2005, the case came to close when Moussaoui admitted his guilt in front of a packed courtroom. His sentencing trial was set for 2006.

ECONOMICS AND BUSINESS

What is **capitalism**?

The cornerstones of capitalism are private ownership of property (capital goods); property and capital create income for those who own the property or capital; individuals and firms openly compete with one another, with each seeking its own economic gain (so that competition determines prices, production, and distribution of goods); and participants in the system are profit-driven (in other words, earning a profit is the main goal). Capitalism is the antithesis of socialism, a theory by which government owns most, if not all, of a nation's capital. There is no pure capitalist system; national governments become involved in the regulation of business to some degree. But the economy of the United States is highly capitalistic in nature, as are the economies of many other industrialized nations, including Great Britain.

What does **laissez-faire** mean?

From the French, *laissez-faire* literally means "to let (people) do (as they choose)." As an economic doctrine, laissez-faire opposes government interference in economic and business matters, or at least desires to keep government's role to an absolute minimum. Laissez-faire favors a free market (a market characterized by open competition). The theory was popularized during the late eighteenth century as a reaction to mercantilism. Noted Scottish economist Adam Smith (1723–1790) was among the advocates of a laissez-faire market.

What is **mercantilism**?

An economic system that developed as feudalism was dissolving (at the end of the Middle Ages [500–1350]), mercantilism advocates strict government control of the national economy. Its adherents believe a healthy economy can only be achieved through state regulation. The goals were to accumulate bullion (gold or silver bars), establish a favorable balance of trade with other countries, develop the nation's agricultural concerns as well as its manufacturing concerns, and establish foreign trading policies.

Who was **Adam Smith**?

Scottish economist Adam Smith (1723–1790) is popular with conservative economists today because of his work titled *The Wealth of Nations* (written in 1776), which proposes a system of natural liberty in trade and commerce; in other words, a free-market economy. Smith, who was teaching at the University of Glasgow at the time, wrote, "Consumption is the sole end and purpose of all production, and the interest of the producer ought to be attended to, only so far as it may be necessary for promoting that of the consumer."

The Wealth of Nations established the classical school of political economy but has been faulted for showing no awareness of the developing Industrial Revolution. While Smith advocated both free-market competition and limited government intervention, he also viewed unemployment as a necessary evil to keep costs—and therefore prices—in check.

What is **Keynesian economics**?

Keynesian economics are the collected theories of British economist and monetary expert John Maynard Keynes (1883–1946), who in 1935 published his landmark work, *The General Theory of Employment, Interest and Money.* A macroeconomist (he studied a nation's economy as a whole), Keynes departed from many of the concepts of a free-market economy. In order to ensure growth and stability, he argued that government needs to be involved in certain aspects of the nation's economic life. He believed in state intervention in fiscal policies, and during recessionary times he favored deficit spending, the loosening of monetary policies, and government public works programs (such as those of President Franklin D. Roosevelt's New Deal) to promote employment. Keynes's theories are considered the most influential economic formulation of the twentieth century.

Having played a central role in British war financing during World War II (1939–45), Keynes participated in the Bretton Woods Conference of 1944, where he helped win support for the creation of the World Bank, which was established in 1945 as a specialized agency of the United Nations. The body aims to further economic development by guaranteeing loans to nations, extending easy credit terms to developing nations, and providing risk capital to promote private enterprise in less-developed nations. It's interesting to note that Keynes was a key representative at the Paris Peace Conference of 1919, where the Treaty of Versailles was drawn up, officially ending

World War I (1914–18). He quit the proceedings in Paris, returned to private life in London, and in 1919 published *The Economic Consequences of Peace,* in which he argued against the excessive war reparations that the treaty required of Germany. Keynes foresaw that the extreme punishment of Germany at the end of World War I would pave the way for future conflict in Europe.

MONEY

When was **money introduced**?

The use of money dates back some 4,000 years, when people began using something of recognized value, such as precious metals including gold and silver, to purchase goods and services. In the absence of money, all transactions were made on the barter system, which is an exchange of goods and services negotiated by the parties involved. The introduction of money simplified the acquisition of products and services. The ancient country of Lydia, in the western part of Asia Minor (modern-day Turkey), is credited with the first use of standardized coins, made of gold and silver, in the seventh century B.C.

When was **paper money** first used?

Paper money first appeared in China during the Middle Ages (500–1350). In the ninth century A.D., paper notes were used by Chinese merchants as certificates of exchange and, later, for paying taxes to the government. It was not until the eleventh century, also in China, that the notes were backed by deposits of silver and gold (called "hard money").

What were **pieces of eight**?

Pieces of eight were Spanish silver coins (pesos) that circulated along with other hard currency in the American colonies. Since the settlements in the New World were all possessions of their mother countries (England, Spain, France, Portugal, and the Netherlands), they did not have monetary systems of their own. England forbade its American colonies to issue money. Colonists used whatever foreign currency they could get their hands on. Pieces of eight (from Spain), *reals* (from Spain and Portugal), and shillings (from England) were in circulation; the pieces of eight were most common. The Spanish silver coin was so named because it was worth eight reals and at one time had an eight stamped on it. To make change, the coin was cut up to resemble pieces of a pie. Two pieces, or "two bits," of the silver coin made up a quarter, which is why Americans still refer to a quarter (of a dollar) as two bits.

There were frequent money shortages in the colonies, which usually ran a trade deficit with Europe. The colonies supplied raw goods to Europe, but finished goods, including manufactured items, were mostly imported, resulting in an imbalance of trade. With coinage scarce, most colonists conducted trade as barter, exchanging goods and services for the same. In 1652 Massachusetts became the first colony to mint its own coins; that year there was no monarch on the throne of England. Although the issue of coinage by colonists was strictly prohibited by England, the Puritans of Massachusetts continued to make their own coins for some 30 years thereafter, stamping the year 1652 on them as a way to circumvent the law.

What were **Continentals**?

Continentals were the paper money issued by the U.S. government during the American Revolution (1775–83). The Second Continental Congress, which governed the new nation after the Declaration of Independence (1776), ran the war effort against Great Britain. The governing body did not have the power to levy taxes, since no constitution had been drawn up yet. So the Congress appealed to each state to contribute to the war fund. However, states that did not face imminent danger—those in which there was no fighting—often did not answer the call. Many of the new nation's most prominent citizens remained loyal to the British and refused to contribute money to the American patriotic cause. Yet money was needed to buy supplies and ammunition, and to pay soldiers. In order to finance the Revolution, Congress was compelled to issue paper bills, which promised holders future payment in silver. But as Congress issued more and more Continentals, the currency became devalued because there was not enough silver to back up the promised payments. By 1780 there were so many Continentals in circulation that they had become almost worthless. The phrase "not worth a Continental" was used by Americans to describe anything that had no value. To help solve the financial crisis, some patriotic citizens contributed sums of money; in exchange, they received interest-bearing securities from the government. But funds continued to be scarce. The problem of funding the revolutionary effort was not solved until foreign powers stepped in to aid the fledgling nation in its fight against the powerful British. European loans to the United States, notably from France, were instrumental in the American victory in the Revolutionary War.

What is **"rag money"**?

Rag money is a derisive term for paper currency. The name comes from the early days of paper money, when paper itself was predominately made with the cotton and linen fibers from rags. Hence, bills were "rag money." Given that valued currency was issued in silver or gold coins by the established governments of Europe, it is not surprising that Americans greeted paper currency—which is nothing more than a promise of future payment in coin—as something to be regarded with skepticism. After the Declaration of Independence (1776), the first bills that were issued by the U.S. government quickly became worthless: In its effort to fund the American Revolution (1775–83), the Second Continental Congress printed so many bills, called Continentals, that there was not enough silver to back them up. The financial crisis that emerged did nothing to inspire American confidence in paper currency. Rag money continued to have its detractors even after the Revolution had been financed by European loans and the U.S. government established the dollar as its unit of currency (1785).

What were **wildcat banks**?

Wildcat banks were state-chartered financial institutions that operated in the United States from the early 1800s until the American Civil War (1861–65). They were known as wildcat banks for their free-lending policies and their issue of paper currency that could not be backed up by gold or silver (called specie). The Second National Bank of the United States operated between 1816 and 1836, during which time the federally controlled bank was able to restrain the wildcat institutions, which predominated in the West and South, requiring them to issue only what currency they could convert to coin. But when the charter of the Second National Bank of the United States was allowed to expire (in 1836), the wildcat banks resumed their unsound banking practices. Paper currency issue and lending went unregulated amidst a rush to buy lands on the frontier. The nation's currency wildly fluctuated as the renegade financial institutions loosened and tightened the money supply to suit their own needs. Further, since there were so many banks issuing their own notes, another problem introduced itself: counterfeiting. No one could tell what was true bank currency and what was the product of a good counterfeiter.

With inflation rampant and land speculation at a new high, on July 11, 1836, President Andrew Jackson (1767–1845), intent on reining in the wildcat banks, issued the Specie Circular, an order that government agents accept nothing but gold or silver as payment for new lands. When prospective land buyers (particularly in the West) took their paper bills to the state-chartered banks to be converted to coin, they found the banks' tills were empty, and the holders were therefore denied the face value of their notes. Bank after bank closed its doors, causing a financial panic in 1837. But many state banks remained in business, and the issue of regulating paper currency continued to trouble the nation.

What was the **National Bank Act**?

The National Bank Act of 1863 was designed to create a national banking system, float federal war loans, and establish a national currency. Congress passed the act to help resolve the financial crisis that emerged during the early days of the American Civil War (1861–65); the fight with the South was expensive, and no effective tax program had been drawn up to finance it. In December 1861 banks suspended specie payments

(payments in gold or silver coins for paper currency)—people could not convert bank notes into coins. The government responded by passing the Legal Tender Act (1862), issuing $150 million in national notes called greenbacks. But bank notes (paper bills issued by state banks) accounted for most of the currency in circulation.

To bring financial stability to the nation and fund the war effort, the National Bank Act of 1863 was introduced in the Senate in January of that year. Secretary of the Treasury Salmon Chase (1808–1873), aided by Senator John Sherman (1823–1900) of Ohio, promoted it to the legislators. The bill was approved in the Senate by a close vote of 23 to 21. The House passed the legislation in February. National banks organized under the act were required to purchase government bonds as a condition of start-up. As soon as those bonds were deposited with the federal government, the bank could issue its own notes up to 90 percent of the market value of the bonds on deposit.

The National Bank Act improved but did not solve the nation's financial problems: Some of the 1,500 state banks, which had all been issuing bank notes, were converted to national banks by additional legislation (passed June 1864 to amend the original bank act). Other state banks were driven out of business or ceased to issue notes because of the 1865 passage of a 10 percent federal tax on notes they issued, making it unprofitable for them to print their own money. The legislation created $300 million in national currency—in the form of notes issued by the national banks. But since most of this money was distributed in the East, the money supply in other parts of the country remained precarious. The West demanded more money—an issue that would dominate American politics in the years after the American Civil War (1861–65). Nevertheless, the nation's banking system stayed largely the same—despite the Panic of 1873—until passage of the Federal Reserve Act in 1913.

What is the **Federal Reserve**?

It is the central banking system in the United States, created by a 1913 act of Congress, the Federal Reserve Act (sometimes called the Glass-Owens Bill). The legislation provided for a stable central banking system after the system set up by the National Bank Act of 1863 proved ineffective in managing the nation's currency, in responding to economic growth, or in exerting a controlling influence on the economy.

The Federal Reserve Act created 12 regional federal reserve banks: in Boston, Massachusetts; New York; Philadelphia, Pennsylvania; Cleveland, Ohio; Richmond, Virginia; Atlanta, Georgia; Chicago; St. Louis, Missouri; Minneapolis, Minnesota; Kansas City, Missouri; Dallas, Texas; and San Francisco, California. These institutions operate as "bankers' banks": member banks (commercial institutions) use their accounts with the Federal Reserve in the same way that consumers use their accounts on deposit at commercial banks. All national banks must be members of the Federal Reserve system; state banks may join the system upon meeting certain requirements. The Federal Reserve Act also established a Federal Reserve Board, now called the

What are greenbacks?

Greenbacks are the paper money printed and issued by the U.S. government during the Civil War. The financial demands of the war quickly depleted the nation's supply of specie (gold and silver). In response, the government passed the Legal Tender Act of 1862, which suspended specie payments and provided for the issue of paper money. (Legal tender is money that must be accepted in payment of any debt.) Since the bills were supported only by the government's promise to pay, it was somewhat derisively observed that they were backed only by the green ink they were printed with. Hence the name greenbacks. The value of the notes depended on the peoples' confidence in the U.S. government—and its future ability to convert the currency to coin. As the fighting between the Union and the Confederacy raged, Americans' confidence in their government fluctuated: When the Union suffered defeat, the value of the greenbacks dropped—one time to as low as 35 cents on the dollar.

Greenbacks remained in circulation after the fighting ended, finally regaining their full value in 1878. After the financial crisis in 1873, many people—particularly western farmers—clamored for the government to issue more. Advocates of the monetary system formed the Greenback Party, which was active in American politics between 1876 and 1884. They believed that by putting more greenbacks into circulation, the U.S. government would make it easier for debts to be paid and prices would go up—resulting in prosperity. The country's present-day system of paper money is based on the government's issue of notes, which was made necessary by the Civil War.

Board of Governors, to supervise the system. The board consists of seven members who are appointed by the president of the United States and are approved by the Senate. To reduce the possibility of nearsighted political influence, members serve staggered 14-year terms (one of the 14 terms expires every other year).

The duties of the Federal Reserve include lending money to commercial (member) banks, directing the reserve banks' purchase and sale of U.S. government securities on the open market, setting reserve requirements (for how much money needs to be in the U.S. Treasury), and regulating the discount rate (the interest rate the Federal Reserve charges commercial banks for loans), which is one of the system's principal influences on the economy. In performing these duties, the Federal Reserve (often called "the Fed" in financial circles) can expand (loosen) or contract (tighten) the supply of money in circulation. The Federal Reserve also issues the national currency and supervises and regulates the activities of banks and their holding companies. It began operation in November 1914.

The central bank systems of other developed nations include the Bank of Canada, Banque de France, and the Deutsche Bundesbank (of Germany).

When were **ATMs** introduced?

The ubiquitous automatic teller machine (ATM) was introduced in 1967 by Britain's Barclays Bank at a branch near London. Two years later, the Chemical Bank opened the first ATM in the United States, at Rockville Centre, New York. Self-service banking grew steadily in the 1980s and took off in the 1990s, when some banks began charging their customers for banking through tellers rather than ATMs.

When was the **euro** introduced?

The euro, the currency of the 12 European Union nations—Belgium, Germany, Greece, Spain, France, Ireland, Italy, Luxembourg, the Netherlands, Austria, Portugal, and Finland—went into circulation January 1, 2002, becoming part of daily life for more than 300 million people. The banknotes and coins replaced national currencies, making the franc, deutschmark, peseta, and lira, among others, history in the participating nations.

The euro's origins can be traced to a series of international agreements, beginning in 1978, which were made among the members of what was then called the European Community, or EC. In February 1986 the framework for the unified monetary system was agreed upon by nations who signed the Single European Act, creating "an area without internal frontiers in which the free movement of goods, persons, services, and capital is ensured." The 1989 Delors Report outlined a plan to introduce the currency in three phases. The final phase of that plan began on January 1, 1999, when the 11 countries (later to become 12) belonging to the European Union established the conversion rates between their respective national currencies and the euro, creating a monetary union with a single currency. A three-year transition phase followed, during which monetary transactions could be made in euro, but there was no requirement to do so. On January 1, 2002, the central banks of the 12 participating countries put into circulation about 7.8 billion euro notes and 40.4 billion euro coins, together worth 144 billion euros. Simultaneously each country began to withdraw its own currency from circulation. By February 28, 2002, the changeover was complete, meaning the national currencies were completely withdrawn and only the euro was in circulation.

When 10 new nations (Cyprus, Czech Republic, Estonia, Hungary, Latvia, Lithuania, Malta, Poland, Slovakia, and Slovenia) joined the EU on May 1, 2004, there was no timetable for their adoption of the euro. Previously, in 2003, Sweden voted against joining the euro area.

COLONIAL AMERICA AND EARLY REPUBLIC

What were **indentured servants**?

During colonial times in America, there were two kinds of indentured servants: voluntary and involuntary. Voluntary servants were people, often trained in a craft or skill, who could not afford passage to the colonies. In exchange for their passage, they agreed to work for a period of four to seven years for a colonial master. At the end of this period, the servant became a freeman and was usually granted land, tools, or money by the former master. Involuntary indentured servants were criminals whose sentence was a period of servitude, the impoverished, or those in debt. Most indentured servants were involuntary. Their period of obligation to a colonial master was longer than that of a voluntary servant, usually 7 to 14 years. But, like their counterparts, the involuntary servants also received land, tools, or money at the end of their contract, and they, too, became freemen.

The arrival of indentured servants in the American colonies addressed a labor shortage that emerged in the early 1600s. In 1618 the Virginia Company, a joint-stock enterprise that encouraged the development of Virginia, adopted a new charter based on the "headright system": Englishmen who could pay their own Atlantic crossing were granted 50 acres of land; each of their sons and servants were also granted an additional 50 acres. Other colonies were also developed under the headright system, with the land amounts varying by colony. Soon there were more farms than there was labor to work the fields. The colonists solved this problem through the system of indentured servitude.

Many indentured servants were drawn from England, Ireland, Scotland, and Germany. In European ports, people contracted themselves or became involuntarily contracted to ship captains, who transported them to the colonies, where their contracts were sold to the highest bidder. Roughly half the colonial immigrants were indentured servants. Colonial laws ensured servants would fulfill the term of their obligation; any servant who ran away was severely punished. Laws also protected the servants, whose masters were obligated to provide them with housing, food, medical care, and even religious training. The system was prevalent in the mid-Atlantic colonies, but it was also used in the South. When the economies of the Caribbean islands failed at the end of the 1600s, plantation owners sold their slaves to the mainland, where they worked primarily on southern plantations, replacing indentured servants by about 1700. In other colonies, the system ended with the American Revolution (1775–83).

What was **triangular trade**?

Triangular trade refers to the various navigation routes that emerged during the colonial period. There were numerous triangular paths that ships traveled, ferrying people, goods (both raw and finished), and livestock. The most common triangular route began on Africa's west coast where ships picked up slaves. The second stop was the Caribbean islands—predominately the British and French West Indies—where the slaves were sold to plantation owners, and traders used the profits to purchase sugar, molasses, tobacco, and coffee. These raw materials were then transported north to the third stop, New England, where a rum industry was thriving. There ships were loaded with the spirits and traders made the last leg of their journey back across the Atlantic to Africa's west coast, where the process began again.

Other trade routes operated as follows: 1) manufactured goods were transported from Europe to the African coast; slaves to the West Indies; and sugar, tobacco, and coffee transported back to Europe, where the triangle began again; 2) lumber, cotton, and meat were transported from the colonies to southern Europe; wine and fruits to England; and manufactured goods to the colonies, where the triangle began again. There were as many possible routes as there were ports and demand for goods.

The tragic result of triangular trade was the transport of an estimated 10 million black Africans. Sold into slavery, these human beings were often chained below deck and allowed only brief if any periods of exercise during the transatlantic crossing, which came to be called the Middle Passage. Conditions for the slaves were brutal and improved only slightly when traders realized that should slaves perish during the long journey across the ocean, it would adversely affect their profits upon arrival in the West Indies. After economies in the islands of the Caribbean crashed at the end of the 1600s, many slaves were sold to plantation owners on the North American mainland, initiating another tragic trade route. The slave trade was abolished in the 1800s, putting an end to the capture of Africans and their forced migration to the Western Hemisphere.

How did the **American tobacco industry** get started?

Tobacco, a member of the nightshade family, is an indigenous American plant. When Christopher Columbus (1451–1506) arrived in the West Indies in 1492, he found the native inhabitants smoking rolls of tobacco leaves, called taino. (The word tobacco is derived from the Spanish *tabaco,* which is probably from *taino.*) The practice of "drinking smoke" was observed to have a relaxing effect. Upon returning to Spain, Columbus took seeds of the plant with him. By 1531 tobacco was being cultivated on a commercial scale in the Spanish colonies of the West Indies. In 1565 English naval commander John Hawkins (1532–1595) introduced tobacco to England, where smoking was condemned as a "vile and stinking custom" by King James I (1566–1625) decades later.

Tobacco was not commercially cultivated on the North American mainland until English colonist John Rolfe (1585–1622) carried seeds from the West Indies to Jamestown, Virginia, where he settled in 1610. By 1612 he had successfully cultivated tobacco and discovered a method of curing the plant, making it a viable export item. Jamestown, Virginia, became a boomtown and England's King James, who collected export duties, changed his mind about the habit of smoking. The coastal regions of Virginia, Maryland, and North Carolina were soon dominated by tobacco plantations, and the crop became the backbone of the economies in these colonies. Cultivation of tobacco did not require the same extent of land or slave labor as did other locally grown crops such as rice and indigo, but it depleted the nutrients of soil more rapidly, causing growers to expand their lands westward into the Piedmont region (the plain lying just east of the Blue Ridge and Appalachian Mountains).

In 1660 British Parliament passed the Second Navigation Act, declaring that tobacco and other articles from England's American colonies could only be exported to the British Isles. Tobacco prices dropped in response to the legislation and the colonial economies were weakened, causing political discontent with the mother country. But European demand was not diminished and the colonists soon resumed exports, despite the Second Navigation Act. By 1765 colonial exports of tobacco were nearly double in value the exports of bread and flour. The crop helped define the plantation economy of the South, which prevailed until the outbreak of the American Civil War (1861–65). During the 1800s companies such as R. J. Reynolds Tobacco and American Tobacco were founded. Despite the highly publicized dangers of using tobacco (smoking or chewing), tobacco has remained an important crop and the manufacture of tobacco products an important industry in the American South.

What were the **Navigation Acts**?

Between 1645 and 1761 British Parliament passed a series of 29 laws intended to tightly control colonial trade, shipping, and industry to the benefit of English interests in America. These acts, which were largely ignored by the American colonists, were intended to ensure that the British colonies in North America remained subservient to the mother country. The initial act of 1645 forbade the import of whale oil into England unless it was transported aboard English ships with English crews. Subsequent laws, those passed in 1651, 1660, and 1663, provided the basis of the Navigation Acts.

The First Navigation Act (1651) resembled the legislation of 1645, but was more far-reaching: It stipulated that goods could only enter England, Ireland, or the colonies aboard English (or English colonial) ships. Further, colonial coastal trade was to be conducted entirely aboard English ships. The Second Navigation Act (1660) reaffirmed that goods could only be transported aboard English ships and established a list of "enumerated articles" that had to be shipped directly to England. The intent was to prevent the colonies from trading directly with any other European country: England required the colonies to sell their materials to English merchants or pay duties on goods sold to other countries. The list of articles included sugar, cotton, tobacco, indigo, rice, molasses, apples, and wool. In 1663 Parliament passed the Staple Act, making it illegal for colonies to buy products directly from foreign countries; European countries would first need to ship their products to England or pay customs fees. Through the Navigation Acts, England tried to establish itself as the gatekeeper of colonial imports and exports. But the laws were difficult to enforce, and the colonists easily circumvented them. Smuggling was rampant. Still, the laws, which continued to be passed until the eve of the American Revolution (1775–83), had little effect on the colonial economy, which grew at twice the rate of England's during the period.

What was **National Road**?

National Road was the first federal road. Today, the path of the great westward route is followed closely by U.S. Highway 40. Congress authorized construction of the road in 1806 to answer the cry of settlers who demanded a better route across the Appalachians into the Ohio River valley. Originally called Cumberland Road, work began in 1811 in Cumberland, Maryland. Progress was slow: The road did not reach present-day Wheeling, West Virginia, a distance of 130 miles, until six years later. But in 1830 President Andrew Jackson (1767–1845) gave the project a boost when he signed an act of Congress appropriating $130,000 to survey and extend the Cumberland Road westward. Jackson called it a "national road" (it was also called the Great National Pike). By the time the route was completed in 1852, it extended westward from Wheeling to cross Ohio, Indiana, and Illinois, where it ended at Vandalia, east of St. Louis. The project cost the government more than $7 million to complete but accomplished what had been hoped: National Road spurred development in the Old Northwest (the present-day states of Ohio, Michigan, Indiana, Illinois, Wisconsin, and part of Minnesota) and the Far West (the territories west of the Mississippi River). The overland route was traversed in covered wagons and Conestogas by pioneers and tradesmen; large quantities of goods, including livestock, grain, and finished products were transported both east and west. Towns along the route boomed. By the end of the century, the road diminished in importance as settlers, new immigrants, and goods were transported along the railroads that had begun to crisscross the nation in 1865. Nevertheless, the National Road heralded the future of federal transportation projects that would knit the nation together.

Why was the completion of the **Erie Canal** important to U.S. development?

Completed in 1825, the Erie Canal joined the Atlantic Ocean to the Great Lakes, linking the East with the West and for the first time allowing freight and settlers to easily move back and forth between the regions. Begun on July 4, 1817, the canal was sponsored by Governor DeWitt Clinton (1769–1828) of New York, who planned and eventually carried out the huge building project. The waterway was funded by the state of New York, which paid just over $7 million to complete it. The original canal was 363 miles long, 40 feet wide at the surface, and 4 feet deep. It had 83 locks, which raised vessels 562 feet between the Hudson River and Lake Erie. (A lock is a section of a canal that can be closed to control the water level and is then used to either raise or lower a vessel to another body of water.) Beginning at Albany, New York, on the Hudson River (which flows into the Atlantic Ocean at New York City), the canal extends west as far as Buffalo, New York, on Lake Erie (one of the five Great Lakes).

The waterway, which was inaugurated by the run of the barge *Seneca Chief* on October 26, 1825, could transport passengers aboard boats and move cargo aboard barges, which were pulled by teams of horses and mules on the ground. In spite of the critics, who dubbed the ambitious project "Clinton's Wonder" and "Clinton's Ditch," the canal's positive impact on the American economy was felt within the first decade of its operation. The new transportation route reduced freight rates both eastward and westward, made Buffalo a major port in the region and New York City a major international port, was a catalyst for population growth in upstate New York and throughout the Old Northwest (the present-day states of Ohio, Michigan, Indiana, Illinois, Wisconsin, and part of Minnesota), and prompted other states (Ohio, Indiana, and Illinois) to build canals, further opening up the country's interior to development and commerce. Since crops could be shipped from these lush farmlands and as more farms came into existence, the Erie Canal helped supply the newly arrived immigrants in the eastern cities with food; in turn, they shipped manufactured goods west to the farming communities. The canal was enlarged several times between 1835 and 1862 to increase its capacity. In 1903 New York voted to link the canal with three shorter waterways in the state to form the New York State Barge Canal, which opened in 1918.

Why was the invention of **barbed wire** important to Western development?

Barbed wire was commercially developed in 1874 by American inventor Joseph Glidden (1813–1906). Consisting of steel wires that are twisted together to make sharp points resembling thorns, the material was quickly implemented in the West to construct fences. With trees scarce on the Great Plains, farmers had lacked the materials to erect wooden fences. Instead they resorted to planting prickly shrubs as a way of defining their lands and confining livestock. However, this method was not always effective. With the advent of barbed wire, farmers were able to fence in their acreage. Cattle owners became angered by small farmers who put up barbed wire: They had previously allowed livestock to roam the open plain. Fearing depletion of grazing lands, ranchers also began using barbed wire to fence tracts, whether or not they could claim legal title to them. Disputes arose between ranchers and between ranchers and farmers. In 1885 President Grover Cleveland (1837–1908) brought an end to illegal fencing, ordering officials to remove barbed wire from public lands and Indian reservations. Legal use of the material to define land claim boundaries brought the demise of the open range and helped speed agricultural development of the prairie.

NATURAL RESOURCES

What was the biggest **gold rush**?

The greatest American gold rush began on January 24, 1848, when James Marshall discovered gold at Sutter's Mill in Coloma, California. Within a year, a large-scale gold rush was on. As the nearest port, the small town of San Francisco grew into a bustling city as fortune seekers arrived from around the world. Due to the influx, by 1850 California had enough people to qualify it for statehood. This pattern repeated itself elsewhere in the American West, including the Pikes Peak gold rush in 1859, which effectively launched the city of Denver, Colorado. The gold rush led to the discovery of copper, lead, silver, and other useful minerals. It also spawned related industry. One of the success stories is that of Levi Strauss (1829–1902), a Bavarian immigrant who in 1853 began making and selling sturdy clothing to miners in San Francisco.

In other countries, gold rushes had the same effect on the growth and development of regions: After the precious metal was discovered in Australia in 1851, the country's population almost tripled over the course of the next decade. The effect of an 1861 gold rush in nearby New Zealand was to double the country's population in six years' time. An 1886 discovery in South Africa led to the development of the city of Johannesburg. Just over a decade later, the infamous Yukon gold rush (in the Klondike region of Canada) spurred development there.

What was the **Comstock Lode**?

The richest silver mine in the United States, the Comstock Lode was also plentiful in gold. The ore deposit was found in 1857 at Mount Davidson in western Nevada, about 16 miles southeast of Reno. The discoverers, Ethan Allen Grosh and Hosea Ballou Grosh, died before they could record the claim. American prospector Henry T. P. Comstock (1820–1870) laid claim to the lode in 1859, but later sold it for an insignificant amount compared to what it was worth. The mine flourished until 1865 and again between 1873 and 1882, when the "Big Bonanza," a super-rich ore vein, yielded more than $100 million. By 1882 the mine had yielded $397 million in ore and had produced half the United States' silver output during the period. Western Nevada had become a hotbed of mining activity, attracting numerous prospectors. Among those who made their fortune from the Comstock Lode was American mining magnate and future senator George Hearst (1820–1891). He used his fortune to buy the *San Francisco Examiner* in 1880, which was taken over by his son, American newspaper publisher William Randolph Hearst (1863–1951), seven years later. Virginia City, established in 1859 at the site of the discovery, became one of the West's boomtowns during the late 1800s. By 1898 the mines at the Comstock Lode were all but abandoned: Wasteful mining methods and the demonetization of silver brought its demise.

Miners in Colorado, 1880. Gold rushes led to a mass migration of prospectors to the American West and spurred the growth of such cities as San Francisco and Denver.

What is **"black gold"**?

Black gold is a term for oil or petroleum—*black* because of its appearance when it comes out of the ground, and *gold* because it made prospectors, drillers, and oil industry men rich. The oil industry in the United States began in 1859 when retired railroad conductor Edwin L. Drake (1819–1880) drilled a well near Titusville, Pennsylvania. His drill, powered by an old steam engine, struck oil. Oil from animal tallow and whales, had been used as a lubricant since colonial times. The discovery of a process for deriving kerosene, a clean-burning and easy-lighting fuel, from coal oil had been patented in 1854. After Drake's Titusville well produced shale oil, the substance was analyzed for its properties and it, too, was determined to be an excellent source of kerosene. Soon others began prospecting for "rock oil." Western Pennsylvania became an important oil-producing region. Wagons and river barges transported barrels to market; later, the railroad reached into the region; and by 1875 a pipeline was built to carry the oil directly to Pittsburgh. Petroleum products soon replaced whale oil as a fluid for illumination. During the 1880s, Ohio, Kentucky, Illinois, and Indiana also produced oil. In 1901 the famous Spindletop Field in eastern Texas produced the nation's first "gusher"—oil literally sprang out of the earth. During the next decade, California and Oklahoma joined Texas to lead the nation's oil production. Between

1859 and 1900, U.S. oil production boomed: Just 2,000 barrels were produced the year it was discovered in Pennsylvania; more than 64 million barrels were produced annually by the turn of the century.

The second half of the 1800s saw the oil industry boom: The fuel was used for lighting, heating, and lubrication (principally of machinery and tools). But the advent of the automobile and its central role in the life of twentieth-century America made the oil industry richer yet. Demand soon exceeded the nation's supply of petroleum, prompting the United States to increasingly rely on imported oil for fuel.

When did **diamond mining** begin in Africa?

An 1867 discovery of a "pretty pebble" along the banks of the Orange River in South Africa led to the finding of a rich diamond field near present-day Kimberley (the city was founded as a result of the mining, in 1871). Similar to the California gold rush roughly a decade and a half earlier, the finding in central South Africa prompted people from Britain and other countries to flock to the area. However, the ultimate outcome was conflict: Since both the British and the Boers (who were Dutch descendants living in South Africa) claimed the Kimberley area, the first Boer War ensued in 1880.

INDUSTRIAL REVOLUTION

How were **finished goods** produced **before the Industrial Revolution**?

Before the factory and machine age ushered in by the Industrial Revolution, people made many of their own finished goods, bought them from small-scale producers (who manufactured the goods largely by hand), or bought them from merchants who contracted homeworkers to produce goods. The putting-out system was a production method that was used in New England from the mid-1700s to the early 1800s. It worked this way: Merchants supplied raw materials (cotton, for example) to families, especially women and young girls, who would make partially finished goods (thread) or fully finished goods (cloth) for the merchant. These manufactured goods were then sold by the merchant. Homeworkers, who "put out" goods, provided the needed manufacturing labor of the day.

How did the **textile industry** begin?

The large-scale factory production of textiles began in the late 1700s, becoming established first in Great Britain, where a cotton-spinning machine was invented in 1783 by

Young girls work in a basket factory, 1908. Child labor was rampant in the United States until the Fair Labor Standards Act of 1938 brought reforms.

Richard Arkwright (1732–1792). Spinning mills were introduced to the United States in 1790 by English-born mechanist and businessman Samuel Slater (1768–1835). The 21-year-old had worked as a textile laborer for more than six years in an English mill where he learned the workings of Arkwright's machine, which the British considered the cornerstone of their booming textiles industry; laws prevented anyone with knowledge of the mill from leaving the country. In 1789 Slater, determined he could recreate the spinning mill and eager to seek his own fortune, disguised himself to evade the authorities and leave the country, sailing from England for American shores. Arriving in Providence, Rhode Island, he formed a partnership with the textiles firm Almy and Brown. Slater began building a spinning mill based on the Arkwright machine. This he did from memory. The spinning mill was debuted December 20, 1790, in the village of Pawtucket, Rhode Island, where the wheels of the mill were turned by the waters of the Blackstone River. The machine was a success and soon revolutionized the American textiles industry, which previously relied on cottage workers (the putting-out system) to manufacture thread and yarn.

Slater's innovation, which would earn him the title Father of the American Textiles Industry, spawned the factory system in the United States. By 1815 there were 165 cotton mills in New England, all working to capacity. The early mills were not large-scale, however, and for a time after Slaters's introductions, New England mills and merchants continued to rely on homeworkers to weave threads (now produced by the mills) into cloth.

In 1813 the Boston Manufacturing Company opened the first textile factory, where laborers ran spinning and weaving machines to produce woven cloth from start to finish. The advent of machinery had given rise to the factory system. And laborers were shifted from working in their homes to working in factories. While native New Englanders continued to provide the labor for the textile industry for the next two decades, an influx of immigrants in the mid-1800s provided the hungry manufacturers with a steady supply of laborers who were willing to work for less money and longer hours. Within the first three decades of the 1800s, New England became the center of the nation's textiles industry. The region's ample rivers and streams provided the necessary water power, and the commercial centers of Boston and New York City readily received the finished products. Labor proved to be in ample supply as well: Since the mill machinery was not complicated, children could operate it and did. Slater hired children ages 7 to 14 to run the mill, a practice that other New England textile factories also adopted. The Jefferson Embargo of 1807, which prohibited importing textiles, also aided the industry. New England's mills provided the model for the American factory system. Slater had brought the Industrial Revolution to America.

How did Eli Whitney invent the **cotton gin**?

American inventor Eli Whitney (1765–1825) is credited with developing the cotton gin, a machine that removes cottonseeds from cotton fibers. A simple cotton gin (called the *churka*) dates back to ancient India (300 B.C.). But Whitney's gin would prove to be far superior. In 1792 Whitney, who had recently graduated from Yale University, was visiting the Georgia plantation owned by Catharine Littlefield Greene, widow of American Revolution (1775–83) hero General Nathanael Greene (1742–1786). Whitney observed that short-staple (or upland) cotton, which has green seeds that are difficult to separate from the fiber, differs from long-staple (also called Sea Island) cotton, which has black seeds that are easily separated. The latter was the staple of American commerce at the time. In 1793 Whitney, who is described as a mechanical genius, completed an invention that could be used to clean bolls of short-staple cotton of their seeds; he patented it the next year.

The machine worked by turning a crank, which caused a cylinder covered with wire teeth to revolve; the teeth pulled the cotton fiber, carrying it through slots in the cylinder as it revolved; since the slots were too small for the seeds, they were left behind; a roller with brushes then removed the fibers from the wire teeth. The cotton gin revolutionized the American textiles industry, which was then but a fledgling concern. The increase in cotton production was as much as fiftyfold: One large gin could process 50 times the cotton that a (slave) laborer could in a day. Soon plantations and farms were supplying huge amounts of cotton to textile mills in the Northeast, where in 1790 another inventor, British-born industrialist Samuel Slater (1768–1835), had built the first successful water-powered machines for spinning cotton. Together the inventions founded the American cotton industry. Whitney struggled to protect his patent, but imitations of his invention were already in production, prompting the U.S. government to allow his patent to expire. Though he did not profit from his cotton gin, he went on to devise a system of interchangeable parts, which introduced the idea of mass production and revolutionized manufacturing.

What is **"king cotton"**?

"King cotton" was an expression of the mid-1800s, when cotton had become so vital to the economies of southern states that it was said to rule them. Until the 1790s growers were limited to producing the quantity of cotton that could be processed by slaves. Separating cotton's fibers from its seeds was a time-consuming and labor-intensive process: the bolls were dried in front of a fire, and the seeds were picked out by hand. In 1793 American inventor Eli Whitney (1765–1825) introduced the cotton gin. In one day, his machine could clean 50 times as much cotton fiber than could a manual laborer. While Whitney patented the machine in 1794, imitations were nevertheless quickly put into production by shrewd businessmen who observed the effect the gin could have on the nation's cotton industry. There was no shortage of demand for the fiber. Just before Whitney developed the gin, another inventor, British-born Samuel Slater (1768–1835), built the first successful water-powered machines for spinning cotton, which he introduced at a Rhode Island mill in 1790. As the 1800s dawned, machinery had made cotton the center of the nation's emerging textiles industry.

Growers in the South stepped up cotton production to keep up with factories' demands. Slave labor and excellent growing conditions in the southern states (predominately Alabama, Mississippi, Georgia, and South Carolina) combined to dramatically increase production. By 1849 cotton exports reached $66 million a year and accounted for roughly two-fifths of total U.S. exports. But cotton came at a dear price. While laborers in the North's textile factories worked under difficult and sometimes dangerous circumstances, in the South cotton crops were planted and harvested by slaves. As abolitionists became increasingly vocal about the immorality of slavery and demanded the U.S. government legislate the abolition of slavery, southern growers defended the system, knowing that their livelihoods and the South's economy depended on it. In 1858 Senator James Henry Hammond (1807–1864) of South Carolina taunted northern sympathizers, saying, "You dare not make war on cotton—no power on earth dares make war upon it. Cotton is king." Hammond was not the first to use the phrase; it was coined three years earlier in the title of a book. Cotton's importance to the South contributed to the deepening North-South divide in the nation. By the time the Civil War (1861–65) began, the southern United States supplied two-thirds of the world's cotton.

Why was the **invention of the reaper** important to the U.S. economy?

Reapers, machines developed in the early 1800s to help farmers harvest grain such as wheat, dramatically increased overall grain production and consumption in the United States and the rest of the world. The first commercially successful reaper was built in 1831 by Virginia-born inventor Cyrus Hall McCormick (1809–1884), who patented it in 1834 and first sold it in 1840 in Virginia. The McCormick reaper was horse-drawn and replaced the use of sickles and scythes in the fields; it also reduced the amount of manual labor required to harvest grain crops. It worked in this way: A straight blade (protected by guards) was linked to a drive wheel; as the drive wheel turned, the blade moved back and forth in a sawing motion, cutting through the stalks of grain, which were held straight by rods; the cut grain stalks then fell onto a platform and were collected with a rake by a worker. The device increased average production from two or three acres a day to ten acres a day. McCormick's reaper was soon in wide use, and the inventor was on his way to becoming an industrialist.

In 1847 he moved his business to Chicago, were he could transport reapers via the Great Lakes and connected waterways to the East and to the South. Within five years McCormick's business had become the largest farm implement factory in the world. Sales and distribution of the equipment increased further during the 1850s as Chicago became a center for the nation's then-expanding rail system. In 1879 Cyrus McCormick's business became the McCormick Harvesting Machine Company, with the inventor himself as president (until 1884, when he was succeeded by his son). The reaper was improved over time: in the 1850s a self-raking feature was added, further reducing the amount of labor required to harvest grain; in the 1870s a binder was added, which bound the sheaves of grain and dropped them to the ground to be collected. In the 1920s the reaper (or harvester) was joined with another invention, the thresher, which separates grains from the stalks. The new reaper-thresher machine was called a combine. Today's combines still use the basic features present in McCormick's revolutionary 1831 invention. His company later became International Harvester (1902) and today is known as Navistar Corporation.

What were **bonanza farms**?

Bonanza farms were extremely successful and large farms, principally on the Great Plains and in the West, that emerged during the second half of the 1800s. The word *bonanza,* which is derived from Spanish and means literally "good weather," was coined in the mid-1800s; it is used to refer to any source of great and sudden wealth—including mines rich in minerals. Large-scale farming had been aided by the development of machinery that greatly increased production, especially of wheat and other grains. The innovations included reapers invented by Cyrus Hall McCormick (1809–1884) and steel plows developed by John Deere (1804–1886). Several events further helped farming interests west of the Mississippi River. To promote westward settlement, Congress passed the Homestead Act (1862), which allowed for ready and cheap acquisition of vast tracts of land: settlers could buy land for as little as $1.25 per acre or they could live on a tract and farm it for a period of five years, at the end of which they were granted 160 acres. The U.S. Army's defeats of rebel Indians were followed by peace treaties that confined Indian agricultural activities to reservations. In 1872 the Northern Pacific Railroad arrived at Fargo, North Dakota, allowing farmers to ship their products greater distances. Finally, dry farming techniques (which allow fields to lie fallow every other year in order to regain their nutrients and moisture to support crops the following year) proved to be a successful method for growing in the Great Plains—previously thought to be too dry for cultivating crops. All of these factors combined to turn some western farms into "bonanzas"—sources of great wealth for their owners. Encouraged by their success, settlers poured into the West. But not all farmers fared as well, and many were severely hit by the Panic of 1873. A drought in the plains states in the 1880s caused farm prices to drop, further hurting western farmers.

When did the American **cattle industry** begin?

As a large-scale commercial endeavor, the beef industry had its beginnings in the decades following the American Civil War (1861–65). Longhorn cattle, a breed of cattle descended from cows and bulls left by early Spanish settlers in the American Southwest, spurred the growth of the industry. Named for their long horns, which span about four feet, by the 1860s they had multiplied and great numbers of them roamed freely across the open range of the West. Ranchers in Texas bred the longhorns with other cattle breeds such as Hereford and Angus to produce quality meat. With beef in demand in the eastern United States, shrewd businessmen capitalized on the business opportunity, buying cattle for $3 to $5 a head and selling them in eastern and northern markets for as much as $25 to $60 a head. Ranchers hired cowboys to round up, sort out, and drive their herds to railheads in places like Abilene and Dodge City, Kansas, which became famous as "cow towns," raucous boom towns where saloons and brothels proliferated. After the long trail drive, the cattle were loaded onto rail cars and shipped live to local butchers who slaughtered the livestock and prepared the beef. For a 20-year period the plentiful longhorn cattle sustained a booming livestock industry in the West: at least 6 million Texas longhorns were driven across Oklahoma to the cow towns of Kansas.

By 1890 the complexion of the industry changed. Farmers and ranchers in the West used a new material, barbed wire, to fence in their lands, closing the open range; railroads were extended, bringing an end to the long, hard, and much-glorified cattle drives; the role of the cowboy changed, making him little more than a hired hand; and big business took over the industry. Among the entrepreneurs who capitalized on beef's place in the American diet was New England-born Gustavus Swift (1839–1903), who in 1877 began a large-scale slaughterhouse operation in Chicago, shipping ready-packed meat via refrigerated railcars to markets in the East.

Why was the **introduction of canning** important?

The advent of canned foods not only created an industry, but it altered the average American diet, helped usher in the consumer age, and saved time. Canning, a process for preserving food (vegetables, fruits, meats, and fish) by heating and sealing it in airtight containers, was developed by French candymaker Nicolas-François Appert (c. 1750–1841) in 1809, though he did not understand why the process worked. Some 50 years later, the pioneering work of French chemist and microbiologist Louis Pasteur (1822–1895) explained that heating is necessary to the canning process since it kills bacteria (microorganisms) that would otherwise spoil the food. Canning was introduced to American consumers in stages. In 1821 the William Underwood Company began a canning operation in Boston, Massachusetts; in the 1840s oyster canning began in Baltimore, Maryland; in 1853 American inventor Gail Borden (1801–1874) developed a way to condense and preserve milk in a can, founding the Borden Company four years later; and in 1858 American inventor John Landis Mason (1832–1902) developed a glass jar and lid suited to home canning.

Though early commercial canning methods did not ensure a safe product and many American women avoided the convenience foods, the canning industry grew rapidly, at least in part due to the male market, which included cowboys in the West. Between 1860 and 1870 the U.S. canning industry increased output from 5 million to 30 million cans. Improvements in the process during the 1870s helped eliminate the chance that cans would burst (a problem early on). And though the canning process changes food flavor, color, and texture, the convenience and long shelf life of canned foods helped them catch on: By the end of the 1800s a wide variety of canned foods, which had also come down in price, were common to the urban American diet. Companies such as Franco-American advertised in women's magazines, promoting their "delicacies in tins." An outbreak of botulism in the 1920s prompted the American canning industry to make further improvements to its process.

What is **Bessemer steel**?

Developed during the early 1850s, the Bessemer process was the first method for making steel cheaply and in large quantities. Named for its inventor, British engineer Henry Bessemer (1813–1898), the process was also developed independently by William Kelly (who patented the process in 1857) in the United States. Bessemer and Kelly experimented with injecting (blowing) air into molten pig iron (crude iron); the oxygen in the air helps rid the iron of its impurities (such as manganese, silicon, and carbon), converting the iron to molten steel, which is then poured into molds. The process was introduced in the U.S. steel manufacturing industry in 1864. Alloys were also added to the refining process to help purify the metal. Within two decades, the method was used to produce more than 90 percent of the nation's steel; it was also implemented throughout the industrialized world.

In the mid-1800s rich iron ore deposits had been discovered in the Upper Peninsula of Michigan, along Lake Superior. The discovery of the minerals and the innovation of the Bessemer process combined to create a thriving steel industry in the United States. At the same time there was a growing market for the material: Railroads needed iron to make rail gauges, while the new auto manufacturing industry used steel to make cars. As a result, between 1880 and 1910 annual U.S. steel production increased by a factor of 20. One of the early industry leaders was Andrew Carnegie (1835–1919), who in 1873 founded the nation's first large-scale steel plant, at Braddock, Pennsylvania. In 1901 Carnegie sold this and other steel mills to the United States Steel Corporation (today the USX Corporation, which is the largest steel producer in the United States). The Bessemer process continued to be used until after World War II (1939–45). The open-hearth method of purification gradually replaced it.

When did **chain stores** begin?

The innovation of the chain store, technically defined as two or more retail outlets operated by the same company and which sell the same kind of merchandise, was made by American businessmen George Gilman (c. 1830–1901) and George Huntington Hartford (1833–1917), who in 1859 set up the Great Atlantic & Pacific Tea Company in New York City. Better known as A&P, the stores proliferated rapidly, and other chain stores, such as W. P. Woolworth (established 1879) and J. C. Penney (1902) opened their doors for business. The early twentieth century saw tremendous growth of the chain stores: Between 1910 and 1931, the number of A&P stores grew from 200 to more than 15,000. While the department stores, also a byproduct of the late 1800s, catered to middle- and upper-class customers, the chain stores, including Woolworth's "five-and-dimes" (which sold many items at such low price points), served lower-income consumers.

Chain stores, which operate within all major retailing categories (including grocery stores, department stores, and drugstores, as well as apparel and food outlets), offer consumers many advantages. Their system of centralized and mass buying allow them to acquire merchandise from manufacturers and wholesalers at reduced costs; this savings is passed along to the consumer, who pays less for the item. Further, they experience economies of advertising: A single ad placement promotes all the stores within the chain. In the 1920s independent retailers rallied against the chain stores, citing they had unfair advantages. This argument has resurfaced off and on throughout the twentieth century, as chain stores entered into more and more retailing sectors, including hardware, jewelry, furniture, music, and books. But the only federal legislation that constructively attempted to regulate the chain stores came in 1936: The Robinson-Patman Act tried to control competition. Today chain stores account for roughly one-third of all American retail sales.

Shoppers in a crowded Woolworth's. Department stores emerged in the mid-1800s.

When did **department stores** begin?

Department stores, which offer a wide variety of goods for sale in various departments, emerged in the mid-1800s. Many evolved out of general stores (which offered a variety of goods but not divided into departments), while others evolved out of dry-goods stores (which sold textiles and related merchandise). The first bona fide department store was established in Paris: the Bon Marché (French meaning "good bargain") opened its doors in 1838. Between the 1850s and 1880s, numerous department stores opened in American cities—including Jordan Marsh, founded 1851 in Boston, Massachusetts; R. H. Macy's, founded 1858 in New York City (the store was known for its creative advertisements); Wanamaker's, founded 1861 in Philadelphia, Pennsylvania (it successfully implemented fixed pricing so that customers no longer haggled over price); and Marshall Field, founded 1881 in Chicago (within 25 years it became the world's largest wholesale and retail dry goods store). These pioneer department stores, multistoried enterprises located in downtown areas, introduced many innovations to merchandising, including the policy of returnable or exchangeable goods, ready-made apparel, clearly marked prices, and window displays. By the early 1900s department stores could be found throughout the country. The timing was right for their emergence. Urban centers grew rapidly at the end of the century, giving department stores a ready clientele; the advent of the telephone, electric lighting, and billing machines helped retailers conduct business efficiently; transportation improvements allowed for the shipment of large quantities of goods; and a variety of finished goods were mass-produced, increasing supply and lowering cost of production as well as the price to the consumer. By the 1910s the stores were part of a new mass culture, which centered in American cities. During the twentieth century, department store sales typically ranged between 6 and 12 percent of total annual retail sales.

What was the first **mail-order** company?

The mail-order business was pioneered by retailer Montgomery Ward & Company, founded in 1872 in Chicago when American merchant Aaron Montgomery Ward (1843–1913) set up shop over a livery stable and printed a one-sheet "catalog" of bargains. Midwestern farmers, hurt by low farm prices and rising costs, were a ready market for the value-priced goods, which were shipped by rail to rural customers. Originally called "The Original Grange Supply House," Montgomery Ward offered 30 dry goods priced at $1 or less and provided special terms of sale for Grange members (the Grange is an association of farmers). Ward bought merchandise directly from wholesalers and, since he did not maintain a store building, overhead was low. By 1876 Ward's catalog had grown to 150 pages; in 1884 it was 240 pages and offered nearly 10,000 products, including household items (such as furniture, cutlery, and writing paper), farm implements (such as harnesses and tools), and fashions (such as ready-made apparel and parasols). Ward offered customers "satisfaction or your money back." In 1886 American Richard W. Sears (1863–1914) entered the mail-order business, opening operations in Minneapolis, Minnesota. He moved the business to Chicago the following year and sold it in 1889. In 1893 he joined with Alvah C. Roebuck (1864–1948) to found Sears, Roebuck and Company. The Sears catalog, which soon consisted of hundreds of pages and thousands of items, became popularly known as the "Wish Book."

Montgomery Ward and Sears Roebuck were aided by the U.S. Postal Service's expansion into remote areas: Beginning in 1896 mail could be delivered via the RFD, Rural Free Delivery. In 1913 parcel post was added to the postal service's offerings, further benefiting the mail-order houses and their growing lists of customers. Montgomery Ward and Sears Roebuck offered rural America more than merchandise; the mail-order houses were farm families' link to the greater consumer society that was emerging at the turn of the century. Regardless of geography, rural Americans could purchase "store-bought" goods, manufactured goods that were mass-produced in factories. The mail-order houses offered customers convenience (since customer purchases no longer had to be deferred for the next trip to a town), variety (since catalogers catered to a nationwide customer base, on-hand inventory included a multitude of products), and low prices (the mail-order houses bought merchandise at reduced rates from the wholesalers). Fashions were no longer restricted to the middle- and upper-class city dwellers who had access to department stores; rural customers

What was Black Friday?

The term refers to one of two Fridays in the second half of the 1800s when severe market drops precipitated financial crises in the United States. The first Black Friday was September 24, 1869: Financiers Jay Gould (1836–1892) and James Fisk (1834–1872) had conspired to raise the market price of gold by buying it in huge quantities (leaving less supply on the open market, which would, theoretically, increase demand and therefore price). Having caused the price to increase, the businessmen planned to sell their gold supplies at a profit. While they did make out handsomely, clearing some $11 million, a panic resulted when the price of gold rose sharply: Businesses that needed gold to meet their obligations were forced to pay exorbitant prices for it. The government responded to the crisis by selling off $4 million of its gold reserves, causing gold prices to tumble. Speculators were hit hard, but Gould and Fisk came out unscathed, selling off their gold supplies before the price plummeted.

Four years later, another Friday turned dark when the investment banking firm of Jay Cooke & Company failed after it had invested too heavily in railroad securities, which had since declined. When news of the company's collapse was released on September 19, 1873, it affected the entire stock market, and prices fell sharply. The so-called Panic of 1873 signaled the beginning of a depression that persisted through most of the 1870s.

became aware of new styles each time the Montgomery Ward and Sears Roebuck catalogs were delivered, which, by the early 1900s, was twice a year.

Though both Montgomery Ward and Sears Roebuck exited the mail-order business to concentrate efforts on their chain store retail operations later in the century, they set the standard for modern mail-order houses through their early policies addressing merchandise returns, competitive pricing, flexible payment methods, and shipping terms.

Who were the **"robber barons"**?

They were the industrial and financial tycoons of the late nineteenth century, the early builders of American business. Some called them the captains of industry. The "robber barons" included bankers J. Pierpont Morgan (1837–1913) and Jay Cooke (1821–1905); oil industrialist John D. Rockefeller (1839–1937); steel mogul Andrew Carnegie (1835–1919); financiers James J. Hill (1838–1916), James Fisk (1834–1872), Edward Harriman (1848–1909), and Jay Gould (1836–1892); and rail magnates Cornelius Vanderbilt (1794–1877) and Collis Huntington (1821–1900). These influential businessmen were hailed for expanding and modernizing the capitalist system and lauded for

Illustration of the "robber barons" of the late nineteenth century: James J. Hill, Andrew Carnegie, Cornelius Vanderbilt, John D. Rockefeller, J. Pierpont Morgan, Edward H. Harriman, and Jay Gould. (Original lithograph by Bernarda Bryson.)

their philanthropic contributions to the arts and education. But they were also viewed as opportunistic, exploitative, and unethical.

Many factors converged to make the robber baron possible. The new nation was rich in natural resources, including iron, coal, and oil; technological advances steadily improved manufacturing machinery and processes; the population growth, fed by an influx of immigrants, provided a steady workforce, often willing to work for a low wage; the government turned over the building and operation of the nation's railways to private interests; and, adhering to the philosophy of laissez-faire (noninterference in the private sector), the government also provided a favorable environment in which to conduct business. Shrewd businessmen turned these factors to their advantage, amassing great empires. Reinvesting profits into their businesses, fortunes grew. The robber barons, especially the railroad men and the financiers who gained control of rail companies through stock buy-outs, hired lobbyists who worked on their behalf to gain the corporations subsidies, land grants, and even tax relief at both the federal and state levels.

The robber barons converted their business prowess into political might. In Washington, politicians grew tired of the advantage-seeking representatives of the nation's business leaders. Reform-minded progressives complained that the robber barons lived in opulent luxury while their workers barely eked out a living, their families teetering on despair.

After dominating the American economy for decades, changes around the turn of the century worked to curb the influence of the robber barons. In 1890 the federal government passed the Sherman Anti-Trust Act, making trusts (combinations of firms or corporations formed to limit competition and monopolize a market) illegal; workers continued to organize in labor unions, with which corporations were increasingly compelled to negotiate; the Interstate Commerce Commission (ICC) was established in 1887 to prevent abusive practices; and in 1913 the Sixteenth Amendment was ratified, allowing the federal government to collect a graduated income tax. Though many American businessmen and women would make great fortunes in the twentieth century, by the end of the 1920s the era of the robber barons had drawn to a close.

MODERN INDUSTRY

When was the **New York Stock Exchange** founded?

The oldest and largest stock exchange in the United States, the New York Stock Exchange (NYSE) had its origins on May 17, 1792, when local brokers who had been buying and selling securities under a designated tree agreed to formalize their business transactions. The NYSE that most people would recognize today opened for business in 1825 at 11 Wall Street, New York City. At the time most shares traded were in canal, turnpike, mining, and gaslight companies. Though a few industrial securities were first traded on the New York Stock Exchange as early as 1831, it was another 40 years before the complexion of trading changed to a more industrial nature. As the nation became increasingly manufacturing oriented, the companies listed on the exchange reflected the economic shift. Today, if corporations wish to list their stocks on the NYSE, they must have a minimum of 2,000 shareholders, each of those original shareholders must have 100 or more shares, the corporation must be able to issue at least 1 million shares of stock, and it must also provide a record of earnings for the previous three-year period. The board of the stock exchange can make exceptions to these guidelines. Corporations may be listed with other stock exchanges (such as the American Stock Exchange) or they may allow stock in their company to be traded as unlisted stocks, which are bought and sold in over-the-counter (OTC) trading. Companies that do not allow shares to be publicly traded are called private corporations.

When was the ***Wall Street Journal*** first published?

The newspaper, considered one of the world's best and certainly the preeminent American financial periodical, was first published in 1889, seven years after the Dow, Jones and Company (the comma between the two names was later dropped) was founded by financial reporters Charles Henry Dow (1851–1902) and Edward Davis Jones (1856–1920). Since the founding of the New York Stock Exchange (NYSE) in 1792, business reporting had been largely based on rumor or speculation. Dow and Jones were determined to provide American businesspeople and investors with up-to-date and accurate reporting on the stock market. In its first seven years of business, their publishing company grew from six employees to a staff of 50. In 1889 Dow and Jones expanded their two-page daily newspaper, titled *Customers' Afternoon Letter,* into the *Wall Street Journal.* The paper's stated goal was to give full and fair information regarding fluctuations in the prices of stocks, bonds, and some commodities. Further, the focus was to be on news rather than opinion. The paper began publishing composite lists of major stocks in 1884. In the decades since, it has expanded coverage to include all facets of the business and economic world and now caters to the leisure interests of businesspeople by publishing reports on the arts, travel, sports, and other recreational activities.

What is the history of the **Dow Jones Industrial Average**?

A measure of stock prices of important industrial companies, the Dow Jones Industrial Average (DJIA) was first printed in the *Wall Street Journal* in 1897. The average is an indicator of the market overall and is used, along with other indexes, by investors, stockbrokers, and analysts to make investment forecasts and decisions. "The Dow," as it has come to be called, was conceived of as a summary measure of the stock market, an index that could be used to analyze past trends, indicate current trends, and even predict future ones. The first DJIA averaged the prices of 12 major companies. The list had been expanded since: in 1916 it averaged the stock prices of 20 companies; in 1928, 30. Adjustments have been made as the result of company mergers and dissolutions. Though it is a measure only of the New York Stock Exchange, the Dow Jones Average has been called a barometer of the stock market. News of fluctuations in the DJIA can affect market prices around the world.

What is the system of **scientific management**?

A system to gain maximum efficiency from workers and machinery, scientific management, also known as Taylorism, was developed by American industrial engineer Frederick Winslow Taylor (1856–1915). As foreman in a steel plant, Taylor undertook time and motion studies and conducted experiments to determine the "one best way" to do any given job, developing detailed systems to yield the highest possible productivity levels. He first presented his theories in 1903 to the American Academy of Mechanical Engineers. Efficiency was the cornerstone of Taylorism: production processes should not waste time or materials. He published his ideas in the landmark work *The Principles of Management* (1911), and became a well-known engineering consultant, contracted by companies eager to maximize their output. The doctrine of scientific management was embraced by American industry: As transportation networks improved and the U.S. population grew rapidly in the early 1900s, markets expanded, placing great demands on industry. Applying Taylor's scientific management, manufacturers were able to boost productivity by as much as 200 percent. Since Taylorism broke production processes into individual tasks, each with its own best practice, new workers could be quickly and easily trained, which adherents believed was another benefit of the concept. Scientific management had many advocates, including engineers Frank (1868–1924) and Lillian (1878–1972) Gilbreth, who furthered Taylor's work, publishing volumes such as *Primer of Scientific Management* (1911), *Psychology of Management* (1912), and studies on motion, fatigue, and time. Among those who applied scientific management were Ford Motor Company (in developing the assembly line for the Model T); Boston retailer Filene's (one of the first commercial enterprises to use the method); and Bethlehem Steel (which conducted experiments in the loading of pig iron).

Scientific management also had its detractors: Taylorism was criticized for having a dehumanizing effect on labor. In making every job routine, some charged that the system separated the minds of workers from their hands, eliminated the need for skilled workers, and gave management absolute control over production processes. Nevertheless, principles of scientific management remain evident in the workplace today, and the adoption of scientific management is credited with boosting American productivity and increasing stockholder profits. Theories concerning worker output were modified during the second half of the twentieth century.

Who invented the **assembly line**?

Ford Motor Company founder Henry Ford (1863–1947) is credited with the creation of the assembly line, an industrial innovation that allowed cars to be produced quickly and efficiently. In 1913, 10 years after founding Ford Motor Company, Henry Ford installed the first moving assembly line in one of his Model T manufacturing plants. The innovation proved to be the beginning of the consumer age: The assembly line allowed greater efficiencies in auto production, which, in turn, reduced the price of a quality car, putting it in reach of the ordinary person. Soon, manufactured goods of every variety would be mass produced.

Why was **Ford's Model T** important?

The enormous success of the Model T, a Ford Motor Company car introduced in 1908 and manufactured until 1927, has been the source of extensive analysis and commentary by historians, sociologists, economists, business writers, and pop culture experts. The Model T has been credited with not only changing America but with defining it. When Ford Motor Company founder and president Henry Ford (1863–1947) unveiled the prize Model T in October 1908, he hailed it as "a motor car

A 1914 Ford Model T. The vehicle, which was nicknamed "Tin Lizzie," became the symbol of low-cost, reliable transportation.

for the great multitude." The product lived up to the promise. The internal combustion vehicle had been in production in the United States only since the 1890s, but in the decade preceding the Model T's debut, manufacturers and consumers alike had come to regard the "horseless carriage" as a luxury item, custom made for wealthy Americans. Ford had conceived of a different and, as the company would advertise throughout the century, a better idea: A car that was simple to operate, easy to service, comfortable, *and* affordable.

The Model T had a wooden body on a steel frame; four-cylinder, 20-horsepower engine; tank capacity of 10 gallons (in the "touring sedan") or 16 gallons (in the "runabout"); and a completely enclosed power plant and transmission. It was also lighter than other models. Through large-scale production, based on a system of interchangeable parts, the Model T took 728 minutes (just more than 12 hours) to build and sold for $850, lower than the price of other automobiles, but still beyond the reach of the average American. Nevertheless, 17,000 Model Ts were bought by American consumers the year they were introduced. Ford improved production methods to realize greater economies and lower the price each year; sales steadily rose. The company raised eyebrows in the business community when it offered workers an eight-hour day for $5 a day—twice what other factory workers were earning. Ford explained that this was merely good business practice. By raising the wages of his factory workers, Ford enlarged the potential market for his Model T.

Ford workers on a flywheel assembly line, 1913. Henry Ford is credited with the creation of the assembly line, an industrial innovation that allowed cars to be produced quickly and efficiently.

In 1914 Ford implemented the moving assembly line. It used the principles of scientific management, where each job has one "best way" of being accomplished, to bring unprecedented efficiency to manufacturing. Assembly time per car dropped to just 90 minutes. That year the Ford plant in Highland Park, Michigan, produced almost 250,000 Model Ts. To keep up with ever-rising demand, operations were sped up and capacity increased to the point that one day in 1925, Ford produced one Model T every 10 seconds. That year the car retailed for just $295, making the so-called "Tin Lizzie" (or the "Flivver") accessible to working-class families. By 1927, when Ford retired the Model T so the company could respond to consumer demand for cars with better performance, power, and styling, the company had turned out 15 million Tin Lizzies. Ford's innovative Model T, a reliable, no-nonsense, mass-produced automobile, manufactured on a moving assembly line, brought mobility within the reach of the average American. It had changed consumer mind-set to view the car as a necessity.

What is a **"Horatio Alger story"**?

It is any story about someone who, through sheer determination and good works, rises from poverty to wealth. During the second half of the 1800s, novels by American clergyman and author Horatio Alger Jr. (1832–1899) were extremely popular: He wrote more than 100 books, including the *Luck and Pluck* and the *Tattered Tom* series. All of the stories center on a boy from inauspicious beginnings who, through hard work, clean living, and a little bit of luck, becomes successful. Alger's real-life experience working with orphans and runaways in New York City provided the foundation for his works, which inspired countless readers and fed into the American Dream: that the United States is a land ripe with possibility. Though dead for more than a century, Alger's name lives on: many Americans still describe an honorable person's rise from rags to riches as "a real Horatio Alger story."

What was **Black Tuesday**?

It was the day the stock market crashed—Tuesday, October 29, 1929, signaling the beginning of the worldwide economic downturn called the Great Depression (1929–39).

On Thursday, October 24, 1929, stock values declined rapidly following a five-year period in which the average price of common stocks on the New York Stock Exchange had more than doubled. The prosperity of the 1920s and the widespread sale and purchase of Liberty Bonds (U.S. government bonds) to help finance World War I (1914–18) had encouraged many Americans to invest in the stock market: With the market robust, timing seemed right for speculation and America's experience with Liberty Bonds had made many people comfortable with and interested in investments. So when the stock market dropped precipitously on that fateful October day in 1929, the effects were felt by many. On the following Monday prices again plummeted; on Tuesday, October 29, stockholders panicked, selling off more than 16.4 million shares, and prices nose-dived. Institutions were also affected: Banks, also investors, lost huge sums of money, forcing many to close their doors. News of the stock market failure and bank closures caused many Americans to try to withdraw their money from their deposit accounts, leading to the famous run on the banks. The late-October financial crises marked the beginning of a decade of hard times.

What was the **New Deal**?

While the Great Depression began with the stock market crash on Black Tuesday, October 29, 1929, many factors contributed to the financial crisis, including overproduction, limited foreign markets (due to war debts that prevented trading), and overexpansion of credit, as well as stock market speculation. Soon the country was in the grips of a severe economic downturn that affected most every American. Some were harder hit than others: many lost their jobs (16 million people were unemployed at the depth of the crisis, accounting for about a third of the workforce); families were unable to make their mortgage payments and lost their homes; hunger was widespread, since there was no money to buy food. The sight of people waiting in bread lines was a common one.

Shacks serve as homes for men and women in a Seattle, Washington, "Hooverville." Such villages popped up during the Great Depression and were sarcastically named after President Herbert Hoover.

It was amidst this crisis, which was soon felt overseas, that Franklin D. Roosevelt (1882–1945) took office as president in 1933. In his inaugural address, he called for faith in America's future, saying, "The only thing we have to fear is fear itself." Roosevelt soon rolled out a program of domestic reforms called the New Deal. For the first time in American history, the federal government took a central role in organizing business and agriculture. Roosevelt initiated aid programs and directed relief in the form of public works programs that would put people back to work. The new government agencies that were set up included the Public Works Administration, Federal Deposit Insurance Corporation, Security and Exchange Commission, National Labor Relations Board, Tennessee Valley Association, the National Recovery Administration, and the Civilian Conservation Corps. These government organizations soon become known by their initials (PWA, FDIC, SEC, NLRB, TVA, NRA, CCC). Roosevelt's critics charged him with giving the federal government too much power and began calling his New Deal "alphabet soup." The president became widely known as FDR.

Though the New Deal measures alleviated the situation and did put some Americans back to work, the country did not pull out of the Depression until industry was called upon to step up production in order to provide arms, aircraft, vehicles, and supplies for the war effort. It was during the early days of World War II (1939–45), the economy buoyed by military spending, that the nation finally recovered. Many New Deal agencies are still part of the federal government today.

Who was **"Rosie the Riveter"**?

The term referred to the American women who worked factory jobs as part of the war effort on the home front, where auto plants and other industrial facilities were converted into defense plants to manufacture airplanes, ships, and weapons. As World War II (1939–45) wore on, more and more men went overseas to fight, resulting in a shortage of civilian male workers. And so, women pitched in. However, at the end of the war, many of these women were displaced as the men returned home to their jobs and civilian life. Nevertheless, the contribution of all the Rosie the Riveters was instrumental to the war effort.

Why did the **auto industry boom** in the postwar era?

In the years following World War II (1939–45), auto ownership in the United States soared from 27.5 million registered vehicles in 1940 to 61.5 million in 1960. Americans had resumed their love affair with the automobile, inextricably linking the car with the U.S. history of the postwar era. Many factors combined to bring about the automobile's widespread popularity.

During World War II the car manufacturers curtailed auto production, converting factories to military production and turning out some $29 billion in materials, including trucks, jeeps, tanks, aircraft, engines, artillery, and ammunition. With the conflict ended, automakers stepped up production to fulfill the unmet demand of the war years, and soon found themselves working to meet new demand, created by an increase in consumer spending and the growth of the suburbs. The overall prosperity of the late 1940s and 1950s produced a new spirit of consumerism: Government regulations (brought about through the efforts of the labor unions) resulted in increased wages and improved benefits—meaning Americans, for the most part, had more disposable income. Advertisers took advantage of the new medium of television to reach wide—and eager—audiences. The housing industry, largely dormant during World War II, built new neighborhoods around the edges of American cities, making the automobile a necessity rather than a luxury. The Big Three (GM, Ford, and Chrysler) increased capacity to meet the tremendous demand, setting new records for production in 1949 and 1950. By 1960 more than three out of every four American families owned at least one car. The infrastructure raced to keep pace with a nation on wheels. Superhighways were built (covering some 10,000 miles of road); motels and fast-food restaurants went up along roadsides; and shopping centers were built outside city centers. While imports would challenge the American automakers in the decades to come, it was the U.S. manufacturers that defined the postwar era.

Why was the **introduction of plastic** important to industry?

Pioneered in the early 1900s, plastic—which is any synthetic organic material that can be molded under heat and pressure to retain a shape—affected every industry and every consumer. As a malleable material, plastic could quite literally be molded for countless uses, both for the production of goods and as a material in finished goods. In 1909 Bakelite plastic was introduced, and over the next three decades the plastics industry grew, developing acrylic, nylon, polystyrene, and vinyl (polyvinyl chloride or PVC) in the 1930s, and polyesters in the 1940s. The applications seemed endless: from household items such as hosiery, clocks, radios, toys, flooring, food containers, bags, electric plugs, and garden hoses, to commercial uses such as automobile bodies and parts, airplane windows, boat hulls, packaging, and building materials. The space industry and medicine have also found critical uses for plastic products. Scientists have continued to find new applications for plastics—in products such as compact disks (CDs), computer diskettes, outdoor furniture, and personal computers (PCs). The material has become essential to modern life.

When did **IBM** enter the personal computer business?

IBM (International Business Machines, organized in 1924) had long been an industry leader in developing and producing computers for business and science, but in August 1981, the company jumped into the consumer business, competing with upstart Apple for a share in the personal computer (PC) business. The PC introduced by IBM used a Microsoft disk-operating system (MS-DOS) and soon captured 75 percent of the market. Observing the company's enormous success, other firms began producing IBM "clones," which could use the same software as the IBM PC.

When was **Microsoft** founded?

It wasn't all that long ago, 1975, that computer whiz Bill Gates (1955–) founded what is now the dominant manufacturer of computer software (so dominant that the company has faced antitrust allegations from the federal government). Gates was only 19 years old when he founded the business with his friend Paul Gardner Allen, and he had dropped out of Harvard to do so. It paid off: Gates was a billionaire by age 30. Though he's undoubtedly a math ace (he scored a perfect 800 in math on his SATs and began writing computer programs when he was all of 13), Gates has more than once credited the success of Microsoft to not his own programming skills—but to hiring the best programming talent for the Redmond, Washington-based company.

U.S. ECONOMIC LEGISLATION

What was the **Embargo Act**?

On December 22, 1807, President Thomas Jefferson (1743–1826) signed the Embargo Act, prohibiting ships that were destined for foreign ports from leaving the United States. The legislation had been drawn up in an effort to pressure France and Britain, which were then at war and had been seizing U.S. merchant ships to prevent each other from receiving American goods. The situation began after the French navy was crushed by the British under Admiral Horatio Nelson (1758–1805) at the Battle of Trafalgar (October 1805). French ruler Napoleon Bonaparte (1769–1821) turned to economic warfare in his long struggle with the British, directing all countries under French control not to trade with Britain. Its economy dependent on trade, Britain struck back by imposing a naval blockade on France, which soon interfered with U.S. shipping. Ever since the struggle between the two European powers had begun in 1793, the United States had tried to remain neutral. But the interruption of shipping to and from the Continent and the search and seizure of ships posed significant problems to the American export business. The Embargo Act was an attempt to solve these problems without getting involved in the conflict. But the effort failed. The embargo made sales of U.S. farm surpluses impossible. New England shippers protested the act and were joined by southern cotton and tobacco planters in their opposition. Nevertheless, the embargo remained in effect for 14 months, during which the American economy suffered and many ships resorted to smuggling. In 1809 Congress passed the Non-Intercourse Act, which limited the shipping embargo to France and Britain; all other foreign ports were again open to U.S. ships. Three years later, the United States was drawn into the conflict, fighting the British in the War of 1812 (1812–14).

What was the **Tariff of Abominations**?

In 1828 the U.S. Congress passed a bill putting high tariffs (government taxes) on imported goods. The measure was intended to protect the burgeoning industries of New England, where numerous factories had opened during the first three decades of the century and the manufacture of finished goods defined the region's economy. Congress figured that by placing high taxes on goods from other countries, Americans would buy American-made products. But southern farmers had come to rely on cheaper imported goods. Believing the 1828 legislation was overly protective of the nation's industrial interests, southerners dubbed it the "tariff of abominations." Vice President John C. Calhoun (1782–1850), from South Carolina, openly and strongly criticized the tax, pronouncing that any state could declare null a federal law it deemed unconstitutional. In response, Congress took measures to lower the tariffs, but not eliminate them. South Carolina remained dissatisfied with the legislation, and in 1832 the state declared the tariff act null and void. Further, it threatened secession from the Union. President Andrew Jackson (1767–1845), unwilling to tolerate such rebelliousness and determined to enforce the federal law at all costs, asked Congress to pass the Force Bill—legislation allowing the nation's armed forces to collect the tariffs. Jackson's move inspired tremendous opposition in Congress. The Senate leader of the anti-Jackson contingency was Henry Clay (1777–1852) of Kentucky. Clay, who had earned himself the nickname "Great Pacificator" for his work in crafting the Missouri Compromise (1820), presented another compromise in 1833. He proposed that duties on certain goods could remain high but others should be gradually reduced over time. The Compromise Tariff authored by Clay averted an all-out conflict in the nation. The measure was passed and thereafter tariffs were adjusted depending on the prevailing economic conditions. But the fury over the Tariff of Abominations further revealed the North-South differences and the federal-government-versus-states'-rights issues that would inspire the southern states—led by South Carolina—to secede from the Union in 1860 and 1861, bringing on the American Civil War (1861–65).

How old is the U.S. **income tax**?

It dates to 1913. Proposed in Congress on July 12, 1909, and ratified February 3, 1913, the Sixteenth Amendment to the U.S. Constitution gives the federal government (specifically, the U.S. Congress) authority to levy and collect income taxes. The language of the amendment states that incomes "from whatever sources derived" may be taxed—and without regard to a census. In other words, it is up to Congress to determine the level at which citizens of the country are taxed, and this may be done without apportionment among the individual states.

One hundred years before the Sixteenth Amendment was approved, Congress had begun eyeing income tax as a way to collect funds for government use. Lawmakers first considered levying an income tax to help pay for the War of 1812 (1812–14), which the new republic fought against Great Britain over shipping disputes. During the American Civil War (1861–65), Congress imposed an income tax for the first time, charging workers and businessmen between 3 and 5 percent of their earnings and establishing (in 1862) a Bureau of Internal Revenue to administer the tax program. Once the war was over, income taxes were phased out. In 1894, responding to increasing economic and political pressures, the legislature again passed an income tax law (2 percent on all incomes over $4,000), as part of the Wilson-Gorman Tariff Act. But it was struck down by the U.S. Supreme Court, which declared it unconstitutional in the case of *Pollock* v. *Farmers' Loan and Trust Company* (1898). In the early 1900s, the idea of an income tax received widespread political support for the first time. Progressive politicians could see that the nation's wealth was poorly distributed, the gap between rich and poor growing wider. Conservative politicians worried that the government would not be able to respond to a national emergency if it lacked resources. These political factions found a single voice in favor of a graduated income tax (a tax based on level of income: those who earn more pay higher taxes). To circumvent the

U.S. Supreme Court, it was necessary for Congress to propose an amendment to the Constitution. In ratifying the amendment, the states gave Congress the authority to set rates and collect income tax.

Tax rates have fluctuated ever since the passage of the Sixteenth Amendment, reaching their highest mark during World War II (1939–45) when the rate soared to 91 percent. The war effort also brought the innovation of automatic withholdings: Taxes were deducted directly from paychecks. In 1953 the Bureau of Internal Revenue was dramatically reorganized to create the Internal Revenue Service (IRS). Over the decades, tax laws (collectively called the Tax Code) have become increasingly complex, prompting a recent movement in favor of a flat (versus the graduated) tax, where all taxpayers are charged at the same rate.

When was the **Interstate Commerce Commission** formed?

The Interstate Commerce Commission (ICC) was established by act of Congress in 1887. The agency is responsible for regulating the rates and services of specified carriers that transport freight (goods, whether raw or finished) and passengers between states. Its jurisdiction, expanded by subsequent acts of Congress, includes trucking, bus services, water carriers, expedited delivery services, and even oil pipelines. The regulatory agency, the nation's first such body, was borne out of necessity in the late 1800s, as farmers in particular charged the railroads with discriminatory freight practices. With rail lines crisscrossing the nation, the question of who would control rates and monitor practices had become an increasingly difficult one to answer. Many states, particularly in the Midwest, set up their own regulatory boards, but because the rail companies operated between states, enforcing state laws on them proved cumbersome and impractical. Meanwhile the railroads, operating without the purview of any effective regulatory body, set their own standards and practices, which resulted in many abuses.

In an 1877 U.S. Supreme Court ruling, in the case of *Munn* v. *Illinois,* the authority of the state boards to regulate the railroads was upheld. But less than a decade later, in the case of *Wabash, St. Louis and Pacific Railway Company* v. *Illinois,* the high court invalidated its earlier decision and proclaimed that only the U.S. Congress has the right to regulate interstate commerce. Citing Section 8 of Article 1 of the U.S. Constitution (1790), which states that "Congress shall have the power…to regulate commerce with foreign nations, and among the several states, and with the Indian tribes," the Interstate Commerce Act was passed in 1887, setting up the Interstate Commerce Commission to regulate the interstate railroads. The agency's purview was later expanded to include all ground and water carriers that operate on an interstate basis. In addition to controlling rates, the agency also enforces laws against discrimination. The ICC's authority was strengthened by congressional legislation including the Hepburn Act (1906) and the Mann-Elkins Act (1910).

What is the **Sherman Anti-Trust Act**?

Passed by Congress in 1890, the Sherman Anti-Trust Act was an attempt to break up corporate trusts (combinations of firms or corporations formed to limit competition and monopolize a market). The legislation stated that "every contract, combination in the form of trust or otherwise, or conspiracy in the restraint of trade" is illegal. While the act made clear that anyone found to be in violation of restraining trade would face fines, jail terms, and the payment of damages, the language lacked clear definitions of what exactly constituted restraint of trade. The nation's courts were left with the responsibility of interpreting the Sherman Anti-Trust Act, and the justices proved as reluctant to take on big business as Congress had been.

The legislation was introduced in Congress by Senator John Sherman (1823–1900) of Ohio, in response to increasing outcry from state governments and the public for the passage of national antitrust laws. Many states had passed their own antitrust bills or had made constitutional provisions prohibiting trusts, but the statutes proved difficult to enforce and big business found ways around them. When the legislation proposed by Sherman reached the Senate, conservative congressmen rewrote it; many charged that the Senators had made it deliberately vague. In the decade after its passage, the federal government prosecuted only 18 antitrust cases and court decisions did little to break up monopolies. But after the turn of the century, a progressive spirit in the nation grew; among progressive reformers' demands was that government regulate business. In 1911 the U.S. Justice Department won key victories against monopolies, breaking up John D. Rockefeller's Standard Oil Company of New Jersey and James B. Duke's American Tobacco Company. The decisions set a precedent for how the Sherman Anti-Trust Act would be enforced and demonstrated a national intolerance toward monopolistic trade practices. In 1914 national antitrust legislation was strengthened by the passage of the Clayton Anti-Trust Act, which outlawed price fixing (the practice of pricing below cost to eliminate a competitive product), made it illegal for the same executives to manage two or more competing companies (a practice called interlocking directorates), and prohibited any corporation from owning stock in a competing corporation. The creation of the Federal Trade Commission (FTC) that same year provided further insurance that U.S. corporations engaging in unfair practices would be investigated by the government.

Between 1880 and the early 1900s corporate trusts proliferated in the United States, becoming powerful business forces. The vague language of the Sherman Anti-Trust legislation and the courts' reluctance to prosecute big business based on that act did little to break up the monopolistic giants. The tide turned against corporate trusts when Theodore Roosevelt (1858–1919) became president in September 1901, after President William McKinley (1843–1901) was assassinated. Roosevelt launched a "trust-busting" campaign, initiating, through the attorney general's office, some 40 lawsuits against American corporations such as American Tobacco Company, Standard Oil Company, and American Telephone and Telegraph (AT&T). Government efforts to break up the monopolies were strengthened in 1914, during the presidency of Woodrow Wilson (1856–1924), when Congress passed the Clayton Anti-Trust legislation and created the Federal Trade Commission (FTC), which is responsible for keeping business competition free and fair. Trust-busting declined during the prosperity of the 1920s, but was again vigorously pursued in the 1930s, during the administration of Franklin D. Roosevelt (1882–1945).

U.S. ECONOMY TODAY

What is **NAFTA**?

NAFTA is the North American Free Trade Agreement, signed on December 17, 1992, by U.S. president George H. W. Bush (1924–), Canadian prime minister Brian Mulroney (1939–), and Mexican president Carlos Salinas de Gortari (1948–). It went into effect January 1, 1994. Inspired by the success of the European Community's open-trade agreement, the architects of NAFTA aimed to create free trade among North America's three largest countries. A 1988 pact between the United States and Canada had already lifted numerous barriers to trade between the two nations; that agreement was expanded to include Mexico through a series of negotiations that were preliminarily approved in August 1992 and were concluded with the signing of NAFTA later that year. The agreement removes trade barriers, including customs duties and tariffs, over the course of 15 years, allowing commodities and manufactured goods to be freely traded among the three nations. NAFTA also includes provisions that allow American and Canadian service companies to expand their markets into Mexico.

What was the **1990s boom**?

It was the longest economic expansion in American history. According to accepted economic indicators, the boom began in March 1991, when the first President Bush was in office, and ended in March 2001, when President George W. Bush was in office. Eight years of the expansion were during the Clinton administration.

The hallmarks of the 1990s boom were the creation of almost 24 million jobs, or an average of 200,000 jobs a month; a national unemployment rate that dropped to around 4 percent for an extended period; productivity gains month over month; gross domestic product (GDP) growth month over month; unprecedented investment in the stock market (Wall Street added $10 trillion in wealth over the decade); a bull market fueled by $100 billion in initial public offerings (IPOs), many of them technology stocks; low interest rates; a low inflation rate averaging 2.6 percent per year; the elimination of the federal budget deficit; and the addition of dollars to the paychecks of many American workers. The last time the economy had seen similar indicators was during the 1960s. But in January 2000 the boom surpassed all others to become the longest sustained expansion in U.S. history.

Economists considered the factors that contributed to the boom. A *Christian Science Monitor* writer credited "a combination of ubiquitous American entrepreneurial spirit, massive amounts of technology, and a man named Greenspan." But Alan Greenspan, the chairman of the Federal Reserve, which controls interest rates, credited information technology as the defining factor of "this special period": "Its major contribution is to reduce the number of worker hours required to produce the nation's output," said Greenspan.

The boom ended in March 2001, along with the end of the dot-com bubble. The nation began a short recession, which ended November 2001, according to economic indicators. The economy then began a slow, and by most indicators, weak recovery.

What was the **dot-com bubble**?

The dot-com bubble was a phenomenon of the late 1990s, when there was unguarded optimism for Internet-based businesses. According to the *Oxford English Dictionary,* a "bubble" has long been defined as "anything fragile, unsubstantial, empty, or worthless"; since the seventeenth century, the word has been applied to "delusive commercial or financial schemes." "Dot-com" refers to a commercial Internet venture, which most often carries the ".com" suffix in its URL, or Internet address.

The dot-com bubble inflated quickly. The graphical user interface (GUI) of the World Wide Web, with point-and-click hyperlinks, was integrated into the previously academic-oriented Internet in the early 1990s, making it more user-friendly for the average person. Public use of the Internet expanded rapidly. Existing businesses realized they needed a presence on the Internet for information or marketing purposes, if not for commerce. The promise of conducting business online, where costs were (wrongly) judged to be low, spurred entrepreneurs. New businesses began popping up to take advantage of the commercial, or e-commerce possibilities, of the "net"; these were the start-ups. They often had no real-world, or bricks-and-mortar, correlation; they strived to make money by reaching consumers only over the Internet. Among the start-ups of the dot-com bubble were Amazon, eBay, eToys, WebMD, HotJobs, and Monster.

Startups were known by several characteristics: Because of investor optimism about e-commerce, startups had quick access to venture capital funds; their managers were usually young, risk-taking Gen-Xers (some of whom were unsalaried workers who signed on for the promise of big earnings through stock options); they spent lavishly on their office spaces and on employee perks; they conducted expensive advertising and marketing campaigns; and, when taken public on the stock market (in an IPO, or initial public offering), the original owners and investors were known to make huge amounts of money, at least on paper. Once the dot-coms were on the stock market (the technology-heavy NASDAQ was home to many), individual investors became enchanted with them, artificially inflating their stock prices, even when many companies had yet to earn a dime. The price-to-earnings ratio (or PE), a measure of performance used by investors, became virtually meaningless when applied to the dot-coms.

The height of the dot-com bubble was the (January) 2000 Super Bowl, when almost 20 dot-com companies paid more than $2 million each for prime advertising spots. On March 10, 2000, the NASDAQ index of leading technology peaked at 5048.62: a year earlier the index was less than half that, right around 2500; and a year later it hovered around 2000, or about 40 percent of its peak. (In spring 2005 the NASDAQ composite index was below 2000.) Some called the March 2000 burst "cataclysmic," but other analysts and investors saw the end of the dot-com bubble as a necessary correction, or thinning, after which earnest players could get on with building Internet-based businesses that would be successful in the long run. Whatever the view, the dot-com bubble was the biggest market bubble ever seen, and many investors lost big.

What was **Y2K**?

Y2K means "year 2000" (K is a metric abbreviation for thousand). In the late 1990s the term was most often coupled with "problem" or "bug" to refer to a potentially disastrous computer programming peculiarity: Over decades of programming and in trillions of lines of code, developers had truncated the four-digit year to two digits as a space-saving mechanism, so that 1999 was rendered as 99. It was therefore feared that many computer programs, as well as simple chips in VCRs, watches, and other consumer devices, might malfunction, reading 00 not as 2000 but as 1900, or not recognize it at all. The problem was a concern for every sector of the global economy, but it was feared to have a particularly ruinous impact on the mainframe computer systems that are the backbone of operations for banks and other financial institutions, electric utilities, water systems, communication systems, oil and gas companies, and government entities (such as the Department of Defense and the Social Security Administration).

In the second half of the 1990s, business and government developed solutions to the Y2K problem, or the "millennial bug"; readiness was gauged periodically. To encourage the sharing of best practices across and between industries, the federal government passed the Year 2000 Information and Readiness Disclosure Act (signed October 1998). Despite assurances that the Y2K bug had been fixed, much of the public met the year 2000 with at least a bit of trepidation, some people having made emergency preparations such as stocking up on food and water, buying generators, and stashing away cash. Though there were isolated glitches, no major problem surfaced. Thanks to rectification efforts, the foretold Y2K crisis never materialized. But the cost of fixing the problem was high: it had an estimated $1.5 trillion price tag.

What was **Enron**?

Enron was a high-flying energy trading and communications company headquartered in Houston, Texas. It was the seventh-largest corporation in the United States, a favorite on Wall Street, and for six years in a row (1996–2001) was named America's Most Innovative Company by *Fortune* magazine. Then Enron filed for Chapter 11 bankruptcy in December 2001, rocking the business world and shocking investors and rank-and-file employees. It was, for a short time, the largest bankruptcy in American history. Federal investigators later learned that the company's collapse was caused by fraudulent accounting practices that allowed Enron to overstate earnings and hide debts: the conglomerate had booked billions in profits that did not really exist and created mythical companies to bury heavy losses. Enron's stock price plummeted, there were massive layoffs, employee retirement accounts (heavily invested in Enron stock) were decimated, executives resigned, and criminal indictments followed. Its accounting firm, Chicago-based Arthur Anderson, collapsed under the weight of its involvement in the scandal.

Enron soon became emblematic of a much larger problem, the so-called "breakdown of corporate America." It was the first of several colossal business failures, the biggest of which was the collapse of telecom giant WorldCom. In July 2002 WorldCom, valued at $180 billion and serving 15 million customers at its 1999 peak, filed for bankruptcy. WorldCom eclipsed Enron to earn the dubious title of largest bankruptcy in U.S. history. Again, fraudulent business practices were to blame. In March 2005 a federal jury convicted former WorldCom Inc. CEO Bernard Ebbers of engineering $11 billion in fraud. He was also found guilty of conspiracy and of filing false financial reports. The conviction was critical to prosecutors in a host of pending cases connected to corporate scandals, including Enron, whose former chairman, Kenneth Lay, and former CEO, Tom Skilling, were awaiting trial (set for 2006).

A *Fortune* magazine writer reflected on the crisis of corporate ethics, saying, "Phony earnings, inflated revenues, conflicted Wall Street analysts, directors asleep at the switch—this isn't just a few bad apples…[it is] a systemic breakdown." Enron happened to be first, and it became the symbol for the many. In 2005 Enron was in the process of distributing remaining assets to creditors and liquidating other operations. WorldCom, meanwhile, emerged from bankruptcy in 2004 as MCI Inc., the name of one of its subsidiaries.

POLITICAL AND SOCIAL MOVEMENTS

How old is **nationalism**?

Nationalism, a people's sense that they belong together as a nation because of a shared history and culture, and often because of a common language and/or religion, emerged at the close of the Middle Ages (500–1350). By the 1700s several countries, notably England, France, and Spain, had developed as "nation-states," groups of people with a shared background who occupy a land that is governed independently. As political and economic entities, the nation-states were preceded by fiefs, tribes, city-states, and empires, which overlapped each other as organizing units, dividing peoples' loyalties.

By the 1800s nationalism had become a powerful force, and the view took hold that any national group has the right to form its own state. Because of this belief, called national self-determination, some nations achieved independence (including Greece, which gained freedom from Turkey in 1829, and Belgium, which won self-rule from the Netherlands in 1830); others formed new, larger countries (both Italy and Germany were created by the unification of numerous smaller states, Italy in 1870 and Germany one year later); and still others carved smaller states out of great empires (for example, the breakup of the Austro-Hungarian Empire following World War I [1914–18] resulted in the formation of the independent countries of Austria, Hungary, Czechoslovakia, Poland, and, later, Yugoslavia).

In the United States, nationalism in the 1800s took the form of Manifest Destiny, the mission to expand the country's boundaries to include as much of North America as possible. By the end of the century the United States had claimed all of its present-day territory.

While nationalism is a source of pride and patriotism and has had many positive results, some leaders (notably German dictator Adolf Hitler [1889–1945]) have carried it to extremes, initiating large-scale movements that resulted in the persecution of other peoples and in the hideous practice of ethnic cleansing.

The boundary lines that were drawn on the world map at the end of the twentieth century were largely the result of nationalistic movements, some of which had resulted in conflicts—and some of which remained unresolved.

What is **Zionism**?

Zionism was founded as the nationalist movement to establish an independent Jewish state; it began in the 1890s, and roughly 50 years later, in 1948, the movement's activism resulted in the proclamation of the state of Israel. Since that time, Zionism has focused its efforts on building bridges between Israel and Jewish people around the world.

The roots of Zionism date to 1882, when a movement began encouraging Jewish settlement of Palestine, the region in the Middle East (in southwest Asia) that borders the Mediterranean Sea to the west, Lebanon to the north, Syria and Jordan to the east, and Egypt (the Sinai Peninsula) to the southwest. Groups advocating immigration to the Jewish homeland in Palestine called themselves Lovers of Zion (*Hoveve Zion*): Mount Zion is the site in Jerusalem where the Temple of David (king of the ancient Hebrews; d. 962 B.C.) was built, and it is therefore considered the center of Jewish spiritual life.

As a political movement, Zionism was founded in the late 1890s by Austrian journalist Theodor Herzl (1860–1904). In 1894 Herzl was among the reporters covering the trial of Alfred Dreyfus, a French army officer falsely convicted of treason. Though the artillery captain, who was Jewish, was later declared innocent (the guilty verdict rendered in his first trial was annulled), many felt the Dreyfus case had exposed a "deep vein of anti-Semitism" in Europe. Herzl's conclusion was that if anti-Semitism could take hold in France, it could prevail anywhere. Based on this belief, he began working for the reclamation of a Jewish state in the Middle East. In 1897 Herzl convened the First Zionist Congress, held in Basel, Switzerland, bringing the movement to worldwide attention. In 1917, against the backdrop of World War I (1914–18), British foreign secretary Arthur James Balfour (1848–1930) issued a declaration vowing his country's support for a national Jewish homeland in Palestine; this came after British troops liberated the Middle East from the control of the Ottoman Empire. In 1920 the Ottoman Empire dissolved as part of the conclusion of World War I and by international agreement the British were given rule over Palestine.

Numerous Jews immigrated to Palestine, where fighting broke out with Arabs who opposed Jewish resettlement. Previously boosted by British support of an independent Jewish state, Zionists received a heavy blow in 1937 when, with another conflict in Europe on the horizon, Britain reversed its policy in Palestine—in an effort to gain Arab support should fighting break out with Germany. At the end of World War II (1939–45), Britain turned over the problem in the Middle East to the newly created United Nations, which decided that out of Palestine both an independent Jewish state and a self-ruling Arab state should be formed. In 1948 the state of Israel was declared by Polish-born and Zionist moderate David Ben-Gurion (1886–1973), who became head of the nation's provisional government. The World Zionist Congress was later separated from the government. The organization has since turned its attention to immigration and cultural activities. German-born scientist Albert Einstein (1897–1955) was among Zionism's most prominent adherents.

What was the nonviolent **Indian reform movement**?

It was the movement led by Indian nationalist leader Mohandas Gandhi (1869–1948), whose methods of protest included staging boycotts, fasting, conducting prayer vigils, and visiting troubled areas in an attempt to end conflicts. Gandhi, whom the people called Mahatma (meaning "great-souled"), was determined to bring about change in India—to bring an end to British control of the country and to topple the ages-old caste system (the strict social structure) there. Gandhi believed that it took great courage to *not* engage in violence, and he began campaigns of passive resistance, which he called *satyagraha* (meaning "firmness in truth"). Gaining a wide following, Gandhi's acts of civil disobedience did bring about changes in his homeland, where he is revered as the founder of an independent India (1947). He remained faithful to his nonviolent beliefs throughout his life. He also adhered to a firm policy of religious tolerance. It was for this reason that the spiritual and nationalist leader was killed by a Hindu extremist in 1948.

What was the **May Fourth movement**?

It was a mass movement that emerged in China after May 4, 1919, when students in Beijing protested one of the outcomes of the peace conference held at Versailles earlier that year to officially settle World War I (1914–18): Japan, which had seized German territories in China during the war, was given control of the holdings. Student demonstrators criticized a weakened Chinese government for allowing the Japanese occupation. Following the death of powerful leader Yuan Shih-k'ai (1859–1916), the country's central government crumbled: In northern China local military leaders (called war lords) rose to power, continually challenging the authority of the capital at Beijing. Meanwhile, revolutionary leader Sun Yat-sen (1866–1925) had begun promoting his three great principles—nationalism, democracy, and people's livelihood—in southern China, where he gained the support of military leaders in the region. At about the same time, Chinese intellectuals had begun attacking traditional culture and society, urging government reforms and the modernization of industry. The May Fourth movement fanned the fires of revolution. The movement would have far-reaching—and unforseen—results. And some might argue that the story has not yet played out.

In 1919 Sun reorganized the Kuomintang (Nationalist) Party and began recruiting student followers. Two years later he became president of a self-proclaimed national government of the Southern Chinese Republic, establishing the capital at

Guangzhou (Canton). His sights were set on conquering northern China. Toppling the northern war lords to reunify the country, in 1924 Sun began cooperating with both the Soviets and the Communist groups that had been formed by students following the 1919 protest. Under Sun's leadership, the Nationalist Party began preparing for war. But Sun, who is regarded as the "father of modern China," would not live to see the culmination of his plans: He died of cancer in 1925. Under military leader Chiang Kai-shek (1887–1975), the Kuomintang turned on its Communist members, whose leaders fled in fear of the generalissimo. In 1928, following a two-year military campaign, Chiang led the nationalists to capture Beijing, reuniting China under one government for the first time in 12 years. His rule of China lasted until 1949, when Communists won control of the mainland and Mao Tse-tung (1893–1976) became the first chairman of the People's Republic of China. The expelled Chiang and his followers established a Chinese nationalist government on the island of Taiwan. Back on the mainland, Mao's Great Leap Forward, his massive collectivization of agriculture and industry, brought economic failure and a two-year famine to China in the late 1950s.

What is **Solidarity**?

It was a worker-led movement for political reform in Poland during the 1980s and it led to the downfall of communism. The movement was inspired by Pope John Paul II's June 1979 visit to his native Poland, where, in Warsaw, he delivered a speech to millions, calling for a free Poland and a new kind of "solidarity." (As scholar and author Timothy Garton Ash noted, "Without the Pope, no Solidarity. Without Solidarity, no Gorbachev. Without Gorbachev, no fall of Communism.")

Shipyard electrician Lech Walesa (1943–) became the leader of Solidarity, formed in 1980 when 50 labor unions banded together to protest Poland's Communist government. The unions staged strikes and demonstrations. By 1981 Solidarity had gained so many followers that it threatened Poland's government, which responded (with the support of the Soviet Union) by instituting martial law in December of that year. The military cracked down on the activities of the unions, abolishing Solidarity in 1982 and arresting its leaders, including the charismatic Walesa. But the powerful people's movement, which had also swept up farmers (who formed the Rural Solidarity), could not be suppressed. Martial law was lifted in mid-1983 but the government continued to exert control over the people's freedom. That year Walesa received the Nobel peace prize for his efforts to gain workers' rights and prevent violence. Solidarity continued its work for reform. In 1989 the collapse of communism on the horizon (people's movements in Eastern Europe had combined with Soviet leader Mikhail Gorbachev's policy of *glasnost* to herald the system's demise), the Polish government reopened negotiations with Solidarity's leadership. Free elections were held that year, with the labor party candidates gaining numerous seats in Parliament. In 1990 Walesa was elected president, at which time he resigned as chairman of Solidarity. Poland's Communist Party was dissolved that year.

Pro-Solidarity rally in Czestochowa, Poland, 1988. Solidarity was a worker-led movement for political reform in Communist Poland; by 1990 the country's Communist government was dissolved.

What was the **antiapartheid movement**?

It was an international movement to throw out the decades-old system of racial segregation in South Africa. (The word *apartheid* means "separateness" in the South African language of Afrikaans.) Under apartheid, which was formalized in 1948 by the Afrikaner Nationalist Party, minority whites were given supremacy over nonwhites. The system further separated nonwhite groups from each other so that mulattoes (those of mixed race), Asians (mostly Indians), and native Africans were segregated. The policy was so rigid that it even separated native Bantu groups from each other. Blacks were not allowed to vote, even though they were and are the majority population. Apartheid was destructive to the society as a whole and drew protest at home and abroad. But the South African government adhered to the system, claiming it was the only way to keep peace among the country's various ethnic groups. In 1961 the government even went so far as to withdraw from the British Commonwealth in a dispute over the issue.

Protesters against apartheid staged demonstrations and strikes, which sometimes became violent. South Africa grew increasingly isolated as countries opposing the system refused to trade with the apartheid government. The no-trade policy had been urged by South African civil rights leader and former Anglican bishop Desmond Tutu (1931–), who led a nonviolent campaign to end apartheid and in 1984 won the Nobel peace prize for his efforts. During the 1980s the economic boycott put pressure on the white minority South African government to repeal apartheid laws. It finally did so, and in 1991 the system of segregation was officially abolished.

White South African leader F. W. de Klerk (1936–), who was elected in 1989, had been instrumental in ending the apartheid system. In April 1994 South Africa held the first elections in which blacks were eligible to vote. Not surprisingly, black South Africans won control of Parliament, which in turn elected black leader Nelson Mandela (1918–) as president; de Klerk was retained as deputy president. The two men won the Nobel peace prize in 1993 for their efforts to end apartheid and give all of South Africa's peoples full participation in government. In 1996 the work of the Truth and Reconciliation Commission, a panel headed by Desmond Tutu and charged with investigating the political crimes committed under apartheid, began work. Its investigations continued in 1999, with many findings proving controversial.

Who was **Biko**?

Stephen Biko (1946–1977) was a black leader in the fight against South African apartheid and white minority rule. In 1969 Biko, who was then a medical student, founded the South African Student's Organization, which took an active role in the black consciousness movement, a powerful force in the fight against apartheid. Preaching a doctrine of black self-reliance and self-respect, Biko organized protests, including antigovernment strikes and marches. Viewing such activities as a challenge to its authority and fearing an escalation of unrest, in August 1977, the white government had Biko arrested. Within one month, he died in prison. Evidence indicated he had died at the hands of his jailers, a revelation that only cemented antigovernment sentiment. Along with Nelson Mandela (1918–), who was imprisoned in South Africa from 1962 to 1990 for his political activities, Biko became a symbol of the antiapartheid movement, galvanizing support for racial justice at home and abroad.

What happened at China's **Tiananmen Square**?

In 1989 Beijing's Tiananmen Square, the largest public square in the world, became the site of a student protest and massacre. Three years before the demonstration, freedom of speech and other democratic beliefs began being espoused on university campuses. In increasing numbers China's youth were demanding political reform. They found a sympathizer in the general secretary (the highest ranking officer) of the Communist Party, Hu Yaobang (1915–1989), who, despite criticism from conservatives in government, adhered to his liberal views, particularly concerning freedom of expression. In January 1987 Hu was removed from his post; he died in April 1989, at which time students organized marches in his honor and demonstrated in favor of democratic reforms. On June 4, 1989, Chinese troops fired on the protesters in Tiananmen Square, killing more than 200 and later arresting anyone thought to be involved in China's pro-democracy movement. The actions raised fury around the world. International observers continued to monitor the tenuous situation in China, where evidence surfaced that the government was continuing its pattern of human rights violations.

THE ANTISLAVERY MOVEMENT

Which U.S. state was the first to **abolish slavery**?

Vermont was first, in 1777. On July 8 of that year Vermont adopted a state constitution that prohibited slavery. The first document in the United States to outlaw slavery, it read in part: "No male person, born in this country, or brought from over sea, ought to be holden by law, to serve any person, as a servant, slave or apprentice, after he arrives to the age of twenty-one years, nor female, in like manner, after she arrives to the age of eighteen years, unless they are bound by their own consent, after they arrive to such age, or bound by law, for the payment of debts, damages, fines, costs, or the like." Vermont's constitution also gave suffrage to all men, regardless of race. Vermonters were the first to put a black legislator in the state house: Alexander Twilight (1795–1857) was elected as a representative in 1836. Twilight also earned another first: In 1823 he graduated from Vermont's Middlebury College to become the first black person in the nation to earn a college degree.

Underground Railroad conductor Harriet Tubman (pictured in 1911) worked against slavery by helping hundreds of blacks head to freedom in northern states and Canada.

When did the **antislavery movement** begin?

In the United States, the campaign to prohibit slavery strengthened in the early 1800s. Across the Atlantic, abolitionists had successfully lobbied for the outlaw of slave trade in Great Britain by 1807. The following year, the U.S. government also outlawed the trade, but possession of slaves remained legal and profitable. In the 1830s the call to abolish slavery and emancipate slaves became an active movement in the United States, precipitated by a revival of evangelical religion in the North. Abolitionists, believing slavery is morally wrong and violates Christian beliefs, called for an end to the system, which had become critical to the agrarian economy of the southern states, where plantations produced cotton, tobacco, and other crops for domestic and international markets.

Who were the **leaders of abolition**?

Leaders of the antislavery movement included journalist William Lloyd Garrison (1805–1879), founder of the influential antislavery journal *The Liberator* and of the American Anti-Slavery Society (established 1833); brothers Arthur (1786–1865) and Lewis (1788–1873) Tappan, prominent New York merchants who were also founders of the American Anti-Slavery Society; and Theodore Dwight Weld (1803–1895), leader of student protests, organizer of the American and Foreign Anti-Slavery Society, and author of *The Bible Against Slavery* (1837) and other abolitionist works.

Underground Railroad conductor Harriet Tubman (c. 1820–1913) worked against slavery by helping to free hundreds of blacks who escaped slavery in the South, heading for northern states and Canada. Writers such as Harriet Beecher Stowe (1811–1896), author of *Uncle Tom's Cabin* (1851–52), helped strengthen the abolitionist cause and were instrumental in swaying public sentiment. In the hands of some activists, the movement became violent: In 1859 ardent abolitionist John Brown (1800–1859) led a raid on the armory at Harpers Ferry (in present-day West Virginia), which proved a failed attempt to emancipate slaves by force.

Who started the Underground Railroad?

American abolitionist, lecturer, and nurse Harriet Tubman (c. 1820–1913) set up the network to emancipate slaves. Tubman was motivated to do so after she had made her way to freedom in 1849, and then wished the same for her family: "I had crossed the line of which I had so long been dreaming. I was free; but there was no one to welcome me to the land of freedom."

For the next 10 years Tubman acted as a conductor on the Underground Railroad, making at least 15 trips into southern slave states, and guiding not only her parents and siblings, but more than 300 slaves to freedom in the North. She was called "the Moses of her people" for her emancipation efforts. The journeys to freedom were demanding and often dangerous missions. Though Tubman was small in stature, she possessed extraordinary leadership qualities. Author, clergyman, and army officer Thomas Wentworth Higginson (1823–1911) called her "the greatest heroine of the age."

What did the **founding of Liberia** have to do with the antislavery movement?

With the goal of transporting freed slaves back to their homeland, members of the American Colonization Society (organized 1816–17), made land purchases on the west African coast. The holdings were named *Liberia,* a Latin word meaning "freedom." The first black Americans arrived there in 1822. But the society's plan was controversial; even some abolitionists and blacks opposed it, as they believed the only answer to the question of slavery was to eradicate it from the United States and extend the full rights of citizenship to the freed slaves in their new American home. Nevertheless, by 1860 11,000 freed black slaves from the United States had been settled there; eventually a total of 15,000 made the transatlantic voyage to a secured freedom in Liberia. The country was established as an independent republic on July 26, 1847.

What did lawmakers do to resolve **the slavery question** before the Civil War?

The mid-1800s were a trying time for the nation—the divide widened between the northern free states and the southern slave states, which were growing increasingly dependent on agricultural slave labor. Government tried but was unable to bring resolution to the conflict over slavery. Instead, its efforts seemed geared toward maintaining the delicate North-South political balance in the nation.

After the Mexican War (1846–48), the issue was front and center as congressmen considered whether slavery should be extended into Texas and the western territories gained in the peace treaty of Guadalupe Hidalgo, which officially ended the war. Lawmakers arrived at the Compromise of 1850, which proved a poor attempt to assuage mounting tensions: The legislation allowed for Texas to be admitted to the Union as a slave state, California to be admitted as a free state (slavery was prohibited), voters in New Mexico and Utah to decide the slavery question themselves (a method called popular sovereignty), the slave trade to be prohibited in Washington, D.C., and for passage of a strict fugitive slave law to be enforced nationally.

Four years later, as it considered how to admit Kansas and Nebraska to the Union, Congress reversed an earlier decision (part of the Missouri Compromise of 1820) that had declared the territories north of the Louisiana Purchase to be free, and set up a dangerous situation in the new states: The slavery status of Kansas and Nebraska would be decided by popular sovereignty (the voters in each state). Nebraska was settled mostly by people opposing slavery, but settlers from both the North and the South poured into Kansas, which became the setting for violent conflicts between proslavery and antislavery forces. Both sides became determined to swing the vote by sending "squatters" to settle the land. Conflicts resulted, with most of them clustered around the Kansas border with Missouri, where slavery was legal. In one incident, on May 24, 1856, ardent abolitionist John Brown (1800–1859) led a massacre in which five proslavery men were brutally murdered as they slept. The act had been carried out in retribution for earlier killings of freemen at Lawrence, Kansas: Brown claimed his was a mission of God. Newspapers dubbed the series of deadly conflicts, which eventually claimed more than 50 lives, "Bleeding Kansas." The situation proved that neither congressional compromises nor the doctrine of popular sovereignty would solve the nation's deep ideological differences.

Why did President Lincoln issue the **Emancipation Proclamation** before the end of the Civil War?

As the war raged between the Confederacy and the Union, it looked like victory would be a long time in the making: In the summer of 1862 things seemed grim for the federal troops when they were defeated at the Second Battle of Bull Run (which took place in northeastern Virginia on August 29–30). But on September 17, with the Battle of Antietam (in Maryland), the Union finally forced the Confederates to withdraw across the Potomac into Virginia. That September day was the bloodiest of the war. President Abraham Lincoln (1809–1865) decided that this withdrawal was success enough for him to make his proclamation, and on September 22, he called a cabinet meeting. That day he presented to his advisers the Preliminary Emancipation Proclamation.

The official Emancipation Proclamation was issued later, on January 1, 1863. This final version differed from the preliminary one in that it specified emancipation was to be effected only in those states that were in rebellion (i.e. the South). This key change had been made because the president's proclamation was based on congressional acts giving him authority to confiscate rebel property and forbidding the military from returning slaves of rebels to their owners.

Abolitionists in the North criticized the president for limiting the scope of the edict to those states in rebellion, for it left open the question of how slaves and slave owners in the loyal (Northern) states should be dealt with. Nevertheless, Lincoln had made a stand, which served to change the scope of the Civil War (1861–65) to a war against slavery.

On January 31, 1865, just over two years after the Emancipation Proclamation, Congress passed the Thirteenth Amendment, banning slavery throughout the United States. Lincoln, who had lobbied hard for this amendment, was pleased with its passage. The Confederate states did not free their 4 million slaves until after the Union was victorious (on April 9, 1865).

When was **slavery outlawed in Europe**?

The slave trade ended in Britain in 1807, when authorities agreed with the growing number of abolitionists (those who argued that slavery is immoral and violates Christian beliefs) and outlawed the trade. In 1833 slavery was abolished throughout the British colonies as the culmination of the great antislavery movement in Great Britain. In the United States, the slave trade was prohibited in 1808, but possessing slaves was still legal. Consequently, trade on the black market continued until Britain stepped up its enforcement of its antislavery law by conducting naval blockades and surprise raids off the African coast, effectively closing the trade. The slave trade as it had been known officially came to an end after 1870, when it was outlawed throughout the Americas. Throughout the world, the United Nations works to abolish slavery and other systems of forced labor.

Is there **slavery today**?

Yes, slavery continues into the twenty-first century. The United Nations Population Fund (UNFPA) has stated: "Although slavery has been formally abolished from the world, the trade in human misery continues." Today it is called "human trafficking." Estimating the size of the problem is difficult, but the UNFPA estimates that about four million people are trafficked across international borders each year. The group also reports that the problem is widespread, but the greatest volume of human trafficking exists in Asia, with Africa and Latin America following close behind. The Asia Pacific region is seen as particularly vulnerable, according to the United Nations Economic and Social Commission for Asia and the Pacific (UNESCAP), because of "its huge population pyramid, growing urbanization, and extensive poverty."

Some human rights groups estimate that the number of slaves in the world today is as high as 27 million people. And experts say that it is a growing problem, fueled by globalization. Men, women, and children, especially in developing countries, are forced into labor in sweatshops and fields, and into prostitution in brothels. In desperately poor regions of the world, families sell their children into slave labor and forced prostitution. Other victims are lured in; according to the United Nations Office on Drugs and Crime, "From Himalayan villages to Eastern European cities, people—especially women and girls—are attracted by the prospect of a well-paid job as a domestic servant, waitress, or factory worker." Traffickers recruit victims through fake advertisements, mail-order bride catalogs, and casual acquaintances. But the victims end up in situations controlled by their traffickers, and they are exploited against their wills to earn illicit revenues.

By the early 2000s, human rights groups and governments were organizing to combat the increase in human trafficking. Several agencies of the United Nations worked to address the roots of the problem and to aid victims. Nongovernment agencies were playing a role as well. One such group is Shared Hope International, founded in 1998 by U.S. Congresswoman Linda Smith (Washington) to "rescue and restore women and children in crisis by providing comprehensive services to meet their needs." Italy's government was at the forefront of the anti-trafficking movement, offering residency permits to victims and funding local shelters through legislation passed in 1999. In 2000 the U.S. Congress passed the Victims of Trafficking and Violence Protection Act (TVPA), declaring that sex trafficking is the "modern day slavery." Government figures estimated that each year 45,000 to 50,000 women and children were trafficked into the United States, where they were trapped in modern-day slavery-like situations such as forced prostitution.

But the trafficking problem in the United States, and elsewhere, is not limited to importing women and children from other countries. According to a September 2001 Justice Department report, 400,000 children are lured or forced into prostitution each year in the United States. Many of the victims are from white, working- and middle-class families, often runaways from troubled homes who end up on the streets.

In September 2004 former representative John R. Miller (Washington) was sworn into the newly created position of ambassador-at-large for the U.S. State Department's anti-trafficking office. In a speech, Miller said, "Today, the slavery is not on plantations and in homes; it is in factories and armies as well, and especially in brothels. But the slave masters use the same tools today as earlier slave masters: kidnapping, fraud, threats, and beatings, all aimed at forcing women, children, and men into labor and sex exploitation."

Experts agreed that ending human trafficking in the twenty-first century would require a coalition of government, special interest groups, human rights organizations, and other nongovernment organizations. Determining the scope of the problem and raising public awareness were important first steps.

THE CIVIL RIGHTS MOVEMENT

What was the **Niagara movement**?

It was a short-lived but important African American organization that advocated "the total integration of blacks into mainstream society, with all the rights, privileges, and benefits of other Americans." Founded in Niagara Falls, Ontario, in 1905, the Niagara movement was led by writer, scholar, and activist W. E. B. Du Bois (1868–1963), who was then a professor of economics and history at Atlanta University. Observers described the organization as the anti-Bookerite camp: Educator Booker T. Washington (1856–1915), who rose from slavery to found Alabama's Tuskegee Institute (1881), believed change for black people should be effected through education and self-improvement—not through demand. Mr. Washington opposed the social and political agitation favored by some reformers; the Niagara movement, on the other hand, placed the responsibility for the nation's racial problems squarely on the shoulders of its white population. The 30 branches of the Niagara movement challenged conservative politics of the so-called "Tuskegee Machine" led by Booker T. Washington. Though the Niagara organization dissolved in 1909, the National Association for the Advancement of Colored People (NAACP) was heir to its ideology and activism. Du Bois helped found that organization, and from 1910 to 1934 edited its official journal, *The Crisis,* in which he published his views "on nearly every important social issue that confronted the black community."

Were activists the only ones who were vocal about **opposing segregation**?

No, segregation was opposed at every level of black society, as well as by many whites. The voices of the civil rights movement included wage laborers, farmers, educators, athletes, entertainers, soldiers, religious leaders, politicians, and statesmen—all of whom had experienced the oppression of Jim Crow laws and policies in the United States.

Before W. E. B. Du Bois (1868–1963) rose to prominence as an educator and writer, he chose to leave the security of his home in Great Barrington, Massachusetts, to attend college at Nashville's Fisk University. There, in 1885, he encountered Tennessee's Jim Crow laws, which strictly divided blacks and whites. He was so intimidated by the "southern system" that he rarely left the campus, and he ultimately returned to New England to complete his studies at Harvard University. He did, however, go back to the South, becoming a professor of economics and history at Atlanta University (1897–1910, 1932–44). As one of the first exponents of full and equal racial equality, in 1909 Du Bois helped found the National Association for the Advancement of Colored People (NAACP), which provided leadership during the civil rights movement.

In 1942 a young Georgia man named John Roosevelt Robinson (1919–1972) was drafted into the military. Robinson applied for Officer's Candidate School at Fort Riley,

Kansas, and although he was admitted to the program, he and the other black candidates received no training until pressure from Washington, D.C., forced the local commander to admit blacks to the base's training school. Later Robinson became a second lieutenant and continued to challenge the Jim Crow policies on military bases. When the army decided to keep him out of a game with the nearby University of Missouri because that school refused to play against a team with black members, Robinson quit the base's football team in protest; at Fort Hood, Texas, Robinson objected to segregation on an army bus. His protests led to court-martial. Acquitted, in November 1944 Robinson was honorably discharged—before the end of World War II (1939–45): The army had no desire to keep this black agitator among the ranks, and, as Robinson later put it, he was "pretty much fed up with the service." In 1947 "Jackie" Robinson became the first black baseball player in the major leagues when he joined the Brooklyn Dodgers, breaking the color barrier in the national pastime.

In the postwar years, American diplomat Ralph Bunche (1904–1971) attracted public attention when he rejected an offer from President Harry Truman (1884–1972) to become an assistant secretary of state. Bunche, a Howard University professor who had worked for the Office of Strategic Services during the war, explained that he declined the position because he did not want to subject his family to the Jim Crow laws of Washington, D.C. Bunche spoke out frequently against racism, and in 1944 he co-authored the book *An American Dilemma,* which examined the plight of American blacks.

These are just a few of the many examples of personal protest that signaled the beginning of the civil rights movement in the United States.

Who was **Emmett Till**?

Emmett Till (1941–1955) was a black 14-year-old from Chicago who was brutally mutilated and killed in the Deep South in August 1955. The young man was visiting relatives in Mississippi when he allegedly whistled at a white female store clerk. Till was sharing a bed with his 12-year-old cousin when two white men came to get him on the morning of August 28; he was not seen alive again. His body was later found in a river, tied to a cotton-gin fan with barbed wire. An all-white jury acquitted the store clerk's husband, Roy Bryant, and half-brother, J. W. Milman, of the crime. The events stirred anger in the black community and among civil rights proponents in general, setting off the civil rights movement.

For four decades, Till's grisly murder continued to deeply trouble many, who believed justice could still be served. Though no one was ever convicted of the crime, and the two men who were tried for it had, by 2005, died, some of Till's family and friends, as well as investigators, believed others who participated in the lynching might still be alive. In a quest for clues, Till's body was disinterred in June 2005 to gather evidence. He was reburied in a quiet funeral. The Till family hoped the pending investigation would yield answers and justice.

How did the **civil rights movement** begin?

It began on Thursday, December 1, 1955, as Rosa Parks (1913–), a seamstress who worked for a downtown department store in Montgomery, Alabama, made her way home on the Cleveland Avenue bus. Parks was seated in the first row that was designated for blacks. But the white rows in the front of the bus soon filled up. When Parks was asked to give up her seat so that a white man could sit, she refused. She was arrested and sent to jail.

Montgomery's black leaders had already been discussing staging a protest against racial segregation on the city buses. They soon organized, with Baptist minister Martin Luther King Jr. (1929–1968) as their leader. Beginning on December 5, 1955, thousands of black people refused to ride the city buses: the Montgomery Bus Boycott had begun. It lasted more than a year—382 days—and ended only when the U.S. Supreme Court ruled that segregation on the buses was unconstitutional. The protesters and civil rights activists had emerged the victors in this—their first and momentous—effort to end segregation and discrimination in the United States.

Parks, who lost her job as a result of the arrest, later explained that she had acted on her own beliefs that she was being unfairly treated. But in so doing Parks had taken a stand and had given rise to a movement.

What was the **nonviolence movement**?

The Reverend Martin Luther King Jr. (1929–1968) was committed to bringing about change by staging peaceful protests; he led a campaign of nonviolence as part of the civil rights movement. King rose to prominence as a leader during the Montgomery Bus Boycott in 1955, when he delivered a speech that embodied his Christian beliefs and set the tone for the nonviolence movement, saying, "We are not here advocating violence.... The only weapon we have...is the weapon of protest." Throughout his life, King staunchly adhered to these beliefs—even after terrorists bombed his family's home. King's "arsenal" of democratic protest included boycotts, marches, the words of his stirring speeches (comprising an impressive body of oratory), and sit-ins. With other African American ministers King established the Southern Christian Leadership Conference (1957), which assumed a leadership role during the civil rights movement.

The nonviolent protest of black Americans proved a powerful weapon against segregation and discrimination: A massive demonstration in Birmingham, Alabama, in 1963 helped sway pubic opinion and motivate lawmakers in Washington to act when news coverage of the event showed peaceful protesters being subdued by policemen using dogs and heavy fire hoses. In response to the outcry over the event in Birmingham, President John F. Kennedy (1917–1963) proposed civil rights legislation to Congress; the bill was passed in 1964. That same year Martin Luther King Jr. received the Nobel peace prize for his nonviolent activism.

The Reverend Martin Luther King Jr. brought change through peaceful protests. A charismatic speaker, here he addresses supporters in Selma, Alabama, in 1965.

King's policy of peace was challenged two years later when the Student Nonviolent Coordinating Committee (SNCC), tired of the violent response with which peaceful protesters were often met, urged activists to adopt a more decisive and aggressive stance and began promoting the slogan "Black Power." The civil rights movement, having made critical strides, became fragmented, as leaders, including the highly influential Malcolm X (1925–1965), differed over how to effect change.

On April 4, 1968, King was in Memphis, Tennessee, to show his support for a strike of black sanitation workers when he was gunned down outside his hotel room shortly after 5:30 in the evening. As news of King's death swept over the nation, blacks in 168 American cities and towns responded with rioting, setting fire to buildings, and looting white businesses. Commenting on the terror, radical African American leader Stokely Carmichael said, "When white America killed Dr. King last night, she declared war on us." The chaos continued for a week: When the rioting ended on April 11, there were 46 dead (most of them black), 35,000 injured, and 20,000 jailed. Nevertheless, the violent crime that claimed the leader's life and the violence that erupted after news spread of his death have not, decades later, overshadowed King's legacy of peace and his message of the brotherhood of all people.

What were the **"freedom rides"**?

The "freedom rides" were a series of bus rides designed to test the U.S. Supreme Court's prohibition of segregation in interstate travel. In 1960, in the case of *Boynton* v. *Virginia,* the Supreme Court ruled in favor of a Howard University student who charged that segregation laws at the Richmond, Virginia, bus station violated federal antisegregation laws. The Congress of Racial Equality (CORE) decided to test the enforcement of the federal law by initiating the freedom rides. On May 4, 1961, 13 people, black and white, boarded a bus for the South. Meant as a nonviolent means of protest against local segregation laws, the riders were nevertheless met with violence: When the bus reached Montgomery, Alabama, on May 20, a white mob was waiting; the freedom riders were beaten. Rioting broke out in the city, and U.S. marshals were sent to restore order. The interracial campaign to desegregate transportation was ultimately successful, but government intervention was required to enforce the laws, as numerous southern whites had demonstrated that they weren't going to comply voluntarily.

When did **Martin Luther King Jr.** give his "I Have a Dream" speech?

The occasion was the March on Washington on August 28, 1963. That summer day more than a quarter million people—lobbying for congressional passage of a civil rights bill—gathered at the Lincoln Memorial to hear speakers, including the charismatic and influential King (1929–1968). His eloquent words defined the movement and still inspire those who continue to work for reforms. Among his words were these: "I have a dream that one day this nation will rise up and live out the true meaning of its creed, 'We hold these truths to be self-evident; that all men are created equal.' "

Congress did pass the Civil Rights Act, in 1964. The most comprehensive American civil rights legislation since the Reconstruction (the 12-year period that followed the Civil War), the act outlawed racial discrimination in public places, assured equal voting standards for all citizens, prohibited employer and union racial discrimination, and called for equality in education.

What does the letter *X* in **Malcolm X**'s name stand for?

The influential but controversial African American leader, who was a staunch defendant of black rights, took the surname X in 1952, upon his release from prison. He explained that the letter stood for the unknown African name of his ancestors. Malcolm X's family's name, Little, was that given to his slave ancestors by their owner. By adopting X as his surname, it was at once a bitter reminder of his family's slavery and an affirmation of his (unknown) African roots.

How were southern **blacks prevented from voting**?

Besides intimidation, there were three different methods used in southern states in the early part of the century to disenfranchise black citizens: 1) the poll tax; 2) literacy tests; and 3) grandfather clauses. The poll tax required a voter to pay a fee in order to exercise the right to vote. Literacy tests were implemented as a prerequisite for voting; this method also kept many poorly educated whites (unable to pass the exam) from casting their ballots as well. Most southern states also adopted legislation by which voting rights were extended only to those citizens who had been able to vote as of a certain date—a date when few if any black men would have been able to vote. Since these laws made provisions for said voters' descendants as well, they were dubbed "grandfather clauses." Such attempts to deny citizens the right to vote were made unlawful in 1964 (by the Twenty-fourth Amendment, which outlaws the poll tax in all federal elections and primaries), in 1965 (by the Voting Rights Act, which outlawed measures used to suppress minority votes), and in 1966 (when poll taxes at the state and local levels were also declared illegal). Literacy tests and grandfather clauses were also struck down as unconstitutional.

What is the history of the **Ku Klux Klan**?

The Ku Klux Klan (KKK) is a white supremacist group originally formed in 1865 in Pulaski, Tennessee, when Confederate Army veterans formed what they called a "social club." The first leader (called the "grand wizard") was Nathan Bedford Forrest (1821–1877), a former general in the Confederate Army, who, on April 12, 1864, in the final days of the Civil War, led a massacre of 300 black soldiers in service of the Union Army at Fort Pillow, Tennessee.

As the unofficial arm of resistance against Republican efforts to restore the nation and make full citizens of its black (formerly slave) population, the Ku Klux Klan waged a campaign of terror against blacks in the South during Reconstruction (1865–77), the 12-year period of rebuilding that followed the Civil War. Klan members, cloaked in robes and hoods to disguise their identity, threatened, beat, and killed numerous blacks. While the group deprived its victims of their rights as citizens, their intent was also to intimidate the entire black population and keep them out of politics. White people who supported the federal government's measures to extend rights to all black citizens also became the victims of the fearsome Klan. Membership in the group grew quickly, and the Ku Klux Klan soon had a presence throughout the South.

In 1871 the U.S. Congress passed the Force Bill, giving President Ulysses S. Grant (1822–1885) authority to direct federal troops against the Klan. The action was successful, causing the group to disappear—but only for a time. In 1915 the society was newly organized at Stone Mountain, Georgia, as a Protestant fraternal organization (called "The Invisible Empire, Knights of the Ku Klux Klan, Inc."), this time widening its focus of persecution to include Roman Catholics, immigrants, and Jews, as well as blacks. Members of all of these groups became the target of KKK harassment, which now included torture, whippings, and public lynchings. The group, which proclaimed its mission of "racial purity," grew in number and became national, electing its own to public office in many states, not just the South. But the society's acts of violence raised the public ire, and by the 1940s, America's attention focused on World War II (1939–45), and the Klan died out or went completely underground. The group had another resurgence during the 1950s and into the early 1970s, as the nation struggled through the era of civil rights. The Klan still exists today, fostering the extremist views of its membership and staging marches to demonstrate its presence on the American landscape. Such demonstrations are often attended by protestors.

What are **reparations**?

Reparations are payments or other compensations made to a group of people who have been wronged or injured. The issue was in the news in the 1990s and early 2000s as lawmakers, academics, and other leaders pressed for a redress for slavery, which some scholars call the American, or black, holocaust. The precedents for making reparations were several: The German government made reparations to survivors and families of victims of the Nazi holocaust. And the American government made reparations to Japanese Americans who had been interned during World War II (1939–45), as well as to Native Americans, for damages done to them.

The recent discussion of reparations began in 1989, when U.S. representative John Conyers (Michigan) introduced a bill, H.R. 40, in Congress to "establish a commission to examine the institution of slavery…and economic discrimination against African Americans" and, if so determined, "to make recommendations to the Congress on appropriate remedies." As the idea of reparations gained currency in the American public in the 1990s, supporters argued that redress for slavery would help heal the "open wound" of race relations and would compensate the descendants of slaves whose ancestors' work had helped build the national economy. They further argued that slavery resulted in long-term discrimination that beleaguered black Americans; they were the victims of a centuries-old, government-sanctioned system that established a legacy of race-based injustices. African American activist and author Randall Robinson explained it this way: "No nation can enslave a group of people for hundreds of years, set them free—bedraggled and penniless—to pit them, without assistance, in a hostile environment against privileged victimizers, and then reasonably expect the gap between the heirs of the two groups to narrow. Lines begun parallel and left alone can never touch." In bolstering support for reparations, Robinson pointed to the consequences of this "massive injustice": that blacks in the United States experience high rates of infant mortality, low incomes, high rates of unemployment, substandard education, high death rates, below-average life spans, and overrepresentation in prisons and on death row.

Critics of reparations said that compensating the descendants of slaves was unrealistic; determining who would be paid would alone constitute an expensive government program. They also questioned why descendants of slaves should be paid by the government a century and a half after the end of the brutal system. Further, they argued that other programs, born of the civil rights movement, have strived to bring equity to African Americans.

Despite criticism, Representative Conyers resolved to reintroduce his bill as often as necessary until Congress would act on it. He emphasized that his goal was to create a commission, informed by town hall meetings, to first determine if there should be reparations—and if so, who should be paid and how much. H.R. 40 had received the support of the city councils of Detroit, Cleveland, Chicago, and Atlanta.

TEMPERANCE

What was the **temperance movement**?

Temperance was an American movement that began in the mid-1800s to outlaw the manufacture and consumption of alcoholic beverages, which were viewed by many to be a corrupt influence on American family life. By 1855 growing public support to ban liquor resulted in 31 states making it illegal to some degree. But a national policy of temperance was still sought by many. During the 1870s temperance became one of the cornerstones of the growing women's movement. As the nation's women, joined by other activists, mobilized to gain suffrage (the right to vote), they also espoused sweeping cultural changes. In 1874 a group of women established the Woman's Christian Temperance Union (WCTU); in 1895 the Anti-Saloon League was formed. Such societies, which grew out of a fundamentalist spirit, found an increasing voice and eventually influenced legislators, many of whom were "dry" candidates that the societies had supported, to take federal action. Even President Woodrow Wilson (1856–1924) supported prohibition, as one of the domestic policies of his New Freedom program.

The movement met with success in January 16, 1919, when the Eighteenth Amendment to the U.S. Constitution (1788) was ratified, forbidding people to make, sell, or transport "intoxicating liquors" in the United States and in all territories within its jurisdiction. Though Congress, which proposed the amendment on December 18, 1917, provided states with a period of seven years in which to ratify the amendment, it took just over a year for it to be approved, such was the prevailing spirit among lawmakers. After the amendment was made, Congress passed the Volstead Act to enforce it. But government nevertheless found prohibition difficult to enforce. Bootleggers (who made their own moonshine—illegal spirits, often distilled at night), rum runners (who imported liquor, principally from neighboring Canada and Mexico), and speakeasies (underground establishments that sold liquor to their clientele) proliferated. Soon organized crime ran the distribution of liquor in the country, whose citizens had not lost their taste for alcoholic beverages. The government now found itself with a bigger problem. As the Federal Bureau of Investigation (FBI) and police worked to control and end mob violence, and as the country suffered through the early years of the Great Depression, lawmakers in Washington reconsidered the amendment. On February 20, 1933, the U.S. Congress proposed that the Eighteenth Amendment be repealed. Approved by the states in December of that year, the Twenty-first Amendment declared the Eighteenth Amendment null, and the manufacture, transportation, and consumption of alcoholic beverages was again legal in the United States, ending the 13-year period of Prohibition. Herbert Hoover (1874–1964), president at the time of repeal, called prohibition a "noble experiment."

Who was **Carry Nation**?

The Kentucky-born Carry Nation (1846–1911) became famous as a temperance agitator in the early 1900s. The saloon was illegal in her resident state of Kansas, and she felt it was her divine duty to take her hatchet to ruining any place that sold intoxicants. Between 1899 and 1909, she went on wrecking expeditions (which she called "hatchetations") throughout the state, incurring the wrath of business owners and government officials. Though many might have favored national prohibition of alcohol, Nation's actions were extreme to say the least, causing her to be arrested, imprisoned 30 times, and even shot at. She persisted, however, buoyed by the belief that she was performing a public—and even divine—service. The propitiously named Carry A. Nation (who tried, it seems, to carry the nation straight to the water fountain) did not live to see prohibition made into a national policy in 1917—nor to see it revoked in 1933.

WOMEN'S RIGHTS

When did the American **suffragist movement** begin?

In the 1840s American women began organizing and, in increasing numbers, demanding the right to vote. The movement was started by women who sought social reforms, including outlawing slavery, instituting a national policy of temperance (abstinence from alcoholic beverages), and securing better work opportunities and pay. These reformers soon realized that in order to make change they needed the power of the vote.

Among the leaders of the suffragist movement was feminist and reformer Elizabeth Cady Stanton (1815–1902). She joined with antislavery activist Lucretia Mott (1793–1880) to organize the first women's rights convention in 1848 in Seneca Falls, New York, launching the woman suffragist movement. In 1869 Stanton teamed with

New York suffragists hold a demonstration outside the White House in February 1917. Ratified in 1920, the Nineteenth Amendment gives women the right to vote in all state and federal elections.

Susan B. Anthony (1820–1906) to organize the National Woman Suffrage Association. That same year, another group was formed: the American Woman Suffrage Association, led by women's rights and antislavery activist Lucy Stone (1818–1893) and her husband Henry Brown Blackwell (1825–1909). In 1870 the common cause of the two groups was strengthened by the passage of the Fifteenth Amendment, which gave all men, regardless of race, the right to vote. When the two organizations joined forces in 1890, they formed the National American Woman Suffrage Association (NAWSA).

The founders of the American women's movement were followed by a new generation of leaders, which included Stanton's daughter, Harriot Eaton Blatch (1856–1940), as well as Alice Paul (1885–1977), who founded the organization that became the National Woman's Party, and organizer and editor Lucy Burns (1879–1966), who worked closely with Paul.

The suffragists appealed to middle-class and working-class women, as well as to students and radicals. They waged campaigns at the state level, distributed literature, organized meetings, made speeches, and marched in parades. They also lobbied federal legislators, picketed, and chained themselves to the White House fence. When jailed, many resorted to hunger strikes and were sometimes met with cruel treatment. The suffragists' fight was a fierce one; the opposition played on the widespread belief that if given the right to vote, women would neglect the traditional duties of wife and mother.

The movement gained strength during World War I (1914–18). As men went off to fight the war in Europe, the women at home demonstrated themselves to be intelligent and involved citizens in the life of the country. A wartime suffragist poster declared in one long column, "As a war measure, the country is asking of women service as…farmers, mechanics, nurses, doctors, munitions workers, mine workers, yeomen, gas makers, bell boys, messengers, conductors, motormen, army cooks, telegraphers, ambulance drivers, advisors to the council of national defense," and in another short column it stated, "As a war measure, women are asking of the country…the vote." By 1918 support for woman suffrage was broad. That year Congress proposed a constitutional amendment stating that the "right of citizens of the United States to vote shall not be denied or abridged by the United States or by any State on account of sex." It was passed, as the proposed Nineteenth Amendment, in the House in 1918 and in the Senate in 1919. The amendment was approved by the required number of state legislatures on August 18, 1920, when Tennessee ratified it.

What was the **Night of Terror**?

It is a little-known episode in the American suffragist movement that took place on November 14, 1917, at the Occoquan Workhouse in Lorton, Virginia.

After President Woodrow Wilson (1856–1924) took office in January 1917, activists began picketing daily outside the White House, demanding the right to vote for women. It was the first time in history that demonstrators had marched at the White House. The suffragists carried banners that read, "Mr. President, how long must women wait for liberty?" and, more radically, "Kaiser Wilson: 20,000,000 American women are not self-governed." Their intent was to expose the government's hypocrisy: In April the United States had entered World War I (1914–18) in an effort to guarantee democracy abroad, yet democracy did not exist at home—where the entire female population remained disenfranchised.

In June police began arresting demonstrators on minor charges, such as obstructing traffic. But the arrests did nothing to deter the suffragists. Upon their release from prison, they returned to protest at the White House gates. In all, 168 women, including Alice Paul (1885–1977) and Lucy Burns (1879–1966) of the National Woman's Party, were arrested.

On the night of November 14, 1917, guards took 33 protesters to the Occoquan Workhouse to be held. Previously, the women had been subject to forced feedings and solitary confinement at Occoquan, but this time new cruelties awaited them. They were beaten, dragged, choked, and handcuffed. Word leaked out about the atrocities committed at the workhouse. Less than two weeks later, a judge ruled that the women had been brutally treated, yet they had done nothing more than exercise their constitutional right to free speech. The women returned to their fight, now with more weight of public opinion behind them. Nevertheless, it was three long years before the Nineteenth Amendment was adopted, guaranteeing women the right to vote.

In 1982 a historical marker was placed on the prison grounds in tribute to the brave women who endured Occoquan's Night of Terror.

Who was **Alice Paul**?

Alice Paul (1885–1977) was a groundbreaking feminist before the word "feminist" came into fashion. The Mount Laurel, New Jersey, institute named in her honor describes her as "the architect of some of the most outstanding political achievements on behalf of women in the twentieth century."

Paul was born in 1885 to Quaker parents who instilled in her a belief in gender equality. After completing high school the top in her class, Paul graduated from Swarthmore College in 1905 and began work toward an advanced degree. In 1906 she traveled to England, where she continued her studies, did social work, and became actively involved in the suffrage movement. She was arrested three times for her involvement in protests.

In 1916, when the American women's suffrage movement was divided and "dead in the water," Paul founded the National Woman's Party (NWP), an organization that spearheaded the campaign for national woman's suffrage and that continued working for women's rights and equality into the twenty-first century. Paul's leadership of the suffrage movement was critical in the passage of the Nineteenth Amendment (1920), which guaranteed women the right to vote; she organized thousands of activists to put enormous pressure on the White House and Congress. Paul employed what was then considered a most unladylike strategy of "sustained, dramatic, nonviolent protest." The suffrage campaign was characterized by national speaking tours, marches, and pickets (including the first ever at the White House). When protesters were arrested, they sometimes endured brutal prison conditions and staged hunger strikes.

After passage of the Nineteenth Amendment, Paul continued her studies, adding to her master's degree in social work (1907) a doctorate in economics (1912). She earned three more advanced degrees, culminating in a doctor of law degree in 1927 from American University. Called a brilliant political strategist, the forward-thinking Paul authored the first equal rights amendment for women, which she introduced to Congress in 1923. In 1942 she became chairperson of the National Woman's Party. She later added language of gender equality to the charter for the United Nations as well as the 1964 Civil Rights Act. After a life of courageous activism on behalf of women, Paul died in 1977.

Who was **Emmeline Pankhurst**?

Pankhurst (1858–1928), a key figure in the women's suffrage movement, was a militant reformer who waged a decades-long battle to win the vote for women in Great Britain.

Pankhurst's sometimes radical campaign greatly influenced her American counterparts. Though she held various municipal offices and was married to an influential barrister (Richard Marsden Pankhurst), she worked for change primarily through the organizations she founded. In 1889 she organized the Women's Franchise League, and five years later the group's work secured the right of all women (married and unmarried) to vote in local elections. She went on to found the Women's Social and Political Union in 1903. The union was known for its extreme tactics. The British suffragist movement culminated in 1928 with the passage of the Representation of the People Act, which gave all women the right to vote in elections. Pankhurst died later that year.

Have all nations of the world granted **women the right to vote**?

No, in a few nations women remain disenfranchised. By the 1990s women had a legal right to vote everywhere in the world except in six Middle Eastern countries (Saudi Arabia, Kuwait, Bahrain, Oman, Qatar, and United Arab Emirates) as well as in Brunei, a small oil-rich country in Southeast Asia. In 2001 Bahrain extended equal voting rights to women, and in 2003 Qatar did the same. But a traditional interpretation of Islamic law kept Muslim women from voting in a few conservative Persian Gulf states. Kuwaiti lawmakers proposed limited women's suffrage in spring 2005, but the measure was not approved.

There was mounting pressure, from inside and outside the Muslim world, for this to change. The issue was an important focal point for the Human Rights Watch, an international watchdog group. In October 2004 a high-ranking Egyptian cleric spoke out on the contentious issue, saying, "It is the right of a Muslim woman to vote for and speak her opinion about whoever serves public or greater interests." He went on to clarify that he was "talking about Muslim women in all Muslim countries, in Egypt, Kuwait, and others."

Suffrage for women has been won country by country and decade by decade. Further, within many countries, rights have been extended only gradually; for example, beginning with local elections. The first nations to extend broad voting rights to women were New Zealand in 1893, Australia and South Wales in 1902, and Finland in 1906. In the 1910s women in several European and Scandinavian nations, including Austria, Denmark, Germany, Luxembourg, the Netherlands, Norway, Poland, and Russia, won the right to vote—largely as a result of World War I (1914–18). The 1920s added not only the United States and the United Kingdom (to a voting status equal to men), but about a dozen other nations, including the former Czechoslovakia and Sweden. Every decade since added more nations to the tally, so that as of 2004 only a few nations denied women the right to vote.

How old is **feminism**?

Feminists—people who believe that women should have economic, political, and social equality with men—have existed throughout history; such women are often described in literature and by history as being "women before their time." But as a movement, feminism, which is synonymous with the women's rights movement, did not get under way until the mid-1800s, when women in the United States and Great Britain began organizing and campaigning to win the vote. Early feminists (and feminists today) were likely influenced by the revolutionary work titled *A Vindication of the Rights of Woman,* published in 1792 by British author and educator Mary Wollstonecraft (1759–1797; her daughter was writer Mary Shelley of *Frankenstein* fame). Wollstonecraft attacked the convention of the day, charging that it kept middle-class and upper-class women in a state of ignorance, training them to be useless. A staunch promoter of education (she was self-educated), Wollstonecraft is credited with being the first major philosophical feminist.

What is the **ERA**?

The ERA stands for the equal rights amendment, a constitutional amendment proposed by Congress in 1972. It stated that "equality of rights under the law shall not be denied or abridged by the United States or any state on account of sex." In proposing the amendment, Congress gave the states 10 years in which to ratify it. But in 1982 only 35 of the necessary 38 states had approved the amendment, which then died. The failure to ratify the ERA was the result of disagreement over how the language would be interpreted. Supporters believed the amendment would guarantee women equal treatment under the law; opponents feared the amendment might require women to forfeit the financial support of their husbands and require them to serve in the military.

How did Betty Friedan's ***Feminine Mystique*** launch the modern women's rights movement?

American author and activist Betty Friedan's (1921–) landmark work was published in 1963. She argued that in the postwar era American society expected women only to be devoted housewives and mothers. This was the ideal, the mystique, that was both promoted and accepted. In the *Feminine Mystique* Friedan challenged this prevailing notion, causing many women to reexamine their lives. After all, during World War II (1939–45) women had ably stepped into the workplace to keep industry running as the men went to war. When the servicemen came back, women returned to the home and were not expected to seek careers. But after nearly two decades of accepting the status quo, Friedan asked the question, "If women could successfully hold jobs, why shouldn't they?" Assessing their happiness, many women opted to pursue work outside the home. The modern women's movement had begun. Soon women were organizing to promote social and political reforms to do away with discrimination in the workplace and eliminate barriers to entry in education and politics. Friedan herself helped found the National Organization for Women (NOW) in 1966; the association grew rapidly and continues to fight for women's equality in the twenty-first century.

What is **Title IX**?

Considered one of the biggest successes of the modern women's movement, Title IX is part of the Education Amendments of 1972, federal legislation that prohibits any school or college that receives federal funds from discriminating on the basis of sex. The law applies to all aspect of education, including admission, athletics, and curriculum.

THE BIRTH CONTROL MOVEMENT

How did the **birth control movement** get started?

The decline in death rates, which has meant an overall increase in the world population, gave rise to the birth control movement. Scientific advances during the eighteenth and nineteenth centuries resulted in better food supplies, the control of diseases, and safer work environments for those living in developed countries. These improvements combined with progress in medicine to save and prolong human lives. During the 1800s, the birth rate, which in earlier times had been offset by the death rate, became a concern to many who worried that population growth would outstrip the planet's ability to provide adequate resources to sustain life.

In 1798 British economist and sociologist Thomas Robert Malthus (1766–1834) published his *Essay on the Principle of Population,* arguing that populations tend to increase faster than do food supplies. He thereby concluded that poverty and suffering are unavoidable. Malthus viewed only war, famine, disease, and "moral restraint" as checks on population growth. In spite of or because of Malthus's assertions, during the 1800s the idea of birth control as a practical method to keep population growth in check gained momentum.

Early in the 1900s the movement found a leader in American Margaret Higgins Sanger (1883–1966), whose personal experience as a nurse working among the poor had convinced her that limiting family size is necessary for social progress. She became convinced that unwanted pregnancy should be avoided by using birth control methods. It was—and remains—controversial. Even though the distribution of birth control information was illegal at the time, Sanger advised people on the subject. In 1914 she founded a magazine called *The Woman Rebel,* and she sent birth control information through the mail. She was arrested and indicted. But she was not deterred. In 1916 in Brooklyn, New York, Sanger founded the first birth control clinic in the United States. In 1921 she organized the first American Birth Control Conference, held in New York. That same year she founded the American Birth Control League, which later became the Planned Parenthood Federation of America. As public support for the movement increased, Sanger succeeded in getting laws passed that allowed doctors to disseminate birth control information to their patients.

In other countries, Sanger's work inspired similar movements, but developed nations continue to have lower birth rates than do developing nations. With the world population exceeding 5.5 billion, the fear of overpopulation had prompted new interest in birth control.

What is the **zero population growth movement**?

Abbreviated ZPG, it's an international movement for population control. It had its beginnings in an organization formed in 1968 called Zero Population Growth, which was established with the help of American biologist Paul Ehrlich (1932–). In his book *The Population Bomb* (1968), Ehrlich argues that the world's population is growing at a pace that will outstrip its supply of natural resources, reaching the conclusion that overpopulation will inevitably cause widespread starvation and death. While his critics charge that this is an oversimplification of the situation, Ehrlich and others who subscribe to his views argue that overpopulation may not be the only factor leading to a severe shortage of natural resources around the globe, but it surely is the most important. The movement urges parents to have no more than two children; at this rate (called a replacement rate) there would be no population growth. In the late 1990s researchers reported that in the United States and in much of Europe, the birth rate is below replacement level, while in some of the world's poorer countries, the average birth rate can be as high as eight children per woman.

POPULISM, PROGRESSIVISM, AND THE LABOR MOVEMENT

What was **populism**?

A commoners' movement, in the United States populism was formalized in 1891 with the founding of the Populist Party, which worked to improve conditions for farmers and laborers. In the presidential election of 1892, the party supported its own political candidate, the former (third-party) Greenback candidate James B. Weaver (1833–1912). Though Weaver lost, the Populists remained a strong force. In the next presidential election, of 1896, they backed Democratic Party candidate William Jennings Bryan (1860–1925), a self-proclaimed commoner who was sympathetic to the causes of the Farmers' Alliances and of the National Grange (reform-minded agricultural organizations) as well as the nation's workers. Bryan lost to William McKinley (1843–

1901), and soon after the election the Populist Party began to fall apart, disappearing altogether by 1908. Nevertheless, the party's initiatives continued to figure in the nation's political life for the next two decades and many populist ideas were made into laws, including the free coinage of silver and government issue of more paper money ("greenbacks") to loosen the money supply, adoption of a graduated income tax, passage of an amendment allowing for the popular election of U.S. senators (the Constitution provided for their election by the state legislatures), passage of antitrust laws (to combat the monopolistic control of American business), and implementation of the eight-hour workday. Since the early 1900s political candidates and ideas have continued to be described as populist, meaning they favor the rights of and uphold the beliefs and values of the common people.

What was the **Progressive movement**?

It was a campaign for reform on every level—social, political, and economic—in the United States. It began during the economic depression that was brought on by the Panic of 1873 and lasted until 1917, when Americans entered World War I (1914–18).

During the first 100 years of the U.S. Constitution (1788), federal lawmakers and justices proved reluctant to get involved in or attempt to regulate private business. This policy of noninterference had allowed the gap to widen between rich and poor. The turn of the century was a time in America when early industrialists built fantastic mansions while many workers and farmers struggled to earn a living; when tenement houses sprang up in urban areas to meet (albeit horribly inadequately) the housing needs made present by a steady stream of immigrants; and when labor unions, which had only recently begun to organize, were beset by outbreaks of violence, hurting their fight for better treatment by employers. Observing these problems, progressive-minded reformers, comprised largely of middle-class Americans, women, and journalists (the so-called "muckrakers"), began reform campaigns at the local and state levels, eventually effecting changes at the federal level.

Progressives favored many of the ideas that had previously been espoused by the Populists, including antitrust legislation to bust up the monopolies and a graduated income tax to more adequately collect public funds from the nation's well-to-do businessmen. Additionally, Progressives combated corrupt local governments; dirty and dangerous working conditions in factories, mines, and fields; and inner-city blight. The minimum wage, the Pure Food and Drug Act, and Chicago's Hull House (which served as "an incubator for the American social work movement") are part of the legacy of the Progressive movement.

When did the **U.S. labor movement** begin?

It began in the early 1800s, when skilled workers, such as carpenters and blacksmiths, banded together in local organizations with the goal of securing better wages. By the time fighting broke out in the Civil War (1861–65), the first national unions had been founded—again, by skilled workers. However, many of these early labor organizations struggled to gain widespread support and soon fell apart. But by the end of the century, several national unions, including the United Mine Workers (1890) and the American Railway Union (1893), emerged. In the last two decades of the 1800s, violence accompanied labor protests and strikes while opposition to the unions mounted. Companies shared blacklists of the names of workers suspected of union activities; hired armed guards to forcibly break strikes; and retained lawyers to successfully invoke the Sherman Anti-Trust Act (of 1890) to crush strikes—lawyers argued that strikes interfered with interstate commerce, which was declared illegal by the Sherman legislation (which had not been the intent of the lawmakers).

In the early decades of the 1900s, unions made advances, but many Americans continued to view organizers and members as radicals. The climate changed for the unions during the Great Depression (1929–39). With so many Americans out of work, many blamed business leaders for the economy's failure and began to view the unions in a new light—as organizations to protect the interests of workers. In 1935 the federal government strengthened the unions' cause in passing the National Labor Relations Act (also called the Wagner Act), protecting the rights to organize and to bargain collectively (when worker representatives, usually labor union representatives, negotiate with employers). The legislation also set up the National Labor Relations Board (NLRB), which still works today to penalize companies that engage in unfair labor practices. The constitutionality of the act was challenged in court in 1937, but the Supreme Court upheld the legislation.

The unions grew increasingly powerful over the next decade: By 1945 more than one-third of all nonagricultural workers belonged to a union. Having made important gains during World War II (1939–45), including hospital insurance coverage, paid vacations and holidays, and pensions, union leaders continued to urge workers to strike to gain more ground—something leaders felt was the worker's right amidst the unprecedented prosperity of the postwar era. But strikes soon impacted the life of the average American: Consumers faulted the unions for shortages of consumer goods, suspension of services, and inflated prices. Congress responded by passing the Labor-Management Relations Act (or the Taft-Hartley Act) in 1947, which limits the impact of unions by prohibiting certain kinds of strikes, setting rules for how unions could organize workers, and establishing guidelines for how strikes that may impact the nation's health or safety are to be handled.

What was **the first big union**?

The first national union of note was the Knights of Labor, founded by garment workers in Philadelphia, Pennsylvania, in 1869. Recruiting women, blacks, immigrants, and unskilled and semiskilled workers alike, the Knights of Labor's open-membership

Who were the Molly Maguires?

They were Irish-American coal miners in eastern Pennsylvania who organized a secret society in 1854. The Molly Maguires aimed to wage a campaign of violence against mine owners and operators. The name of the group came from a society in Ireland that used physical force to fight ruthless landlords. The American miners became determined to defeat their oppressors at all costs. Their numbers grew, and in the decade following the Civil War (1861–65), the Molly Maguires were active both as agitators, and it would later be revealed, as assassins. In 1875 the group incited a coal miners strike, which was broken by the detective work of Irish-American James McParlan (1844–1919), a Pinkerton guard hired by Philadelphia and Reading Coal and Iron Company to infiltrate the Molly Maguires. McParlan revealed the identities of gunmen responsible for the deaths of nine mine company foremen. Several members of the secret society were arrested, tried and convicted (in 1876), and hanged (in 1877) for the crimes. American sympathies for the plight of the miners were diminished by the headlines proclaiming the terrorist activities of the Molly Maguires. The society dissolved by 1877. Their presence, however, was long felt in the anthracite coal fields of Pennsylvania, where company police monitored activities in the mines, and effectively intimidated many miners from organizing.

policy provided the organization with a broad base of support, something previous labor unions, which had limited membership based on craft or skill, lacked. The organization set its objectives on instituting the eight-hour workday, prohibiting child labor (under age 14), instituting equal opportunities and wages for women laborers, and abolishing convict labor. The group became involved in numerous strikes from the late 1870s to the mid-1880s.

At the same time, a faction of moderates within the organization was growing, and in 1883 it elected American machinist Terence Powderly (1849–1924) as president. Under Powderly's leadership, the Knights of Labor began to splinter. Moderates pursued a conciliatory policy in labor disputes, supporting the establishment of labor bureaus and public arbitration systems; radicals not only opposed the policy of open membership, they strongly supported strikes as a means of achieving immediate goals—including a one-day general strike to demand implementation of an eight-hour workday. In May 1886 workers demonstrating in Chicago's Haymarket Square attracted a crowd of some 1,500 people; when police arrived to disperse them, a bomb exploded and rioting ensued. Eleven people were killed and more than 1,000 were injured in the melee. For many Americans, the event linked the labor movement with anarchy. That same year several factions of the Knights of Labor seceded from the

Striking mill workers demonstrate in Gastonia, North Carolina, 1929.

union to join the American Federation of Labor (AFL). The Knights of Labor remained intact for three more decades, before the organization officially dissolved in 1917, by which time the group had been overshadowed by the AFL and other unions.

How old is the **AFL-CIO**?

The roots of the American Federation of Labor-Congress of Industrial Organizations (AFL-CIO), today a federation of national unions, date to 1881 when the Federation of Organized Trade and Labor Unions was formed in Pittsburgh, Pennsylvania, by trade union leaders representing some 50,000 members in the United States and Canada. Reorganizing in 1886, the association of unions changed its name to the American Federation of Labor (AFL) and elected Samuel Gompers (1850–1924) president.

Unlike the open-membership policy of the Knights of Labor (from whom the AFL gained numerous members in 1886), the AFL determined to organize by craft: At the outset, its member unions included a total of 140,000 skilled laborers. Gompers, who immigrated from England in 1863 and became the first registered member of the Cigar-Makers' International Union in 1864, had been active in labor for more than two decades. Once chosen as president of the AFL, Gompers remained in that office, with the exception of only one year, until his death in 1924. During the nearly 40-year period, he shaped the labor federation and helped it make strides by determining a general policy that allowed member unions autonomy. Unlike the Knights of Labor, which pursued long-term goals such as Knights leader Terence Powderly's abstract objective of making "every man his own master—every man his own employer," the AFL focused its efforts on specific, short-term goals such as higher wages, shorter hours, and the right to bargain collectively (when an employer agrees to negotiate with worker/union representatives).

In the 1890s the AFL was weakened by labor violence, which evoked public fears. A July 1892 strike at the Carnegie Steel plant in Homestead, Pennsylvania, turned into a riot between angry steelworkers and Pinkerton guards. The militia was called in to monitor the strike, which five months later ended in failure for the AFL-affiliated steelworkers. Nevertheless, under Gompers's leadership, membership of the AFL grew to more than 1 million by 1901 and to 2.5 million by 1917, when it included 111 national unions and 27,000 local unions. The federation collected dues from its members, creating a fund to aid striking workers. The organization avoided party politics, instead seeking out and supporting advocates regardless of political affiliation. The AFL worked to support the establishment of the U.S. Department of Labor (1913), which administers and enforces statutes promoting the welfare and advancement of the American workforce, and the passage of the Clayton Anti-Trust Act (1914), which strengthened the Sherman Anti-Trust Act of 1890, eventually delivering a blow to monopolies.

The CIO was founded in 1938. In the early 1930s several AFL unions banded together as the Committee for Industrial Organization and successfully conducted campaigns to sign up new members in mass-production facilities such as the automobile, steel, and rubber industries. Since these initiatives (which resulted in millions of new members) were against the AFL policy of signing up only skilled laborers by craft (the CIO had reached out to all industrial workers, regardless of skill level or craft), a schism resulted within the AFL. The unions that had participated in the CIO membership drive were expelled from the AFL; the CIO established itself as a federation in 1938, officially changing its name to the Congress of Industrial Organizations.

In 1955, amidst a climate of increasing anti-unionism, the AFL and CIO rejoined to form one strong voice. Today the organization has craft and industrial affiliates at the international, national, state, and local levels, with membership totaling in the millions.

Who were the **Wobblies**?

The Wobblies were the early radical members of the Industrial Workers of the World (IWW), a union founded in 1905 by the leaders of 43 labor organizations. The group pursued short-term goals via strikes and acts of sabotage as well as the long-term goal of overthrowing capitalism and rebuilding society based on socialist principles. One IWW organizer proclaimed that the "final aim is revolution." Their extremist views and tactics attracted national attention, making IWW and Wobblies household terms during the early decades of the twentieth century.

Founded and led by miner and socialist William "Big Bill" Haywood (1869–1928) and mine workers agitator Mary "Mother" Jones (1830–1930), the IWW aimed to unite all workers in a camp, mine, or factory for the eventual takeover of the industrial facility. The union organized strikes in lumber and mining camps in the West, in the steel mills of Pennsylvania, and in the textile mills of New England. The leadership advocated the use of violence to achieve its revolutionary goals and opposed mediation (negotiations moderated by a neutral third party), collective bargaining (bargaining between worker representatives and an employer), and arbitration (third-party mediation). The group declined during World War I (1914–18), when the IWW led strikes that were suppressed by the federal government. The organization's leaders were arrested and the organization weakened. Haywood was convicted of sedition (inciting resistance to lawful authority) but managed to escape the country. He died in the Soviet Union, where he was given a hero's burial for his socialist views.

The IWW never rose again to the prominent status of its early, controversial days. Many accounts of the group's history cite its demise in the 1920s. But, according to its own statement, the organization continued to "enjoy a more or less continuous existence" into the twenty-first century. As the IWW prepared to celebrate its one-hundredth anniversary in 2005, it continued to promote its original goal of organizing workers by industry rather than trade. Under the IWW's scheme, workers around the world would organize into one big union divided into six camps (or "departments"): agriculture and fisheries, mining and minerals, general construction, manufacture and general production, transportation and communication, and public service. In the early 2000s the IWW had a few dozen member unions in the United States, as well as branches in Australia, Japan, Canada, and the British Isles.

Who was **Eugene Debs**?

Debs (1855–1926) was a radical labor leader who in 1893 founded the American Railway Union (ARU), an industrial union for all railroad workers. Debs was a charismatic speaker, but he was also a controversial figure in American life around the turn of the century. In 1894 workers at the Pullman Palace Car Company, which manufactured railcars in Pullman, Illinois (near Chicago), went on strike to protest a significant reduction in their wages. Pullman was a model "company town" where the railcar manufacturer, founded by American inventor George W. Pullman (1831–1897) in 1867, owned all the land and buildings, and ran the school, bank, and utilities. In 1893, in order to maintain profits following declining revenues, the Pullman Company cut workers wages by 25 to 40 percent, but did not adjust rent and prices in the town, forcing many employees and their families into deprivation. In May 1894 a labor committee approached Pullman Company management to resolve the situation. The company, which had always refused to negotiate with employees, responded by firing the labor committee members. The firings incited a strike of all 3,300 Pullman workers. In support of the labor effort, Eugene Debs assumed leadership of the strike (some Pullman employees had joined the ARU in 1894) and directed all ARU members not to haul any Pullman cars. A general rail strike followed, which paralyzed transportation across the country. In response to what was now being called "Deb's rebellion," a July 2, 1894, federal court order demanded all workers to return to the job, but the ARU refused to comply. U.S. president Grover Cleveland (1837–1908) ordered federal troops to break the strike, citing it interfered with mail delivery. The intervention turned violent. Despite public protest, Debs, who was tried for contempt of court and conspiracy, was imprisoned in 1895 for having violated the court order. Debs later proclaimed himself a socialist and became leader of the American Left, running unsuccessfully for president as the Socialist Party candidate five times, in 1900, 1904, 1908, 1912, and 1920. He actively supported the causes of the International Workers of the World (IWW), a radical labor organization founded in 1905.

Why was the fire at the **Triangle Shirtwaist Factory** important to the labor movement?

The March 25, 1911, blaze, which killed 146 people (most of them women), prompted public outrage and led to the immediate passage of fire safety legislation and became a rallying cry for labor reforms.

The Triangle Shirtwaist Factory occupied the top three floors of a Manhattan office building. It was one of the most successful garment factories in New York City, employing some 1,000 workers, mostly immigrant women. But the conditions were hazardous: The space was cramped, accessible only via stairwells and hallways so narrow that people had to pass single-file; only one of the four elevators was regularly in service; the cutting machines in the workroom were gas-powered; scraps of fabric littered the work areas; the water barrels (for use in case of fire) were not kept full; and the no-smoking rule was not strictly enforced. In short, it was an accident waiting to happen.

When the fire broke out on a weekend (the cause is unknown since the building was charred so badly), about half of the employees were there. Smoke and fire, however, were not the only causes of death: In the panicked escape, people were trampled, fell in elevator shafts, jumped several stories to the pavement below, and were killed when a fire escape melted and collapsed.

While the fire happened during a time of labor reform, those reforms had not come soon enough to save the lives of the Triangle Shirtwaist Factory employees, who had been subjected to extremely poor and dangerous working conditions. The disaster became a rallying cry for the labor movement: Tens of thousands of people marched in New York City in tribute to those who had died, calling attention to the grave social problems of the day.

The World.

"Circulation Books Open to All"

154 KILLED IN SKYSCRAPER FACTORY FIRE; SCORES BURN, OTHERS LEAP TO DEATH.

700 WORKERS, MOSTLY GIRLS, TRAPPED; BODIES OF DEAD HEAP THE STREETS;

News of the March 1911 Triangle Shirtwaist Company fire made the front page of *The New York World;* 146 people died in the workplace disaster.

In New York State, the fire safety reforms for factories came right away: The legislature appointed investigative commissions to examine factories statewide, and 30 ordinances in New York City were enacted to enforce fire prevention measures. One of the earliest was the Sullivan-Hoey Fire Prevention Law of October 1911, which combined six agencies to form an efficient fire commission. Soon factories were required to install sprinkler systems.

The Triangle Shirtwaist Factory fire became an object lesson for the entire nation, prompting the consolidation of reform efforts. The much-needed labor reforms, which addressed the miserable working conditions, did not come until years later.

Who was **Cesar Chavez**?

Mexican American farm worker Cesar Chavez (1927–1993) was a labor union organizer and spokesperson of the poor. Born in Arizona, his family lost their farm when he was just 10 years old; they became migrant workers in California, where farm production—particularly of grape crops—depended on the temporary laborers. Chavez knew the migrant worker's life intimately, and as a young man he began working to improve conditions for his people. In 1962 he organized California grape pickers into the National Farm Workers Association. Four years later this union merged with another to form the United Farm Workers Organizing Committee, or UFWOC (the name was changed to the United Farm Workers of America, or UFW, in 1973). An impassioned speaker known for squishing bunches of grapes in his hands as he delivered his messages, Chavez went on to lead a nationwide boycott of table grapes, since growers had refused to accept the collective bargaining of the UFWOC. By the close of the 1970s, California growers of all crops had accepted the migrant workers' union, now called the United Farm Workers. Like Martin Luther King Jr., Chavez maintained that nonviolent protest was the key to achieving change.

COUNTERCULTURE, CONSUMERISM, AND THE ENVIRONMENT

What was the **beat movement**?

The post-World War II era bred unprecedented prosperity and an uneasy pace in the United States. Out of this environment rose the beat generation, alienated youths who rejected society's new materialism and threw off its "square" attitudes to reinvent "cool." The beat generation of the 1950s bucked convention, embraced iconoclasm, and attracted attention. Mainstream society viewed them as anarchists and degenerates. But many American youths listened to and read the ideas of its leaders, including writers Allen Ginsberg, Jack Kerouac (whose novel *On the Road,* 1957, was the bible of the Beat movement), William Burroughs, and Lawrence Ferlinghetti. The "beatniks," as they were dubbed by their critics, believed in peace, civil rights, and radical protest as a vehicle for change. They also embraced drugs, mystical (Eastern) religions, and sexual freedom—all controversial ideas during the postwar era. Beat writers and artists found their homes in communities like San Francisco's North Beach, Los Angeles's Venice Beach, and New York City's Greenwich Village. The movement merged—or some would argue, gave birth to—the counterculture movements of the 1960s, including the hippies. Beat literature is the movement's legacy.

Who were the **hippies**?

Most hippies of the 1960s and 1970s were young (15 to 25 years old), white, and from middle-class families. The counterculture (antiestablishment) movement advocated peace, love, and beauty. Having dropped out (of modern society) and tuned in (to their own feelings), these flower children were as well known for their political and social

Vietnam War protesters gather on 5th Avenue in New York City in 1968.

beliefs as they were for their controversial lifestyle: They opposed American involvement in the Vietnam War (1954–75) and rejected an industrialized society that seemed to care only about money; they favored personal simplicity, sometimes living in small communes where possessions and work were shared, or living an itinerant lifestyle, in which day-to-day responsibilities were few if any; they wore tattered jeans and bright clothing usually of natural fabrics, grew their hair, braided beads into their locks, walked around barefoot or in sandals, and listened to a new generation of artists including the Beatles, the Grateful Dead, Jefferson Airplane, Bob Dylan, and Joan Baez. Some hippies were also known for their drug use: Experimenting with marijuana and LSD, some hoped to gain profound insights or even achieve salvation through the drug experience—something hippie guru Timothy Leary told them was possible. New York City's East Village and San Francisco's Haight-Ashbury neighborhood became havens of the counterculture. The movement began on American soil, but was soon embraced elsewhere as well—principally Canada and Great Britain.

What happened to the hippies? The conflict in Vietnam ended, flower children grew older, drugs took their toll on some, and by 1980 there was no such thing as free love. Still, a few continued to lead an alternative lifestyle, while at the opposite end of the spectrum others bought back into the establishment. Still others adapted their flower-child beliefs to the ever-changing world around them and got on with their lives, working at jobs and raising children in the most socially and politically conscious way possible.

What are **"Nader's Raiders"**?

They were (and are) investigators working with American lawyer and consumer advocate Ralph Nader (1934–). With the help of his research team, Nader wrote the landmark work *Unsafe at Any Speed* in 1965; it charged that many automobiles weren't as safe as they should be or as consumers had the right to expect them to be. In part the book was responsible for passage in 1966 of the National Traffic and Motor Vehicle Safety Act, which set motor vehicle safety standards. Nader has continued his watchdog work, founding the Public Citizen organization, which researches consumer products, promotes consumer awareness, and works to influence legislators to improve consumer safety.

While Nader may be the most recognizable face of consumer advocacy, the movement's roots predate his activism. As the consumer age dawned at the end of the 1800s and early 1900s, when mass-production techniques came into wide use, some observers decried industry standards (or lack thereof) that put the public who used their products at risk. The muckraking journalists of the early twentieth century disclosed harmful or careless practices of early industry, raising awareness and bringing about needed reforms. Upton Sinclair (1878–1968), for example, penned the highly influential novel *The Jungle* (1906), revealing scandalous conditions of meat-packing plants. The public was outraged; upon reading Sinclair's work, President Theodore Roosevelt (1858–1919) ordered an investigation. Finding the novelist's descriptions to be true to life, government moved quickly to pass the Pure Food and Drug Act and the Meat Inspection Act that same year (1906). Industry watchdogs continued their work in the early decades of the century: In 1929 Consumers' Research Inc. was founded, and in 1936, the Consumers Union was formed; both of these independent organizations test and rate consumer products. (Consumers Union publishes its reports in the monthly magazine *Consumer Reports.*)

Such consumer advocacy has served to heighten public awareness, compelling industry to make changes and improving the safety of products in general. Abuses still occur, but at roughly 100 years into the consumer age, the work of consumer watchdogs like Nader's Raiders has made the consumer experience far less risky than it once was.

What did ***Silent Spring*** have to do with the environmental protection movement?

The 1962 work by American ecologist Rachel Carson (1907–1964) cautioned the world on the ill effects of chemicals on the environment. Carson argued that pollution and the use of chemicals, particularly pesticides, would result in less diversity of life. The best-selling book had wide influence, raising awareness of environmental issues and launching green (environmental protection) movements in many industrialized nations.

What is the **Kyoto Protocol**?

It is an environmental agreement signed by 141 nations that agree to work to slow global warming by limiting emissions, cutting them by 5.2 percent by 2012. Each nation has its own target to meet. The protocol was drawn up December 11, 1997, in the ancient capital of Kyoto, Japan, and went into effect on February 16, 2005. The United States is not among the signatories: American officials said the agreement is flawed because large developing countries including India and China were not immediately required to meet specific targets for reduction. Upon the protocol's enactment, Japan's prime minister called on non-signatory nations to rethink their participation, saying that there was a need for a "common framework to stop global warming." Environmentalists echoed his call to action.

NATURAL AND MAN-MADE DISASTERS

MOTHER NATURE

How did ancient societies interpret **catastrophic weather** events?

Different cultures developed wholly unscientific explanations for dramatic weather events or other natural phenomena—explanations typically rooted in the existing mythology or folklore of its people. For example, the ancient Maya (in Mexico's Yucatan Peninsula and in parts of Central America) believed that earthquakes were the gods' way of thinning out an overcrowded population. Indians in central Mexico are believed to have worshiped the grasshopper—or locust—after swarms destroyed their crops. One Japanese myth maintained that the entire island string rested on the back of a giant catfish who would grow restless and flop around when the gods were displeased, resulting in an earthquake. According to Hawaiian myth, the volcano goddess Pele causes Mount Kilauea to erupt whenever she has a temper tantrum.

What are the **largest known volcanic eruptions** in history?

Scientists measure volcanic eruptions by the amount of material that a volcano ejects into the atmosphere. Based on this measurement system, the largest eruptions include (in descending order of strength) one at Yellowstone Park in the United States, c. 600,000 B.C.; another at Toba, Indonesia, about 74,000 B.C.; a Tambora, Indonesia, eruption in A.D. 1815; Santorini, Greece, in 1470 B.C.; Laki, Iceland, in A.D. 1783 (which also produced the largest known lava flow in recorded history); and another in Indonesia, at Krakatau, in A.D. 1883.

The eruption in Yellowstone is hard to fathom: The volcano (which would have been located in present-day Wyoming) left a crater that measures 30 by 45 miles and released about 10,000 cubic kilometers of material into the atmosphere. To put this into perspective, consider that the next largest eruption, that at Toba, released one-tenth that amount, or 1,000 cubic kilometers. The one at Tambora released one-tenth of the Toba amount, 100 cubic kilometers. All of the others released about 10 cubic kilometers of earth debris into the atmosphere.

The May 18, 1980, eruption of Mount St. Helens in southwestern Washington State is also considered among the largest known eruptions in history and is the largest eruption in the modern history of the 48 contiguous Unites States. Mount St. Helens released a comparatively small amount of material, one cubic kilometer, but the damage was great after the volcano erupted. Much of the region was blanketed in ash, miles of forest were devastated, and the North Fork of the Toutle River was laden with ash and other volcanic debris up to 600 feet deep. The eruption claimed 57 lives.

What were the **deadliest volcanic eruptions** in history?

The April 5, 1815, eruption of Mount Tambora in Indonesia was the deadliest yet, killing some 92,000 people. Another Indonesian eruption later in the nineteenth century—this time at Karkatoa—claimed 36,417 lives in late August 1883.

On the French island of Martinique in the West Indies, Mount Pelee erupted on August 30, 1902, and more than 29,000 people perished. A relatively recent and deadly eruption was that on November 13, 1985, at Nevada del Ruiz, Columbia; it claimed 23,000 lives.

Iceland's Skaptar volcano erupted in 1783. While the number of lives lost may not qualify it for inclusion on a short list of "deadliest" volcanoes, the human toll was great indeed: 20 percent of the country's population died.

Because of population growth, today more people live closer to volcanoes—both active and inactive. As a result, the number of volcano-related deaths has increased: Between the years 1600 and 1900, the estimated average death toll per year due to volcanoes was 315. Since the beginning of the twentieth century, an average of 845 people have died each year because of volcanic eruptions.

What is the **strongest earthquake ever** measured?

It was one that shook Chile on May 22, 1960; it measured 9.5 on the Richter scale. Two thousand died, 3,000 were injured, and 2 million were left homeless. Damage was $550 million. The quake also spawned tsunamis (seismic waves), which claimed 61 lives in Hawaii, 138 in Japan, and 32 dead or missing in the Philippines.

What was the **most damaging earthquake** in recent history?

It was a July 28, 1976, quake that rocked the Chinese city of Tangshan at four o'clock in the morning. In less than a minute, 89 percent of the homes and 78 percent of the indus-

What is the strongest earthquake recorded in an urban area?

It is believed to be a November 1, 1755, earthquake that struck Lisbon, Portugal. The quake may have registered at least 9.0 on the Richter scale and lasted six or seven minutes. The port city was demolished, and more than 60,000 people perished. It was felt as far away as Sweden and generated a giant wave (tsunami) that struck the West Indies in the Caribbean Ocean. The catastrophe in Lisbon generated an intense debate among European philosophers who tried to explain why God destroyed that particular city, then the seat of the Holy Inquisition, during High Mass on All Saints' Day.

trial buildings were destroyed, killing 250,000 people, according to the official reports. However, international observers believe the death toll was even higher, about 750,000, which means that the quake claimed three-fourths of the area's total population.

A quake had not occurred in that region in six centuries, and the area was considered to be at low risk for earthquakes. Consequently, the building codes in the region were not stringent enough for the structures to withstand the force of the quake.

How disastrous was the **San Francisco earthquake**?

The quake of 1906 struck at 5:12 A.M. on April 18, and registered 8.3 on the Richter scale. Twenty seconds of trembling were followed by 45 to 60 seconds of shocks. The quake cracked water and gas mains, which resulted in a fire that lasted three days and destroyed two-thirds of the city. The destruction and loss of lives were great: As many as 3,000 (of San Francisco's 400,000 people) were killed; the entire business district was demolished; three out of five homes had either crumbled or burned; 250,000 to 300,000 people were left homeless; and 490 city blocks were destroyed.

The quake was a milestone for American journalism: The offices of the city's newspapers, the *Examiner* (owned by William Randolph Hearst; 1863–1951), the *Call,* and the *Chronicle* had all burned. But the first day after the disaster, the three papers joined forces across the bay in Oakland to print a combined edition, the *California Chronicle-Examiner.* Across the country, Will Irwin (1873–1948) of the *New York Sun,* who had been a reporter and editor at the *San Francisco Chronicle* from 1900 to 1904, wrote a story titled "The City That Was," which he completed from memory alone. It was picked up by papers around the country and became a classic of journalism. The San Francisco tragedy demonstrated the newfound ability of the American press to create an instant national story out of a local event.

The Bay Area was hit again by a sizeable quake in 1989. As millions tuned in to watch the World Series at Candlestick Park outside San Francisco, the TV cameras

People on Sacramento Street, in San Francisco, watch smoke rise from fires after a severe earthquake hit the city in 1906.

began to shake. Because of media coverage of the baseball game, the earthquake had literally been broadcast live around the world. Once again, fires resulted from broken gas mains, and the damage was extensive. The so-called Loma Pietra quake registered 7.1 on the Richter scale, claimed 67 lives, and damaged $15 billion worth of property. San Francisco's Marina District was particularly hard hit—at least in part due to the fact that the area was built largely on landfill, including debris from the 1906 quake. The San Francisco earthquake of 1906 remains the worst to ever hit an American city.

What were the worst **earthquakes of the past century**?

In order of magnitude on the Richter scale they were: Chile, 1960, 9.5; Prince William Sound, Alaska, United States, 1964, 9.2; Andreanof Islands, Alaska, United States, 1957, 9.1; Kamchatka, northeast Russia, 1952, 9.0; off the coast of northern Sumatra, Indonesia, 2004, 9.0; off the west coast of Ecuador, 1906, 8.8; Rat Islands, Alaska, United States, 1965, 8.7; Assam, India, and Tibet, 1950, 8.6; Kamchatka, northeast Russia, 1923, 8.5; Banda Sea, Malay Archipelago, 1938, 8.5; and Kuril Islands, off the east coast of Asia (extending from Russia in the north to Japan in the south), 1963, 8.5.

How frequently have **tsunamis** occurred throughout the **world**?

Tsunamis typically occur about every six years in the Pacific Ocean, and most often during March, August, and November. Although sometimes called tidal waves, tsunamis are created not by tides but by seismic movements (earthquakes), which produce chains of waves that move across the water at terrific speeds of more than 500 miles per hour. Upon reaching shallow water, the waves grow in height, sometimes to 100 feet or more, as was the case in 1883 when tsunamis reaching up to 130 feet hit an Indonesian island, destroying more than 150 villages and claiming some 36,000 lives.

In ancient times, it is believed that a tsunami destroyed the Minoan Greek culture, that of a people who lived on the island of Crete (in the Mediterranean Sea). In about 1450 B.C. Crete was struck by a 200-foot tsunami, which either demolished the island or weakened the population such that they could be taken over by the Mycenaeans, who were Greek mainlanders.

While tsunamis are known to strike along the Pacific Rim, damage has been minimized by sophisticated instruments that help meteorologists monitor and predict disastrous weather, alerting the public to evacuate from areas of possible danger. Such systems did not exist for the Indian Ocean when a 9.0 earthquake off the coast of the Indonesian island of Sumatra struck on December 26, 2004.

How do the 2004 **Southeast Asia tsunamis** rank among natural disasters?

The Southeast Asia tsunamis killed more people than any tsunami ever recorded. The series of seismic waves that rushed across the Indian Ocean on December 26, 2004, caused damage of biblical proportions and prompted a humanitarian rescue and aid effort on an unprecedented scale. That morning a 9.0 earthquake occurred off the northwestern tip of the island of Sumatra, Indonesia. Witnesses to the tsunamis reported that following the earthquake, ocean waters receded from shorelines hours before the giant waves roared in, washing over islands and sweeping through coastal villages in 12 countries, including Indonesia, Myanmar, India, and Sri Lanka. The waves struck as far west as the coast of Africa.

More than 150,000 people died in the disaster; Indonesia's death toll alone surpassed 85,000. The international response was immediate and reached into the billions of dollars. Nonetheless relief efforts were hampered by remote island locations, the destruction of infrastructure, and ongoing conflicts in some areas. In the weeks following the tsunamis, officials recognized that the true death toll would take time to be known, since survivors had yet to be interviewed about relatives and friends who remained missing. It was expected that many had been washed out to sea and thus had not been counted in the initial death toll, which was based on body counts. A preliminary report from the World Bank put the damages at $4.5 billion in Indonesia alone. But officials acknowledged that it would take months to calculate damages.

Survivors wander amidst the rubble of Banda Aceh, Indonesia, which was devastated in late December 2004 when earthquake-triggered tsunamis swept across Southeast Asia.

The earthquake that struck the morning of December 26, 2004, was the third-biggest earthquake in the past 100 years (and the biggest since 1964, when a 9.2-magnitude temblor occurred off Alaska). Scientists believe that the Southeast Asia quake occurred about 6.2 miles beneath the ocean floor and caused a great protrusion in the sea bed, generating waves that moved across the ocean in the early morning hours. Though probably not huge when they were out at sea, the waves grew higher as they approached shore, as tremendous volumes of water were forced to the surface.

Has a **tsunami** ever hit the **United States**?

Yes, in fact the tallest tsunami ever recorded hit Alaska in 1964; it was 220 feet high. The March 28 event was brought on by an earthquake—the Great Alaskan Earthquake in Prince William Sound—measuring 9.2 on the Richter scale. The seismic wave (tsunami) hit the southwest part of the state and claimed 107 lives.

Hawaii also sees an occasional tsunami, or giant wave, though the most remarkable one hit the islands before they became a state. On April 1, 1946, all of the water drained from the three-mile-wide harbor at Hilo, which was immediately followed by a tsunami that rushed onshore, destroying the waterfront. The process repeated itself twice and resulted in the death of more than 150 people in Hawaii.

What have been the deadliest **tropical storms** in the world?

The deadliest tropical storms are not hurricanes but rather the unnamed "super cyclones" that sweep out of the Bay of Bengal (in the Indian Ocean), striking the densely populated Indian subcontinent. These storms have been known to kill 100,000 people or more. The fatalities numbered more than 300,000 when a 1970 cyclone struck East Pakistan (Bangladesh). That storm is still the deadliest cyclone to hit the region. But more recent Indian cyclones, which usually occur in April through June and September through November, have sometimes reached "super cyclone" status. An April 1991 storm packing 160-mile-per-hour winds and 20-foot waves swept over Bangladesh's low-lying coastal plain. There was no place for residents to seek shelter from the advancing sea. An estimated 140,000 people perished and 10 million were left homeless. Property damage climbed to more than $2 billion. In October 1999 another devastating cyclone struck the Bay of Bengal region; it was the strongest and deadliest since the 1991 disaster: 10,000 died and more than a million people lost their homes.

What was the **worst hurricane in U.S.** history?

Hurricane Katrina, which hit the Gulf Coast in late August 2005, was not only the most disastrous hurricane in U.S. history, it was the nation's worst single weather disaster. Though not the strongest possible hurricane when it made landfall (Katrina had weakened from a category 5 storm to a category 4 storm just before it struck the Gulf

Coast), Katrina was a monster: The storm stretched about 200 miles in diameter, packed winds up to 145 miles per hour, produced torrential rain and huge waves, spawned twisters throughout the region, and pushed up a 28-foot storm surge—a surge usually found only in category 5 hurricanes.

Katrina moved ashore on the Gulf Coast on Monday, August 29. In anticipation of the hurricane, New Orleans, which sits below sea level, had been evacuated. But there were still tens of thousands who stayed—out of necessity (such as law enforcement and healthcare workers), because they were unable to evacuate, or because they chose not to leave. Some 23,000 of those who stayed holed up in the aging Superdome sports arena, which was set up as an emergency shelter. The structure barely withstood the lashing winds of Katrina, which blew off portions of the dome. On Tuesday morning, after Katrina ravaged Louisiana, Mississippi, and Alabama, before weakening as it moved inland, officials and news reporters generally agreed that there was unbelievable damage all along the coast, but New Orleans had "dodged a bullet": the Big Easy had not taken the worst of it. Gulfport and Biloxi, Mississippi, appeared the hardest hit. The devastation there was astonishing. As reports of damage began to be made, it became clear that about 90 percent of the structures along the Gulf Coast were destroyed and hundreds of thousands of people were displaced by Katrina. The death toll was not known, and officials conceded it would take time to determine. The storm had been so ruinous—of biblical proportions, some said—that rescue and recovery would take weeks and months. Later, when the full extent of damage began to be discovered, the recovery estimates were revised to years.

Later Tuesday, New Orleans' fate changed: levees that protect the city could not hold back a swollen Lake Pontchartrain; 80 percent of the city filled with water, 25 feet deep. Officials and volunteers could not get flood victims out fast enough, usually plucking them from rooftops or finding them in attics, where survivors sought refuge as waters rose. The city descended into chaos and lawlessness. Heart-wrenching images of human despair filled the media, touching people around the nation. Americans responded with donations of money, goods, and time. The American Red Cross launched the largest mobilization effort in its history. The Federal Emergency Management Agency (FEMA, a part of the Department of Homeland Security [DHS]), the Coast Guard (also part of the DHS), and the U.S. military struggled to keep pace with Katrina's aftermath all along the coast.

For all the effort, it was widely acknowledged that the government's response to the disaster was inadequate and late. On Friday, September 2, President George W. Bush said that the results had been unacceptable; he further promised to "make it right." While politicians and the people spoke out on the subject, some experts said the catastrophe had simply overwhelmed the system: Emergency programs across the nation had been set up to rely on local and state response first, backed by the federal government. In Katrina's case, the devastation was so great that the localities and states either were not able to respond at all or could not respond with enough help;

federal intervention had been needed sooner and in greater measure to alleviate human suffering and protect lives. Government officials all seemed of one accord, however: The fact-finding could wait; the victims could not. Cities and states across the nation sent resources to the Gulf Coast and set up emergency centers to receive storm refugees, whose needs were immediate (water, food, clothing, shelter, medicine, healthcare, and counseling) and long term (jobs, schools, and permanent housing).

On Wednesday, September 7, nine days after Katrina struck, the situation continued to unfold. The size and scope of the tragedy remained to be fully understood. New Orleans ordered a forced evacuation of the holdouts; the city remained flooded with toxic floodwaters as repairs were made in the levees and the hazardous water began to be pumped out. Efforts to reunite families, separated in the chaos, were ongoing. For many days, rescue workers all along the Gulf Coast had moved from house to house to find survivors; now it was a matter of trying to identify and count the dead. The death toll was expected to be in the thousands. The affected area was 90,000 square miles, or about the size of Minnesota. Property damage was projected to be at least $26 billion in insured losses and perhaps twice that amount in uninsured losses. (Katrina had also caused damages as it struck Florida as a category 1 hurricane on August 26; it later moved into the warm waters of the Gulf of Mexico and strengthened before striking the Gulf Coast.)

Katrina's aftermath was felt in every state of the union: volunteers shored up efforts to assist survivors who were being relocated in an effort to ease the burden on the afflicted region; schools across the nation opened doors to displaced students; fuel prices skyrocketed and natural gas and heating oil prices promised to follow, as offshore rigs and Gulf Coast refineries suffered; and most Americans worried about the government's response to the catastrophe. The disaster made an impact around the world as well, with some 95 countries offering assistance.

Before Katrina, the deadliest hurricane to strike the United States was an unnamed category 4 storm that struck Galveston, Texas, in 1900. The September 8 storm claimed at least 8,000 lives (some estimates place the number as high as 12,000).

Hurricane strength is measured on the Saffir-Simpson scale, where category 1 is the weakest (with sustained winds of at least 74 miles per hour and a storm surge of 4 to 5 feet above normal) and category 5 is the strongest (with sustained winds of more than 155 miles per hour and a storm surge higher than 18 feet). The United States has been hit by three category 5 storms since record-keeping began. The first was a Labor Day hurricane, which struck the Florida Keys in 1935; 408 people died. The second was Camille, which hit Mississippi and southeast Louisiana in 1969, claiming 256 lives. Hurricane Andrew, which struck south Miami-Dade County, Florida, in late-August 1992, was measured as a category 4 storm at the time, but the National Oceanographic and Atmospheric Agency (NOAA) reclassified it in 2002 as a category 5 storm. Andrew claimed more than 100 lives and devastated a wide area, mostly around the town of Homestead, Florida.

How long have **hurricanes** been given **names**?

The practice of naming hurricanes has a long history. According to the National Hurricane Center, for hundreds of years Caribbean storms were named according to the saint whose liturgical day it was when the storm hit. This became confusing when hurricanes struck on the same day but in different years, leading to such references as Hurricane San Felipe the Second. During World War II, the U.S. military began naming storms, giving them women's names. In 1951 the American weather services began naming Atlantic Ocean storms according to a phonetic alphabet (Able, Baker, Charley, etc.). Just a few years later forecasters returned to using women's names, and a new list of names was created for each Atlantic hurricane season (June 1–November 30). Storms in some areas of the Pacific began being named in 1959, and by 1964, all regions of the Pacific were using the naming convention. For each year, there is a list for each region of the world where tropical cyclones occur (Atlantic, Eastern North Pacific, Central North Pacific, Western North Pacific, etc.). The lists of names are agreed upon at international meetings of the World Meteorological Association. Establishing names for storms helps meteorologists track more than one storm at any given time, makes clear the communication of warnings, and facilitates study since the names of major hurricanes are retired to avoid confusion later. In 1979 equality was brought to the naming process, introducing men's names as well as multicultural names to each season's list.

Since each season's storms are named in alphabetical order, there is a preponderance of storms beginning with the letters A, B, and C. However, in 1995, Hurricane Opal ravaged the Florida Panhandle: Not since forecasters began naming hurricanes had they reached the letter O. The letters Q, U, X, Y, and Z are not used because there are so few names beginning with those letters.

What was the **Johnstown Flood**?

The city of Johnstown, in southwest Pennsylvania (east of Pittsburgh), has been the site of numerous floods, but the most disastrous one occurred on May 31, 1889, when the South Fork Dam on the Conemaugh River gave way, releasing a torrent of water. Since the dam was located some 14 miles into the Allegheny Mountains, the waters rushed into Johnstown at a rate of 50 miles per hour. The water hit with a force strong enough to have tossed a 48-ton locomotive one mile. It killed more than 2,000 people; some sources place the estimate as high as 5,000 lives lost. At the time of the flood, the city's population was about 30,000, meaning between 6 and 16 percent of the population died as a result of the disaster. In 1977 Johnstown was again the site of a disastrous flood, though advance warning systems helped minimize the loss of life to 77.

How does the **Great Flood of 1993** compare with other floods?

The flood, which occurred in the summer of 1993, was immense: So much of Iowa was under water that a satellite image (monitoring moisture on the earth) made the flood-

People stand on rooftops in Johnstown, Pennsylvania, after a massive flood destroyed most of the town and killed at least 2,000 people in May 1889.

ed area look like it was the size of Lake Michigan or Lake Superior. The area that was under water was roughly equivalent to twice the size of Massachusetts. Even so, the Great Flood resulted in a smaller water-covered area than did floods earlier in the century (in 1926 and 1973).

Due to heavy rainfall during the spring and summer of 1993, the Mississippi River widened to as much as seven miles at some points, and the Missouri River also overflowed its banks—despite the levee system that had been put in place by the federal government earlier in the century. The flood took 50 lives, displaced 85,000 people from their homes, destroyed 8,000 homes, damaged the contents of another 20,000 homes, stranded 2,000 loaded barges, and resulted in just more than 400 counties being declared disaster areas. The property and crop losses totaled $15 billion.

What is the **worst flood** in history?

The worst flood to date occurred in China in 1887, when more than 900,000 people died as the Huanghe (Yellow) River overflowed. China is particularly susceptible to severe and regular flooding; the Huanghe River alone typically floods two out of every three years. And because river valleys are densely populated, the human toll is often great. A 1939 flood in North China claimed more than half a million lives.

What were the harshest **blizzards** to hit the **United States**?

Regions of the United States—particularly the Great Plains, Midwest, and New England—typically experience extreme winter weather, but some storms do stand out. In March 1888 the northeast was hit by a blizzard dubbed the Great White Hurricane. After a warm spell that had caused the buds to open on trees in New York's Central Park, on March 12 the temperature in the city plummeted to 10 degrees Fahrenheit, and winds off the Atlantic built up to 48 miles per hour, bringing unpredicted snow that continued intermittently until the early morning of March 14. The three-day accumulation totaled 20.9 inches, and snowdrifts 15 to 20 feet high halted traffic. The snowfall was even greater elsewhere, averaging 40 inches or more in parts of southeastern New York and southern New England. The storm extended down into Chesapeake Bay, isolating the nation's capital from the world for more than a day. Two hundred ships were lost or grounded, and at least 100 died at sea. A total of at least 400 people died, half of them in New York City alone. Just two months before this nor'easter, another blizzard had swept through the Great Plains, moving eastward into Minnesota. There, high winds, blowing snow, and sudden drops in temperature combined to make it a dangerous storm, killing many people and thousands of cattle.

The Great Blizzard of 1993 caused loss of life and extensive damage all along the eastern seaboard, from Maine to Florida. More than 300 people died, almost 50 of them at sea, and economic losses totaled \$3 to \$6 billion. While several so-called "100-year storms" have bombarded the East Coast in the 1990s alone (there had been another just a few months earlier, in December 1992), the statistics of the March 1993 storm are impressive indeed, probably qualifying it as the storm of the century. Wind gusts exceeded 75 miles per hour all along the East Coast, with winds exceeding 100 miles per hour were measured at various points, including Flattop, North Carolina. Tennessee saw the highest snowfall of the storm, with 56 inches at Mount LeConte. Snowfall amounts were also heavy in the northeast, but snow accumulated as far south as the Florida Panhandle. Experts estimated that the amount of water that fell (in the form of snow) was equivalent to 40 days' flow of the mighty Mississippi River past New Orleans. According to record-low barometer readings, this storm surpassed Hurricanes Hugo (1989) and Hazel (1954).

However, midwesterners and residents of the northern Great Plains could argue that storms such as the Great Blizzard of 1993 are relatively common in their regions. To wit: Another storm that could easily vie for the title storm of the century occurred January 10–11, 1975, in the upper Midwest. The blizzard was accompanied by winds of 90 miles per hour and wind chills as low as minus 80 degrees Fahrenheit. Trains were stranded in snowdrifts, and at least 80 people died. Ranchers and farmers were hard hit, losing some 55,000 head of livestock.

What was the **dust bowl**?

The dust bowl was the most severe drought in U.S. history. In the spring of 1934, the country in the grips of the Great Depression, farmers across the Great Plains of the United States witnessed two great dust storms. First, in mid-April, after days of hot, dry weather and cloudless skies, 40- to 50-mile-per-hour winds picked up and took with them the dry soil, resulting in thick, heavy clouds. In Texas and Oklahoma, these dirt clouds engulfed the landscape. The next month was extremely hot, and on May 10 a second storm came up as the gales returned, this time creating a light brown fog.

On May 11 experts estimated that 12 million tons of soil fell on Chicago as the dust storm blew in off the Great Plains, and the same storm darkened the skies over Cleveland. On May 12 the dust clouds had reached the eastern seaboard. Between the two storms, 650 million tons of topsoil had blown off the plains.

The resulting dust bowl covered 300,000 square miles across New Mexico, eastern Colorado, Texas, western Oklahoma, and Kansas. The damage was great: Crops, principally wheat, were cut off at ground level or torn from their roots; cattle that ate dust-laden grass eventually died from "mud balls"; dust drifted, creating banks against barns and houses, while families tried to keep it from penetrating the cracks and crevices of their homes by using wet blankets, oiled cloths, and tape, only to still have everything covered in grit. Vehicles and machinery were clogged with dirt. In addition to the farmers who died in the fields, suffocated by the storm, hundreds of people suffered from "dust pneumonia."

What were the **effects of the dust bowl**?

After the dust had settled in the spring of 1934, the reaction among many Great Plains farm families was to flee the devastation: More than 350,000 people packed up their belongings and headed west, their lives forever changed by the disaster. In his 1939 novel, *The Grapes of Wrath,* American writer and Nobel laureate John Steinbeck (1902–1968) chronicled the harrowing and sorrowful westward journey of one Oklahoma family that was among the so-called "Okies" who deserted their farmlands in the devastated area of the Great Plains in search of a better life elsewhere.

Nature alone was not to blame for the dust bowl: By the end of the nineteenth century, farmers, aided by the advent of large tractors and reapers (harvesting machines), were cultivating the Great Plains, uprooting the native buffalo grass, which holds moisture in the soil, keeping it from blowing away. Even strong winds and extended droughts had not disturbed the land when it was covered by the grassland. When the demand for wheat increased after World War I (1914–18), farmers responded by planting more than 27 million new acres of the grain. By 1930 there were almost three times as many acres in wheat production as 10 years earlier: most of the buffalo grass that had prevented the earth from blowing had been removed. When the next dry period came (in spring 1934) and the wind picked up, the dust bowl resulted.

The government stepped in to remedy the problem: Soil conservation became the focus of federal agencies, and the U.S. Forest Service undertook a project to plant a "shelter belt" of trees within a 100-mile-wide zone, from Canada to the Texas Panhandle. Recovery was aided by the return of the rains. Soon the buffalo grass had grown back, helping to ensure that the dust bowl would not recur.

FIRES

How much damage was done by the **Great Fire of London**?

The fire, which began early in the morning on Sunday, September 2, 1666, and burned for four days and nights, consumed four-fifths of the city (which was then walled), plus 63 acres lying just outside the city walls. The blaze began in Pudding Lane near London Bridge and quickly spread through crowded wooden houses to the Thames wharf warehouses. The destruction included London's Guildhall, the Custom House, the Royal Exchange, and St. Paul's Cathedral. Additionally, 44 livery company halls, 86 churches, and more than 13,000 houses were destroyed.

Though the fire was unquestionably disastrous, London soon rebuilt and became one of Europe's most modern cities. The fire also destroyed thousands of old buildings where lice-infested rats had lived—they were partly responsible for spreading the plague through the English city.

What was the impact of the **fire at the MGM Grand** in Las Vegas?

The November 21, 1980, blaze, which killed 85 people and injured more than 600, led to a nationwide revision of local fire codes, giving the tragic event large-scale political significance. The MGM Grand Hotel had, in fact, passed fire inspections, but the building, which was then the world's largest gambling casino, had been eight years in the making. Between the time it was designed (fire protection systems included) and the time it was built, the building no longer complied with the always-improving safety standard for high-rise buildings. A short circuit started the blaze, which sent thick black smoke through the air ducts and escape stairwells in the 21 floors of guest rooms. Since more people were harmed or killed by smoke inhalation than by the fire itself, the American public became aware of the danger of smoke—over and above that of fire.

The event was a catalyst for change: Prior to the November blaze, most communities had not required existing buildings to be retrofitted every time fire safety codes changed and improved. After the fire, many communities chose to require building owners to comply with current protection capabilities.

Was the Chicago Fire really started by a cow?

According to legend, the Great Chicago Fire, which burned from October 8 to 9, 1871, was started by a cow (usually described as belonging to a Mrs. O'Leary) kicking over a kerosene lantern on De Koven Street. But the exact cause is unknown, and many theories exist about how it started—from one of the O'Leary's cows, who all perished in the blaze, to speculation that a meteor broke apart, raining fiery particles in the region.

The Chicago Historical Society cataloged the damage: "The so-called 'Burnt District' encompassed an area four miles long and an average of three-quarters of a mile wide—more than two thousand acres—including more than 28 miles of streets, 120 miles of sidewalks, and more than 2,000 lampposts, along with countless trees, shrubs, and flowering plants in 'the Garden City of the West.' Gone were 18,000 buildings and some $200 million dollars in property, about a third of the valuation of the entire city. Around half of this was insured, but the failure of numerous companies cut the actual payments in half again. One hundred thousand Chicagoans lost their homes, an uncounted number their places of work."

Chicago resident Julia Newberry described the aftermath in her diary in an entry dated October 17 (published in 1933 by W. W. Norton): "The fire began at twelveth (sic) street on Sunday night Oct. It swept the two magnificent avenues, & every building on the South side from twelveth street to the river. The Court House, with the original copy of Father's will & no one knows how many invaluable papers, legal documents, records, the beautiful Crosbie Opera house, a perfect bijou (sic) of a theatre, all the banks, insurance offices, railway depots, churches, & block after block of stores, unequalled any where. And then oh misery, the fire, the red, angry, unrelenting fire, leapt across the [Chicago] river, & burnt & burnt, till Mr. Mahlon Ogden's house was the only one left standing up to Lincoln Park. Yes the whole North Side is in ashes.... "

Though the United States has suffered other disastrous fires, including an 1835 blaze in New York City, which destroyed some 500 buildings, the Chicago Fire is the worst fire tragedy in the recorded history of North America. The damage was not limited to the city of Chicago: Sparks lit forest fires that destroyed more than a million acres of Michigan and Wisconsin timberland, burning from October 8 to October 14. These fires were responsible for the loss of more than 1,000 lives in the logging town of Peshtigo, Wisconsin, and in 16 surrounding communities.

Like the Great Fire of London 200 years earlier, a flurry of construction activity followed in the city, making Chicago one of the United States' most architecturally impressive urban centers. In fact, the fire was the impetus for the development of the Chicago School of Architecture, also called the commercial style (since most of it was devoted to office buildings, warehouses, and department stores). The Chicago School was instrumental in establishing the modern movement of architecture in the United States.

A wildfire burns along the edge of a highway in Lake Arrowhead, California, 2003. The fires that raged in southern California that year were the most damaging in the state's history.

How damaging were the **California wildfires of 2003**?

The wildfires that raged in southern California in late 2003 were the most damaging in the state's history: 15 fires destroyed 3,000 homes, burned 750,000 acres, caused more than 200 injuries, and claimed 24 lives. The fires produced one of the worst disasters the state had ever seen. The wildfires began October 21 and, driven by Santa Ana winds, swept through wooded canyons and mountain communities in Los Angeles, Riverside, San Bernardino, San Diego, and Ventura counties. On October 27 President Bush declared a disaster area, clearing the way for federal assistance. Firefighters worked into early November to bring the blazes under control. The fires had forced mandatory evacuations and caused widespread air pollution. Damages were estimated between $2.5 and $3 billion.

Although southern California was devastated by the emergency, and the western wildfires were in the national headlines, the 2003 fire season was, on the whole, about average. According to the National Interagency Fire Center (NIFC), 85,943 fires burned in U.S. wildlands during the 2003 season; the 10-year average (1993–2002) was 101,575 fires per season. The number of acres burned in 2003 was 4,918,088, with a 10-year average of 4,663,081 acres per season. (The 2004 fire season burned almost 2 million acres more than in 2003.) But the NIFC acknowledged that the number of structures destroyed was above average in 2003: a total of 5,781 structures were burned, including 4,090 primary residences, leaving many homeless. Federal agencies alone spent $1.3 billion to suppress wildfires across the United States in 2003—considerably higher than the 10-year average.

ACCIDENTS AND TECHNOLOGICAL FAILURES

What happened to the **Tacoma-Narrows Bridge**?

In 1940, the new 2,800-foot suspension bridge, carrying traffic across Washington's Puget Sound, was hit by high winds, causing it to buckle and undulate. In the simplest of terms, an engineering error allowed one of the suspensions to give way in the wind, and the bridge became ribbonlike, moving in waves. It was 10 years before a second span was opened over the body of water. The 1940 accident prompted engineers and bridge designers to be more cautious in the design of suspension bridges. The first wire suspension bridge in the U.S. was built in 1842: The 358-foot-long and 25-foot-wide bridge spanned the Schuylkill River, near Philadelphia, Pennsylvania. It was supported by five wire cables on either side, and was built by U.S. civil engineer Charles Ellet Jr. (1810–1862). The first chain suspension bridge in the U.S. was built in 1800.

Illustration depicting the struggle of survivors to get away from the sinking *Titanic.* Only 711 of the 2,224 passengers survived the 1912 disaster.

What are the facts about the *Titanic*?

As the brainchild of Lord William James Pirrie and J. Bruce Ismay, *Titanic* was a marriage of British technology and American money: Pirrie was head of Harland & Wolff, a firm known for building the sturdiest and best ships in the British Isles; Ismay was chairman of the White Star Line, owned by American financier J. Pierpont Morgan's (1837–1913) International Mercantile Marine.

In 1907 Pirrie and Ismay came up with a plan to compete with the top-notch Cunard liners by surpassing them both in size and luxury. The ship they planned, *Titanic,* was built in Belfast along with her sister ship, *Olympic,* which *Titanic* exceeded in gross tonnage but not in length. *Titanic* was 882 feet long, 92 feet wide, and weighed 46,328 gross tons; 9 steel decks rose as high as an 11-story building. Registered as a British ship and manned by British officers, *Titanic* was launched on May 31, 1911.

The ship was everything Pirrie and Ismay had planned. *Titanic's* size not only allowed more room to accommodate the increasing number of steerage (cheapest-fare) passengers who were immigrating to the United States, but also featured lavish elegance for first- and second-class travelers. Creature comforts included the first shipboard swimming pool, Turkish bath, gymnasium, and squash court. First-class cabins were nothing short of opulent, including coal-burning fireplaces in the sitting rooms and full-size, four-poster beds in the bedrooms. Additionally, there was a loading crane and a compartment for automobiles. The ship's hospital even featured a modern operating room.

With her steerage full and some of society's most prominent individuals on board, the RMS *Titanic* left the docks at Southampton, England, on April 10, 1912; New York Harbor was her final destination. On April 14, the ship was traveling in the exceptionally calm and icy waters of the North Atlantic, near Newfoundland. At 11:40 P.M., *Titanic* scraped an iceberg, sustaining damage along the starboard (right) side, from the bow to about midship. The *Titanic,* which immediately began taking on water, sank in 2 hours and 40 minutes in the early morning hours of April 15.

Only 711 of the 2,224 aboard survived; the 1,513 lost included American industrialists and businessmen John Jacob Astor IV, Isidor Straus (of R. H. Macy's), Benjamin Guggenheim, and Harry Elkins Widener. Survivors—mostly women and children who had been traveling as first-class passengers—were picked up by the *Carpathia,* which was 58 miles away when it received *Titanic's* distress signals. It took three and a half hours for *Carpathia* to reach the site of the disaster, by which time the *Titanic* was gone.

Why was the ***Titanic*** thought to be **unsinkable?**

The RMS *Titanic* was state of the art, a huge and luxurious ocean liner equipped with the latest and best. The ship's size afforded it great stability; its structure included more steel than had been used in previous ships; it was built with a double bottom—both skins were heavier and thicker than those of other ships. The hull was divided by 15 bulkheads (upright partitions) that rose five decks forward and aft (back), and four decks midship. These transverse bulkheads divided the ship into 16 compartments—"watertight" chambers—any two of which could take on water without sinking the ship. This marvel of modern technology, which was to be the jewel in the crown of the White Star Line, was given a fitting name: *titanic* is a Greek word meaning "having great force or power." And it was described as "practically unsinkable."

However, the ship designer did not—and could not—prepare the ship for what happened on the night of April 14, 1912. Just before midnight, the *Titanic* was speeding—at 21 knots—through the North Atlantic, even though the crew had been warned by other ships that the unusually calm waters were full of ice. When the *Titanic's* two watchmen, who were not using binoculars, sighted an iceberg in the ship's path, it was only a quarter mile away. The ship was turned to the port (left), but it was too late. The underwater shelf of the ice tore through the plating on the starboard (right) side of the ship. Thin slits developed at the seams in the ship's hull, allowing seawater to enter.

The effect was similar to filling an ice tray with water: Once one "watertight" chamber had filled, the rushing water spilled over the top and into the next.

Titanic came to symbolize human arrogance: The ship owners and operators believed the *Titanic* was impervious to nature. Consequently, the ocean liner had not been equipped with the number of lifeboats needed to rescue everyone on board: *Titanic's* lifeboats had room for about half the passengers. Since there had been no safety drills on board, many lifeboats were launched only half full. The enormous loss of life, which included society's most prominent individuals as well as ordinary families who were immigrating to America, stands out as one of the great tragedies in the history of transportation.

What **effect** did the sinking **of the *Titanic*** have on sea travel?

The sinking of the *Titanic* brought about new regulations to increase the safety of sea travel. First, and perhaps most simply, all ships are required to carry enough lifeboats such that there is one spot for each person on board. (When *Titanic* sailed, the number of required lifeboats was based on the ship's tonnage, not on the number of passengers and crew.) Also, new rules required lifeboat drills to be held soon after a ship sails.

Shipping lanes were moved farther south, away from the ice fields, and are monitored by a patrol. Ships approaching ice fields are required to slow their speed or alter their course.

Until 1912 most ships employed only one wireless operator. Such was the case on the *California,* which was less than 20 miles from *Titanic* when wireless operator Jack Phillips sent out the distress signal. However, the operator on the *California* was not on duty at that hour. Phillips stayed at his station, desperately trying to reach a nearby ship, and eventually went down with *Titanic.* In the aftermath of the disaster, the U.S. Congress moved quickly to pass the Radio Act of 1912, which required that radios be manned day and night, that they have an alternate energy source (besides the ship's engine), and that they have a range of at least 100 miles. Further, operators must be licensed, adhere to certain bandwidths, and observe a strict protocol for receiving distress signals. (This was the beginning of the Federal Communications Commission, or FCC.) These measures were meant to rid the airwaves of those amateur operators who had confused official operators the night of April 15, 1912. One erroneous wireless message transmitted by amateurs that night had the *Titanic* moving safely toward Halifax, Nova Scotia.

Was the *Titanic* the **most disastrous shipwreck** of all time?

Though it is certainly the most famous, it is not the most disastrous. According to shipping registries, three wrecks were worse than the *Titanic.* In April 1865 the sidewheel steamboat *Sultana* exploded on the Mississippi River, killing 1,653 of the estimated 2,300 people on board. The packet steamboat had routinely carried passengers

The German dirigible *Hindenburg* crashes to earth after exploding over the U.S. Naval Station in Lakehurst, New Jersey, in 1937.

and cargo between St. Louis and New Orleans. In 1917 the *Mont Blanc* exploded in the harbor at Halifax, Nova Scotia, claiming 1,635 lives and severely injuring more than 1,000. The ship, which was a French munitions carrier (World War I was raging at the time), was struck by a Norwegian relief ship, the *Imo.* The *Mont Blanc* was laden with thousands of tons of TNT, acid, and other explosives, which were ignited in the collision. The explosion was so terrific that it laid waste to much of Halifax and generated a tsunami that swept through the city. Most recently, in 1987, the *Doña Paz* collided with another ship off the Philippines; 1,840 died.

What happened to the ***Hindenburg***?

The image of the large airship bursting into flames is familiar to many: *Hindenburg,* a German vessel and the largest airship ever built, exploded while it was trying to land at Lakehurst, New Jersey, at about 7:25 P.M. on May 6, 1937. *Hindenburg* had just completed a transatlantic flight and had dropped its mooring lines to the ground crew when the hydrogen gas that kept the airship afloat caught fire. Within 32 seconds, *Hindenburg* was nothing but smoldering rubble on the ground. Sixty-two of the 97 people on board survived the crash. In addition to the 35 passengers and crew who lost their lives, one member of the American ground crew also perished. Though the cause of the fire has never been conclusively determined, it is believed that an atmospheric electrical spark—not sabotage—ignited hydrogen gas that was flowing from a leak. The fact that the outer cover of the tail section had been observed to flutter just seconds before the explosion lends credence to the explanation that there had been a gas leak.

The crash was thoroughly documented. Though travel by airship had been going on for more than 25 years and some 50,000 passengers had been transported without a single fatality, the *Hindenburg's* landing in New Jersey was still an event for which many spectators turned out. Airships were a marvel of technology, and the *Hindenburg* in particular was worth seeing since it was the largest afloat. Even though the airship was more than 12 hours behind schedule (due to weather over the Atlantic), the arrival was eagerly anticipated. The entire event was caught on film, and that documentary was widely shown in movie newsreels. Newspaper and radio coverage also helped link the *Hindenburg*—and airship travel on the whole—with terrifying technological disaster.

The highly publicized crash effectively ended airship travel. A sister ship, the *Graf Zeppelin,* was en route from Rio de Janeiro back home to Germany when news of the *Hindenburg* disaster came in. Upon arrival, the *Graf Zeppelin* was grounded until the cause of the *Hindenburg's* crash was known. No passenger airships took flight again. Two years later, an airplane carried its first paying passenger across the Atlantic.

Today, airships, or "blimps," are used by major corporations such as Goodyear during national events, primarily sporting events. Some airships are also used for reconnaissance and patrol.

Why did the ***Hindenburg*** use **hydrogen** to keep afloat?

The fact that *Hindenburg* used hydrogen might have been the airship's only flaw; and it was made necessary by the political climate of the time. *Hindenburg* was the fulfillment of German airship designer Hugo Eckener (1868–1954), whose Zeppelin Company had enjoyed years of experience and success even as other airship companies folded. By 1934 Eckener felt that his successful *Graf Zeppelin,* which had made several transatlantic trips, was not well suited to such long-distance flights. Eckener envisioned a larger and speedier vessel. In the *Hindenburg,* which took her maiden flight on March 4, 1936, Eckener's vision was made real. Named for the German war hero and politician Paul von Hindenburg (1847–1934), the immense airship measured 803 feet in length and had a diameter of 135 feet, allowing it to hold nearly twice as much gas as other airships. The vessel was equipped with the latest technology, including four Daimler-Benz diesel engines that allowed it to travel as fast as 85 miles per hour.

Hindenburg was also a luxury liner: it featured private cabins, showers, dining room, promenade decks, picture windows, and even a pressurized and sealed smoking room. (Cigarettes, pipes, and cigars had to be lit using an electric lighter; matches were strictly forbidden on board.)

But there was one problem: *Hindenburg* had been designed to be lifted by helium. However, the gas was scarce at the time, and the United States refused to sell any to Germany, which had been taken over by national extremist Adolf Hitler (1889–1945). The American government suspected the Germans might soon have military plans for their airships. Thus, the *Hindenburg* was forced to use hydrogen—7 million cubic feet of the flammable gas.

Were there any **other airship disasters** before *Hindenburg*?

Yes, as Hugo Eckener (1868–1954) and his Zeppelin Company laid plans in 1934 to build the large and luxurious *Hindenburg,* most other nations with airship programs had either abandoned them or were about to, since all had experienced disastrous and fatal crashes. One of these was when a British dirigible R-101 burned on October 5, 1930, northwest of Paris while on her maiden voyage to Australia. That disaster claimed 54 lives.

What is the **worst airplane accident** in history?

With thousands of accidents since the beginning of aviation history, records differentiate among ground collision, midair collision, and single-aircraft accidents. (Some records differentiate by cause, including pilot error, weather, and fuel starvation.) The worst ground collision—and the deadliest airplane accident in history—was the Tenerife disaster of March 27, 1977, which killed 583 people. Two Boeing 747 airliners ran into each other on Tenerife, in the Canary Islands (Atlantic Ocean). One was a Pan Am flight, "The Clipper Victor," which originated at Los Angeles International Airport, made a stop at New York's JFK Airport, and was headed for the Canary Islands; it was diverted to Tenerife at the last minute due to a bomb threat at its destination airport on neighboring Las Palmas Island. The other 747 was a KLM flight, "The Flying Dutchman," originating in Amsterdam; it, too, was diverted to Las Palmas's Los Rodeos Airport because of the threat there. On takeoff, the KLM plane slammed into the taxiing Pan Am plane. Heavy fog on the runway contributed to the disaster, but there were communication problems as well: According to tower records, the KLM flight had not yet been cleared for takeoff. Upon collision, the jumbo jets burst into flames; there were only 61 survivors (54 passengers and 7 crew members), all from the Pan Am flight.

The worst midair collision happened on November 12, 1996, over Charkhi Dadri, India: 349 people perished when a Saudi Arabian Airlines Boeing 747 collided with a Kazakh Ilyushin Il–76 aircraft. There were no survivors. The worst single-plane accident happened on August 12, 1985, when a Japan Air Lines Boeing 747 crashed into a mountain on a domestic flight, killing 520 people; there were four survivors, all of them passengers.

What happened on ***Apollo 13***?

On April 13, 1970, a damaged coil caused an explosion in one of the oxygen tanks on the moon-bound U.S. spacecraft, leaving astronauts Jim Lovell, Jack Swigert, and Fred

Haise in a disastrous situation. The explosion damaged the fuel cells as well the craft's heat shield, which was needed to protect the vessel upon re-entry into Earth's atmosphere. While the National Aeronautics and Space Administration (NASA) had experienced a previous disaster—in 1967, when three astronauts died in a fire on the launch pad—mission control had not faced anything like this before. And no Americans had ever been lost in space.

After hearing a loud bang and seeing an oxygen tank empty, the *Apollo 13* astronauts reported to mission control at the Johnson Space Center, "OK, Houston, we've had a problem." The ensuing real-life drama proved that to be an understatement. The crew moved into the craft's tiny lunar module, designed to keep two men alive for just two days. With the astronauts four days from home, NASA engineers had their work cut out for them. Among other measures, the temperature in the module was lowered to 38 degrees Fahrenheit to conserve oxygen and electricity. The world was waiting and watching as the module splashed down in the South Pacific, just barely ahead of the failure of the oxygen. All three astronauts survived the disaster, which came to be known as the "successful failure." *Apollo 13* never reached its destination but, despite the odds, made it back to Earth safely.

What happened to ***Challenger***?

On January 28, 1986, at Cape Canaveral, Florida, the National Aeronautics and Space Administration (NASA) launched the twenty-fifth mission of its space shuttle program. The *Challenger* carried a crew of seven, including Christa McAuliffe (1948–1986), who was to be the first schoolteacher in space. She was slated to broadcast a series of lessons to schoolchildren throughout America. The crew's commander was Francis Scobee (1939–1986), who had piloted a 1984 shuttle mission; the pilot was Michael Smith (1945–1986), who was making his first flight in space; mission specialists Ellison Onizuka (1946–1986), Ronald McNair (1950–1986), and Judith Resnik (1949–1986) were all experienced space travelers; and payload specialist Gregory Jarvis (1944–1986) was making his first space flight.

That cold and clear January morning, the *Challenger's* takeoff was delayed by two hours. Freezing temperatures overnight had produced ice on the shuttle and launch pad, which prompted NASA to conduct inspections to assess the condition of the craft. At 11:38 A.M., *Challenger* was launched into space. Just 73 seconds later and at an altitude of 48,000 feet—the craft still in view of the spectators on the ground—*Challenger* burst into flames. While NASA controllers were aware of what had happened (they had, among other things, heard Smith utter, "Uh oh," just one second prior to the explosion), it took a moment for the spectators to understand. But as the fireball grew bigger and debris scattered, the spectators, including family and friends of the crew, fell silent.

The crew, inside a module that detached from the shuttle during the blowup, evidently survived the explosion but died upon impact after a nine-mile free-fall into the

The space shuttle *Challenger* explodes 73 seconds after takeoff on January 28, 1986. One of the shuttle's booster rockets, whose faulty O-rings were blamed for the disaster, shoots off to the right.

Atlantic Ocean. Six weeks after the disaster, the crew module was recovered from the ocean floor; all seven astronauts were buried with full honors.

Investigations into the crash revealed that the O-rings (seals) on the shuttle's solid rocket boosters had failed to work; due to the low temperatures, the O-rings had stiffened and thereby lost their ability to act as a seal. A government commission recommended a complete redesign of the solid rocket booster joints, a review of the astronaut escape systems (to work toward achieving greater safety margins), regulation of the rate of shuttle flights to maximize safety, and a sweeping reform of the shuttle program's management structure. The space agency retrenched. It was almost three years later—on September 29, 1988—before another American shuttle flew in space.

Is it true that the engineers of the ***Challenger's* O-rings** warned NASA that the devices might fail?

Yes, but sadly the advice of the engineers went unheeded: the O-ring manufacturer, Morton Thiokol, gave the National Aeronautics and Space Administration (NASA) the go-ahead in the hours before *Challenger's* takeoff. On January 27, 1986, the night before the planned takeoff, the temperature at Cape Canaveral, Florida, dropped to well below freezing. Since no shuttle had been launched in temperatures below 53 degrees Fahrenheit, NASA undertook a late-night review to determine launch readiness. As a contractor, Morton Thiokol participated in this process, with their engineers expressing concerns about the O-rings on the shuttle's solid rocket boosters. They feared the rings would stiffen in the cold temperatures and lose their ability to act as a seal. Since the space agency was under pressure to launch the shuttle on schedule, NASA managers pushed the manufacturer for a go or no-go decision. The managers of Thiokol, who were aware that the O-rings had never been tested at such low temperatures, signed a waiver stating that the solid rocket boosters were safe for launch at the colder temperatures.

Challenger was launched the next morning, at 11:38 A.M. About one minute into the flight, a flame became evident, and seconds later, the spacecraft exploded. All seven crew members died. Investigators later concluded that the tragic accident had been caused by the failure of the O-rings.

What happened in the ***Columbia*** space shuttle disaster?

The U.S. space shuttle *Columbia* was lost upon its reentry into Earth's atmosphere on the morning of February 1, 2003. All seven crew members died.

The *Columbia* was in the skies over Texas about 15 minutes before its scheduled landing at Florida's Kennedy Space Center when, shortly before 9:00 A.M. (EST), ground controllers lost data from temperature controllers on the spacecraft. Over the next several minutes, National Aeronautics and Space Administration (NASA) ground control lost all flight data. At about the same time, witnesses in Texas reported the sound of rolling thunder and debris falling from the sky. Heat-detecting weather radar showed a bright red streak moving across the Texas sky. The shuttle was 40 miles above Earth and traveling at 18 times the speed of sound when it disintegrated, leaving a trail of debris from eastern Texas to western Louisiana. The investigation later revealed that damage to the spacecraft had gone unseen during the mission, causing the *Columbia* to break apart upon reentry.

The shuttle was commanded by Rick Husband and piloted by William McCool. The mission specialists were Michael Anderson, Kalpana Chawla, David Brown, and Laurel Clark. The payload specialist was Israeli astronaut Ilan Ramon. In President George W. Bush's remarks to the nation that day, he said, "These men and women assumed great risk in the service to all humanity. In an age when space flight has come to seem almost routine, it is easy to overlook the dangers of travel by rocket.... These astronauts knew the dangers, and faced them willingly."

The *Columbia* tragedy occurred within a week of the anniversaries of two other deadly NASA disasters: the *Challenger* explosion on January 28, 1986, and the launchpad fire that killed three *Apollo* astronauts on January 27, 1967. After investigating the cause of the *Columbia* disaster, NASA focused on implementing a new system of sensors to detect potentially fatal damage to spacecrafts while in orbit. NASA relaunched its space shuttle program in late July 2005 with the *Discovery*.

What happened in the **blackout of 2003**?

A three-month investigation found that line failures and system errors combined to cause the power outage that hit much of the Northeast and Great Lakes portions of the United States, as well as parts of eastern Canada, on August 14, 2003. The blackout affected 50 million people, 40 million of them in the United States.

A task force headed by the U.S. Department of Energy and its Canadian counterparts delved into the sequence of events that caused the massive power failure. Investigators concluded that mistakes made by FirstEnergy Corporation of Akron, Ohio, had combined with heavy loads of power to stress the interstate electrical grid and cause its failure through a chain reaction. Just after 4:00 P.M. (EDT) on Friday, August 14, power was knocked out from Detroit to Toronto to New York City.

The widespread outage immediately affected water supplies, halted air transportation and subway systems, snarled ground transportation, and brought most distribution lines to a standstill—affecting the flow of food and other supplies. Though there were isolated reports of looting, crime was minimal, with New York City reporting a lower crime rate for the night of August 14 than usual. Eight deaths were related to the blackout.

Power began to be restored to some areas later that evening, but it was August 16 before all affected areas were back on line.

Two weeks later, on August 28, London experienced a blackout at the peak of evening rush hour. The outage paralyzed the British capital and left a half million commuters stranded. Power was restored within an hour. A utilities official said that the "freak event" was caused by two faults that happened in quick succession.

In late September 2003 there were also widespread power outages in Denmark, Sweden, and Italy. The blackout in Italy affected the entire nation (with the exception of Sardinia), or 57 million people. Because it happened overnight and on a weekend, the impact was minimal.

The blackouts drew attention to the need for system upgrades, improved maintenance, and more sophisticated alert systems to prevent such grid failures, which take both an economic and human toll.

INDUSTRIAL ACCIDENTS

What was **Love Canal**?

When Love Canal, a community east of Niagara Falls, New York, made international headlines in August 1978, it was only after the neighborhood had already been the subject of local newspaper stories since 1976. And sadly, more headlines followed, into 1980. What had become clear during these years was that Love Canal was toxic. Community residents had experienced unusually high incidences of cancer, miscarriages, birth defects, and other illnesses. There were also reports that foul odors, oozing sludge, and multicolored pools of substances were emerging from the ground; and children and animals returned from outdoor play with rashes and burns on their skin.

Unbeknownst to the residents, all of these problems were attributable to the history of the site upon which their community had been built. Beginning in 1947 the Hooker Electrochemical Company had used Love Canal, with its clay walls, to dump 21,800 tons of chemical waste. In 1953 the company sold the canal to the Niagara School Board for the sum of one dollar. The deed acknowledged the buried chemicals, although it did not disclose their type or toxicity. A disclaimer protected the firm from future liability. The canal pit was subsequently sealed with a clay cap designed to prevent rainwater from disturbing the chemicals. Grass was planted. Soon Love Canal had become a 15-acre field. The following year, a school was under construction on the site. In 1955 400 elementary school children began attending classes there and playing on the surrounding fields. Development happened fast: roads, sewers, and utility lines crisscrossed the site, disrupting the soil.

While residents began to discern problems as early as 1958, when they complained of nauseating smells and incidences of skin problems, it was not until the mid-1970s that the extent of the hazard became evident. It was then that unusually heavy rainfalls caused chemicals to surface. A portion of the schoolyard collapsed, strange substances seeped into basements, and trees and gardens died. In October 1976 the *Niagara Gazette* began investigating these problems, but an official investigation did not begin until the following April. By this time, the site was a disaster: toxins were found in storm sewers and basements, exposed chemical drums leaked substances, and air tests detected dangerously high chemical levels in homes. Further testing identified more than 200 different compounds at the site, including 12 carcinogens (cancer-causing agents) and 14 compounds that can affect the brain and central nervous system.

The residents of Love Canal organized, forming citizen groups including the Love Canal Homeowners Association. These groups succeeded in getting media coverage and in pressuring public officials to act. Finally on August 2, 1978, the New York State Health Commissioner declared Love Canal unsafe. Six days later, President Jimmy Carter (1924–) approved emergency assistance and New York governor Hugh Carey announced that funds would be used to purchase homes nearest the canal.

While more than 200 families that were perceived to be in danger were moved, in 1980 problems resurfaced when researchers found that blood tests of residents showed abnormally high chromosome damage. The state recommended that pregnant women and infants be removed from homes—even those that had been certified as safe. In May 1980 conflict ensued between 300 Love Canal homeowners and officials from the Environmental Protection Agency (EPA). On May 21, President Carter declared a second emergency at Love Canal. This time the actions were more comprehensive: Almost 800 families were evacuated, and their homes were either destroyed or declared unsafe until further clean-up could be done. Four years later, a new clay cap was installed over the canal. It was also in 1984 that Occidental Petroleum, parent company of the firm that had dumped chemicals in Love Canal, reached a $20-million settlement with residents.

What **impact** did **Love Canal** have?

The effect of the crisis was felt on many levels: by area residents whose lives were forever changed by the hazards, by residents nearby Love Canal who feared for their own safety, by Americans across the country who lived near other chemical waste sites, and by Americans for whom Love Canal had become synonymous with the problems posed by hazardous waste.

At the government level, the tragic events at Love Canal helped to speed the passage of the Comprehensive Environmental Response, Compensation, and Liability Act of 1980. Also known as the Superfund, the legislation set up a multibillion-dollar fund to clean up the nation's worst toxic disasters. The Environmental Protection Agency (EPA) assigned clean-up priority to some 1,200 abandoned and potentially contaminated waste sites.

Along with the chemical plant explosion at Bhopal, India, in 1984, Love Canal also contributed to a "community-right-to-know" provision, which was part of the 1985 Superfund Amendments and Reauthorization Act. The new legislation gave all citizens the right to know what chemicals are produced, stored, or buried in their neighborhoods.

What happened at **Three Mile Island**?

The March 1979 accident—a near meltdown—at the nuclear power station at Three Mile Island, outside Middletown, Pennsylvania (near Harrisburg), was eventually contained. Had it not been, the damage would have been on a level with that of the Chernobyl (Ukraine) disaster, which happened some seven years later. Instead, Three Mile Island served as a wake-up call, reminding the American public and its utility companies of the potential risks involved in nuclear energy.

The sequence of events at the nuclear power plant, which is located on an island in the Susquehanna River, was as follows: At 4:00 A.M. on Wednesday, March 28, an overheated reactor in Unit II of the power plant shut down automatically (as it should have); Metropolitan Edison Company operators, guided by indicators that led them to believe water pressure was building (and an explosion was therefore imminent), shut down those pumps that were still operating; the shut-down of all the pumps caused the reactor to heat further; then, tons of water poured out through a valve that was stuck open; this water overflowed into an auxiliary building through another valve that was mistakenly left open. This final procedure, which took place at 4:38 A.M., released radioactivity.

Since there was no cooling system in operation, the reactor in Unit II was damaged. But this was not the end of it: The radiation within the buildings was released into the atmosphere, and at 6:50 A.M. a general emergency was declared. Early that afternoon, the hydrogen being created by the uncovered reactor core accumulated in a containment building and exploded. Since hydrogen continued to be emitted, officials feared another—catastrophic—explosion. Worse yet, they feared the reactor would become so hot that it would melt down. The effect of a meltdown would be that the superheated material would eat its way through the bottom of the plant and bore through the ground until it hit water, turning the water into high-pressure steam, which would erupt, spewing radioactivity into the air.

As technicians worked to manage the crisis, radiation leaked into the atmosphere off and on through Wednesday and Thursday. On Friday the governor of Pennsylvania ordered an evacuation: Some 144,000 people were moved from the Middletown area. The situation inside the plant remained tenuous as a hydrogen bubble developed and increased in size, again raising fear of explosion. Meantime, public alarm was mounting as the media attempted to monitor the ongoing crisis. Finally, on Sunday, April 1, the plant was visited by President Jimmy Carter (1924–). At about the same time, the hydrogen bubble began to decrease in size, ending the crisis.

What was the **impact of** the accident at **Three Mile Island**?

Prior to the March 1979 events at Three Mile Island, it was thought that the danger of a nuclear meltdown was almost negligible. Though there were safety systems in place, none of them would have prevented a complete catastrophe. Since the accident, the American Nuclear Regulatory Commission (NRC) and utility companies have worked together to resolve the problems that were revealed. Among the efforts and requirements put into place were: more stringent licensing procedures for operators; better training of plant operators in the event of an emergency; wider sharing of information on emergency management systems; effort to locate new plants outside of densely populated areas; more rigid quality assurance standards at all plants; strict implementation of the standards, which are subject to review by the NRC; and emergency evacuation plans that must be approved by the Federal Emergency Management Agency (FEMA). Even with these improvements to safety programs, the accident at Chernobyl in 1986 again produced worldwide concern over the hazards of nuclear power.

What was the **worst industrial accident**?

It was the gas leak at a Union Carbide chemical plant in Bhopal, India, on December 3, 1984. At about 12:30 A.M., methyl isocyanate (MIC), a deadly gas, began escaping from the pesticide plant, and it spread southward, eventually covering approximately 15 square miles. Within a few hours, thousands of Bhopal residents were affected by the asphyxiating gas. General symptoms included severe chest congestion, vomiting, paralysis, sore throat, chills, coma, fever, swelling of legs, impaired vision, and palpitations. Estimates of the total death toll range from the official government estimate between 3,000 and 10,000, a figure based on what medical professionals described. In total, 200,000 people were directly or indirectly affected by the poisonous gas.

Within hours of the accident Bhopal police moved into action, closing the plant and arresting its manager and four of his assistants. The five men were charged with "culpable homicide through negligence." Union Carbide dispatched a team of technical experts from its Danbury, Connecticut, headquarters, but upon arrival at the plant, they were turned away by local authorities. Meanwhile, the Indian Central Bureau of Investigation seized the plant's records and log books and ordered an inquiry into the accident. Union Carbide's chief executive officer Warren M. Anderson flew to Bhopal, but he was promptly arrested, along with two officials of the company's Indian subsidiary. The corporate executives were charged with seven offenses including criminal conspiracy, culpable homicide not amounting to murder, making the atmosphere noxious to health, and causing death by negligence. Anderson was later released on bond.

Upon learning of the horrific accident, U.S. president Ronald Reagan (1911–2004) sent a message conveying the grief shared by him and the American people. Multinational corporations, including Union Carbide, were vilified in the press; the Soviet news agency accused such companies of marketing "low-quality products and outdated technology to developing countries." Prime Minister Rajiv Gandhi (1944–1991) of India visited the disaster site and announced immediate creation of a $4-million relief fund for victims; he also vowed that he would prevent multinational corporations from setting up "dangerous factories" in India.

The implications of the industrial accident were many. It prompted public scrutiny of safety systems at chemical plants around the globe. Given the number of plants where poisonous chemicals are produced and stored, some observers believe chemical accidents could happen as often as once in every 10 years. Union Carbide, of course, suffered financially; the stock dropped more than 12 points, wiping out 27 percent, or almost $1 billion, of its market value, in about one week. Damage claims were filed in behalf of the victims, with noted American criminal attorney Melvin Belli filing one of them in the amount of $15 billion.

In addition to the thousands who died in Bhopal, others suffered from long-term effects including chronic lesions of the eyes, permanent scarring of the lungs, and injuries to the liver, brain, heart, kidneys, and the immune system. In the years after the accident, studies showed that the rate of spontaneous abortions and infant deaths in Bhopal were three to four times the regional rate.

What caused the nuclear accident at **Chernobyl**?

The damaged reactor at the Chernobyl nuclear power plant, Ukraine; a 1986 explosion and fire at the antiquated facility released radioactive material into the atmosphere.

The April 1986 accident—the world's worst nuclear power plant disaster—was caused by explosions at the Soviet power plant, sending radioactive clouds across much of northern Europe. According to the World Nuclear Association, the accident was the result of "a flawed reactor design that was operated with inadequately trained personnel and without proper regard for safety."

The trouble began at 1:24 A.M. on Saturday, April 26, when Unit 4 of the Chernobyl Nuclear Power Plant, about 70 miles outside of the Ukrainian capital of Kiev, was rocked by two enormous explosions. The roof was blown off the plant and radioactive gasses and materials were sent more than a half-mile into the atmosphere. Though two workers were killed instantly, there was no official announcement about the hazardous blast. It was the Swedes who detected a dramatic increase in wind-borne radiation, and on April 28—two full days after the accident—news of the event was briefly reported by the Soviet news agency Tass.

Two weeks later, on May 14, First Secretary Mikhail Gorbachev (1931–) went on national television and explained what officials knew about the accident. More details were revealed over the following months. The explosions were caused by an unauthorized test carried out by plant operators, who were trying to determine what would happen in the event of a power outage. There were six critical errors made by workers during the testing, which combined to spell disaster. Perhaps the most significant of these mistakes was turning off the emergency coolant system: Once the test was under way, further mistakes caused the core to heat to more than 9000 degrees Fahrenheit, producing molten metal that reacted with what cooling water was left to produce hydrogen gas and steam, resulting in a powerful explosion. What caused the second explosion is not clear, and experts disagree on what might have happened. Some theorize that it was a pure nuclear reaction.

What was the **impact of the disaster at Chernobyl**?

As the worst nuclear power plant disaster to date, the Chernobyl accident in 1986 had far-reaching effects. Total fallout from the accident eventually reached a level 10 times that of the atomic bomb dropped on Hiroshima, Japan, on August 6, 1945, at the end of World War II (1939–45). Some 30 firefighters and plant workers died just after the accident.

Plants and animals in the immediate area and downwind of the plant were heavily contaminated with radioactive fallout. More than 10 years after the accident, food crops still could not be planted in the region.

The immediate after-effects were felt in Europe as well: Some Italian vegetables were found to be contaminated; reindeer meat in Lapland (a region above the Arctic Circle and extending over northern Norway, Sweden, Finland, and Russia) was declared unfit for human consumption, also due to radioactive contamination; and for a time, fresh meat from eastern Europe was banned by the EC (European Community).

After the accident, some experts predicted disastrous long-term effects, estimating that between 6,500 and 45,000 people could die as a result of cancer caused by exposure to radiation. But a report issued by the World Health Organization (WHO) in 2000 concluded that "there is no evidence of a major public health impact attributable to radiation exposure 14 years after the accident. There is no scientific evidence of increases in overall cancer incidence or mortality or in non-malignant disorders that could be related to radiation exposure."

Since the type of nuclear power plant (called RMBK) in use in the former Soviet Union is no longer in operation elsewhere in the world, non-Soviet scientists had few lessons to learn from the event. One American nuclear expert remarked that "most of the lessons from Chernobyl have been learned already and applied in the United States." However, opinion remained divided over the relative safety of nuclear power in general.

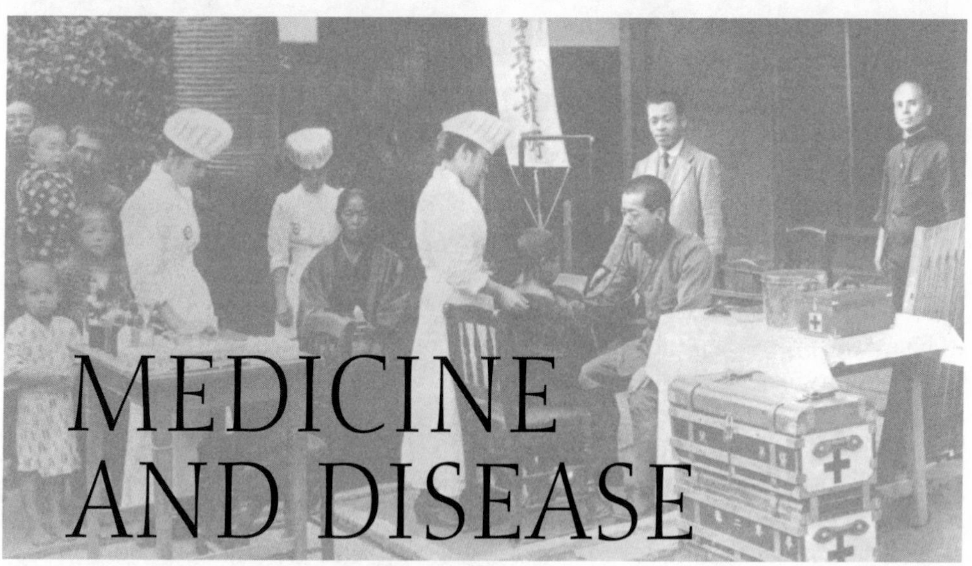

MEDICINE AND DISEASE

How long have people been **getting sick**?

Humankind has suffered from illnesses since humans first appeared on Earth millions of years ago. But in the absence of medical knowledge, which only began about 5,000 years ago, ailments and sickness were assumed to have been brought on by angered gods or evil spirits. Even today, some primitive groups believe they must appease the gods and spirits through offerings and sacrifices in order to stave off or rid themselves of illness; these are practices carried out by witch doctors or healers. In prehistoric times such superstitious beliefs resulted in the first "medical" procedure, in which a hole was drilled into the patient's skull—this practice was believed to allow evil spirits to escape the body and thereby rid the body of sickness. Fossil findings showing this method of treatment date back about 10,000 years. One vestige of ancient treatments of illness that still has relevance today is the use of plants and plant substances for medicinal purposes.

Who was the **first physician** in history?

The first physician known by name was Imhotep, an Egyptian who lived about 2600 B.C. Also considered a sage, Imhotep lived at a time when the Egyptians were making progress in medicine. The advances included a textbook on the treatment of wounds, broken bones, and even tumors. Imhotep was later worshiped as a god by the Egyptians.

What is the **Hippocratic Oath**?

The Hippocratic Oath is the pledge taken by many medical students upon graduation or upon entering into practice. While the text of the oath varies by translation, one important line reads, "I will prescribe regiment for the good of my patients according

What are the four humors?

The four humors are the bodily fluids: blood, phlegm, yellow bile, and black bile, originating in the heart, brain, liver, and spleen, respectively. One work assigned to Greek physician Hippocrates (c. 460–c. 377 B.C.), *Nature of Man,* asserts that illness is caused by an imbalance of the four humors (fluids) in the body. The presence of these humors was thought to determine the health and personality of a person. This belief prevailed for centuries but was finally discredited by modern science.

During the Middle Ages (500–1350) each of the humors was assigned certain characteristics. Someone of ruddy complexion was believed to have an excessive amount of blood in his or her system; that person would be sanguine (cheerful and optimistic) in character. (The word sanguine is derived from the Latin word *sanguiss,* meaning "blood.") Someone who had an imbalance resulting in more phlegm was considered phlegmatic, and would have a slow and impassive temperament. An individual who had excessive yellow bile was considered hot-tempered. And a person who had more black bile in his or her physiological system was believed to be melancholic.

to my ability and my judgment and never to harm anyone." The vows are attributed to the Greek physician and teacher Hippocrates (c. 460–c. 377 B.C.), who practiced on the island of Cos. Unlike his predecessors, who relied on superstitious practices in their treatment of patients, Hippocrates believed that diseases were brought on not by supernatural causes but by natural ones. He further believed that disease could be studied and cured; this assertion forms the basis of modern medicine, which is why Hippocrates is called the "father of medicine."

It is largely owing to another prominent Greek physician that the oath was handed down through history: Galen (A.D. 129–c. 199) was physician to Roman emperors Marcus Aurelius (121–80) from 161 and Lucius Commodus (161–92) from 168. He demonstrated that arteries carry blood, not air (as had been thought), and, like Hippocrates, Galen believed in the four humors of the body. He left medical texts that for centuries were considered the authoritative works on medical practice. Galen's writings reveal his high regard for Hippocrates, who lived and worked many centuries earlier.

How old is **biological warfare**?

Biological or germ warfare has a long history. For example, in the year 1343, Tatars (originally a nomadic tribe of east-central Asia) became sick with the bubonic plague. The disease, which is carried by fleas and rats, was called the Black Death because nearly all who became afflicted died. Invading the Crimea (in present-day Ukraine), the marauding Tatars encountered a group of Genoese (Italian) merchants at a trading post. Besieging them, the Tatars catapulted their dead at their enemy, many of whom became infected, carrying the plague to Constantinople (present-day Istanbul, Turkey) and to the western European ports where they traveled.

In the twentieth century, the use of microorganisms or toxins that produce sickness in people or in animals, or that cause destruction to crops, was outlawed by the Geneva Gas Protocol of 1925. In 1972 the Biological and Toxin Weapons Convention was simultaneously opened for signature in Moscow, Washington, and London, and the agreement entered into force on March 26, 1975. Signed by more than 162 nations, the convention bans "the development, production, stockpiling, acquisition, and retention of microbial or other biological agents or toxins, in types and in quantities that have no justification for prophylactic, protective or other peaceful purposes." Nevertheless, several nations have conducted further research into defense against biological warfare, including developing microorganisms suitable for military retaliation. The existence of such biological weapons—including anthrax and smallpox—remains a concern today. The possibility that Iraq possessed biological and chemical weapons of mass destruction was the primary reason for the U.S.-led invasion of that country in 2003.

Is **anthrax a new disease**?

No, the disease dates back thousands of years, at least to biblical times. But its potential use as a bio-terrorism weapon is relatively recent.

Anthrax is caused by the *bacillus anthracis* bacterium, spores that can survive in soil for years. It is mainly a disease of grass-eating livestock, but humans who work with herd animals may become infected through exposure. In humans, anthrax occurs as a cutaneous (skin) form, as a pulmonary (inhaled) form, or as an intestinal infection after the consumption of contaminated meat. The fifth and sixth plagues on Egypt, as described in Exodus chapters 9 ("The Pestilence") and 10 ("The Boils"), are consistent with anthrax in livestock and humans. In the late 1800s scientists made several important discoveries regarding anthrax: The anthrax germ, *bacillus anthracis,* was the first germ linked to a particular disease. In 1881 French scientist Louis Pasteur developed an inoculation to protect animals from the disease. Anthrax emerged as a potential weapon of bio-terrorism during the twentieth century. Several countries, including the United States, the United Kingdom, Germany, Japan, Iraq, and the former Soviet Union experimented with the bacterium. Beginning in the 1990s, U.S. troops headed for combat in the Persian Gulf were vaccinated for anthrax.

THE DEVELOPMENT OF MODERN MEDICINE AND SURGERY

What advances were made in **medicine during the Middle Ages**?

During the Middle Ages (500–1350) medicine became institutionalized; the first public hospitals were opened and the first formal medical schools were established, making health care (formerly administered only in the home) more widely available and improving the training of doctors. These developments had been brought on by necessity: Europe saw successive waves of epidemics during the Middle Ages. Outbreaks of leprosy began in the 500s and peaked in the 1200s; the Black Death (the bubonic plague) killed about a quarter of the European population; and smallpox and other diseases afflicted hundreds of thousands of people. Consequently, many hospitals—meant to serve the poor—were established, as were the first medical schools, some of them associated with universities that were then forming, such as the University of Bologna (Italy) and the University of Paris (France). In 900 the first medical school was started in Salerno, Italy.

European physicians during the period were greatly influenced by the works of Persian physician and philosopher Rhazes (or Razi; c. 865–c. 930). Considered the greatest doctor of the Islamic world, Rhazes's works accurately describing measles and smallpox were translated into Latin and became seminal references in the Christian world as well. Another prominent Islamic, the scientist Avicenna (or Ibn Sina; 980–1037), produced a philosophical-scientific encyclopedia, which included the medical knowledge of the time. In the West, the work became known as *Canon of Medicine* and with its descriptions of many diseases, including tetanus and meningitis, it remained influential in European medical education for the next 600 years.

Were there **hospitals before the Middle Ages**?

Public hospitals emerged during the Middles Ages (500–1350), as Christianity spread and religious orders set up the facilities to serve the poor. Still, most people received a doctor's care in the privacy of their own homes. The concept of a public health care facility originated in India as early as the third century B.C. when Buddhists established hospital-like installations.

The Middle Ages saw the establishment of facilities more closely resembling modern hospital including Paris's Hotel Dieu (founded in the seventh century); today it is the oldest hospital still in operation. In 970 a hospital in Baghdad (in present-day Iraq) divided physicians into the equivalent of modern-day interns and externs. Its pharmacy disseminated drugs (as well as spices deemed to have medicinal value) from all over the known world.

When was the **first hospital** established **in North America**?

It was in 1503 when the Spanish built a hospital in Santo Domingo in the Dominican Republic (then known as Hispaniola). It is no longer in existence, but ruins remain.

On the North American mainland, the first hospital was opened in Quebec, Canada, in 1639. The first incorporated hospital in the United States was the Pennsylvania Hospital in Philadelphia, chartered in 1751 with the support of statesman Benjamin Franklin (1706–1790).

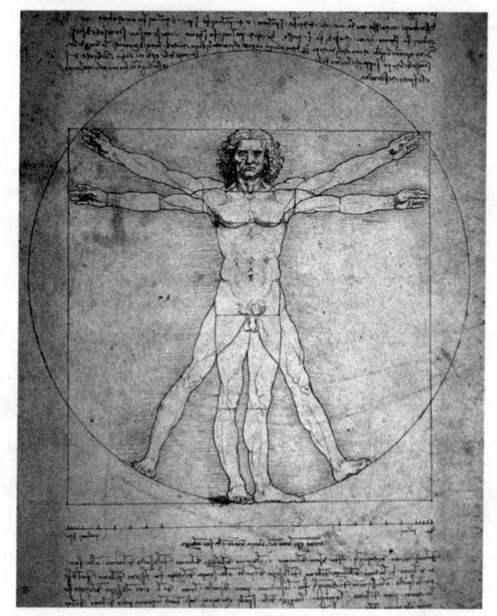

Leonardo da Vinci's study of the human body contributed greatly to the understanding of anatomy. He produced more than 750 anatomical drawings, such as this one on human proportion.

What advances were made in **medicine during the Renaissance**?

The chief advance of the Renaissance (1350–1600) was an improved understanding of the human anatomy. This knowledge was the direct result of dissection, which was prohibited during the Middle Ages (500–1350). The scientific spirit of the Renaissance saw those laws relaxed, and researchers were free to dissect human corpses for study.

Among those who practiced dissection was Leonardo da Vinci (1452–1519). While the Italian artist may be better known for the *Mona Lisa,* he also contributed greatly to the understanding of human anatomy, producing more than 750 anatomical drawings as a result of his studies in dissection.

What was the **first scientific textbook on human anatomy**?

It is a work titled *On the Structure of the Human Body,* written by Belgian physician and professor Andreas Vesalius (1514–1564) and published in 1543, when he was in his late twenties. Like other anatomists during the Renaissance (1350–1600), Vesalius conducted numerous dissections of human cadavers. Publishing his findings and drawings, his textbook soon became the authoritative reference, overturning the works of Greek physician Galen (129–c. 199).

What is ***Gray's Anatomy***?

It is the popular name for *Anatomy of the Human Body, Descriptive and Surgical,* written by English physician Henry Gray (1825 or 1827–1861). First published in

1858, the tome is still considered the standard work on anatomy, and it is in print today in several editions, including *The Concise Gray's Anatomy.* Gray was a lecturer in anatomy at London's St. George's Hospital and was a fellow of Britain's Royal College of Surgeons. He was 33 years old when he compiled the book, which went on to be used by medical students for more than a century.

When did **modern medicine** begin?

The practices of modern medicine have their roots in the 1600s. It was early in the century when the work of English physician William Harvey (1578–1657) demonstrated to the science community that effective medicine depends on knowledge of the body's structure. From 1597 to 1602 Harvey studied medicine at Padua (Italy) under Italian surgeon Fabricius (or Fabrici; 1537–1619) and went on to perform numerous experiments to learn how blood circulates through the body. In his studies, Harvey discarded the accepted method of studying parts of a problem and then filling in the gaps with theory; instead he aimed to understand the entire circulatory system, studying the pulse and heartbeat, and performing dissections on cadavers. He accurately concluded that the heart pumps blood through the arteries to all parts of the body and that the blood returns through the veins to the heart. Putting his discovery into writing, Harvey published *An Anatomical Study of the Motion of the Heart and of the Blood in Animals* in 1628.

Another medical development during the 1600s came not at the hands of a physician or surgeon, but rather a naturalist, Antoni van Leeuwenhoek (1632–1723). A surveyor to the court of Holland, van Leeuwenhoek began making his own microscopes and used them to study organisms invisible to the naked eye—he had discovered microorganisms. Leeuwenhoek also observed (but did not name) bacteria, and he accurately described red blood corpuscles, striated muscle fibers, and the lens of the eye. This amateur scientist also disproved the theory of spontaneous generation, the belief that living organisms could be generated by lifeless matter.

How long did people believe in the theory of **spontaneous generation**?

Spontaneous generation, the theory that living things can develop from nonliving things, originated in prehistoric times and held sway throughout the Middle Ages (500–1350). One of the first scientists to test this theory was Italian physician Francesco Redi (1626–1697). In 1668 Redi demonstrated that as long as meat was covered, maggots would not "form" on it. (When left uncovered, flies would land on the meat, lay eggs, and thus, produce maggots.) Despite Redi's findings, the theory of spontaneous generation continued to influence scientists and physicians for centuries. It was ultimately discredited by the successive experiments of Dutch naturalist Antoni van Leeuwenhoek (1632–1723), French chemist and microbiologist Louis Pasteur (1822–1895), and German physician and pioneer bacteriologist Robert Koch (1843–1910), who together proved that bacteria cause infectious diseases.

What was the **"germ theory"**?

The germ theory, established in the mid-1800s, posited that certain germs cause diseases, refuting the ages-old notion of spontaneous generation. The idea was first put forth by French chemist and microbiologist Louis Pasteur (1822–1895) in a paper he published in 1861. His research, and that of German physician Robert Koch (1843–1910), eventually substantiated the germ "theory" as fact: They proved that the microbe, or germ, is a living organism that can cause disease. Koch was even able to isolate certain bacteria as the causes of particular diseases, including anthrax (for which he published a method of preventive inoculation), tuberculosis, cholera, and rinderpest (a cattle disease). The anthrax germ was the first germ linked to a particular disease—by Koch in 1876. By the end of the 1800s researchers had discovered the kinds of bacteria and other microbes responsible for the plague, diphtheria, dysentery, gonorrhea, leprosy, malaria, pneumonia, tetanus, and other infectious diseases.

Who invented the **vaccine**?

English physician Edward Jenner (1749–1823) is credited with inventing the vaccine; however, evidence suggests that vaccination (inoculation of a substance into the body for the purpose of producing active immunity against a disease) was used in China, India, and Persia (present-day Iran) in ancient times.

In modern times, Jenner pioneered the science of immunology by developing a vaccination against smallpox. The English physician was practicing medicine in rural Gloucestershire in 1796 when he observed that dairymaids who had been sick with cowpox did not contract smallpox, suggesting that they had developed an immunity to the often fatal disease, which then occurred in epidemics. Jenner must have been quite certain of his theory: He chose to test it on an eight-year-old schoolboy, James Phipps, whom Jenner vaccinated with matter from cowpox vesicles from the hands of a milkmaid, Sarah Nelmes. Jenner then allowed the boy's system to develop the immunity he had previously observed in the dairymaids. Several weeks later, Jenner inoculated Phipps with smallpox, and the boy did not become even the least bit ill. The experiment was a success. Jenner continued his experiments for two years and then published his findings, officially announcing his discovery of vaccination in 1798.

As Jenner suspected, vaccines provide immunity by causing the body to manufacture substances called antibodies, which fight a disease. Over the course of the twentieth century, vaccination programs have greatly reduced disease, particularly in developed nations where childhood immunization programs are very effective. By 1977 vaccination had virtually wiped out smallpox.

When did **modern surgery** begin?

Modern surgical techniques were developed during the late Renaissance, largely owing to the work of one man, French surgeon Ambroise Paré (1510–1590), called the "father of modern surgery." Prior to Paré's lifework, physicians had regarded surgery as something lowly. They left this "dirty work" to barber-surgeons. As a young man living in the French countryside, Paré became apprenticed to one such barber-surgeon. When he was only 19 years old, Paré entered Paris's Hotel Dieu hospital to study surgery. Becoming a master surgeon by 1536, he later served as an army surgeon and then as physician to four sixteenth-century French kings—Henry II, Francis II, Charles IX, and Henry III. Paré also built a flourishing surgical practice and authored works on anatomy, surgery, the plague, obstetrics, and the treatment of wounds. Opposing the common practice of cauterizing (burning) wounds with boiling oil to prevent infection, he introduced the method of applying a mild ointment and allowing the wound to heal naturally. Paré was renowned for his patient care, which he based on his personal credo, "I dressed him, God cured him."

When was **anesthesia** first used?

The first use, which was the subject of an embittered debate, was determined to have been in 1842, when Georgia physician Crawford Williamson Long (1815–1878) became the first doctor to use ether as an anesthetic. He went on to use ether in seven more operations before 1846, when he made a public demonstration of anesthesia. Long published his accounts of the experiences in December 1849.

But Boston dentist William T. G. Morton (1819–1868) disputed Long's claim to have been first in using anesthesia. Morton had begun experimenting with anesthetics at about the same time as Long, and on October 16, 1846, he had arranged the first hospital operation using ether as an anesthetic: A tumor was removed from the neck of a patient at the Massachusetts General Hospital in Boston. Nevertheless, it is Long who gets credit for being the first doctor to use ether during an operation.

When were **antiseptics** introduced to surgery?

Antiseptics, which prevent infections, were introduced in the middle of the nineteenth century, and by the end of the century were in widespread use. The introduction in 1846 of anesthetics such as ether and chloroform handled the problem of pain during surgery. But even after successful operations, patients were dying or becoming permanently disabled from infections contracted while in the hospital. These infections, which often became epidemic inside the medical facilities, included tetanus, gangrene, and septicemia. In 1846 Hungarian obstetrician Ignaz Philipp Semmelweis (1818–1865), who was practicing at a Vienna hospital, concluded that infection, in this case puerperal (or childbirth) fever, was coming from inside the hospital ward. His analysis was met with strong rebuttal. While he began practicing antisepsis (cleanliness to reduce infection) and his statistics showed a decrease in mortality rates, the methods did not gain acceptance in the medical community.

Nearly two decades later, in 1864, English surgeon Joseph L. Lister (1827–1912) became interested in French chemist and microbiologist Louis Pasteur's (1822–1895) work with bacteria. While practicing surgery in Glasgow, Lister replicated Pasteur's experiments and concluded that the germ theory applied to hospital diseases. In order to stave off inflammation and infections in his patients, Lister began working with solutions containing carbolic acid, which kills germs. Observing favorable results, Lister reported his findings in 1867 in the British journal of medicine *Lancet*. Many physicians still rejected Lister's claims that antisepsis could reduce the danger of infection. Nevertheless, the medical community began adopting these methods. By the turn of the century, not only had these principles saved lives, they had transformed the way doctors practice medicine: Since doctors could not ensure necessary cleanliness in their patients' homes, hospitals became the preferred place to treat all patients—not just the poor or the very sick.

What were **Pasteur's discoveries on disease**?

Louis Pasteur (1822–1895) may be best known for developing the process that bears his name, pasteurization, but the French chemist and microbiologist made other important contributions to public health, including the discovery of vaccines to prevent diseases in animals and the establishment of a Paris institute for the study of deadly and contagious diseases.

In the 1860s the hard-working Pasteur was asked to investigate problems that French winemakers were having with the fermentation process: Spoilage of wine and beer during fermentation was resulting in serious economic losses for France. Observing wine under a microscope, Pasteur noticed that spoiled wine had a proliferation of bacterial cells that produce lactic acid. The chemist suggested gently heating the wine to destroy the harmful bacteria, and then allowing the wine to age naturally. Pasteur published his findings and his recommendations in book form in 1866. The idea of heating edible substances to destroy disease-causing organisms was later applied to other perishable fluids—chief among them milk.

Pasteur later studied animal diseases, developing a vaccination to prevent anthrax in sheep and cattle. The deadly animal disease is spread from animals to humans through contact or the inhalation of spores. In 1876 German physician Robert Koch (1843–1910) had identified the bacteria that causes anthrax, and Pasteur weakened this microbe in his laboratory before injecting it into animals, which then developed an immunity to the disease. He also showed that vaccination could be used to prevent chicken cholera.

In 1881 Pasteur began studying rabies, an agonizing and deadly disease spread by the bite of infected animals. Along with his assistant, Pierre-Paul-Émile Roux (1853–1933), Pasteur spent long hours in the laboratory, and the determination paid off: Pasteur developed a vaccine that prevented the development of rabies in test animals. On July 6, 1885, the scientists were called on to administer the vaccine to a small boy who was bitten by a rabid dog. Pasteur hesitated to provide the treatment, but as the boy faced certain and painful death from rabies, Pasteur proceeded. Following several weeks of painful injections to the stomach, the boy did not get rabies: Pasteur's treatment was a success. The curative and preventive treatments for rabies (also called hydrophobia) we know today are based on Pasteur's vaccination, which has allowed officials to control the spread of the disease.

In 1888 the Institut Pasteur was established in Paris to provide a teaching and research center on contagious diseases; Pasteur was director of the institute until his death in 1895.

When were **antibiotics invented**?

The idea of antibiotics, substances that destroy or inhibit the growth of certain other microorganisms, dates back to the late nineteenth century, but the first antibiotics were not produced until well into the twentieth century.

The great French chemist Louis Pasteur (1822–1895) laid the foundation for understanding antibiotics when in the late 1800s he proved that one species of microorganisms can kill another. German bacteriologist Paul Ehrlich (1854–1915) then developed the concept of selective toxicity, in which a specific substance can be toxic (poisonous) to some organisms but harmless to others. Based on this research, scientists began working to develop substances that would destroy disease-spreading microorganisms. A breakthrough came in 1928 when Scottish bacteriologist Alexander Fleming (1881–1955) discovered penicillin. Fleming observed that no bacteria grew around the mold of the genus *Penicillium notatum,* which had accidentally fallen into a bacterial culture in his laboratory.

But penicillin proved difficult to extract. It was not until more than a decade later (in 1941) that the substance was purified and tested, by British scientist Howard Florey (1898–1968). Another British scientist, Ernst Boris Chain (1906–1979), developed a method of extracting penicillin, and under his supervision the first large-scale penicillin production facility was completed, making the antibiotic commercially available in 1945. That same year, Fleming, Florey, and Chain shared the Nobel prize in physiology or medicine for their work in discovering and producing the powerful antibiotic, still used today in the successful treatment of bacterial diseases, including pneumonia, strep throat, and gonorrhea.

The term "antibiotic" was coined by American microbiologist Selman A. Waksman (1888–1973), who tested about 10,000 types of soil bacteria for antibiotic capability. In

1943 Waksman discovered a fungus that produced a powerful antibiotic substance, which he called streptomycin. The following year, the antibiotic was in production for use in treating tuberculosis, typhoid fever, bubonic plague, and bacterial meningitis. Although streptomycin was later found to be toxic, it saved countless lives and led to the discovery of many other antibiotics, which have proven both safe and effective.

In 1860 Florence Nightingale (shown in an undated photo) established a training institution for nurses, which marked the beginning of professional education in nursing.

Who was **Florence Nightingale**?

The English nurse, hospital reformer, and philanthropist is considered the founder of modern nursing. The daughter of well-to-do British parents, Florence Nightingale (1820–1910) was born in Florence, Italy. Though she was raised in privilege on her family's estate in England, Nightingale had a natural and irrepressible inclination toward caring for others. Despite her parents' wishes, Nightingale—who, in accordance with the social standards of her set and day had already been presented to the queen—entered a training program for nurses near Dusseldorf, Germany. She went on to study in Paris. In 1853 Nightingale became superintendent of a hospital for invalid women in London.

In 1854 Nightingale took 38 nurses with her to the city of Üsküdar, near Istanbul, Turkey. There, despite great obstacles, she set up a barrack hospital to treat soldiers who were injured in the Crimean War (1853–56), then being fought between Russian forces and the allied armies of Britain, France, the Ottoman Empire (present-day Turkey), and Sardinia (part of present-day Italy). Nightingale set about cleaning the filthy hospital facility; established strict schedules for the staff; and introduced sanitation methods that reduced the spread of infectious diseases such as cholera, typhus, and dysentery. While her methods were considered controversial at first (doctors initially found Nightingale to be demanding and pushy), they got results. Before long, Nightingale was put in charge of all the allied army hospitals in the Crimea.

During the fighting Nightingale visited the front and caught Crimean fever, which threatened her life. By this time she had become so well known that Queen Victoria (1819–1901) was aware of and deeply concerned about Nightingale's illness. By the end of the war, Nightingale's care of the sick and wounded was legendary: Known for

Red Cross workers provide relief to flood victims in Japan, 1916. The society was formed in 1863 to provide care to the wounded and sick in times of war, and later expanded its mission to tend to victims of natural disasters.

walking the floor of the hospital at night, tending her patients, she became known as "the Lady with the Lamp."

After the war Nightingale returned to London and in 1860, with 50,000 pounds sterling, she established a training institution for nurses in London. In 1873 Massachusetts General Hospital in Boston, Bellevue Hospital in New York City, and New Haven Hospital in Connecticut opened the United States' first nursing schools; all of them were patterned after the London program founded by Nightingale.

Nightingale's fierce determination, which ran contrary to her parents' wishes for her as well as to the social standard of the day, made her a legend. And rightly so: Because of her concern for the sick, the standard of care of all patients improved.

When was the **Red Cross** founded?

The Red Cross was founded in Switzerland in October 1863 when the delegates from 16 nations met in Geneva to discuss establishing "in all civilized countries permanent societies of volunteers who in time of war would give help to the wounded without regard for nationality." The idea had been described in a pamphlet published in 1862 by Swiss philanthropist Jean Henri Dunant (1828–1910). In 1859 Dunant was in Italy when French and Italian troops under Napoleon III fought Austrians under Emperor Francis Joseph in an indecisive battle in Lombardy (northern Italy). At Solferino, Dunant observed the suffering of the wounded and immediately organized a group of volunteers to help them.

At the Geneva conference in 1863, the delegates decided the organization symbol and name: The name of the organization comes from its flag showing a red cross on a white background—the inverse of the flag of Switzerland, where the organization was founded. The following August (1864), European delegates met again; this time they were joined by two American observers. The meeting gave rise to the first Geneva Convention, which determined the protection of sick and wounded soldiers, and of medical personnel and facilities during wartime. The Red Cross was adopted as a symbol for neutral aid. In Muslim countries the organization is known as the Red Crescent.

Who was **Clara Barton**?

The American humanitarian was called "the Angel of the Battlefield" for her work during the Civil War (1861–65). Clara (Clarissa Harlowe) Barton (1821–1912) was a nurse in army camps and on battlefields, where she cared for the wounded. When the fighting ended, Barton formed a bureau to search for missing men. This demanding work left her exhausted. Recuperating in Switzerland in 1869, Barton learned of the newly formed International Red Cross (established 1863). She rallied to the aid of that volunteer organization, tending the needs of those wounded in the fighting of the Franco-Prussian War (1870–71), which German chancellor Otto von Bismarck (1815–1898) had provoked in his attempt to create a unified German empire.

In 1877 Barton began working to form the American Red Cross. Her efforts came to fruition in 1881 with the establishment of the first U.S. branch of the International Red Cross. She became the organization's first president, a post she held from 1882 to 1904. When Johnstown, Pennsylvania, experienced a devastating flood in 1889, Barton took charge of relief work there. She subsequently advocated a clause be added to the Red Cross constitution, stating that the organization would also provide relief during calamities other than war. She was successful. It is because of Barton that the Red Cross has become a familiar and welcome site in times of disaster.

When was **insulin** discovered?

Insulin, a hormone that regulates sugar levels in the body, was first discovered in 1889 by German physiologist Oskar Minkowski (1858–1931) and German physician Joseph von Mering (1849–1908). They observed that the removal of the pancreas caused diabetes in dogs. Researchers set about isolating the substance but it was not until 1922 that insulin was used to treat diabetic patients. The first genetically engineered human insulin was produced by American scientists in 1978.

Who invented the **X-ray**?

German physicist Wilhelm Conrad Roentgen (1845–1923) discovered X-rays in 1895—but did not understand at first what they were—which is how they got their name: In science and math, *X* refers to an unknown. By the end of the decade, hospitals had put X-rays to use, taking pictures (called radiographs) of bones and internal organs and tissues to help diagnose illnesses and injuries. Using the new technology, doctors could "see" the insides of a patient. In 1901 Roentgen received the first Nobel prize in physics for his discovery of a short-wave ray.

What did the **Curies contribute to medicine**?

In 1898 French chemists-physicists and husband-and-wife team Pierre (1859–1906) and Marie Curie (1867–1934) discovered radium, the first radioactive element, which proved to be an effective weapon against cancer. They conducted further experiments in radioactivity, a word that Marie Curie coined, distinguishing among alpha, beta, and gamma radiation. Upon Pierre's death in 1906, Marie succeeded him as professor of physics at the Sorbonne. During World War I (1914–18), Curie organized radiological services for hospitals. From 1918 to 1934 she went on to become director of the research department of the Radium Institute of the University of Paris. The Curies's daughter, Irène (1897–1956), followed in her parents' footsteps, becoming a physicist, and marrying (in 1926) another scientist, Frédéric Joliot (1900–1958), who served as director of the Radium Institute for 10 years beginning in 1946. The pair, who were known as the Joliot-Curies, contributed to the discovery and development of nuclear reactors. The Curies and the Joliot-Curies were all Nobel laureates.

What is **Jonas Salk** known for?

American physician Jonas Edward Salk (1914–1995) is familiar to many as the inventor of the polio vaccine. In 1952 more than 21,000 cases of paralytic polio—the most severe form of polio—were reported in the United States. An acute viral infection, poliomyelitis (also called polio or infantile paralysis) invades the central nervous system; it is found worldwide and mainly in children.

In 1953, after years of research that included sorting through all the studies done on immunology since the mid-1800s, Salk announced the formulation of a vaccine, which contained all three types of polio known at the time. Salk tested it on himself first, and then on his wife and three children. Experiencing no side effects and finding the vaccine to be effective, it was then tested on 1.8 million schoolchildren, in a program sponsored by the March of Dimes (then called the National Foundation for Infantile Paralysis). In April 1955 the vaccine was pronounced safe and effective. Salk was duly honored, including with a congressional gold medal and a citation from President Dwight D. Eisenhower (1890–1969). Four years later, American physician

Albert B. Sabin (1906–1993) developed an effective polio vaccine that could be taken orally (versus via injection)—it is the sugar cube so well known to people around the world. That vaccine contains live viruses (Salk's was a killed-virus vaccine). The two vaccines virtually eradicated polio from developed nations.

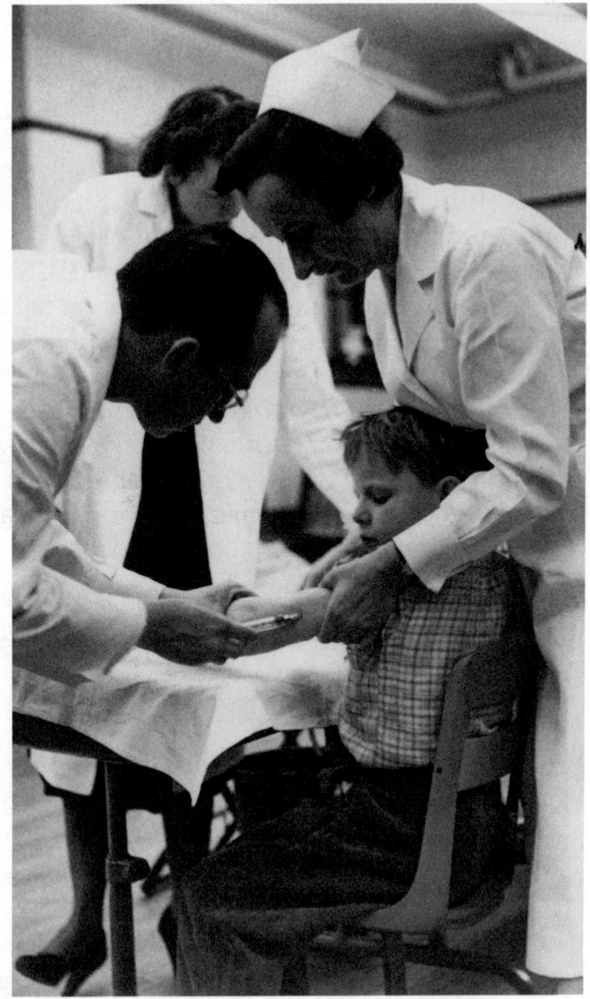

American physician Jonas Salk inoculates a schoolboy during the polio vaccine field trials of 1954.

When was the first human **organ transplant**?

The first human organ transplant occurred on June 17, 1950, at the Little Company of Mary Hospital in Evergreen Park, Illinois. The suburban Chicago hospital, better known as the "baby hospital" for the high number of births there each year, was an unlikely place for this landmark in medical history. And the doctors who took part in the transplant tried to keep the highly experimental procedure quiet. The subject was a 44-year-old woman who suffered from polycystic kidney disease. She received a donor organ, a kidney, from a cadaver, making the procedure even more controversial for the Catholic hospital. (At the time, the church was opposed to the idea that tissue could be taken from a dead person and put into a living person, and that the tissue would then come to life again.) But the three doctors who performed the procedure had the confidence and trust of the sisters running the hospital. Doctors James W. West, Richard H. Lawler, and Raymond P. Murphy were surgeons on the faculty at Loyola's Stricht School of Medicine and the Cook County Hospital but also practiced at Little Company of Mary. The operation was the last resort for the patient, who had seen her mother, sister, and uncle die from the same disease. Word leaked about the operation, and several days after the procedure, when the patient was doing well, the hospital and doctors went public with their breakthrough, making headlines around the world. The transplanted kidney functioned in the patient for about six weeks—enough time for her other kidney to begin working again; she lived another five years before finally succumbing to the disease.

On December 23, 1954, Harvard University physicians led by surgeon Joseph E. Murray (1919–) performed the world's first successful transplant from a living donor, the patient's identical twin brother. The operation took place at Peter Bent Brigham Hospital (now Brigham and Women's Hospital). Since the patient and the donor had the same genetic makeup, organ rejection was not an issue. The procedure saved the patient's life, and the well-publicized breakthrough immediately opened up the possibility for similar transplants (between identical twins) as well as for the transplantation of other organs. Dr. Murray and other Harvard researchers continued working on the problem of rejection, eventually developing new drugs that reduce the possibility that a recipient would reject an organ from a non-relative. In 1990 Murray was awarded the Nobel prize for his pioneering work. He shared the prize with his friend and colleague E. Donnall Thomas (1920–), an innovator in bone marrow transplant.

Today tens of thousands of organs are transplanted each year in the United States. In October 2004 doctors performed the first organ transplant arranged and brokered over the Internet.

When was the first **heart transplant**?

The world's first heart transplant took place on December 3, 1967, in Cape Town, South Africa. Surgeon Christiaan Barnard (1922–2001) conducted the operation; the patient lived for 18 days. Over the next two years, more than a hundred heart transplant operations were performed, but the survival rate was not encouraging. Surgeons have continued the practice with moderately improved results: While some heart recipients have lived as long as six years after the procedure, only 20 percent of the recipients survive more than one year.

How old is **animal experimentation**?

Scientific experimentation using animals—including mice, rats, rabbits, guinea pigs, monkeys, and dogs—dates back to ancient times. But the practice did not become widespread until the late nineteenth century. Clinical experimentation, which includes vivisection (surgery on live animals), has yielded benefits to human health, but because it often results in the suffering and death of the animals, many people are against the practice. Tens of millions of animals are used for experimentation in the United States today; official estimates cite that mice and rats account for some 90 percent of this number. The practice remains controversial as people grapple with the issues surrounding animal rights and weigh these considerations against improved scientific understanding of illnesses.

When was the first **test-tube baby** born?

The process of in-vitro (artificial) fertilization (IVF), in which doctors retrieve an egg from the mother and mix it with the father's sperm in a petri dish or test tube to achieve fertilization, made possible the birth of Louise Brown on July 25, 1978, in Bristol, England. She became the world's first "test-tube baby." The scientific and medical advance of IVF gave parents who were otherwise unable to conceive, another chance at procreation. The procedure has since resulted in numerous successful births: Ten years after Louise Brown was born, an infertile couple had a one-in-ten chance to procreate using IVF technology; and 20 years later the chances had increased to one in five. But as Louise Brown and her parents celebrated her twentieth birthday in 1998, news stories abounded discussing the ethics of in-vitro fertilization. With scientists now able to clone sheep and mice, public opinion often veered toward fear. Would scientists soon be able to clone humans? While technological advances continue to be made, government leaders around the world grapple with how to regulate the use of new "life-giving" technologies such as IVF.

Why is **stem cell research** controversial?

Stem cell research raises important bioethical issues. Stem cells have the potential to develop into all body tissues, and they may be able to replace diseased or defective human tissue. The best source for these cell clusters is human embryos, which are destroyed when the stem cells are extracted. Opponents to the research, including any on the religious right who also oppose abortion, argue that the embryo is a potential human life and therefore should not be destroyed for the sake of science. But proponents of the controversial research say that a variety of treatments and cures for diseases could be gained through scientific advancements made because of stem cell research. Supporters add that the embryos cannot develop on their own, and therefore should be put to use for the sake of better medicine—which could help people who suffer from many different diseases, including diabetes, Alzheimer's, and Parkinson's, thus improving and extending human life. (It is important to note that the embryos exist in laboratories because of advances previously made in reproductive science.)

In August 2001 the George W. Bush administration moved cautiously forward on the issue by allowing stem cell research as long as it is limited to existing cells, the embryos having already been destroyed. In other words, new stem cells cannot be created strictly for the purpose of laboratory work. Bush said he concluded that federal funding should be used to support research on 60 existing genetically diverse stem cell lines, which have the ability to regenerate themselves indefinitely. The president acknowledged the complexity of the issue, saying in a radio address, "At its core, this issue forces us to confront fundamental questions about the beginnings of life and the ends of science. It lies at a difficult moral intersection, juxtaposing the need to protect life in all its phases with the prospect of saving and improving life in all its stages."

But, he added, for the existing stem cell lines, "the life and death decision has already been made." Over the next few years, state legislatures took up the issue, creating a patchwork of policies across the nation by 2005.

EPIDEMICS

How old is the concept of **public health**?

Public health is an old concept, dating back to when people first began living in communities. Through the ages, governments have shown varying degrees of concern for the public health. The ancients Greeks, and the Romans after them, tried to ensure the health of their citizens by providing a supply of clean water (via aqueducts and pipelines), managing the disposal of waste, and working to control disease by hiring public physicians to treat the sick. These measures may have helped prevent the spread of certain diseases, but epidemics still occurred. After the fall of the Roman Empire (c. 476), Europe's civilizations largely ignored matters of public health. Once disease was introduced to a community, it would spread quickly. Epidemics of leprosy, the plague, cholera, and yellow fever ensued.

During the late 1800s European governments began turning their attention to matters of public health in an effort to control the spread of disease. In the United States, the public health became an official concern when in 1866 a cholera epidemic struck the nation—for the eighteenth consecutive year. It was part of a worldwide epidemic that persisted for 12 years. Though governments set up health facilities, including laboratories for the study of infectious disease, by 1893 another cholera pandemic (widespread epidemic) began. During the twentieth century, the measures taken by national governments to safeguard their citizens from health risks have been strengthened by the establishment of regional and local laboratories, public education programs, and the research conducted at universities and other institutions. These combined efforts have made outbreaks of diseases such as diphtheria, dysentery, typhoid fever, and scarlet fever increasingly less common in developed nations. In developing nations, public health officials continue working with international agencies (such as the World Health Organization and other United Nations agencies) to reduce instances and the spread of infectious disease.

When was **leprosy** first diagnosed?

Leprosy is an ages-old disease, described in many historical texts. Mentioned in the Bible, leprosy was introduced in Europe in the 400s B.C., probably by the troops of the Persian ruler Xerxes (c. 519–465 B.C.) as they moved westward. By the twelfth century

What is the plague?

The plague is a general term that refers to any contagious epidemic disease, but usually refers specifically to bubonic plague (which gets its name from the swelling of the lymph nodes, or buboes). A bubonic plague epidemic spread throughout Europe and Asia in the middle of the fourteenth century, killing as much as 75 percent of the population in 20 years; that epidemic came to be known as the Black Death.

An acute infectious disease, the bubonic plague is carried to humans by fleas that have bitten infected rats and other rodents. Human symptoms include high fever, chills, swelling of the lymph nodes, and hemorrhages. Once the bacteria spreads to the lungs, it is quickly fatal. (This form of the disease is called pneumonic plague and can be transmitted from person to person via droplets.)

Improved sanitation, chiefly in developed nations, has reduced the occurrence of the disease. Bubonic plague still occurs, but the development of antibiotics in the twentieth century has greatly reduced the mortality rate.

leprosy had reached epidemic proportions in western Europe, even claiming the lives of rulers (Portugal's Alfonso II died from it in 1223, and Robert I, King of Scots, in 1329). Explorers and settlers from the European continent later carried the infectious chronic skin disease to the New World, where it was previously unknown.

The cause of leprosy was unknown. While some theorized it was contagious, others asserted that it was hereditary or was caused by eating certain foods (even potatoes were at one time blamed for originating the affliction). The disease gradually disappeared from Europe, attributable to improved living conditions, better nutrition, and, later, the advent of drugs that are effective in treatment.

The first clinical description was not made until 1874 when Norwegian physician Gerhard Henrik Hansen (1841–1912) discovered the leprosy bacterium. Since then the disease has also been called Hansen's disease. Today, leprosy afflicts about 5 million people worldwide. It is endemic (native) to tropical or subtropical regions, including Africa, Central and South America, India, and Southeast Asia. Most cases of leprosy that occur in the United States are among immigrants from areas where the disease is endemic. Beginning in the mid-1950s, the Roman Catholic nun Mother Teresa (1910–1997) of Calcutta ministered to those afflicted with leprosy, setting up colonies for their care.

Does **yellow fever** still exist?

Yellow fever, an acute infectious disease, does still exist in some select areas of the world. Outbreaks still occur in jungle areas. The disease was once widespread, afflicting people in tropical climates such as Central and South America, Africa, and Asia. But with exploration during the 1500s and 1600s, and the opening of trade routes during the 1700s, the disease spread to North America by 1699, when there were epidemics in Charleston, South Carolina, and Philadelphia, Pennsylvania; three years later, an epidemic broke out in New York City. Yellow fever first materialized in Europe in 1723. An epidemic in Philadelphia in 1793 was determined to have been carried there aboard a ship from the West Indies; nearly all of the city's people were afflicted by the fever, and more than 4,000 people died in what has been called the worst health disaster ever to befall an American city.

Breakthroughs in controlling yellow fever came in the late 1800s and early 1900s. In 1881 Cuban physician Carlos Finlay (1833–1915) wrote a paper suggesting that yellow fever was transmitted by mosquitoes. This was proved to be true by U.S. Army surgeon Walter Reed (1851–1902), who in 1900 headed a commission sent to Cuba to investigate the cause and mode of transmission of yellow fever. With this knowledge, U.S. Army officer and physician William Gorgas (1854–1920) applied strict measures to destroy mosquitoes in Havana, eventually eliminating yellow fever from the Cuban port city. Serving as chief sanitary officer of the Panama Canal Commission from 1904 to 1913, Gorgas implemented similar measures in the Panama Canal Zone, where the disease had been a menace. Again his methods proved effective, greatly reducing the instances of yellow fever, which allowed the canal to be completed.

In 1937 the 17-D vaccine was developed by American physician and bacteriologist Max Theiler (1899–1972). The vaccine was found to be effective in combating yellow fever. In 1951 Theiler was awarded the Nobel prize in physiology or medicine for his discoveries concerning the infectious disease. Conquering yellow fever was one of the great achievements of modern medicine.

What was the **first disease conquered** by human beings?

Smallpox was the first disease eradicated by medicine. Caused by a virus spread from person to person through the air, smallpox was one of the most feared diseases and there was no treatment for it. Before the discovery of the New World, smallpox epidemics swept across Africa, Asia, and Europe, leaving victims scarred and/or blind, and killing countless millions. When explorers set out to find new trade routes and landed in North and South America, they brought the disease with them, infecting the indigenous peoples.

But once a person had the disease, he or she would not contract it again. This and other observations led British physician Edward Jenner (1749–1823) to develop a successful vaccine against the disease. Prior to the vaccine, the only preventive method was inoculation of the disease itself, which sometimes led to further spread of the disease. For example, in 1777 American general George Washington (1732–1799) obtained congressional approval to inoculate the entire Continental army against smallpox, but the results were mixed.

After its discovery in 1798 the use of Jenner's vaccine quickly spread. The first vaccine given in the United States was in 1799 by a Harvard physician. During the 1800s many countries passed laws requiring vaccination. Improvements in the vaccine resulted in the elimination of smallpox from Europe and North America by the 1940s. When the World Health Organization (WHO) was created by the United Nations in 1946, one of its aims was to reduce the instances of smallpox around the world. Immunization programs brought this about: The last natural occurrence of the disease was reported in October 1977 in Somalia, Africa. When no further cases were documented within the next two years, the disease was considered eradicated.

Who was **Typhoid Mary**?

Typhoid Mary was the name given to Mary Mallon (c. 1870–1938), the first known carrier of typhoid fever in the United States. Though Mallon had recovered from the disease, as a cook in New York City area restaurants she continued to spread typhoid fever germs to others, infecting more than 50 people between 1900 and 1915. The New York State sanitation department connected her to at least 6 typhoid fever outbreaks there. Officials finally—and permanently— institutionalized her in 1914 to prevent further spread of the acute infectious disease.

When was **AIDS** first diagnosed?

The first AIDS (acquired immunodeficiency syndrome) cases were identified in 1981 by physicians in Los Angeles and New York City. Since that time researchers traced possible cases of the disease back to 1969. The human immunodeficiency virus (HIV), which severely damages the body's ability to fight disease, is transmitted through sexual contact, shared drug needles, and infected blood transfusions. While the disease was believed to have been transmitted somehow to humans from monkeys (since research shows HIV to be similar to simian immunodeficiency viruses), HIV has never been isolated in any wild animal. While the source of the deadly disease has not been definitively determined, scientists believe that infection began in Africa during the 1960s and 1970s when significant numbers of people migrated from rural areas to cities. The overcrowding and unemployment that resulted contributed to the spread of sexually transmitted diseases.

AIDS is now considered endemic to many developing nations, where it is spread mostly among heterosexual men and women. In developed nations, education programs have made the public aware of how the disease is transmitted, helping curb the spread of HIV. Drug treatments are still being developed to treat HIV/AIDS; no cure has been discovered. A 2004 report from the United Nations stated that there were 38 million people living with HIV in the world, almost 70 percent of them in sub-Saharan Africa. The same report states that more than 20 million people had died since the first cases were identified in 1981.

Has **the flu** reached epidemic proportions?

Yes, influenza (also called the grippe), a contagious virus, has reached worldwide epidemic proportions three times in the twentieth century alone. An epidemic occurred in 1918 and lasted into the next year; 20 million people died across the globe, including half a million Americans. Many of these deaths occurred as a result of secondary infections in patients whose immune systems were weakened by the flu. The advent of antibiotics, which became commercially available in 1945, helped prevent any subsequent flu outbreak from becoming so deadly. From 1957 to 1958 the Asian flu caused a worldwide epidemic, and from 1968 to 1969, the Hong Kong flu spread quickly to cause an epidemic. But dramatically fewer deaths were caused in these years due to the availability of antibiotics to control secondary infections.

BEHAVIORISM, PSYCHOLOGY, AND MENTAL HEALTH

Why are **Pavlov's dogs** well known?

Russian physiologist Ivan Pavlov (1849–1936) carried out famous experiments with dogs, which were intended to demonstrate conditioned reflex. Noticing that the laboratory dogs would sometimes salivate merely at the approach of the lab assistants who fed them, Pavlov, who was already a Nobel laureate for his research on digestion, set out to determine whether he could turn normally "unconditioned" reflexes or responses of the central nervous system into conditioned reflexes. He demonstrated that if a bell is rung every time a dog is fed, eventually the dog becomes conditioned to salivate at the sound of a bell, even if there is no food present. In this way, Pavlov substituted artificial stimulus (the ringing of the bell) for natural or environmental stimulus (food) to prompt a physiological reaction (salivation). Based on these experiments, Pavlov concluded that all acquired habits depend on chains of conditioned reflexes. This conclusion contributed to the development of behaviorism.

What is **behaviorism**?

Behaviorism is a school of psychology that attempts to explain human behavior in terms of responses to environmental stimuli. Influenced by the conditioned reflex demonstrated by Russian physiologist Ivan Pavlov (1849–1936), American psychologist John Broadus Watson (1878–1958), of Johns Hopkins University, codified and popularized the theory, which discards introspection and consciousness as influences on human behavior. Behaviorism was further studied by another American psychologist and Harvard professor B. F. Skinner (1904–1990). Skinner focused his work on patterns of responses to observable stimuli (versus unobservable stimuli such as introspection and conscience) and external rewards. Applied to human learning, Skinner's theories on behaviorism affected educational methods, which tangibly reward good behavior.

Sigmund Freud (pictured in 1920) believed that human behavior and all mental states are influenced by repressed and forgotten impressions, many from childhood.

What did **Freud** believe?

The Austrian neurologist believed that human behavior and all mental states are influenced by repressed and forgotten impressions, many of them from childhood. Sigmund Freud (1856–1939) further believed that by uncovering these impressions (collectively called a complex), he could effect a cure for his patient. Freud regarded infantile mental processes, including infantile sexuality, of particular importance to the unconscious, and therefore influential to human behavior.

While he initially used hypnosis (a sleeplike state in which the patient is open to suggestion) as a method of revealing the unconscious, Freud later turned to a new form of treatment called free association. By this method, a patient talks about whatever is on his or her mind, jumping from one idea to the next. The memories and feelings that surface through free association are then analyzed by the therapist to find the root of the patient's mental or emotional problem. Freud also interpreted his patients' dreams, which he believed are unconscious representations of repressed desires. Free association and dream analysis are the cornerstones of psychoanalysis.

In analyzing human behavior, Freud came to the conclusion that the mind (or psyche) is divided into three parts: the id, the ego, and the superego. The id is the source of instincts; the ego is the mediator between those instincts and reality; and the superego is the conscience. The superego functions to reward or punish through a system of moral attitudes and a sense of guilt. The theories of psychoanalysis hold that if the parts of the mind oppose each other, a mental or emotional disorder (called a neurosis) occurs.

Freud's theories revolutionized the fields of psychiatry and psychology. They also influenced methods and philosophies of child-rearing and education. While psychoanalysis has been credited with helping millions of mentally ill patients, Freud's theories have also been rejected or challenged by many.

What does **"Jungian"** mean?

"Jungian" refers to the analytical psychology founded by Swiss psychiatrist Carl Gustav Jung (1875–1961). Early in his career Jung conducted experiments in mental association and through this work came into contact with famed psychoanalyst Sigmund Freud (1856–1939) in 1907. While initially in harmony with each other, Jung later broke with Freud's theories, establishing his own doctrines of human behavior.

Like Freud, Jung believed that the unconscious (that part of the mind of which a person is unaware) affects human behavior. But unlike his Austrian colleague, Jung denied that neuroses have any sexual basis. Instead, Jung believed that many factors influence human behavior, including the personalities of one's parents. He also believed in something he described as the "collective unconscious": In his revolutionary work *Psychology of the Unconscious,* published in 1912, Jung asserted that there are two dimensions of the unconscious: the personal and the collective. The collective unconscious, according to Jung, is made up of those acts and mental patterns that are shared by members of a culture or are perhaps universally shared by all humankind. He theorized that the collective unconscious manifests itself in archetypes—images, patterns, and symbols that appear in dreams and fantasies as well as in mythology, religion, and literature. Jung believed that the collective unconsciousness can serve as a guide to humanity and, therefore, he taught that therapy should make people aware of it. Jung's theories of archetypes, or universal symbols, have influenced such diverse fields as anthropology, art, filmmaking, and history.

Jung later developed a system for classifying personalities (into introverted and extroverted types) and distinguishing among mental functions (classifying them as thinking, feeling, sensing, or intuitive). Jung taught that therapists should help their patients balance introversion (relying only on oneself for personal fulfillment) with extroversion (relying on others for personal fulfillment). Jung's system of classifications, or "typology," has been used to develop theories of personality types and their influences on human behavior.

Who was **Dorothea Dix**?

Dorothea Lynde Dix (1802–1887) was a philanthropist and among the first American women to become active in social reform. Having been headmistress of her own school for girls in Boston from 1821 to 1836, in 1841 Dix toured Massachusetts state correctional institutions, where she was shocked to see deplorable treatment of the mentally ill. Thereafter Dix became an impassioned advocate for the mentally ill. Leading a drive to build hospitals for the specialized care of those afflicted with mental illnesses, Dix appealed to the consciences of legislators and philanthropists. She was successful in establishing mental hospitals throughout the United States, Canada, and Europe, many of which still bear her name. Dix's campaign for humane treatment of the mentally ill transformed American attitudes and institutions in the two decades that led up to the Civil War (1861–65). During the war she acted as superintendent of the U.S. Army nurses. She also worked to improve prison conditions during her lifetime.

PHILOSOPHY

What is **philosophy**?

From the Greek *philo,* meaning "love of," and *sophia,* meaning "wisdom," philosophy is literally a love of wisdom. In practice, it is the pursuit of understanding the human condition—how, why, and what it means to exist or *to be.* Philosophers use methods such as observation and questioning to discern the truth. Philosophy is traditionally divided between Eastern thought and Western thought. Further, Western thought consists of five branches: metaphysics (concerned with the nature of the universe or of reality); logic (the laws of reasoning); epistemology (the nature of knowledge and the process by which knowledge is gained); ethics (the moral values or rules that influence human conduct); and aesthetics (the nature of beauty or the criteria for art).

How **old is philosophy**?

Philosophy, apart from religion, emerged in the East and the West at about the same time—roughly 600 years before Christ (before the common era). It was then that thinkers in Greece began questioning the nature of existence, and it was c. 500 B.C. when Confucianism was formulated in China. Philosophy has long been intertwined with religion, including Hinduism, which developed about 1500 B.C. But as a pursuit of wisdom in and of itself, philosophy is roughly 2,600 years old.

Western philosophy is divided into three major periods: ancient (c. 600 B.C.–c. A.D. 400); medieval philosophy, often called Scholasticism, which also included Eastern thinkers (400s–1600s); and modern philosophy (since the 1600s).

How old is **Taoism**?

It dates back to the sixth century B.C. when it was founded by Chinese philosopher Lao-tzu (c. 550 B.C.–?). Master Lao, as he was known, believed in inaction and simplic-

Confucius was a highly revered Chinese philosopher who believed that the family is the model for all human relationships.

ity, which he combined with religious practices to form the mystical philosophy of the Tao (Dao), the path of virtuous conduct. Lao-tzu reasoned that since humans face a "cloud of unknowability," they ought not to react to things at all: He viewed the world as a pendulum, with the Tao as its hinge. Anyone who struggles against the current of life is like an insect caught at the end of a pendulum—swinging back and forth, and suffering with each movement. But by crawling along the hinge (Tao) to reach the top, a place of complete stillness is found. Lao-tzu advised that people do away with their desires, avoiding that which is extreme, extravagant, or excessive, and steer clear of any competition. Many of these ideas are embodied in a work usually ascribed to him, *Tao-te Ching* (Classic of the Way of Power). However, modern scholars now believe that tome to be the work of his followers.

Taoism is still relevant to many today. When it was developed some 26 centuries ago, the philosophy filled a spiritual void that was not addressed by the practical doctrines of traditional Confucianism. One legend has it that Lao-tzu rebuked a young Confucius (551–479 B.C.) for his pride. The Tao also contributed greatly to Buddhism, especially in its emphasis on meditation and sudden enlightenment.

One of the great thinkers of the Chinese Taoist school was teacher and philosopher Chuang-tzu (fourth century B.C.), who constructed an nonpolitical, transcendental philosophy that promoted an individual's spiritual freedom. His self-titled work (*Chuang-tzu*) is another classic of Taosim.

Why was **Socrates** condemned to death?

Socrates (c. 470–399 B.C.), the Greek philosopher who is credited, along with philosophers Plato (c. 428–347 B.C.) and Aristotle (384–322 B.C.), for laying the foundations of Western thought, had many followers in his own time. However, his ideas and methods were controversial, too, which led him to be tried before judges and sentenced to death, which he carried out by drinking hemlock (poison). He had been charged for not worshiping the Athenian gods and for corrupting the young.

Who was Confucius and who do people still quote him?

Confucius (551–479 B.C.) was a Chinese philosopher whose real name was K'ung Ch'iu; Confucius is the Latinized version. Born into a class of lesser nobility in the province of Lu, his father died before Confucius had turned three and he was raised in humble circumstances by his mother. He lived in the middle of China's feudal period, when there were enormous problems, including famine and poverty, which had been brought on by weak emperors and, consequently, chronic warfare among rival feudal states. Because of his upbringing, Confucius possessed a profound sympathy for the common people. In his view, the feudal princes, only interested in their own personal gain, were responsible for the suffering of the people. Confucius set out on a reform mission: Believing that good government can only be achieved by ethical leaders, Confucius endeavored to train a new generation of them. He taught literature and music (important in building character), conduct, and, most importantly, ethics, to anyone who wanted to learn. He is regarded as the first Chinese teacher to offer education freely, that is to say, to all comers, rather than just the privileged. The great philosopher is revered for his belief that the family is the model for all human relations. Confucianism regards the chief relations in life to be those between ruler and subject, father and son, elder and younger brother, husband and wife, and friends. Most importantly, he taught students that rulers are responsible for the happiness of their subjects. He believed that government leaders need not be expert administrators. Instead, they must be humane, honest, and above corruption and personal gain. Some of his students went on to hold positions of power in city governments.

Over the centuries, he became the most venerated person in Chinese history, but his teachings transcend cultural lines, which is why Confucian wisdom, including the principle "What you do not want done to yourself, do not do to others," is often quoted. The philosophy's maxims are set forth in the work *Lun-Yü* (Analects), recorded by his followers.

Except for his time spent in military service, Socrates lived his entire life in Athens, where he was as well known for his disheveled appearance as for his moral integrity, self-control, and quest for wisdom. He lived during a time when attention was turning away from the physical world (of the heavens) and toward the human world (of the self, the community, and the law). He participated in this turning point by walking the streets of Athens, engaging people—including rulers who were supposed to be wiser than he—in conversation. In these conversations, he employed what came to be known as the "Socratic method" or dialectic, a series of seemingly simple questions designed to elicit a rational response. Through the line of questioning, which usually centered around a moral concept such as courage, the person being questioned was intended to realize that he did not truly know that which he thought he knew. Socrates's theory was that once the person being questioned realized his weak understanding, he could divest himself of false notions, and was then free to participate in the quest for knowledge. These philosophical "disputes," however, gained Socrates many enemies.

Though he left no writings, Socrates's student, Plato, documented his recollections of dialogues with his teacher. A staunch believer in self-examination and self-knowledge, Socrates is credited with saying that "the unexamined life is not worth living" (some ascribe the quote to Plato). Socrates also believed that the psyche (or "inner self") is what should give direction to one's life—not appetite or passions. A seminal figure in Greek (and Western) thought, philosophy that predates him is termed "pre-Socratic."

What was **Plato**'s relationship to Socrates and Aristotle?

The Athens-born Plato (originally, Aristocles) was Socrates's disciple and Aristotle's teacher. The philosophies of these three men combined to lay the foundations of Western thought.

With the death sentence of his spiritual guide, Socrates, in 399 B.C., Plato's (c. 428–347 B.C.) dissatisfaction with the Athenian government reached its peak. Traveling throughout the Mediterranean after the death of Socrates, Plato returned to Athens in 387 B.C., and one mile outside of the city he established the Academy, a school of philosophy supported entirely by philanthropists; students paid no fees. One of the pupils there was young Aristotle (384–322 B.C.), who remained at the Academy for 20 years before venturing out on his own.

Plato wrote a series of dialogues in which Socrates figures prominently. The most highly regarded of these is the *Republic,* in which Plato discusses justice and the ideal state. It was his belief that people would not be able to eliminate injustice from society until rulers became philosophers: "Until all philosophers are kings, or the kings and princes of this world have the spirit and power of philosophy, and political greatness and wisdom meet in one, and those commoner natures who pursue either to the exclusion of the other are compelled to stand aside, cities will never have rest from their evils—no, nor the human race." Also on the subject of the ideal state, Plato wrote but did not finish *Laws.* His other works include *Symposium,* which considers ideal love; *Phaedrus,* which attacks the prevailing notions about rhetoric; *Apology,* which is a rendering of the speech Socrates delivered at his own trial in 399 B.C.; and *Phaedo,* which discusses the immortality of the soul and which is supposed to be a record of Socrates's last conversation before he drank hemlock and died.

What is Plato's **theory of forms**?

The theory, or doctrine, of forms (also called the theory of ideas) is Greek philosopher Plato's (c. 428–347 B.C.) expression of his belief that there are forms that exist outside the material realm, and therefore are unchanging—they do not come into existence, change, or pass out of existence. It is these ideas that, according to Plato, are the objects or essence of knowledge. Further, he posited that the body, the seat of appetite and passion, which communes with the physical world (rather than the world of ideas or forms), is inferior to the intellect. He believed the physical aspect of human beings to be irrational while the intellect, or reason, was deemed to be rational.

The origins of Plato's theory can be traced to Socrates (c. 470–399 B.C.), who believed that the psyche (inner spirit) has intuitive access to divinely known principles or truths, which he attempted to formulate through his conversations with others. Indeed, the Socratic dialogues, written by Plato, reveal that Socrates was striving to define the exact nature of the traditional Greek moral virtues of piety, temperance, and courage.

Did **Aristotle develop his own philosophy**?

A student of Plato's for 20 years, Aristotle's ideas were unquestionably influenced by his teacher. However, Aristotle developed his own doctrine, which he applied to many subjects.

Aristotle rejected Plato's theory of forms (or theory of ideas). While Aristotle, too, believed in material (the physical being) and forms (the unchanging truths), unlike his teacher, he believed that it is the concrete (material) that has substantial being. Aristotle viewed the basic task of philosophy as explaining why and how things are, or how they become what they are. It is for this reason that Aristotle had not only a profound and lasting influence on philosophy but on scientific spirit.

What does **"epicurean"** mean?

While "epicurean" has come to refer to anything relating to the pleasure of eating and drinking, it is an oversimplification of the beliefs of the Greek philosopher Epicurus (341–270 B.C.), from whose name the word was derived. While Epicurus did believe that pleasure is the only good, and that it alone should be humankind's pursuit, later scholars misinterpreted the philosophy as a license for sensory excess. In actuality, Epicurus defined pleasure not as unbridled sensuality but as freedom from pain and as peace of mind, which can only be obtained through simple living.

In about 306 B.C., Epicurus established a school in Athens, which came to be known as the Garden School because residents provided for their own food by gardening. There he and his students strived to lead lives of simplicity, prudence, justice, and honor. In this way, they achieved tranquility—the ultimate goal in life, according to the

Why is Aristotle considered one of the greatest minds in Western history?

The system of philosophy that Aristotle (384–322 B.C.) developed became the foundation for European philosophy, theology, science, and literature. The Aristotelian system may be so much a part of the fabric of Western culture that the only effective way to describe his philosophy is through example.

Among his writings on logic is *Organon,* meaning "tool" or "instrument." Here he defines the fundamental rules for making an argument. While other thinkers may well have formulated the argument before Aristotle, no one had made a systematic study of it. In *Organon,* Aristotle puts forth a method for coming to a conclusion based on circumstantial evidence and prior conclusions rather than on the basis of direct observation. This deductive scheme, called a syllogism, is made up of a major premise, a minor premise, and a conclusion. For example: every virtue is laudable (major premise); courage is a virtue (minor premise); therefore courage is laudable (conclusion). (It is worth noting, however, that the belief in deductive logic was later rejected by English philosopher Sir Francis Bacon [1561–1626] in 1620, in favor of an inductive system, or one that is based on observation.)

In *Poetics,* Aristotle expounded upon his literary views. He maintained that epic and tragedy portray human beings as nobler than they truly are, while comedy portrays them as less noble than they are. In order to explain how tragedy speaks to the emotions of the spectator, Aristotle introduced the idea of catharsis. He separated tragedy from epic with the distinction that tragedy maintains unity of plot (later translated as unity of plot, time, and place), while the epic does not. Because of the keen understanding evident in *Poetics,* the work has illuminated literary criticism since antiquity.

In addition to logic and rhetoric, Aristotle wrote on natural science (*Physics, On the Heavens, Parts of Animals,* and *On Plants*) and on ethics and politics (*Politics*). His great philosophical work was *Metaphysics,* so named because, in the body of his works, it comes after (the Greek word for which is *meta*) the work *Physics.* Metaphysics as a philosophy is the study of substance, or the nature and structure of reality. It is considered one of five major branches of Western philosophy. In modern thought, metaphysics can include many disciplines, such as cosmology (the study of the origins and structure of the universe) and theology (the study of religion). Most of the great philosopher's writings are compilations of notes from lectures he delivered to his students at the Lyceum, also called the Peripatetic School, in Athens. Among his pupils there were Greek leaders, including Alexander the Great (356–323 B.C.).

philosophy of Epicureanism. He further believed that intellectual pleasures are superior to sensual pleasures, which are fleeting. In fact he held that one of the greatest and most enduring pleasures is friendship. These ideas were put forth by the Greek philosopher and writer Lucretius (c. 99–c. 55 B.C.) in his poem *On the Nature of Things.*

While the Epicurean school endured for several centuries, ultimately Christian leaders deemed the philosophy a pagan creed. However, some critics have posited that the writer of Ecclesiastes in the Old Testament of the Bible was likely a member of the Garden School and that the Epistles of Saint Paul in the New Testament were strongly influenced by Epicurean thought. In more recent times, Thomas Jefferson (1743–1826), author and signatory of the Declaration of Independence and third president of the United States, was a self-proclaimed Epicurean.

What was the **philosophy of the Middle Ages**?

During medieval times (500–1350) philosophers concerned themselves with applying the works of ancient Greek thinkers, such as Aristotle (384–322 B.C.) and Plato (c. 428–347 B.C.), to Christian thought. This movement, which spanned most of the Middle Ages and reached its high-water mark in the thirteenth century, was called Scholasticism since its proponents were often associated with universities: the word *scholastic* is derived from the Greek *scholastikos,* meaning "to keep a school." In the simplest terms, the goal of Scholasticism has been described as "the Christianization of Aristotle." Indeed, medieval philosophers strived to use reason to better understand faith. Scholasticism was, therefore, both rational and religious. The movement was also an interesting occasion of East meets West: The commentaries of Islamic philosophers, principally Avennasar (c. 878–950), Averroës (1126–1198), and Avicenna (980–1037), figured prominently in Scholasticism. Theologians, including St. Anselm and St. Thomas Aquinas, used the non-Christian philosophy—both of the ancient Greeks and of Muslim thinkers—to better understand their own Christian faith.

Who were the great **Islamic philosophers of the Middle Ages**?

Three thinkers of the Islamic world stand out as important interpreters of Greek thought, and therefore, as a bridge between ancient philosophy and the Scholasticism of the Middle Ages: their Latin names are Avennasar, Averroës, and Avicenna.

Avennasar (c. 878–950), who studied with Christian Aristotelians in Baghdad (Iraq), proved so adept at applying the teachings of Aristotle (384–322 B.C.) to Muslim thought that he became known as "the second Aristotle" or the "second teacher." He posited that philosophy and religion are not in conflict with each other; rather, they parallel one another. Also known for his work in interpreting the great Aristotle for the Muslim world, Avicenna (980–1037) is sometimes referred to as the "third teacher." He was also the first to expand the distinction between essence and existence. Averroës (1126–1198) also was no stranger to Aristotle, writing commentaries on him as well as Plato (specifically, the *Republic*); he also wrote on religious law and philosophy as well as religion and logic.

Who were the great thinkers of **Scholasticism**?

Just as Islamic philosophers reinterpreted faith by applying reason, subordinating revelation to reason, Western philosophers endeavored to incorporate the doctrines of Greek philosophy into the theology of the Christian church. Leaders in this movement included St. Augustine (Augustine of Hippo), St. Anselm, and St. Thomas Aquinas.

Augustine of Hippo (354–430) lived during a time when the last vestiges of the pagan world of the Romans was giving way to Christianity. His theological works, including sermons, books, and pastoral letters, reveal a Platonic influence, foreshadowing the movement of Scholasticism that emerged more than six centuries later (during the eleventh century). Augustine believed that understanding can lead one to faith and that faith can lead a person to understanding. He also argued that Christians can understand the nature of the Trinity by examining their own nature (through introspection).

One of Scholasticism's founders, Anselm (c. 1033–1109) was a Benedictine monk who in 1093 became archbishop of Canterbury. He became famous for writing about the attributes of God (in his work *Monologion*) and for trying to prove the existence of God (in *Proslogion*) by rational means alone, arguing that God is that of which nothing greater can be thought; that of which nothing greater can be thought must include existence (if it did not, then something greater could be thought); and therefore God necessarily exists.

But the greatest figure of Scholasticism was St. Thomas Aquinas (1225–1274), who is also one of the principal saints of the Roman Catholic Church. In 1879 his philosophical works were declared the official Catholic doctrine by Pope Leo XIII (1810–1903). While he was teaching at universities in Cologne (Germany) and Paris between 1248 and 1272, Thomas Aquinas penned his major works, *Summa contra gentiles* (1259–64) and *Summa theologica* (1266–73). He discarded the Platonic leanings of St. Augustine (to whom truth was a matter of faith), interpreting Aristotle's naturalistic philosophy: Similar to the Islamic philosopher Avennasar (c. 878–950), who argued that religion and philosophy are not in conflict with each other, Thomas Aquinas believed faith and reason are in harmony with each other. His work is considered the greatest achievement of medieval philosophy, making the thirteenth century Scholasticism's golden age. Thomas Aquinas was canonized in 1323 and was proclaimed a doctor of the Catholic Church in 1567.

What were **Sir Francis Bacon**'s beliefs?

The English philosopher, author, and statesman was one of the great minds of the Scientific Revolution of the 1500s and 1600s, during which the way that Europeans viewed themselves and the universe underwent a dramatic change. Bacon (1561–1626) believed that humankind's accepted notions about nature should be aggressively challenged. As a young man studying at Trinity College, he concluded that the Aristotelian system (or deductive logic) was without merit; Bacon favored observation (or inductive logic) as a system for interpreting and understanding nature. He argued that the understanding of nature was being held back by the blind acceptance of the beliefs of ancient philosophers such as Aristotle (384–322 B.C.) and Plato (c. 428–347 B.C.). A religious person, Bacon maintained that theology should *not* be questioned: He believed that rational inquiry can unlock the secrets of nature—but not of the human soul. Bacon therefore insisted on the separation of philosophy and theology, an idea that ran counter to the academic traditions of the time. Consequently he was a staunch proponent of educational and scientific reform.

English philosopher Sir Francis Bacon believed that nature was best understood by direct observation. (Original engraving by S. Freeman.)

Trained in law, Bacon served as a royal diplomat in France, was admitted to the bar, elected to Parliament, and served in public office (including the jobs of solicitor general and attorney general). He penned several seminal works, including *Essayes* (1597), which consists of practical wisdom and observations; *Advancement of Learning* (1605), a survey of the state of knowledge (Bacon was attempting to enlist the support of the king in the total reform of education and science in England); and *Novum Organum* (1620), in which he put forth his method for understanding nature by an inductive system, based on direct observation (versus Aristotle's deductive method, which was based on circumstantial evidence and prior conclusions).

What is the **"doctrine of idols"**?

This was a phrase used by English philosopher Sir Francis Bacon (1561–1626) in his written attack on the widespread acceptance of the thinking of ancient philosophers such as Aristotle (384–322 B.C.) and Plato (c. 428–347 B.C.) and the founder of modern astronomy, Copernicus (1473–1543). In his 1620 work, *Novum Organum,* Bacon vehemently argues that human progress is held back by adherence to certain concepts, which it does not question. By hanging on to these concepts, or "idols," humankind may proceed in error in its thinking. The double edge is that in holding to notions accepted as true, we run the danger of dismissing any new notion, a tendency Bacon characterized as arrogance. A quality that goes hand in hand with arrogance is skepticism: In adhering to that which we know, we are likely to dismiss any new ideas. To combat these obstacles, Bacon advocated a method of persistent inquiry. He believed that humans can understand nature only by carefully observing it with the help of instruments. He went on to describe scientific experimentation as an organized endeavor that should involve many scientists and which requires the support of leaders. Thus, Bacon is credited with no less than formulating modern scientific thought.

Why is **Descartes** considered the **"father of modern philosophy"**?

French mathematician and philosopher René Descartes (1596–1650) was living in Holland in 1637 when he published his first major work, *Discourse on Method.* In this treatise, he extends mathematical methods to science and philosophy, asserting that all knowledge is the product of clear reasoning based on self-evident premises. This idea, that there are certitudes, provided the foundation for modern philosophy, which dates from the 1600s to the present.

Descartes may be best known for the familiar phrase "I think, therefore I am" (*Cogito ergo sum,* in Latin). This assertion is based on his theory that only one thing cannot be doubted, and that is doubt itself. The next logical conclusion is that the doubter (thinker) must, therefore, exist. The correlation to the dictum (I think, therefore I am) is dualism, the doctrine that reality consists of mind and matter: Since the thinker thinks and is, he or she is both mind (idealism) and body (matter, or material). Descartes concluded that mind and body are independent of each other, and he formulated theories about how they work together. Modern philosophers have often concerned themselves with the question of dualism.

Descartes's other major works include *Meditations on First Philosophy* (1641), which is his most famous, and *Principles of Philosophy* (1644). His philosophy became known as Cartesianism (from *Cartesius,* the Latin form of his name).

What is **empiricism**?

Empiricism is the philosophical concept that experience, which is based on observation and experimentation, is the source of knowledge. According to empiricism, the information that a person gathers with his or her senses is the information that should be used to make decisions, without regard to reason or to either religious or political authority. The philosophy gained credibility with the rise of experimental science in the eighteenth and nineteenth centuries, and it continues to be the outlook of many scientists today. Empiricists have included English philosopher John Locke (1632–1704), who asserted that there is no such thing as innate ideas—that the mind is born blank and all knowledge is derived from human experience; Irish clergyman George Berkeley (1685–1753), who believed that nothing exists except through the perception of the individual, and that it is the mind of God that makes possible the apparent existence of material objects; and Scottish philosopher David Hume (1711–1776), who evolved the doctrine of empiricism to the extreme of skepticism—that human knowledge is restricted to the experience of ideas and impressions, and therefore cannot be verified as true.

Why are **Kant's philosophies** still relevant?

Immanuel Kant (1724–1804) remains one of the great modern thinkers because he developed a whole new philosophy, one that completely reinterpreted human knowledge. A professor at Germany's Königsberg University beginning in 1755, Kant lectured widely and was a prolific writer. His most important work came somewhat late in life—after 1775. It was in that year that he undertook "a project of critical philosophy," in which he aimed to answer the three questions that, in his opinion, have occupied every philosopher in Western history: What can I know? What ought I do? For what may I hope?

Kant's answer to the first question (What can I know?) was based on one important conclusion: What a person can know or make claims about is only his or her experience of things, *not* the things in themselves. The philosopher arrived at this conclusion by observing the certainty of math and science: He determined that the fundamental nature of human reality (metaphysics) does not rely on or yield the genuine knowledge of science and math. For example, Newton's law of inertia—a body at rest tends to remain at rest, and a body in motion tends to remain in motion—does not change based on human experience. The law of inertia is universally recognized as correct and as such, is a "pure" truth, which can be relied on. But human reality, argued Kant, does not rest on any such certainties. That which a person has not experienced with their senses cannot be known absolutely. Kant therefore reasoned that free will cannot be proved or disproved—nor can the existence of God.

Even though what humans can know is extremely limited, Kant did not become skeptical. On the contrary, he asserted that "unknowable things" require a leap of faith. He further concluded that since no one can disprove the existence of God, objections to religion carry no weight. In this way, Kant answered the third question posed by philosophers: For what may I hope?

After arriving at the conclusion that each person experiences the world according to his or her own internal laws, Kant began writing on the problem of ethics, answering the second question (What ought I do?). In 1788 he published the *Critique of Practical Reason,* asserting that there is a moral law, which he called the "categorical imperative." Kant argued that a person could test the morality of his or her actions by asking if the motivation should become a universal law—applicable to all people: "Act as if the maxim from which you act were to become through your will a universal law." Kant concluded that when a person's actions conformed with this "categorical imperative," then he or she was doing his or her duty, which would result in goodwill.

German philosopher Georg Hegel (in an 1884 portrait) theorized that at the center of the universe there is an absolute spirit that guides all reality. (Original painting by Ernst Hader.)

Kant's theories have remained relevant to philosophy for more than two centuries: modern thinkers have either furthered the school of thought that Kant initiated or they have rejected it. Either way, the philosopher's influence is still felt. It's interesting to note that among his writings is an essay on political theory (*Perpetual Peace*), which first appeared in 1795: In it, Kant described a federation that would work to prevent international conflict; the League of Nations and the United Nations, created more than a century after Kant, are the embodiments of this idea.

What is the **Hegelian dialectic**?

It is the system of reasoning put forth by German philosopher Georg Hegel (1770– 1831), who theorized that at the center of the universe there is an absolute spirit that guides all reality. According to Hegel, all historical developments follow three basic laws: Each event follows a necessary course (in other words, it could not have happened in any other way); each historical event represents not only change but progress; and one historical event, or phase, tends to be replaced by its opposite, which is later replaced by a resolution of the two extremes. This third law of Hegel's dialectic is the "pendulum theory" discussed by scholars and students of history: that events swing from one extreme to the other before the pendulum comes to rest at middle. The extreme phases are called the thesis and the antithesis; the resolution is called the synthesis. Based on this system, Hegel asserted that human beings can comprehend the unfolding of history. In this way, he viewed the human experience as absolute and knowable.

What is **existentialism**?

Existentialism is not a single school of thought but rather a label applied to several systems that are influenced by the theories of Danish philosopher Soren Kierkegaard (1813–1855). Existentialist thinkers consider one problem: human existence in an unfathomable universe. However, in considering this "plight," philosophers have arrived at different conclusions.

The founder of existentialism, Kierkegaard rejected the principles put forth by traditional philosophers such as Georg Hegel (1770–1831), who had considered philosophy as a science, asserting that it is both objective and certain. Kierkegaard overturned this assertion, citing that truth is not objective but rather subjective; that there is no such thing as universal truths; and that human existence is not understandable in scientific terms. He maintained that human beings must make their own choices, based on their own knowledge. When he wrote on the subject, Kierkegaard frequently used pseudonyms, a practice he defended by intimating that he was putting the onus on his readers to determine what is true—that they shouldn't rely on the "authority" of his philosophies.

In the twentieth century, heirs to Kierkegaard's school of thought included German philosopher Martin Heidegger (1889–1976), who rejected the label "existentialist," and the French writer Jean-Paul Sartre (1905–1980), the only self-proclaimed existentialist. They grappled with the dilemma that human beings must use their free will to make decisions—and assume responsibility for those decisions—without knowing conclusively what is true or false, right or wrong, good or bad. In other words, there is no way of knowing absolutely what the correct choices are, and yet individuals must make choices all the time, and be held accountable for them. Sartre described this as a "terrifying freedom." However, theologians such as American Paul Tillich (1886–1965) reconsidered the human condition in light of Christianity, arriving at far less pessimistic conclusions than did Sartre. For example, Tillich asserted that "divine answers" exist. Similarly, Jewish philosopher Martin Buber (1878–1965), who was also influenced by Kierkegaard, proposed that a personal and direct dialogue between the individual and God yields truths.

What was **Nietzsche's philosophy** about the "will to power"?

The German philosopher Friedrich Nietzsche (1844–1900) developed many theories of human behavior, and the will to power was one of these. While other philosophers (including the ancient Greek Epicurus) argued that humans are motivated by a desire to experience pleasure, Nietzsche asserted that it was neither pleasure nor the avoidance of pain that inspires humankind, but rather the desire for strength and power. He argued that in order to gain power, humans would even be willing to embrace pain. However, it's critical to note that he did not view this will to power strictly as a will to dominate others: Nietzsche glorified a superman or "overman" (*ubermensch*), an individual who could assert power over himself (or herself). He viewed artists as one example of an overman—since that person successfully harnesses his or her instincts through creativity and in so doing has actually achieved a higher form of power than would the person who only wishes to dominate others. A notable exception to Nietzsche's esteem for artists was the composer Richard Wagner (1813–1883), whom the philosopher opposed. Since Wagner led an immoral lifestyle, unlike the ubermensch, Nietzsche maintained that the composer had not gained power over his own instincts.

Nietzsche was a professor of classics at the University of Basel in Switzerland from 1868 to 1878. Retiring due to poor health, he turned to his writing, which included poetry. In 1889 he suffered a mental breakdown and died the next year. After his death, his sister, Elisabeth Förster-Nietzsche (1846–1935), altered her brother's works in editing, changing their meaning. In 1895 she married an anti-Semitic agitator, Bernhard Förster (1843–1889), who, with his wife, attempted to establish a pure Aryan (a non-Jewish Caucasian) colony in Paraguay. The effort failed, and Förster took his own life. These events and, more importantly, the changes to the philosopher's own words resulted in the popular misconception that Nietzsche's philosophies had given rise to Nazism.

PHILOSOPHY AND GOVERNMENT

What is **natural law**?

Natural law is the theory that some laws are fundamental to human nature, and as such they can only be known through human reason—without reference to manmade law. Roman orator and philosopher Cicero (106–43 B.C.) insisted that natural law is universal, meaning it is binding to governments and people everywhere.

What is the **social contract**?

The social contract is the concept that human beings have made a deal with their government, and within the context of that agreement, both the government and the people have distinct roles. The theory is based on the idea that humans abandoned a natural (free and ungoverned) state in favor of a society that provides them with order, structure, and, very importantly, protection.

Through the ages, many philosophers have considered the role of both the government and its citizens within the context of the social contract. In the theories of English philosopher John Locke (1632–1704), the social contract was inextricably tied to natural law (the theory that some laws are fundamental to human nature). Locke argued that people first lived in a state of nature, where they had no restrictions on their freedom. Realizing that conflict arose as each individual defended his or her own rights, the people agreed to live under a common government, which offers them protection. But in doing so they had *not* abandoned their natural rights. On the contrary, argued Locke, the government should protect the rights of the people—particularly the rights of life, liberty, and property. Locke put these ideas into print, publishing his two most influential works in 1690: *Essay Concerning Human Understanding* and *Two Treatises of Government*. These works firmly established him as the leading "philosopher of freedom." His writings profoundly influenced Thomas Jefferson (1743–1826), author of the Declaration of Independence (1776), which asserts that there are "self-evident truths" (natural laws), that people are "endowed by their Creator with certain unalienable rights" (natural rights), and that among these are "Life, Liberty, and the pursuit of Happiness."

French philosopher Jean-Jacques Rousseau (1712–1778), one of the great figures of the Enlightenment (a cultural period of the seventeenth and eighteenth centuries, during which reason was celebrated as a superior human virtue), later published a book on the subject of the social contract. In his book titled *Social Contract* (1762), he wrote that people enter into a binding agreement among themselves, and it is incumbent upon them to establish their government and government systems. According to Rousseau, people "have a duty to obey only legitimate powers," meaning that only the people can decide who governs them. Rousseau's ideas helped promote the causes of the French Revolution (1789–99) and the American Revolution (1775–83). The concept of the social contract as defined by Rousseau materialized in the Declaration of Independence (1776), which proclaims that "Governments are instituted among Men, deriving their just powers from the consent of the governed," and "whenever any Form of Government becomes destructive of these ends, it is the Right of the People to alter or to abolish it." The well-known words of the "American's Creed," written in 1917 by William Tyler Page (1868–1942) of Maryland, also asserts these principles: "I believe in the United States of America as a government of the people, by the people, for the people; whose just powers are derived from the consent of the governed.... "

Why are **Thomas Paine**'s philosophies important to democratic thought?

English political philosopher and author Thomas Paine (1737–1809) believed that a democracy is the only form of government that can guarantee natural rights. Paine arrived in the American colonies in 1774. Two years later he wrote *Common Sense,* a pamphlet that galvanized public support for the American Revolution (1775–83), which was already underway. During the struggle for independence, Paine wrote and distributed a series of 16 papers, called *Crisis,* upholding the rebels' cause in their fight. Paine penned his words in the language of common speech, which helped his message reach a mass audience in America and elsewhere. He soon became known as an advocate of individual freedom. The fight for freedom was one that he waged in letters: In 1791 and 1792 Paine, now back in England, released *The Rights of Man* (in two parts), a work in which he defended the cause of the French Revolution (1789–99) and appealed to the British people to overthrow their monarchy. For this he was tried and convicted of treason in his homeland. Escaping to Paris, the philosopher became a member of the revolutionary National Convention. But during the Reign of Terror (1793–94) of revolutionary leader Maximilien Robespierre (1758–1794), Paine was imprisoned for being English. An American minister interceded on Paine's behalf, insisting that Paine was actually an American. Paine was released on this technicality. He remained in Paris until 1802, and then returned to the United States. Though he

Political philosopher and author Thomas Paine believed that a democracy is the only form of government that can guarantee natural rights. (Original painting by George Romney.)

played an important role in the American Revolution by boosting the morale of the colonists, he nevertheless lived his final years as an outcast and in poverty.

What is **Marxism**?

Marxism is an economic and political theory named for its originator, Karl Marx (1818–1883). Marx was a German social philosopher and revolutionary who in 1844 in Paris met another German philosopher, Friedrich Engels (1820–1895), beginning a long collaboration. Four years later they wrote the *Communist Manifesto,* laying the foundation for socialism and communism. The cornerstone of Marxism, to which Engels greatly contributed, is the belief that history is determined by economics. Based on this premise, Marx asserted that economic crises will result in increased poverty, which in turn, will inspire the working class (proletariat) to revolt, ousting the capitalists (bourgeoisie). According to Marx, once the working class has seized control, it will institute a system of economic cooperation and a classless society. In his most influential work, *Das Kapital* (The Capital), an exhaustive analysis of capitalism published in three volumes (1867, 1885, and 1894), Marx predicted the failure of the capitalist system, based on his belief that the history of society is "the history of class struggle." He and Engels viewed an international revolution as inevitable.

While Marxism still has followers in the late twentieth century, most scholars have discredited Marx's predictions, citing improved conditions for workers in industrialized nations, which has been brought about by the evolution of capitalism.

What is the difference between **socialism and communism**?

In practice, there is little distinction between the two systems, which both rely on the elimination of private property and the collective ownership of goods. But in theory, there are distinctions between the two. According to Marxism, socialism is a transition state between capitalism and communism: In socialism, the state (or government) still exists, and is in control of property and the programs for collectivization. Marxist theory holds that communism is the final stage of society—after the state has dissolved. In a Communist society economic goods and property are distributed equally among the people.

What is **fascism**?

Fascism is an extreme political philosophy that holds nation and race above the individual and supports the establishment of an authoritarian government, where absolute power is vested in the leader. In post-World War I Italy, a fascist movement developed, led by Benito Mussolini (1883–1945), paralleling the rise of Nazism in Germany. Mussolini took power of the Italian government in 1922, instituting programs of economic and social regimentation. Opposition to his dictatorship was forcibly suppressed. Italy became allied with Germany (under German chancellor and führer Adolf Hitler [1889–1945]) and Japan in 1936 to form the Axis powers, an alliance that was ultimately defeated by the Allied nations (chiefly Great Britain, the United States, and Soviet Union) in World War II (1939–45). Mussolini died by execution in 1945. Italy's fascist movement dramatically declined at the end of the war.

SCIENCE AND INVENTION

Who was **Pythagoras**?

Known by students today for the Pythagorean theorem (the square of the length of the hypotenuse of a right triangle is equal to the sum of the squares of the lengths of the other two sides), Pythagoras (c. 580–c. 500 B.C.) was a Greek philosopher and mathematician who lived in the sixth century B.C. and whose followers kept Pythagoreanism alive into the middle of the fourth century B.C. Religious in nature (some have referred to the society as a "cult"), Pythagoreans believed in the functional and even mystical significance of numbers and made considerable advances in mathematics and astronomy. Pythagoras left no writings, which has prompted some scholars to believe that no such person ever existed, but rather, any doctrines ascribed to him are actually attributable to a group of people. Whatever the case, the Pythagorean legacy is real. It includes: the word *calculus* (Pythagoreans used lines, triangles, and squares made of pebbles to represent numbers; the Latin word for pebble is *calculus*); the rendering of astronomy and music in numerical patterns (which were studied as mathematical subjects); and the first suggestion (as early as the sixth century B.C.) that Earth is spherical (not flat) and that Earth, the moon, and the planets revolve around the sun.

Who was **Euclid**?

The Greek mathematician Euclid (330?–270? B.C.) is considered the "father of geometry." He used axioms (accepted mathematical truths) to develop a deductive system of proof, which he wrote in his textbook *Elements.* This book proved to be a great contribution to scientific thinking and includes Euclid's proof of the Pythagorean theorem.

Euclid's first three postulates, with which he begins his *Elements,* are familiar to anyone who has taken geometry: 1) it is possible to draw a straight line between any two points; 2) it is possible to produce a finite straight line continuously in a straight line; and 3) a circle may be described with any center and radius.

What was the **Ptolemaic System**?

It was a scheme devised by the ancient Greek astronomer Ptolemy (c. 170–c. 100 B.C.). He proposed a system that placed Earth directly at the center of the universe—with the sun, the moon, and the planets all orbiting around Earth. However, Ptolemy observed that the movement of the planets did not match his scheme and so he added small orbits (called epicycles) to the model to try to make it work. Even though it was erroneous—and complicated—the Ptolemaic system was functional enough to make predictions of planetary positions. The system took hold, influencing thinking for 1,400 years. The Roman Catholic Church adopted the system as part of its doctrine, which the church hierarchy held to even when Polish astronomer Nicholas Copernicus (Mikolaj Kopernik; 1473–1543) refuted it in 1543, arguing that the sun, not Earth, is the center of the universe. In the 1570s accurate measurements of planet positions that had been taken by Danish astronomer Tycho Brahe (1546–1601) proved that the Ptolemaic system was inaccurate. But it was not until 1609, when German astronomer Johannes Kepler (1571–1630) devised a better explanation of planetary orbits, that the Ptolemaic system was put to rest.

What is the **Copernican view** of the universe?

The Copernican view of the universe, proposed in 1507, argued that Earth was only one of several galactic bodies that orbit the Sun. This theory, put forth by Polish astronomer Nicholas Copernicus (Mikolaj Kopernik; 1473–1543), was controversial in its day because it ran counter to the astronomical beliefs that had held sway for some 1,400 years—those of the Ptolemaic system, which maintained that Earth is the center of the universe and that the sun and the planets all revolve around it.

Copernicus devised his scheme out of necessity, really: He had found that using the Ptolemaic system to predict the positions of the planets over long periods of time yielded haphazard results. Once he made the assumption that the sun, rather than Earth, is the center of the solar system and that all the planets orbit the sun, Copernicus realized that tables of planetary positions could be calculated much more easily—and accurately.

However, Copernicus was not the first to put forth such a radical idea: the Greek astronomer Aristarchus of Samos (c. 310–230 B.C.) was the first to maintain that Earth rotates on an axis and revolves around the Sun. But it was Ptolemy's ideas that took hold; not Aristarchus's. Copernicus did, however, take the argument a step further, averring that Earth itself is small and unimportant compared with the rest of the universe.

But the Copernican view had its problems, too: Copernicus assumed that the planetary orbits were perfectly circular. Because of this error, he found it necessary to use some of Ptolemy's cumbersome epicycles (smaller orbits centered on the larger ones) to reduce the discrepancy between his predicted orbits and those that he observed. It was not until the early 1600s that the elliptical orbit of the planets was put forth, by German astronomer Johannes Kepler (1571–1630).

Italian physicist and astronomer Galileo was the first scientist to try to prove or disprove theory by conducting tests and observing the results.

What were **Galileo's contributions** to science and mathematics?

Galileo Galilei (1564–1642) is credited with establishing the modern method of experimentation. He was the first scientist and thinker to try to prove or disprove theory by conducting tests and observing the results. Prior to Galileo, scientific theory was purely based on hypothesis and conjecture. It was in the interest of conducting accurate tests and in making precise observations that Galileo developed a number of inventions, including the hydrostatic balance (a device designed to measure the density of objects) in about 1586, and the thermometer (one of the first measuring devices used in science) in 1593.

The invention most widely credited to Galileo is the telescope; however, he did not originate the instrument, but rather improved it (in 1609). He was also the first to use a telescope to study the skies, which led him to make a series of discoveries, all in 1610: the Moon shines with reflected light; the surface of the moon is mountainous; the Milky Way is made up of countless stars; and Jupiter has four large satellites. He was even able to correctly estimate the period of rotation of each of these moons, which he named "Medicean stars" (for his benefactor, Cosimo de Medici). Galileo was also the first to observe the phases of Venus, which are similar to the moon's, and to discover sunspots.

Prior to these astronomical discoveries, Galileo had already made significant contributions to science. In 1589, when he was only 25 years old, he published a treatise on the center of gravity in solids. From 1602 to 1609 he studied the motion of pendulums and other objects along arcs and inclines. From these observations, he concluded that falling objects accelerate at a constant rate. This law of uniform acceleration later helped Sir Isaac Newton (1642–1727) derive the law of gravity. Galileo also demonstrated that the path of a projectile is a parabola.

Galileo was a professor of mathematics at Pisa (1589–1591) and at Padua (1592–1610), Italy. In 1610 he was appointed philosopher and mathematician extraordinary to the grand duke of Tuscany, Cosimo de Medici.

Why was Galileo tried for heresy by the Inquisition?

The trouble began for Galileo (1564–1642) in 1613 when he published *Letters on the Solar Spots,* in which he advocated the Copernican system of the universe, which proposed that Earth (along with other galactic bodies) revolves around the sun. This view ran contrary to the accepted beliefs of the Roman Catholic Church, whose doctrine was based on Ptolemy's theory that Earth was the center of the universe and that all the planets (including the sun) revolved around it. Thus, in 1616 the Pope issued a decree declaring the Copernican system to be "false and erroneous," and Galileo was ordered not to support it.

When a new pope, Urban VIII, was coronated in 1624, Galileo traveled to Rome to make an appeal that the edict against the Copernican theory be revoked. The pope declined to do so, but he did give Galileo permission to write about the Copernican system under the condition that he not give it preference over the church-sanctioned Ptolemaic model. So, in 1632, Galileo published again: *Dialogue Concerning the Two Chief World Systems,* however, contained unconvincing objections to the Copernican view. The church saw through it and summoned the author to Rome to stand before the Inquisition (church interrogators). Galileo was accused of violating the original edict of 1616, put on trial for heresy, and found guilty. Though he was ordered to recant, at some point he uttered the famous statement: "And yet it moves," a reference to the Copernican theory that Earth rotates on its axis.

Galileo was supposed to be imprisoned, but the pope commuted this sentence to house arrest at Galileo's home near Florence, where he died blind at the age of 78.

Why is **Johannes Kepler** important to modern astronomy?

German astronomer Johannes Kepler (1571–1630) put forth the theory that planets, including Earth, rotate around the sun in elliptical orbits. But he had the help of research conducted by astronomer Tycho Brahe (1546–1601) to use as a basis for his conclusion.

In 1600 Kepler moved to Prague (Czech Republic), where he began working as an assistant to the flamboyant Brahe, a Danish aristocrat who a few years earlier (in 1576) had set up the first real astronomical observatory in history. Brahe's benefactor in this work was none other than King Frederick II (1534–1588), ruler of Denmark and Norway, who was a patron of science. In the observatory, Brahe had made and recorded extraordinarily accurate observations of planetary positions. Even though he rejected the Copernican (sun-centered) view of the universe because it violated church beliefs, he also realized that his observations of the planets could not be explained by the Ptolemaic system. He soon put forth his own theory of planetary orbit. The Tychonic theory was something of a compromise between the two existing models (the sun-centered system of Copernicus and Earth-centered system of Ptolemy): Brahe's model followed Copernicus's theory in that it, too, had the planets orbiting around the sun; but it kept to the Ptolemaic belief that the sun orbited Earth (in this way, he accounted for a year). The theory was ignored.

But when Brahe hired Kepler in 1600, he turned over his observations to him and charged him with the task of devising a theory of planetary motion. As a mathematician, Kepler was the right man for the job, and he devoted himself to the effort for the next 20 years. At one point, Kepler had devised a scheme that almost matched Brahe's observations, but not quite. Believing Brahe's observations were perfectly accurate, Kepler threw out the scheme and started again. Finally, he gave up on using circular orbits and epicycles (smaller orbits centered on the larger ones) and began working with ellipses (ovals). When Kepler charted the planets' orbits as ellipses, the results matched Brahe's data. In 1609 Kepler published his first two laws of planetary motion (in the work *Astronomia nova*): a planet orbits the sun in an ellipse, not a circle (as Copernicus had believed); and a planet moves faster when near the sun and slower when farther away.

We also have Kepler to thank for a word that is in everyday use: satellite. After Galileo discovered the moons of Jupiter, Kepler used a telescope to view them for himself. He dubbed them satellites (*satelles* is from the Latin meaning "attendant"), a name that stuck. The celebrated astronomer also did pioneer work leading to the invention of calculus (late 1600s).

When did **Halley's comet** first appear?

Its first noted appearance was in 239 B.C., but it was British astronomer Edmund Halley (1656–1742) who noted that the bright comet he observed in 1682 followed roughly the same path as those that had been observed in 1531 and 1607. He suggested that they were all the same comet and that it would reappear in 1758. It did reappear, and it was therefore named for Halley, who was England's second astronomer royal. Thanks to Halley, the common man could rest a bit easier at night: Before the British astronomer proved, through his observations and previous astronomical data, that comets are natural objects subject to the laws of gravity, people had viewed occurrences of comets as harbingers of doom. Halley's comet has been observed by astronomers every time it has appeared since 239 B.C. Most recently seen in 1985 and 1986, the comet will make another appearance in 2061, and roughly every 76 years thereafter.

It's also worth noting that it was Halley who encouraged his friend and fellow scientist Sir Isaac Newton (1642–1727) to write his theory of gravity, which he did: *Principia mathematica,* which Halley used his own money to publish, appeared in 1687 and is considered a seminal work of modern science.

Did **Newton** really formulate the laws of motion after observing an apple falling from a tree?

While it may sound more like legend than fact, Newton maintained that it was true. In 1665 Sir Isaac Newton (1642–1727), who was newly graduated from Cambridge University, escaped bubonic-plagued London and was visiting the family farm. There he saw an apple fall to the ground, and he began considering the force that was responsible for the action. He theorized that the apple had fallen because all matter attracts other matter, that the rate of the apple's fall was directly proportional to the attractive force that Earth exerted upon it, and that the force that pulled the apple was also responsible for keeping the moon in orbit around Earth.

But he then set aside these theories and turned his attention to experimenting with light. In the 1680s Newton revisited the matter of the apple, taking into consideration Galileo's (1564–1642) studies of motion (1602–09), from which the Italian scientist had concluded that falling objects accelerate at a constant rate. In 1687 Newton, with the considerable support of his friend Edmund Halley (of Halley's comet fame), published *Principia mathematica* (Mathematical Principles of Natural Philosophy), which outlined the laws of gravity and planetary motion. Newton arranged Galileo's findings into three basic laws of motion: 1) a body (any object or matter in the universe) that is at rest tends to remain at rest, and a body in motion tends to remain in motion—moving in the same direction, unless acted upon by an outside force (this is the law of inertia); 2) the force to move a body is equal to its mass times acceleration (F = MA, where F is force, M is mass, and A is acceleration); and 3) for every action, there is an equal and opposite reaction. These three laws allowed Newton to calculate the gravitational force between Earth and the moon.

Did Newton invent **calculus**?

Yes, Newton did invent calculus. But so did German mathematician Gottfried Leibniz (1646–1716), independently of Newton. Both men had developed calculus in the context of trying to explain the laws of physics. Newton's development of calculus predated that of Leibniz, but he failed to publish it. Leibniz published his results in 1684; Newton followed suit in 1693. Each used different symbols and notations, but Leibniz's were considered superior and were more widely adopted, causing friction between the men. Their conflict became a matter of national pride, with English scientists refusing to accept Leibniz's version. Nevertheless, since Newton's system predated that of Leibniz, he is credited as the originator.

How extensive were **George Washington Carver**'s agricultural discoveries?

American botanist and agricultural chemist George Washington Carver (c. 1864–1943) won international fame for his research, which included finding more

African American botanist and agricultural chemist George Washington Carver revolutionized agricultural development by developing new products from peanuts, sweet potatoes, and soybeans.

than 300 uses for peanuts and more than 100 uses for sweet potatoes. The son of slave parents, Carver was born near Diamond Grove, Missouri, and through his own efforts obtained an education, earning a bachelor's degree in 1894 and his master of science in agriculture in 1896 from Iowa State University. That year he joined the faculty of Alabama's Tuskegee Institute (now Tuskegee University), where he served as director of agricultural research until his death in 1943. His first research projects centered on soil conservation and agricultural practices. Carver gave lectures and made demonstrations to southern farmers, particularly black farmers, to help them increase crop production. He then turned his attention to finding new uses for two southern staple crops: peanuts and sweet potatoes. Carver found that peanuts could be used to make a milk substitute, printer's ink, and soap. He also found new uses for soybeans and devised products that could be made from cotton waste. His efforts were all intended to improve the economy in the American South and better the way of life of southern black farmers.

Carver was lauded for his accomplishments: he was named a fellow of the Royal Society of Arts of London (1916); he was awarded the Spingarn Medal for distinguished service in agricultural chemistry (1923); and he was bestowed with the Theodore Roosevelt Medal for his valuable contributions to science (1939).

What did **Alfred Nobel** do?

The Swedish chemist whose name is known around the world because of the Nobel prize, invented dynamite (1866). Even though dynamite improved the safety of explosives, Alfred Nobel (1833–1896) became concerned with how his invention would be used. Nobel was a pacifist; he was involved in the explosives industry because it was his family's business. In his will, he set up a fund (bequeathing a sum of $9.2 million) to reward people who make strides in the sciences, literature, and promoting international peace. He died in 1896, and the Nobel prize has been awarded annually (except for 1940–42) since 1901. Recipients in any of five categories—physics, chemistry, medicine/physiology, literature, and peace—are presented with a gold medal, a diploma, and a substantial monetary award (in the hundreds of thousands of U.S. dollars for each laureate). A sixth related award is the Prize in Economic Sciences in Memory of Alfred Nobel, which was established in 1968 by the Swedish national bank and was first awarded in 1969. The laureates are announced each year in October, and the prizes are handed out in ceremonies on December 10, the anniversary of Nobel's death.

Why was the development of the **quantum theory** important?

German physicist and professor Max Planck (1858–1947) originated and developed the quantum theory (from 1900), for which he was awarded the Nobel prize in physics in 1918. The basic theory is that energy and some other physical properties can exist in tiny, finite amounts (called quanta). Before Planck's work, theories of classical physics held that energy and physical properties varied continuously. Planck experimented with black-body radiation (a black body is any substance that absorbs all of the radiant energy that falls on it, reflecting none of it). He concluded that radiant energy can be divided, and the particles (quanta) would have values proportional to those of the energy source; Planck determined the relationship between the amount of energy that light has and its frequency. Along with Albert Einstein's (1897–1955) theory of relativity, the quantum theory forms the basis of modern physics. Since it was developed, the quantum theory has been applied to numerous processes involving the transfer of energy in an atomic or molecular scale, including in 1913 when it was used by Danish physicist and Nobel laureate (1922) Niels Bohr (1885–1962) to explain atomic structure. The theory has been used to explain how electrons move though the chips in a personal computer, the decay of nuclei, and how lasers work.

Why is the equation **$E = mc^2$** historically significant?

The famous equation was put forth in 1905 by German-born physicist and Nobel laureate (1921) Albert Einstein (1879–1955) and became important to history largely because it, along with the quantum theory (developed by Max Planck in 1900) laid the foundation for nuclear energy. It is part of Einstein's theory of relativity (1905, 1916).

In the formula, E is energy, m is mass, and c-squared is a constant factor equal to the speed of light squared. The equation illustrates the relationship between, and exchangeability of, energy and matter. In the 1930s, when scientists discovered a way to split atoms (the minute particles of which elements are made), they learned that the subatomic particles that were created have a total mass less than the mass of the original atom. In other words, when the atom was split, part of the mass of the atom had been changed into subatomic particles but some of it had been converted into energy. Using Einstein's formula $E = mc^2$, the scientists calculated how much energy was produced by splitting an atom. This atom-splitting method for creating energy is the basis for nuclear energy. The term nuclear refers to the atomic process, which uses the nucleus or central portion of an atom to release energy. Today, we live in the nuclear age, where power and weapons are produced using the atomic process for creating energy.

Albert Einstein writes an equation for the density of the Milky Way in 1931. Einstein is best known for his theory of relativity.

Why did **Einstein** write to President Roosevelt urging U.S. development of the atom bomb?

Albert Einstein, who was born in Germany in 1879 and was educated in Switzerland (where he also became a naturalized citizen), was an ardent pacifist. But, being a brilliant observer, he quickly perceived the threat posed by Nazi Germany. Einstein was visiting England in 1933 when the Nazis confiscated his property in Berlin and deprived him of his German citizenship. Some might call it lucky that the Nobel prize-winning (1921) scientist was out of the country when this happened—in the coming years, other Jews certainly suffered worse fates as Nazi Germany, under despotic ruler Adolf Hitler (1889–1945), persecuted and killed more than 6 million Jews.

Einstein moved to the United States, where he took a position at the Institute for Advanced Study in Princeton, New Jersey. There he settled and later became an American citizen (1940). In August 1939, just before Adolf Hitler's German troops invaded Poland and began World War II (1939–45), Einstein wrote a letter to President Franklin D. Roosevelt (1882–1945), urging him to launch a government program to study nuclear energy. He further advocated the United States build an atomic bomb, cautioning that such an effort might already be underway in Germany.

The United States did in fact begin development of the atomic bomb, which releases nuclear energy by splitting heavy atomic particles. The program was called the Manhattan Project, and it was centered at Oak Ridge, Tennessee, and Hanford, Washington, where scientists worked to obtain sufficient amounts of plutonium and uranium to make the bombs. The bombs themselves were developed in a laboratory in Los Alamos, New Mexico. The project was funded by the government to the tune of $2 billion. The first atomic device (made of plutonium) was tested in Alamogordo, New Mexico, on July 16, 1945. And less than a month later the United States dropped atomic bombs on Hiroshima (August 6) and Nagasaki (August 9), Japan, forcing the Japanese to surrender.

After the war, Einstein, who was always interested in world and human affairs and had regretted the death and destruction in Japan, advocated a system of world law that he believed could prevent war in the atomic (or nuclear) age. A Zionist Jew, in 1952 he was offered the presidency of the relatively new state of Israel (founded 1947), but he declined the honor. He died in New Jersey in 1955.

Who was **Enrico Fermi**?

The Italian-born physicist Enrico Fermi (1901–1954) was one of the chief architects of the nuclear age. In 1934 Fermi announced that he had discovered elements beyond uranium; but what he had really done, and which was later proved, was split the atom. In 1938, just before World War II (1939–45), this process was named nuclear fission. And one year later, German-born physicist Albert Einstein (1879–1955) would write a letter to President Franklin D. Roosevelt (1882–1945), urging the American government to study this process, which releases energy. Also in 1938 Fermi was awarded the Nobel prize in physics and he escaped Italy, where the fascist regime of Benito Mussolini (1883–1945) had taken hold. Fermi became a professor of physics at Columbia University in 1939, where he taught for three years. In 1942 he got involved in the Manhattan Project, which was developing the atomic bomb. In that capacity, he directed the first controlled nuclear chain reaction.

After World War II, Fermi, who had become an American citizen in 1944, taught at the University of Chicago and continued his research on the basic properties of nuclear particles. In 1953 he had an element named after him, fermium, an artificially created radioactive element. The U.S. Atomic Energy Commission (AEC) honors accomplishments in physics with the Enrico Fermi Award.

Why was **Oppenheimer** investigated for disloyalty to the United States?

American physicist J. Robert Oppenheimer (1904–1967), who had directed the Los Alamos, New Mexico, laboratory where the first atomic bomb was developed and built, was investigated by the government in 1953 and 1954 because of his opposition to the United States' development of a hydrogen bomb. He was also suspected of having ties with the Communist Party, and therefore was viewed as a security risk.

Following World War II (1939–45), Oppenheimer, seeing the devastating and awesome power of the atomic bomb his laboratory had created, became a vocal advocate of international control of atomic energy. When the United States began developing the hydrogen bomb (also called a "thermonuclear bomb" because of the high temperatures that it requires in order to create a reaction), Oppenheimer objected on both moral and technical grounds: The hydrogen bomb is a far more destructive weapon than the atomic bomb.

In 1953 Oppenheimer was suspended from the U.S. Atomic Energy Commission (AEC) because he was believed to pose a threat to national security. Hearings were held, but the New York-born scientist was cleared of charges of disloyalty. A decade later, in 1963, the organization gave Oppenheimer its highest honor, the Enrico Fermi Award, for his contributions to theoretical physics. Indeed, Oppenheimer had done much to further the science during his lifetime: As a member of the faculty of the University of California at Berkeley, he established a center for research in theoretical physics; he also taught at the California Institute of Technology; and he served as director of the Institute for Advanced Study in Princeton, New Jersey, from 1947 to 1966. There he knew German-born physicist Albert Einstein (1879–1955), who accepted a position at the institute in 1933 and remained there until his death in 1955.

What is the **doomsday clock**?

The clock represents the threat of nuclear annihilation. It was created by the board of directors of *Bulletin of the Atomic Scientists* and first appeared on the cover of that magazine in 1947—two years after the United States had used two nuclear weapons against Japan (at Hiroshima and Nagasaki) to end World War II (1939–45). The atomic scientists developed the idea in order to illustrate the threat of total destruction posed by nuclear weapons. On the clock, midnight is the time of destruction. When the clock first appeared, the scientists had set the time at seven minutes before midnight. In the decades since, the clock had been adjusted based on the proliferation of or agreements to limit nuclear weapons.

The closest it ever came to "doomsday" was two minutes until midnight. This was in 1953, shortly after the United States and the Soviet Union each tested hydrogen bombs. The farthest the minute hand has ever been from striking the hour of midnight was in 1991, when the United States and the Soviet Union signed the Strategic Arms Reduction Treaty (START) and announced cuts in nuclear weapons. The scientists moved the clock to read 17 minutes until midnight.

In the late 1990s the clock read 14 minutes to midnight, but the 1998 testing of nuclear weapons in Pakistan and India, neighboring countries long at odds with each other, resulted in the clock being forwarded to nine minutes before midnight. In 2002 the *Bulletin of Atomic Scientists* informed the world that the clock had been adjusted to seven minutes to midnight, saying that not only had little progress been made on global nuclear disarmament, but the United States had rejected a series of arms control treaties and announced that it would withdraw from the Anti-Ballistic Missile Treaty. Further, terrorists sought to acquire and use nuclear and biological weapons. All of this added up to a greater threat of nuclear annihilation.

What is the **big bang theory**?

It is a theory of the origin of the universe. According to the big bang theory, the universe began as the result of an explosion that occurred between 15 and 20 billion years ago. Over time, the matter created in the big bang broke apart, forming galaxies, stars, and a group of planets we know as the solar system. The theory was first put forth by Edwin Hubble (1889–1953), who observed that the universe is expanding uniformly and objects that are greater distances are receding at greater velocities. In the 1960s Bell Telephone Laboratories scientists discovered weak radio waves that are believed to be all that remains of the radiation from the original fireball. The discovery further supported Hubble's theory, which puts the age of the universe between 15 and 20 billion years.

Astronomers have observed that the galaxies are still moving away from each other and that they'll probably continue to do so forever—or at least for about the next 70 billion years. If the galaxies did come together again, scientists believe that all of the matter in the universe would explode again (in other words, there would be another big bang), and the result would be consistent with that of the first—it would produce a universe much like the one people live in today.

Another supporter of the big bang theory is British theoretical scientist Stephen Hawking (1942–). In 1988 Hawking, who is known for his theories on black holes (gravitational forces in space, made by what were once stars), published his ideas in the best-selling book *A Brief History of Time: From the Big Bang to Black Holes.*

How did **Carl Sagan** popularize science?

Carl Sagan (1934–1996), a Cornell University astronomy and space science professor, became known to many Americans via his 13-part television program, *Cosmos,* which first aired on Public Broadcasting Service (PBS) affiliates in the fall of 1980. The show covered a variety of science topics, including the origin and evolution of life on Earth, the evolution of the human brain, black holes, time travel, space exploration, and the ultimate fate of the universe. The program did so well—for a while ranking as the highest-rated regular series in public television history—that it also spun off a book by the same name. *Cosmos,* the book, became a best-seller and is still in print.

Who was Dolly the sheep?

Dolly was the first mammal cloned from the DNA of an adult animal: She was a Finn Dorset sheep born in 1996 and was hailed as a monumental scientific breakthrough when her birth was announced in early 1997. Scientists at Scotland's Roslin Institute used somatic cell nuclear transfer (SCNT), a reproductive cloning method, to produce the lamb, which carried the same nuclear DNA as the donor sheep (the cells were taken from the donor's udders). Dolly made headlines around the world and launched a public debate about the possibilities—and ethics—of cloning. Over the years, research groups around the world reported the cloning of mice, rats, cows, goats, rabbits, pigs, a horse, a mule, and a dog.

In 2003 Dolly was put to sleep. Though she lived only about half the expected 10- to 12-year life span for a Finn Dorset sheep, scientists who conducted a postmortem examination of her found that other than her ailments (arthritis and lung cancer), she appeared to be normal. The celebrity sheep was the mother of six lambs, which were brought into the world the old-fashioned way.

NATURAL HISTORY

How did Charles Darwin develop his **theory of evolution**?

The English naturalist Charles Darwin (1809–1882) was attending Cambridge when, through a friend, he gained appointment as naturalist aboard the HMS *Beagle*. The around-the-world voyage began in 1831, and Darwin was able to gather data on flora, fauna, and geology in the world's southern lands—the South American coasts, the Galapagos, the Andes Mountains, Australia, and Asia. The trip lasted into 1836, and in the years that followed Darwin published and edited many works on natural history—though none of them put forth his theory of evolution. (It's interesting to note that Darwin's paternal grandfather, Erasmus Darwin [1731–1802], was a physician, botanist, and poet, who in 1803 wrote verse that he titled, "The Temple of Nature or the Origin of Society," which anticipated the theory of evolution that his grandson would put forth.)

Encouraged by Sir Charles Lyell (1797–1875), who is considered the "father of modern geology," Charles Darwin wrote out his theory. Coincidentally, he received the abstract of an identical theory of natural selection, formulated by English naturalist Alfred Russell Wallace (1823–1913)—who had arrived at his conclusions independently of Darwin. Wallace had made his formulation of evolution based on his study of comparative biology in Brazil's Amazon River and the East Indies (Malay Archipelago). In 1858 Darwin published both his work as well as the abstract written by Wallace. The following year he backed up his theory in *On the Origin of Species by Means of*

British naturalist Charles Darwin (pictured in 1875) laid the foundation of modern evolutionary theory with his concept of natural selection.

Natural Selection, which was found to be scientifically credible. It was widely read by the public, which snatched up so many copies that the first printing of the first edition sold out in only one day. Darwin added to this work with *The Descent of Man* in 1871.

What Darwin put forth was that all life originated with a simple, primordial protoplasm (the fundamental material of which all living things are composed). In his theory of natural selection, Darwin posited that those species that are best adapted to the environment are the species that survive and reproduce.

While organic evolution and natural selection had a profound impact on the world of science (which further studied and supported the concepts), they jarred much of the religious community and created a controversy. Creationists, who believe the origin story put forth in the Bible—the story of Adam and Eve in the Garden of Eden, objected to Darwin's theories. Debate between evolution and creationism ensued, most notably in the so-called Scopes "monkey trial" of July 1925: Dayton, Tennessee, public schoolteacher John T. Scopes (1900–1970) was charged with violating a state law that prohibited the teaching of evolution. He was defended by one of the most acclaimed attorneys of all time, Clarence Darrow (1857–1938), but the prosecution was led by another prominent lawyer, William Jennings Bryan (1860–1925), and Scopes lost the case. But Scopes was later released on a technicality. The law was eventually repealed (in 1967).

Though scientific research has largely substantiated the theory of evolution described by Charles Darwin, the theory of creation still has its followers, mostly religious fundamentalists.

Who are the **Leakeys**?

The prominent British family has included four scientists who have made significant anthropological findings in East Africa. Family patriarch Louis S. B. Leakey (1903–1972) was born near Nairobi, Kenya, the oldest child of British missionaries. There he grew up, learning the tribal language of the Kikuyu people before he learned English and wandering the countryside, where he discovered primitive stone arrow-

When was the structure of DNA discovered?

In 1953 American biologist James Dewey Watson (1928–) and British biophysicist Francis Crick (1916–2004) developed a model of the structure of DNA (short for deoxyribonucleic acid), the acid found in cell nuclei. The scientists posited that DNA is constructed of a double helix (a spiral ladder) held together by hydrogen bonds. In reaching this conclusion, the pair relied on data gathered by British biophysicist Maurice Wilkins (1916–2004). In the Watson-Crick model, each rung of the DNA ladder consists of two pairs of chemicals. When DNA is replicated (which it is during reproduction), the ladder rungs are divided and the legs form new ladders, identical to the original. The model has helped scientists understand how genetic traits are passed from parent to offspring: Cells in the human body have 46 chromosomes, arranged in 23 pairs; children inherit half a set of chromosomes (threadlike bodies in the nucleus of a cell) from each parent; and different combinations of the parents' DNA (a process called recombination) produces offspring of different, though related, inherited characteristics. Each person's genetic information is carried in his or her DNA—between 10 and 20 billion miles of it, which is distributed among trillions of cells in the average human body. The study of DNA's structure has proved invaluable to scientists working in the fields of evolution, pathology, forensics, and many others. DNA's "fingerprinting" ability is so powerful that forensic scientists can use the DNA found on a single strand of human hair to identify the owner. In 1962 Watson, Crick, and Wilkins shared the Nobel prize for their groundbreaking work on DNA.

heads and tools. While attending Cambridge University, Leakey determined that he would pursue a career in archeology, and he went on to earn his doctorate degree.

Louis Leakey married archeologist and artist Mary Douglas (1913–1996) in 1936. Returning to Leakey's boyhood home to conduct their work, the husband-and-wife team made their first discovery of note in 1948. Near Lake Victoria, Kenya, they found more than 30 fragments of the skull of an apelike creature. Scientists concluded that the animal was a common ancestor of humankind and apes—and had lived between 25 and 40 million years ago.

The Leakeys made their most well-known discoveries in neighboring Tanzania during the late 1950s and into the 1960s, proving that human evolution was centered in Africa. At the Olduvai Gorge, a 35-mile-long ravine, the archaeologists discovered layers of Earth's history, including almost 100 forms of extinct animal life. They also unearthed the fossils of a near-man, *Zinjanthropus,* who possessed a brain about half the size of the modern human and who walked upright at a height of about 5 feet, roughly 1.75 million years ago. Because he lived on a diet of nuts and meat, the discovery came to be called "Nutcracker Man." Subsequent findings at the gorge included that of *Homo habilis,* called "Able Man," since it is believed that he made use of the stone tools found nearby. Louis Leakey later decided the two humanlike creatures, Able Man and Nutcracker Man, had actually lived in the same place at the same time—meaning that the evolution of humankind was not along the linear path that had been thought.

While Leakey's controversial conclusion challenged the scientific community, so would the finds of their scientist son Richard (1944–): In the decades that followed his parents' discoveries at Olduvai Gorge, Richard pursued his own projects at Lake Turkana in north-central Kenya. There Richard discovered more than 200 early-man fossils. Like his father, Richard Leakey is part of a husband-and-wife team of scientists. In 1971 he married British-born Meave Epps (1942–), a zoologist and paleontologist who had been hired by Louis Leakey in 1965 to work on his African digs. Together Richard and Meave Leakey, along with American anthropologist Alan Walker, have discovered and identified some of the oldest known humanlike fossils. In 1994 and 1995, near Lake Turkana, the team found prehistoric fossils, identified as *Australopithecus anamnesis,* humanlike creatures that lived about 4 million years ago.

INVENTION

How old is the **compass**?

The first compass dates back to the first century B.C., when the Chinese observed that pieces of lodestone, an iron mineral, always pointed north when they were placed on a surface. There is evidence that Arab sailors were using compasses as early as A.D. 600, and as Arab influence spread north into Europe, so did the compass. By the fourteenth century, European ships carried maps that were charted with compass readings to reach different destinations. Portugal's Prince Henry (1394–1460), also called Henry the Navigator, advanced the use of compasses in navigation by encouraging sailors and mapmakers to coordinate their information to make more accurate maps of the seas. Also in the fifteenth century, an important observation was made by none other than Genoese navigator and explorer from Spain Christopher Columbus (1451–1506), who noticed that as he sailed to the New World, his compass did not align directly with the North Star. (The difference between magnetic north and true north is called declination.) In the sixteenth and seventeenth centuries scientists began to better understand Earth's magnetic fields.

American Elmer Sperry (1860–1930) built the first gyrocompass, a device that works day or night, anywhere on Earth—even at the poles, where lines of force are too close together for magnetic devices to function properly. When the gyrocompass is pointed north, it holds that position.

Before the compass, which simply indicates north by a means of a magnetic needle or needles that pivot, sailors used the sun, the moon, and the stars to determine direction and navigate their ships.

When did sailors begin using **latitude and longitude** to navigate?

It was after English inventor John Harrison (1693–1776) presented his ship's chronometer to London's Board of Longitude in 1736. The instrument was accurate to within one-tenth of a second per day (1.3 miles of longitude). Since it was set to the time of zero degrees longitude (Greenwich time), it enabled navigators to fix longitudinal position by determining local time. Even though Harrison's award-winning invention was heavy (weighing 65 pounds), complicated, and delicate, it was subsequently improved upon so that it could be used on any sea-faring vessel in any weather conditions.

How long have **sundials** been in use?

Sundials, which indicate the time of day by the shadow cast by a stick, pin, or other object, usually on a horizontal plate, have been in use since before the sixth century B.C., when both the ancient Chinese and Egyptians used the device to tell time. Sundials proved to be a fairly accurate indicator of the passage of time. But it has its problems: A sundial can be difficult to read, the markings have to be adjusted according to latitude, and the readings differ with the seasons. They remain popular as garden ornaments today.

How old is the **calendar**?

The calendar that is in general use today is the Gregorian calendar; it dates to 1582, when Pope Gregory XIII (1502–1585) asked for a revision of the Julian calendar. That calendar is named for its initiator, Julius Caesar (100–44 B.C.), who in about 46 B.C. commissioned the astronomer Sosigenes of Alexandria to develop a universal solar calendar to be used throughout the Roman Empire (as Roman armies conquered more and more territory, the empire included many peoples and differing calendars, including the lunar-based Roman calendar). The Julian calendar consisted of a year of 365 days, with one day added every fourth year (leap year, when the year is divisible by four) to compensate for the fact that the solar year is really 365.25 days. It had 12 months, each of 30 or 31 days except February, which had 28, and the new year began on January 1. The Gregorian calendar retained these features but revised the Julian to bring the Christian celebration of Easter in alignment with the vernal equinox (first day of spring). It also dropped leap years for any century year not divisible by 400—an effort to keep the solar calendar in line with the seasons: For example, 1900, though divisible by four, was not a leap year since it was a centenary year not divisible by 400; the year 2000, divisible by 400, was a centenary leap year.

How old is the oldest clock?

The first mechanical timekeeping device was a water clock called a clepsydra, which was used from about 1500 b.c. through the Middle Ages (500–1350). One very elaborate clepsydra was constructed for Holy Roman Emperor Charlemagne (742–814) in a.d. 800: Upon the hour, it dropped a metal ball into a bowl. Because of problems with water (it evaporated, froze, and eroded the surfaces of its container), a more accurate device was needed. It is believed that the first completely mechanical clock was developed by a monk around 1275: The clock was driven by the slow pull of a falling weight that had to be reset to its starting position after several hours. The clocks in monasteries were among the first to be fitted as alarm clocks: striking mechanisms were added to the timekeeping devices so the monks would know when to ring the monastery bell.

Other calendars remain in use in the world today, including the lunar Babylonian, Chinese, and Muslim calendars; the Jewish calendar, which is a combination of solar and lunar; and the solar Coptic, Japanese, and Hindu calendars. Secular calendars include the Julian Day, used by astronomers; and the perpetual calendar, which gives the days of the week for the Julian and the Gregorian calendar, and therefore is used by historians and other scholars to reconcile world events along a single timeline.

How old is **standard time**?

Standard time was introduced in 1884; it was the outcome of an international conference held in Washington, D.C., to consider a worldwide system of time. By international agreement, Earth was divided into 24 different "standard" time zones; within each time zone, all clocks are to be set to the same time. The device of standardized time zones was necessitated by the expansion of industry: businesses, particularly those in the transportation industry, could not coordinate schedules when each community used its own solar time (the local time as determined by the position of the sun). Railroad schedules had been extremely complicated before the establishment of standard time zones, which the railroads readily adopted.

Each time zone spans 15 degrees of longitude, beginning at zero longitude (called the "prime meridian"), which passes through the observatory at Greenwich (a borough of London), England. Time kept at the observatory is called Greenwich mean time (GMT). Time zones are described by their distance east or west of Greenwich. The model also dictates that each time zone is one hour apart from the next. However, the borders of the time zones have been adjusted throughout the world to accommodate national, state, and provincial boundaries.

The contiguous United States has four time zones: Eastern, Central, Mountain, and Pacific. Waters off the eastern seaboard are in the Atlantic time zone; Alaska, Hawaii, Samoa, Wake Island, and Guam each have their own time zones. Congress gave the Interstate Commerce Commission (ICC) authority to establish limits for U.S. time zones in 1918. This authority was transferred to the Department of Transportation (DOT) in 1967.

When was the **flush toilet** invented?

The invention dates to the 1590s and is credited to Sir John Harington (1561–1612), hence its nickname, "the john." A courtier and godson of England's Queen Elizabeth I (1533–1603), Harington installed a flush lavatory in one of the queen's palaces. Though he was a serious scholar and translator, Harington was also a rebel who wrote controversial satire, leading to his banishment. His invention of the so-called "water closet" was not taken seriously in its day. But over the following two centuries various inventors worked to improve it, ultimately developing the plumbed sanitary toilet, a flush commode that is connected to plumbing and sewers or septic tanks.

Who invented the **thermometer**?

While the Greeks made simple thermometers as early as the first century B.C., it wasn't until Galileo (1564–1642) that a real thermometer was invented. It was an air thermometer, in which a colored liquid was driven *down* by the expansion of air, so that as the air got warmer (and expanded), the liquid dropped. This is unlike ordinary thermometers in use today, which rely on the colored liquid of mercury to rise as it gets warmer.

In 1612 Italian physician Santorio Santorio (1561–1636), a friend of Galileo, adapted the device to measure the body's change in temperature due to illness. (The clinical thermometer wasn't Santorio's only invention: As the first doctor to use precision instruments in the practice of medicine, Santorio also developed the pulse clock.)

It was a full century though before thermometers had a fixed scale. This was provided by German physicist Daniel Fahrenheit (1686–1736), who in 1714 invented the mercury thermometer.

Is **Thomas Edison** the greatest inventor in history?

Some believe that he is, and for good reason: Thomas Alva Edison (1847–1931), the so-called "Wizard of Menlo Park," registered 1,300 patents in his name during his lifetime, more than have been credited to any other individual in American history. His best-known inventions include an automatic telegraphy machine, a stock-ticker machine, the phonograph (1877), the incandescent light bulb (1878), and the motion picture machine (kinetoscope; 1891). Edison also made one major scientific find during his research, when he observed that electrons are emitted from a heated cathode (the con-

Thomas Alva Edison registered 1,300 patents in his name during his lifetime, more than have been credited to any other inventor in American history.

ductor in an electron tube)—the phenomenon is known as the "Edison effect."

When Edison was still in his twenties, he set up a laboratory where 50 consulting engineers worked with him on various inventions. By his own description, the Newark, New Jersey, plant was an "invention factory." It operated for six years, during which Edison was granted about 200 new patents for work completed there. The laboratory is regarded by most as the first formally organized nonacademic research center in the United States. By 1876 Edison had outgrown his Newark facilities and arranged for the construction of a new plant at Menlo Park, New Jersey. His most productive work was accomplished at this location over the next decade.

Is it true that **Edison had no formal education**?

It's almost true: Thomas Edison (1847– 1931) had no formal education to speak of. Born in Milan, Ohio, Edison's father moved the family to Port Huron, Michigan, in 1854. There young Edison attended school for the first time—in a one-room schoolhouse where he was taught by the Reverend and Mrs. G. B. Engle. But this arrangement lasted just a few months: The boy grew impatient with his schooling—behavior his teachers interpreted as a sign of mental inferiority. When Edison overheard Mrs. Engle refer to him as "addled," he reported it to his mother, Nancy, who promptly withdrew him from the school. From then on his mother taught him at home, introducing young Edison to natural philosophy—a mixture of physics, chemistry, and other sciences. He showed an inclination toward science, and by the age of 10 he was conducting original experiments in the family's home.

Edison furthered his education through voracious reading. He sought, and was granted, permission to sell periodicals, snacks, and tobacco to passengers on the train between Port Huron and Detroit, Michigan, some 60 miles away. During the layover in downtown Detroit, Edison spent his time at the public library, where, according to his own recollection, he read not a few books, but the entire library. Even though he lacked a formal education, Edison possessed a keen mind and a natural curiosity. Further, he had the benefit of the schooling provided by his mother as well as access to a library, of which he took full advantage.

Then something happened that changed Edison's life: While still a young man, he lost his hearing. Biographer Matthew Josephson explains that the deafness had two effects on Edison: Not only did he become "more solitary and shy," but Edison turned with an even greater intensity toward his studies and began to "put forth tremendous efforts at self-education, for he had absolutely to learn everything for himself."

TRANSPORTATION

Who invented the **steam engine**?

Like many other modern inventions, the steam engine had a long evolution. It was first conceived of by Greek scientist Hero of Alexandria in the first century A.D. The mathematician invented many "contrivances" that were operated by water, steam, or compressed air. These included not only a fountain and a fire engine, but the steam engine. Many centuries later, Englishman Thomas Newcomen (1663–1729) developed an early steam engine (about 1711) that was used to pump water. He was improving on a previous design, which had been patented by another inventor in 1698.

But it was Scottish inventor James Watt (1736–1819) who substantially improved Newcomen's machine, patenting his own steam-powered engine in 1769. It was the first practical steam engine, and Watt's many improvements to the earlier technology paved the way for the use of the engine in manufacturing and transportation during the Industrial Revolution (c. 1750–c. 1850); Britain was just on the cusp of this new age when Watt patented his engine. The steam engine was eventually replaced by more efficient devices such as the turbine (developed in the 1800s), the electric motor (also developed in the 1800s), the internal-combustion engine (first practical engine built in 1860), and the diesel engine (patented 1892). Nevertheless, James Watt's steam engine played a critical role in moving society from an agricultural- to industrial-based economy. Watt's legacy also includes the use of "horsepower" and "watts" as units of measure.

How long have **trains** been in use?

Trains date as far back as the sixteenth century when crude railroads operated in the underground coal and iron ore mines of Europe. These systems consisted of two wooden rails that extended into the mines and across the mine floors. Wheeled wagons were pulled along the rails by men or by horses. Early in the eighteenth century, mining companies expanded on this rail system, bringing it above ground to transport the coal and iron ore. Workers found that they could cover the wooden rails with iron so they wouldn't wear out as quickly. Before long, rails were made entirely of iron.

Meanwhile, the steam engine had been developed. An engineer in the mines of Cornwall, England, Richard Trevithick (1771–1833), constructed a working model of a

Railroad officials and employees celebrate the completion of the first transcontinental railroad in Promontory, Utah, 1869.

locomotive engine in 1797. Three years later, he built the first high-pressure steam engine. He made quick progress from there, building a road carriage, which on Christmas Eve 1801 became the first vehicle to convey passengers by steam. Two years later, the inventor had built the world's first steam railway locomotive.

In 1825 progress in rail transportation was made by another English inventor, George Stephenson (1781–1848), who, after patenting his own locomotive engine (1815), finished construction on the world's first public railroad. The train ran a distance of about 20 miles, conveying passengers from Stockton to Darlington, England. In 1830 Stephenson completed a line between Liverpool and Manchester. Rail travel caught on quickly—and remains an efficient means of transport today, with commuters around the world relying on trains to get them to work each day.

While Stephenson went on to build more railways, and build a family business in the process, Trevithick did not fare nearly as well: Though he later found other uses for the high-pressure steam engine (including rock boring, dredging, and agriculture), he died penniless.

When did **rail service** in the United States reach from **coast to coast**?

On May 10, 1869, the last tracks of the United States' first cross-country railroad were laid, making North America the first continent to be spanned from coast to coast by a rail line. The event was the fulfillment of a great national dream to knit the vast country closer together. Short-run rail lines had been in use since the 1840s, but the nation lacked a method for transporting people, raw materials, and finished goods between distant regions. In the early 1860s U.S. Congress decided in favor of extending the railroad across the country. The federal government granted land and extended millions of dollars in loans to two companies to complete the project. After a long debate that had become increasingly sectional, the legislature had earlier determined the railroad should run roughly along the 42nd parallel—from Omaha, Nebraska, to Sacramento, California. This route was settled on for its physical properties: The topography of the landscape would best allow the ambitious project.

The Union Pacific Railroad was to begin work in Omaha and lay tracks westward; the Central Pacific Railroad was to begin in Sacramento and lay tracks eastward, crossing the Sierra Nevada Mountains. Work began in 1863; and six years later the two projects met at Promontory in north-central Utah, northwest of Ogden. The Central Pacific Railroad laid 689 miles of track eastward from Sacramento, while the Union Pacific Railroad laid 1,086 miles of track westward from Omaha. Six hundred workers, including numerous Irish, Chinese, and Mexican immigrant workers, attended the May 10, 1869, ceremony celebrating the accomplishment. The last steel spike that was driven into the railroad was wired to a telegraph line; when the spike was pounded, a signal was sent around the world to announce the completion of the first transcontinental railroad. By the end of the 1800s, 15 rail lines crossed the nation.

When was the **bicycle** invented?

A series of inventions during the 1800s resulted, in 1876, in the introduction of the safety bicycle, the direct ancestor of the modern bike and the first commercially successful bicycle. It had wheels that were equal in size, making it easier and safer to ride than its "high-wheeler" predecessor. The bikes proliferated: By 1900 more than 10 million Americans owned bicycles.

As with other inventions, the bicycle was the result of the work of several innovators. In 1817 German baron Karl Friedrich Freiherr von Drais de Sauerbronn (1785–1851) developed a device that resembled a scooter. Named for its inventor, the *drasienne* later improved by Scotsman Kirkpatrick Macmillan (1813–1878), who in 1839 added pedals to the vehicle, giving the world the first real bicycle. In 1870 English inventor James Starley (1830–1881) designed a bicycle with a large front wheel and small rear wheel. Named the Ariel, the invention was also called a "penny-farthing" (after two very different-sized British coins), the "high-wheeler," and the "ordinary." Though the bicycle was easier to pedal and faster (one revolution of the pedals turned the front wheel once), its high center of gravity made it unstable and even dangerous. The innovation of the tricycle, or velocipede, improved the design of the Ariel by giving it the added stability of the third wheel. But it was not until the safety bike was developed in 1876 that the bicycle's popularity took off. Invented by Englishman Henry John Lawson (1852–?), the bicycle had wheels of equal size and a bike chain (to drive the rear wheel). This practical design was improved again in 1895 when air-filled (pneumatic) tires were added. Mass production of the safety bicycle began in 1885.

Even after the advent of the automobile, the bicycle continued to figure prominently in everyday life. In the United States, bicycle riding became a leisure pursuit that rivaled baseball in popularity. Cycling clubs emerged. The tandem, the bicycle built for two, allowed American youths an opportunity for courtship. Further, the bicycle industry produced some of the great innovators in transportation, including bicycle designer Charles Duryea (1861–1938), who, with his brother Frank (1869–1967), demonstrated the first successful gas-powered car in the United States; and brothers Wilbur (1867–1912) and Orville (1871–1948) Wright, who were the first to successfully build and fly an airplane.

Did **Henry Ford** invent the automobile?

No, while Henry Ford (1863–1947) transformed American industry and changed the way we travel, live, and work, he did not invent the automobile. Just before the turn of the century, there were several inventors who were tinkering with gas-powered vehicles, and by the time Ford had finished his first working car, the Duryea brothers (Charles, 1861–1938; Frank, 1869–1967) had demonstrated the first successful gas-powered car in the United States, and Ohio-born inventor Ransom Eli Olds (1864–1950) already had a car, the Oldsmobile, in production. Even prior to the work of these American inventors and entrepreneurs, Europeans had made strides in developing the automobile.

The automobile is the result of a series of inventions, which began in 1769 when French military engineer Nicolas-Joseph Cugnot (1725–1804) built a steam-powered road vehicle. In the early 1800s other inventors also experimented with this idea. And the steam-powered vehicle was put into production, both in Europe and the United States. In 1899 William McKinley (1843–1901) became the first U.S. president to ride in a car—a Stanley Steamer, built by twin brothers Francis (1849–1918) and Freelan (1849–1940) Stanley.

A breakthrough in developing gas-powered automobiles came in 1860, when an internal combustion engine was patented by Frenchman Étienne Lenoir (1822–1900). But the car as we know it was born in 1885 when Germans Gottlieb Daimler (1834–1900) and Carl Benz (1844–1929), working independently of each other, developed the forerunners of the gas engines used today. In 1891 to 1892 French company Panhard et Levassor designed a front-engine, rear-wheel drive automobile. This concept remained relatively unchanged for nearly 100 years. Until 1900 Europeans led the world in the development and production of automobiles. In 1896 the Duryea Motor Wagon Company turned out the United States' first production motor vehicle. The gas-powered cars were available for purchase that same year.

The development of the car was helped by advances in rubber and in the development of pneumatic tires during the nineteenth century, with names like Charles Goodyear (American; 1800–60), John Boyd Dunlop (Scottish; 1840–1921), and the Michelin brothers (French), Andre (1853–1931) and Edouard (1859–1940), figuring prominently in automotive history.

Even after the gas engine was invented, which was ultimately more efficient than the steam engine, the car's development continued along parallel tracks: About 1891 American William Morrison successfully developed an electric car. Electric cars were soon put into production, and by the turn of the century they accounted for just less than 40 percent of all American car sales.

Who invented the **hot-air balloon**?

French papermakers Joseph-Michel (1740–1810) and Jacques-Étienne (1745–1799) Montgolfier built the first practical balloon—filled with hot air. On June 5, 1783, the Montgolfiers launched a large balloon at a public gathering in Annonay, France. It ascended for 10 minutes. Three months later, they sent a duck, a sheep, and a rooster up in a balloon—and the animals were landed safely. This success prompted the Montgolfiers to attempt to launch a balloon carrying a human. In October 1783 French scientist Jean-François Pilâtre de Rozier (1756–1785) became the first person to make a balloon ascent, but the balloon was held captive (for safety). The following month he became one of two men to make the first free flight in a hot-air balloon, which ascended to a height of about 300 feet over Paris on November 21, 1783, and drifted over the city for about 25 minutes. Hot-air ballooning, which proved to be better than the rival hydrogen-filled balloon developed in France at about the same time, became very popular in Europe. In January 1785 hot-air balloonists successfully crossed the English Channel, from Dover, England, to Calais, France. Across the Atlantic, hot-air balloons made their debut in the United States in Philadelphia in 1793—before a crowd that included George Washington (1732–1799), who was then president of the United States.

Who invented the **airship**?

The airship, a lighter-than-air aircraft that has both propulsion (power) and steering systems, had a long evolution. The first successful power-driven airship was built by French engineer Henri Giffard (1825–1882), who in September 1852 flew his craft a distance of 17 miles from Paris to Trappes, France, at an average speed of five miles per hour. The airship was cigar-shaped with a gondola that supported a three-horsepower steam engine. Though it included a rudder, the craft proved difficult to steer. Austrian David Schwarz (1845–1897) is credited with designing the first truly rigid airship, a craft that he piloted—unsuccessfully—in November 1897; the airship crashed.

The inventor whose name is most often associated with airships (also called "dirigibles" after 1885) is Ferdinand von Zeppelin (1838–1917), who designed, built, and flew the first successful rigid airship in 1900. The Zeppelin (as the aircraft came to be known) flew at a top speed of about 17 miles per hour. The German aeronaut steadily improved his craft in the years that followed and in 1906 set up a manufacturing plant where the Zeppelins were built. In 1909 Zeppelin helped establish the world's first commercial airline; the transport was wholly via airship. The crafts saw military use during World War I (1914–18), but after the *Hindenburg* crashed on landing in New Jersey in 1937, the airships declined dramatically in use. Their decline largely paralleled advances in the development of airplanes.

Did the **Wright brothers** invent the airplane?

The Wright brothers were the first to successfully build *and* fly an airplane; and both events went virtually unnoticed at the time. The owners of a bicycle shop in their hometown of Dayton, Ohio, Wilbur (1867–1912) and Orville (1871–1948) Wright were interested in mechanics from early ages. After attending high school, the brothers went into business together and, interested in aviation, began tinkering with gliders in their spare time. The brothers consulted national weather reports to determine the most advantageous spot for conducting flying experiments. Based on this data, they concluded it was Kitty Hawk, North Carolina. There in 1900 and 1901, on a narrow strip of sand called Kill Devil Hills, they tested their first gliders that could carry a person. Back at their bicycle shop in Ohio, they constructed a small wind tunnel (about six feet in length) in which they ran experiments using wing models to determine air pressure. As a result of this research, the Wright brothers were the first to write accurate tables of air pressures on curved surfaces.

Based on their successful glider flights and armed with their new knowledge of air pressure, Orville and Wilbur Wright designed and built an airplane. The returned to Kitty Hawk in September 1903 to try the craft. But weather prevented them from doing so until December. It was days before Christmas when, on December 17, 1903, the Wright brothers made the world's first flight in a power-driven, heavier-than-air machine. Orville piloted the craft a distance of 120 feet and stayed in the air 12 seconds. They made a total of four flights that day, and Wilbur made the longest: 59 seconds of flight time that covered just more than 850 feet.

It was not a news event: The brothers had witnesses (a few spectators on the beach in North Carolina), and there were a handful of newspaper accounts of the Wrights's marvelous feat, but some were inaccurate. After they made a public announcement in January 1904, *Popular Science Monthly* published a report (in March), as did another magazine. Other than these scant notices, the Wrights received no attention for their accomplishments. Many were trying to do what the Wrights had done, but the public was skeptical that any heavier-than-air manmade machine could take flight. The doubt played a role in the lack of acclaim. Meanwhile, the brothers continued their experiments at a field near Dayton: In 1904 and 1905 they made 105 flights, but totaled only 45 minutes in the air.

Wilbur and Orville Wright flew the first airplane in 1903 at Kitty Hawk, North Carolina. Here the brothers are shown near New York Harbor in September 1909.

The Wright brothers persisted, and in spite of public skepticism, which initially included that of the U.S. government, in 1908 Orville and Wilbur Wright signed a contract with the Department of War to build the first military airplane. Only then did they receive the media attention they deserved. A year later, they set up the American Wright Company to manufacture airplanes. In spring 1912 Wilbur became sick and died; three years later Orville sold his share in the company and retired. The plane piloted by the two brothers in December 1903 near Kitty Hawk is on display at the National Air and Space Museum in Washington, D.C.

When did the first **jet airplane** take flight?

The aviation event took place in 1939 in Germany, just as World War II was beginning. The development of the jet aircraft was made possible by British inventor Frank Whittle (1907–1996), who built the first successful jet engine in 1937 (the Germans copied Whittle's design). The engine propels an object forward by discharging a jet of heated air or exhaust gases rearward. Whittle's company, Powerjets Limited, built the engine for Britain's first jet plane in 1941, which became the model for early U.S. jets. During World War II, Great Britain, the United States, and Germany all employed jets (though in limited numbers) in their military operations.

After the war ended (1945), aircraft manufacturers began developing jet airliners. The innovator in the field was the De Havilland Aircraft Company (founded 1920), which produced the Comet, the first commercial jet. The aircraft was used by British Overseas Airways Corporation (now British Airways), which in 1952 initiated passenger flights. But flaws in the Comet's structure were later discovered to be the cause of several midair explosions. The craft was redesigned, and in 1958 British Airways launched transatlantic passenger service using the improved Comets. U.S. jet airplane passenger service was introduced by American Airlines in 1959; the airline used the Boeing Company's 707 to transport passengers from New York City to Los Angeles.

COMMUNICATIONS

Who invented the **telegraph**?

Though the invention came as the result of several decades of research by many people, Samuel F. B. Morse (1791–1872) is credited with making the first practical telegraph, the first instrument that could send messages across wires via electricity, in 1837. Morse was a portrait painter in Boston when he became interested in magnetic telegraphy in about 1832. With technical assistance from chemistry professor Leonard Gale (1800–1883) and financial support from Alfred Vail (1807–1859), Morse conducted further experiments. He also developed Morse code, a system of variously arranged dots and dashes, which can be used to transmit messages. (For example, the most frequently used letter of the alphabet is *e*, which is rendered in Morse code by using one dot; the less frequently used *z* is rendered by two dashes followed by two dots.) By 1837 Morse had demonstrated the telegraph to the public in New York, Philadelphia, and Washington, D.C. He received a patent for his invention in the United States in 1840. In 1843 his invention got a boost when the U.S. Congress approved an experimental line, to be built between Washington, D.C., and Baltimore, Maryland. The following year, on May 24, 1844, Morse sent his first message across that line: "What hath God wrought!" Vail was on the receiving end of the wire.

By 1861 most major U.S. cities were linked by telegraph wires. The first successful transatlantic cables were laid in 1866. Morse code transmissions, called telegraphs when transmitted via aboveground wires and cablegrams (or cables) when transmitted via underwater cables, were translated by operators or mechanical printers on both the sending and receiving ends of the message. The introduction of the telegraph marked the beginning of modern communications: When the first transcontinental telegraph line in the United States was completed on October 24, 1861, it eliminated the need for the Pony Express, which had briefly enjoyed the status of the fastest way to transmit a message—about eight days from St. Louis, Missouri, to Sacramento,

California, a distance that could be bridged by telegraph lines within minutes. The telegraph became the chief means of long-distance communication. The telephone (invented 1875), which allows voice transmission over electrical wires, gradually replaced the telegraph. But for many decades the two technologies were both in use.

Who invented the **telephone**?

For more than 100 years Alexander Graham Bell (1847–1922) was credited with inventing the telephone. But in 2002 the U.S. Congress officially recognized a previously unknown Italian American inventor, Antonio Meucci (1808–1889), as the "father of the telephone."

Meucci was born in Florence, Italy, and arrived in the United States in 1845. His focus was the use of electricity in medicine. But in carrying out his experiments, he realized that the voice could be transmitted over wire. By 1862 he had developed dozens of models of his invention but could not afford to protect his prototypes with patents. In 1870 illness forced the poor inventor to sell his early models. By 1874 he had assembled new models, which he gave to an executive at Western Union Telegraph Company. Two years later, the announcement was made that Scottish-born inventor Alexander Graham Bell had pioneered the technology.

Bell came to the United States in 1871 as a teacher of speech to the deaf. He believed that sound wave vibrations could be converted into electric current at one end of a circuit, and the current could be reconverted into identical sound waves at the other end of the circuit. He had described this idea to his father as early as 1874, and some believe that he may have conceived of it as early as 1865. Still, Meucci's prototypes would have predated these ideas.

On June 3, 1875, while trying to perfect a method for simultaneously carrying more than two messages over a single telegraph line, Bell heard the sound of a plucked spring along 60 feet of wire. In 1876 the first voice transmission via wire was made when Bell had a laboratory accident and called out to his assistant, "Watson, please come here. I want you." Thomas Watson (1854–1934) was on another floor of the building, with the receiving apparatus, and distinctly heard Bell's message. That same year, the Bell telephone was patented in the United States and exhibited at the Philadelphia Centennial Exposition. Meucci died in poverty in 1889. His recognition as the "father of telephony" was the result of persistence on the part of Italian Americans who wished to set the record straight. In fact, in 1871, five years earlier than Bell's first voice communication, Meucci had secured a certificate stating his paternity to the invention of the telephone, which he called the tele*tro*phone.

When was the **radio** invented?

The radio, or "wireless," was born in 1895 when Italian physicist and inventor Guglielmo Marconi (1874–1937) experimented with wireless telegraphy. The following year he transmitted telegraph signals, through the air, from Italy to England. By 1897 Marconi founded his own company, Marconi's Wireless Telegraph Company, Ltd., in London and began setting up communication lines across the English Channel to France, which he accomplished in 1898. In 1900 Marconi established the American Marconi Company. He continued making improvements, including those that allowed for sending out signals on different wavelengths so that multiple messages could be transmitted at one time, without interfering with each other. The first transatlantic message, from Cornwall, England, to Newfoundland, Canada, was sent and received in 1901.

At first radio technology was regarded as a novelty and few understood how it could work. But in January 1901 a Marconi wireless station at South Wellfleet, Massachusetts (on Cape Cod), received Morse code messages as well as faint music and voices from Europe. That event changed the perception of radio: Before long, Americans had become accustomed to receiving "radiograms," messages transmitted via the wireless. In 1906 the first radio broadcast of voice and music was made: The event originated at Brant Rock, Massachusetts, on Christmas Eve—and the program was picked up by ships within a radius of several hundred miles. That accomplishment resulted from the invention of another radio pioneer, American engineer Reginald Fessenden (1866–1932), who patented a high-frequency alternator (1901) capable of generating continuous waves rather than intermittent impulses; it was the first successful radio transmitter.

In 1910 American inventor Lee De Forest (1873–1961), the "father of radio," broadcast opera singer Enrico Caruso's (1873–1921) tenor voice over the airwaves. In 1916 De Forest transmitted the first radio news broadcast. Westinghouse station KDKA in Pittsburgh, Pennsylvania, was the first corporately sponsored station and the first broadcast station licensed on a frequency outside amateur bands. Within three years of its first commercial radio broadcast, which announced the election returns in the presidential race on November 2, 1920 (Warren G. Harding won), there were more than 500 radio stations in the United States.

Who invented the **television**?

The television, which may seem to many to be a decidedly American invention, was actually the outcome of a series of inventions by a cast of international characters. As early as 1872 British engineer Willoughby Smith (1828–1891), inspired by an experiment on selenium rods, imagined a system of "visual telegraphy." Five years later, the tube technology that would make television possible was developed in Strasbourg by German physicist Karl Ferdinand Braun (1850–1918). He invented a cathode-ray tube (also known as the Braun tube), which improved the Marconi wireless (radio) technology by increasing the energy of sending stations, and arranged antennas to control the direction of radiation.

In 1907 Russian physicist Boris Rosing proposed using Braun's tube to receive images—something he called "electric vision." One year later, Alan Campbell Swinton

(1863–1930) suggested using the cathode-ray tube to both receive and transmit images. That same year, the idea of using cathode-ray tubes to scan images for the purpose of television was published, and by 1912 it was being worked on by Rosing and his former pupil, Vladimir Zworykin (1889–1982), in Russia.

In 1923 a competing technology, which was entirely mechanical, reached an early milestone when British inventor John Logie Baird (1888–1946) demonstrated an electrified hatbox with disks, which constituted the world's first working television set. But the race was still on and in that same year, Zworykin, who had moved to the United States in 1919 and was hired by Westinghouse Electronic Corporation (in 1920), advanced the tube-based technology when he patented the iconoscope, which would become the television camera. In 1929 Zworykin, now a U.S. citizen, invented the kinescope (television tube). Zworykin's inventions together comprised the first all-electronic television.

Regularly scheduled U.S. television began on April 30, 1939, when President Franklin D. Roosevelt (1882–1945) opened the New York World's Fair, billed as "The World of Tomorrow," giving a speech that was the first televised presidential talk. The National Broadcasting Company's (NBC) coverage of the fair's opening initiated its weekly television scheduling, a victory for parent company RCA, whose president, David Sarnoff (1891–1971), founded NBC and is considered a broadcasting pioneer.

When was **color television** invented?

In 1940 Hungarian American engineer Peter Carl Goldmark (1906–1977), the head of the Columbia Broadcasting System's (CBS) research and development laboratory, came up with a technology that broke down the television image into three primary colors through a set of spinning filters in front of black and white, causing the video to be viewed in color. His system gave way in the 1950s to an RCA system whose signals were compatible with conventional black-and-white TV signals.

In September 1962 American Broadcasting Corporation (ABC) began color telecasts, for three and a half hours a week. By this time, competitor National Broadcasting Corporation (NBC) was broadcasting 68 percent of its prime-time programming in color, while CBS had opted to confine itself to black and white after having transmitted in color earlier. By 1967 all three networks were broadcasting entirely in color.

It was not until 1967 that color television was broadcast in England: On July 1 the BBC-2 transmitted seven hours of programming, most of it coverage of lawn tennis from Wimbledon.

When was the first **fax machine** developed?

The fax (facsimile) machine may seem like a recent invention, but it was developed long ago—it took more than 100 years for the machines to become part of everyday life. In 1842 to 1843 Scottish philosopher and psychologist Alexander Bain (1818–1903) invented the first, albeit crude, fax machine. The scanning technology was improved enough by 1924 that newspapers began using the device to transmit photographs. By the 1930s wirephotos were an important component of newspaper reports. It was not until the 1980s that faxes came into widespread use, as manufacturers produced the more compact and affordable machines that are visible in most every place of business today.

Who invented the **computer**?

English mathematician Charles Babbage (1792–1871) is recognized as the first to conceptualize the computer. He worked to develop a mechanical computing machine called the "analytical engine," which is considered the prototype of the digital computer.

While attending Cambridge University in 1812, Babbage conceived of the idea of a machine that could calculate data faster than could humans—and without human error. These were the early years of the Industrial Revolution, and the world Babbage lived in was growing increasingly complex. Human errors in mathematical tables posed serious problems for many burgeoning industries. After graduating from Cambridge, Babbage returned to the idea of a computational aid. He spent the rest of his life and much of his fortune trying to build such a machine, but he was not to finish. Nevertheless, Babbage's never-completed "analytical engine" (on which he began work in 1834) was the forerunner of the modern digital computer, a programmable electronic device that stores, retrieves, and processes data. Babbage's device used punch cards to store data and was intended to print answers.

More than 100 years later, the first fully automatic calculator was invented; development began in 1939 at Harvard University. Under the direction of mathematician Howard Aiken (1900–1973), the first electronic digital computer, called Mark I, was invented in 1944. (The Mark II followed in 1947.) In 1946 scientists at the University of Pennsylvania completed ENIAC (Electronic Numerical Integrator And Calculator), the first all-purpose electronic digital computer. Operating on 18,000 vacuum tubes, ENIAC was large, required great deal of power to run, and generated a lot of heat. The first computer to handle both numeric and alphabetical data with equal facility was the UNIVAC (UNIVersal Automatic Computer), developed between 1946 and 1951, also at the University of Pennsylvania.

Who wrote the **first computer program**?

The first functional computer program was written by Grace Murray Hopper (1906–1992), an admiral of the U.S. Navy. She wrote a program for the Mark I computer (developed in 1944), the first fully automatic calculator. During the 1950s Hopper directed the work that developed one of the most widely used computer programming

Two men work in 1959 on UNIVAC, the first computer to handle both numeric and alphabetical data with equal facility.

languages, COBOL (Common Business Oriented Language). She is also credited with coining the slang term *bug* to refer to computer program errors. The story goes that her machine had broken down, and when she looked into the problem, she discovered a dead moth in the computer. As she removed it, she reportedly announced that she was "debugging the machine." Hopper served the U.S. Navy for 43 years, from 1943 to 1986, and retired as its most senior officer. She was also a professor at Vassar College and a programmer for the Sperry Rand Corporation from 1959 to 1971. She is one of the pioneers of computer science.

The very first computer program written, though never used, was also by a woman: the English baroness Augusta Ada Byron (the poet Lord Byron's daughter, born 1815) wrote it for Charles Babbage's "analytical engine," which was never completed, and so the program was not tested.

When was the **computer chip** developed?

The computer chip, or integrated circuit, was developed in the late 1950s by two researchers who were working independently of each other: Jack Kilby (1923–) of Texas Instruments (who developed his chip in 1958) and by Robert Noyce (1927–1990)

Why was the invention of the transistor important?

Invented by Bell Laboratories scientists in 1947 and announced in 1948, the transistor allowed electronic equipment to become much, much smaller. The transistor is a device that transfers electronic signals across a resistor (it gets its name from *trans*ferring over a res*istor*). The device is smaller than the vacuum tube it replaced, requires less energy, generates less heat, and is more dependable and faster. The transistor radio was among the early innovations made possible by the 1948 invention, making radios truly wireless—and portable. Transistors, of which many types have been designed since their invention, are also used in computers (on the computer chip), and in numerous other consumer electronics. Another principal use is in the space and military industries where transistors are used in automatic control instrumentation for flight and guided missiles.

of Fairchild Semiconductor (in 1959). The chip is an electronic device made of a very small piece (usually less than one-quarter inch square) of silicon wafer, and today has typically hundreds of thousands miniature transistors and other circuit components that are interconnected. Since its development in the late 1950s, the number of tiny components a chip can have has steadily risen, improving computer performance, since the chips perform a computer's control, logic, and memory functions. A computer's microprocessor is a single chip that holds all of the computer's logic and arithmetic. It is responsible for interpreting and executing instructions given by a computer program (software). The microprocessor can be thought of as the brain of the computer's operating system.

Many other consumer electronic devices rely on the computer chip as well, including the microwave, the VCR, and calculators.

Who invented the **first personal computer**?

Development of the personal computer (PC), a microcomputer designed to be used by one person, was first developed for business use in the early 1970s. Digital Equipment Corporation developed the PDP-8, which was predominately used in scientific laboratories. The credit for development of a computer for home use goes to Steve Wozniak (1950–) and Steve Jobs (1955–), college dropouts who founded Apple Computer in 1976. They spent six months working out of a garage, developing the crude prototype for Apple I, which was bought by some 600 hobbyists—who had to know how to wire, program, and set up the machine. Its successor, Apple II, was introduced in 1977 as the first fully assembled, programmable microcomputer, but it still required customers to use their televisions as screens and to use audio cassettes for data storage. It retailed for just less than $1,300. That same year Commodore and Tandy introduced affordable personal computers. In 1984 Apple Computer introduced the Macintosh (Mac), which became the first widely used computer with a graphical user interface (GUI). By this time, International Business Machines (IBM) had introduced its PC (1981), which quickly overtook the Mac, in spite of the fact that IBM was behind in developing a user-friendly graphical interface.

Why was the **Internet** invented?

The computer network was invented in the late 1960s so that U.S. Department of Defense researchers could share information with each other and with other researchers. The Advanced Research Projects Agency (ARPA) developed the Internet; its users, who were mostly scientists and academics, saw the power of the new technology: Wires linking computer terminals in a web of networks allow people anywhere in the world to communicate with each other over the computer. Even though it was developed by the government, the Internet is not government-run. The Internet Society, comprised of volunteers, addresses usage and standards issues.

The technology caught on, made more accessible by the innovation of the user-friendly World Wide Web. In spring 2005 there were an estimated 888 million Internet users around the world, about 35 percent of them in Asia, 30 percent in Europe, and 25 percent in North America (about 200 million of those in the United States). The powerful network had become part of everyday life in the developed world.

How old is the **World Wide Web**?

The Web, which adds an ease-of-use layer to the Internet by providing a graphical user interface (GUI), was developed in 1990 by English computer scientist Tim Berners-Lee (1955–), who wrote the Web software at the CERN physics laboratory near Geneva, Switzerland. Berners-Lee wrote a program defining hypertext markup language (HTML), hypertext transfer protocol (HTTP), and universal resource locators (URLs). The Web became part of the Internet in 1991 and has played a major role in the growing popularity of the international computer network, making information more accessible to the user via multimedia interfaces, which allow the presentation of graphics (formatted text and hyperlinks, photos, and illustrations) as well as streaming or downloadable audio and video.

When was **e-mail** invented?

Short for "electronic mail," e-mail was invented in 1971 by computer engineer Ray Tomlinson (1941–) who developed a communications program for computer users at the Advanced Research Projects Agency (ARPA). The result was ARPAnet, a program that allowed text messages to be sent to any other computer on the local network. ARPAnet is now hailed as the Model T of the Information Superhighway. The technology expanded in the 1970s with the use of modems, which connect computers via telephone lines. Within a decade of its introduction, e-mail had become widely used as a communications mode in the workplace. In the 1990s usage expanded rapidly to Internet users at home, schools, and elsewhere. Some technology analysts call e-mail the "killer app" of the Internet, the most powerful tool on the worldwide computer network.

When did **mobile phones** first come into use?

Mobile communication dates back to radiophones used in the 1940s and 1950s. They were two-way radio systems that were powered by car batteries and required operator assistance; they were not very reliable, and the phones were anchored to a place, not a person. The first truly mobile phone call, in that it used a portable handset, was manufactured on April 3, 1973. The caller was Dr. Martin Cooper of Motorola, who, from the streets of Manhattan, called rival researcher Joel Engel at Bell Laboratories (AT&T's research arm); the two companies were in a heated race to develop mobile telephony. The device used by Cooper that day was called the Dyna-Tac; it weighed two pounds and had simple dial, talk, and listen features.

The first generation of mobile phones began to be widely used in the 1980s. These phones were large by today's standards and were usually installed in a car or briefcase. Transmission was via clusters of base stations, or cellular networks. The next generation of mobile phones appeared in the 1990s; the handset and battery technology improved, allowing for more features in smaller-sized phones and greater mobility; these were reliable phones that people could carry with them. As more users adopted the technology, cellular providers expanded transmitting systems. In some areas of the world, usage took off to the point of near universality by 2000. Usage in the United States, though strong, lagged behind the rest of the developed world. Some analysts believed that this was due to relatively high service fees, while others cited a lack of reliability, especially in rural areas. The land-based telephone system in the United States was designed to nine nines of reliability (meaning it can be counted on to function 99.9999999 percent of the time), a standard as yet unmet by cellular technology.

When was **instant messaging** introduced?

The ability to send instant text messages over computers was introduced to the public in 1996. Internet service provider America Online (AOL) launched instant messaging, or IM, as another way for its members to communicate with each other. By logging onto a home or work computer (or cell phone with Internet capabilities), users could view their "buddy list" and see which of their AOL contacts were online at the time. IM users could then send messages back and forth in real time, next door or across thousands of miles. In 1997 AOL expanded the service to non-AOL users with a utility called AOL Instant Messaging (AIM), and in 1998 the company acquired another IM utility, ICQ. By 1999 MSN and Yahoo rolled out their instant messenger services, and others followed. The concept caught on, particularly with young users who embraced the concept of a private chat room. In 2004 a Pew Internet and American Life Project report estimated that 53 million Americans, or 4 out of 10 Internet users, were instant messaging. IM had become the primary form of communication for many, replacing telephone calls and emails. The mode of communication also brought about a new shorthand, or subculture language, with a heavy reliance on abbreviations and icons. The use of the technology continued to grow rapidly, expanding to practical business use. In 2004 analysts estimated that there were about 600 million registered IM users worldwide.

How long have humans been producing **art**?

The first true art was originated by *Homo sapiens sapiens* (called "man the double wise") in Europe about 35,000 years ago (during the Stone Age). Man the double wise painted his own hand prints, warrior images, and animals (including bison, horses, and reindeer) on the walls and ceilings of uninhabited caves in France and Spain between 35,000 B.C. and 8000 B.C. He used red, black, and yellow paints, which were made by mixing powdered earth and rock pigments with water. Among the most famous paintings are those in the caves at Lascaux (in Dordogne, France), Niaux (Ariège, France), Pech-Merle (Lot, France), Gasulla (Castellón, Spain), and Altamira (Cantabria, Spain).

These early modern humans—who, if dressed in contemporary clothing, would be nearly indistinguishable from anyone on a modern city street—also decorated tools and created lifelike sculptures of animals and women. European man of this period, who had a fully developed human brain, is also referred to as Cro-Magnon man for a shallow rock shelter near Les Eyzies in the Dordogne region of southwestern France, where, in 1868, skeletal remains of the tall, erect-walking species were found.

WRITTEN LANGUAGE

When did **people begin to write**?

If writing is viewed as a means of communicating between humans, using conventional, visible marks, then writing spans the entire history of visual communication—

from early pictographic (picture writing) beginnings to alphabetical writing. The oldest picture writing identified thus far was found in Mesopotamia (present-day Iraq), the valley between the lower Tigris and Euphrates rivers. This writing, called cuneiform, dates back to the last centuries of the fourth millennium B.C. (c. 3700 B.C.). The finding consisted of about a dozen pictures inscribed on both sides of a limestone tablet. The characters are made up of wedge-shaped strokes; the system was probably created by the Sumerians, a non-Semitic people whose origins are unknown; they probably immigrated to southwest Asia (Mesopotamia) from the East. Cuneiform pictographs closely resemble Egyptian hieroglyphics, picture script developed by ancient Egyptian priests and perfected by the first Egyptian dynasty (3110 to 2884 B.C.). Egyptian hieroglyphics consisted of some 600 symbols (phonograms). Cuneiform and hieroglyphic systems were the predecessors to the alphabet.

When was the **first alphabet** created?

The earliest form of an alphabet (a set of letters with which one or more languages are written) was developed between 1800 and 1000 B.C. by a Semitic people of unknown identity. In 1928 in the northern Syrian city of Ras Shamra, clay tablets were discovered with a cuneiform (wedge-shaped) alphabet of characters. According to most scholars, these early Semitic inventors got the idea of developing an alphabet because of their contact with Egyptian hieroglyphics. Although some of the Semitic symbols were Egyptian, the system of writing was distinctly their own, consisting of 22 characters representing only consonants. (There were no written directions for vowels; the reader had to supply the vowels from his or her knowledge of the language.)

The characters were developed this way: Important Semitic words were selected so that each would begin with a different consonant. Then, stylized pictures portraying the words (mostly nouns) were assigned the phonetic value of the initial sound of each word. For example, the first character in the Semitic (Hebrew) alphabet, *aleph,* originally meant and was the symbol for an oxhead. The second, *beth,* which came to represent the sound "b," meant "house" and renderings of the symbol reveal a shelter with a roof. The Phoenician alphabet (developed c. 1000 B.C.) is believed to have been derived from this early Semitic alphabet; in turn, the Greeks (in about 500 B.C.) adapted the Phoenician alphabet for their own use.

How was **paper invented**?

The oldest writing surfaces in existence include Babylonian clay tablets and Indian palm leaves. Around 3000 B.C. the Egyptians developed a writing material using papyrus, the plant for which paper is named. Early in the Holy Roman Empire, the long manuscript scrolls that were made of fragile papyrus and used by Egyptians, Greeks, and Romans were replaced by the codex, separate pages bound together at one side and having a cover (like the modern book). Eventually, papyrus was replaced by vellum (made of a fine-grain lambskin, kidskin, or calfskin) and parchment (made of sheepskin or goatskin), both of which provided superior surfaces for painting.

The wood-derived paper we know today was developed in A.D. 105 by the Chinese, who devised a way to make tree bark, hemp, rags, and fishnets into paper. The process used then contained the basic elements that are still found in paper mills today. The Moors introduced papermaking to Europe (Spain) in about 1150; by the 1400s paper was being made throughout Europe. But it was not until the late 1700s that paper was produced in continuous rolls. In 1798 a French paper mill clerk invented a machine that could produce a continuous sheet of paper in any desired size from wood pulp. The machine was improved and patented by English papermakers Henry (1766–1854) and Sealy Fourdrinier (d. 1847) in 1807. The invention spurred the development of newspapers.

What was the **first book**?

It was *The Diamond Sutra,* published in the year A.D. 868. Archaeologists discovered it in the Caves of the Thousand Buddhas, at Kansu (Ganzu), China.

Who developed **printing**?

If printing is defined as the process of transferring repeatable designs onto a surface, then the first known printing was done by the Mesopotamians, who as early as 3000 B.C. used stamps to impress designs onto wet clay.

Printing on paper developed much later; Chinese inventor Ts'ai Lun (A.D. c. 50–c. 118) is credited with producing the first paper in A.D. 105. During the T'ang dynasty (618–906) Chinese books were printed with inked wood blocks, and it was the Chinese—not German printer Johannes Gutenberg (c. 1390–1468), as is widely believed—who developed movable type, allowing printers to compose a master page from permanent, raised characters. However, movable-type printing did not catch on in medieval China because the Chinese language has some 80,000 characters; printers found it more convenient to use carved blocks.

How old is the oldest **illuminated manuscript**?

The oldest painted manuscript known is the *Vatican Vergil,* which dates to the early fifth century A.D. The content is pagan, representing a scene from Roman poet Virgil's (70–19 B.C.) *Georgica,* his verses idealizing country life and nature. But illuminated manuscripts were typically renderings of sacred texts. When the fathers of the Eastern Church advised that pictures could be used in books and art to teach people, the decorations and pictures (called illuminations since they were meant to illuminate, or make clear, the text) took on great importance. Since the beauty of the manuscript

What was the first American newspaper?

America's first regular newspaper was the weekly News-Letter published by Boston postmaster John Campbell (1653–1728). It was published for nearly two decades, between 1704 and 1722, and consisted of a single 7-by-11-inch sheet of paper covered on both sides with news (not all of it substantiated) from post riders, sea captains, and sailors. By 1765 there were more than 20 newspapers in the American colonies.

was meant to represent the spiritual beauty of the text (usually gospels, psalms, prayers, or meditations), sacred books became increasingly ornamental. Some "books of hours," devotional books developed during the 1300s, are considered masterpieces of illumination. The works (c. 1409–16) of Belgian illuminators Pol, Hermann, and Jehanequin Limburg represent the height of painted religious manuscripts.

Why is **Gutenberg** considered the pioneer of **modern printing**?

Johannes Gutenberg (c. 1390–1468), a German who built his first printing press around 1440 to 1450, is considered the inventor of a lasting system of movable-type printing. Printing technology had only a brief existence in Europe before Gutenberg, whose process, culminating in the publication of the Gutenberg Bible (1452–1456), made printed material available to everyone, not just the clergy or the privileged class. Gutenberg is credited with helping spread the ideals of the Renaissance (1350–1600) throughout Europe.

When was the **first newspaper** published?

Newspapers, publications that are issued regularly (usually daily or weekly) to notify readers of current events, had their first recorded appearance about 59 B.C., when Romans posted a daily handwritten news report in public places (the report was called the *Acta Diurna,* or "Daily Events"). In about A.D. 700 the Chinese developed the world's first printed newspaper, called the *Dibao;* it was printed from carved wooden blocks. The invention of the moveable-type printing press (c. 1450) spurred the development of the newspaper as we know it. In Germany regularly published newspapers began being circulated in the early 1600s. Newspapers proliferated in the 1800s after the development of a machine that allowed paper to be produced in sheets of any size.

What were **penny newspapers**?

Although the number of newspapers published in the United States climbed quickly during the first decades of the new republic (the decades following independence,

1783), they were not generally accessible to the common man. Priced at an average six cents a copy, the cost was outside the reach of the average American. The labor-saving machinery developed during the Industrial Revolution (late 1700s and early 1800s) was instrumental in lowering the cost of the newspaper: Because production costs were lower, publishers could charge less per copy. The first so-called penny newspaper (or one-cent paper) was published by Benjamin H. Day (1810–1889): the debut issue of the daily *The New York Sun* appeared in 1833. The American newspaper industry was off and running. Now reaching a mass audience, publishers worked feverishly to outdo each in order to keep their readers. By the late 1800s news reporting had become increasingly sensationalistic. Population growth (spurred by increased immigration at the end of the nineteenth century and early in the twentieth century) meant there were plenty of readers for the now thousands of newspapers. During the first decade of the 1900s, before the proliferation of radio (invented 1895), the number of American newspapers peaked at about 2,600 dailies and 14,000 weeklies.

What is **muckraking** journalism?

Muckrakers are journalists who seek out and expose the misconduct of prominent people or of high-profile organizations; they emerged on the American scene in the late 1800s and early 1900s. Crusaders for social change, muckraking journalists wrote articles not about news events, but about injustices or abuses, bringing them to the attention of the American public. Published in newspapers and magazines, the articles exposed corruption in business and politics. While the early muckrakers were sometimes criticized for their tactics, their work succeeded in raising widespread awareness of social, economic, and political ills, prompting a number of reforms, including passage of pure food laws and antitrust legislation. American politician Theodore Roosevelt (1858–1919) dubbed the controversial journalists "muckrakers," a reference to a character in *Pilgrim's Progress* (by English preacher John Bunyan [1628–1688]) who rejects a crown for a muckrake, a tool used to rake dung.

EDUCATION

When were the **first schools** established?

The first formal education began shortly after the development of writing (c. 3000 B.C.), when both the Sumerians (who had developed a cuneiform system of pictographics) and the Egyptians (who developed hieroglyphics) established schools to teach students to read and write the systems. After the development of the first alphabet (between 1800 and 1000 B.C.) by Semitic people in Syria, religious schools were set up. Priests taught privileged boys to read sacred Hebrew writings (the Torah). The first

Typical American schoolroom of the late nineteenth century. The U.S. public school system traces its roots to colonial times.

school that was open to everyone, not just the upper classes, may well have been that established by Chinese philosopher Confucius (551–479 B.C.), who taught literature and music, conduct, and ethics to anyone who wanted to learn.

The western model of education is based on the ancient Greek schools, which were founded about the fifth century B.C. In the city-state of Sparta, boys were not only trained for the military; they also learned reading and writing and studied music. In Athens, boys learned to read and write, memorized poetry, and learned music as well as trained in athletics. In the second half of the fifth century B.C., the Sophists (ancient Greek teachers of rhetoric and philosophy) schooled young men in the social and political arts, hoping to mold them into ideal statesmen.

How old is the concept of **public schools**?

It dates at least as far back as ancient China. The philosopher Confucius (551–479 B.C.) was among the first in China to advocate that primary school education should be available to all. He averred that "in education there should be no class distinctions." He never refused a student, "even though he came to me on foot, with nothing more to offer as tuition than a package of dried meat." Confucius asserted that any man—including a "peasant boy"—had the potential to be a man of principle.

However, it was not until the Age of Enlightenment that public schools were widely instituted. In Prussia (present-day Germany), Frederick the Great (1712–1786) was considered an enlightened ruler for, among other things, founding a public education system (which became established during the early 1800s). After Prussia united with Germany to form a powerful state, other European countries began instituting systems of public education—which were credited by many as an important factor in Prussia's rise. By the early twentieth century, public elementary schooling was both free and compulsory in most of Europe. Free secondary education was also offered in some nations.

In the United States, public schools had their beginnings during colonial times: In 1647 Massachusetts passed a law requiring the establishment of public schools.

When were **schools** in the United States **desegregated**?

On May 17, 1954, in the case of *Brown* v. *Board of Education,* the Supreme Court ruled (nine to zero) that racial segregation in public schools is unconstitutional. The court overturned the "separate-but-equal" doctrine laid down in the 1896 case *Plessy v. Ferguson.* Chief Justice Earl Warren (1891–1974) ordered the states to proceed "with all deliberate speed" to integrate educational facilities. Also in 1954, on November 7, the Supreme Court ordered desegregation of public golf courses, parks, swimming pools, and playgrounds. In the aftermath of these rulings, desegregation proceeded slowly and painfully. In the early 1960s sit-ins, "freedom rides," and similar expressions of nonviolent resistance by blacks and their sympathizers led to a decrease in segregation practices in public facilities.

How has the **U.S. separation of church and state** affected the public schools?

Religion in American public schools continued to be a hot topic throughout the 1900s. But the Supreme Court rulings in the middle of the twentieth century proved to have the most bearing on religious practices in state-supported schools. On June 17, 1963, in an eight-to-one ruling, the Supreme Court decided that prayer and Bible reading in U.S. public schools were unconstitutional. The decision, in the case of *Schempp* v. *Abington Township,* culminated a series of high court rulings over the course of almost 20 years, which gradually removed the practice of religious activities from public schools.

The rulings began in 1947 with the New Jersey case of *Everson* v. *Board of Education,* in which the court (in a five-to-four vote) defended the use of state funds to transport children to parochial schools, but warned that "a wall of separation between church and state" must be maintained. In 1948, in *McCollum* v. *Board of Education,* the court banned a program of religious instruction from the schools of Champaign, Illinois. In *Engel* v. *Vitale* (1962) the justices of the Supreme Court ruled that the state-composed prayer recited in New York classrooms was unconstitutional.

African American students at Central High School under trooper escort in Little Rock, Arkansas, 1957, after the Supreme Court ruled that racial segregation in public schools is unconstitutional.

When was the **first kindergarten**?

The world's first kindergarten opened in 1837 in Blankenburg, Germany, under the direction of educator Friedrich Froebel (1782–1852). Froebel went on to establish a training course for kindergarten teachers, and he introduced the schools throughout Germany. Such schools and classes for children ages four to six are the norm today in much of the world.

How did **Montessori schools** get started?

The schools, evident throughout the United States, as well as Great Britain, Italy, the Netherlands, Spain, Switzerland, Sweden, Austria, France, Australia, New Zealand, Mexico, Argentina, Japan, China, Korea, Syria, India, and Pakistan, carry the name of their founder, Maria Montessori (1870–1952). She was the first woman in Italy to earn a medical degree and to practice medicine. In 1900 Montessori pioneered teaching methods to develop sensory, motor, and intellectual skills in retarded kindergarten and primary school students. Under her direction, these "unteachable" pupils not only mastered basic skills, including reading and writing, but they passed the same examinations given to all primary school students in Italy.

Montessori then spent time in the country's primary schools, where she observed the educators' practice of teaching by rote (by using repetition and memory) and their reliance on restraint, silence, and a system of reward and punishment in the classroom. She believed her system, called "scientific pedagogy," which was based on noncoercive methods and self-correcting materials (such as blocks, graduated cylinders, scaled bells, and color spectrums), would yield better results in students. Montessori theorized that children possess a natural desire to learn and, if put in a prepared environment, their "spontaneous activity" would prove educational. Instead of lecturing to their students, Montessori encouraged educators to simply demonstrate the correct use of materials to students who would then teach themselves and each other. She also believed in community involvement in schools, encouraging parents and other community members to take active roles in the education of the children. When Montessori put these principles into action, it was to highly favorable results.

In 1909 Montessori published *The Montessori Method,* which was made available in English three years later and became an instant best-seller in the United States. Her method, which she believed "would develop and set free a child's personality in a marvelous and surprising way," caught on. For Montessori, who has been called a "triumph of self-discipline, persistence, and courage," spreading the message about her teaching method became her life's work. She was still traveling, speaking to enthusiastic crowds the world over, when she died in the Netherlands at the age of 81. Montessori's beliefs—which were both scientific and spiritual—had a profound effect not only on students in Montessori schools, but on primary education in general.

When did **higher education** begin?

About the sixth century B.C. schools of medicine existed on the island of Cos, Greece, where philosophers theorized on the nature of man and the universe. The Pythagoreans (followers of Greek philosopher and mathematician Pythagoras, c. 580–500 B.C.) began the first schools of higher education in southern Italy, where philosophy and mathematics were taught in Greek. The great philosophers Socrates, Plato, and Aristotle carried on the Pythagorean tradition, as did Epicurus and Zeno in the fourth century B.C. Universities have a long history in the Arab world; for example, the Al-Azhar University in Cairo was founded in about A.D. 970 and is one of the oldest universities in the world.

When was the **first university** established in the **West**?

The first modern western university was established in the Middle Ages—1158 to be exact—in Bologna, Italy. It was in that year that Frederick I (c. 1123–1190), Holy Roman emperor, asserted his authority in Lombardy. He granted the first university charter for the University of Bologna, authorizing its students to organize. The universities that were set up in Europe during the Middle Ages (500–1350) were not necessarily places or groups of buildings; they were more often groups of scholars and students. The University of Paris, which today includes the renowned Sorbonne (the university's liberal arts and sciences division), soon became the largest and most famous university in Europe. The Sorbonne itself was founded in 1250 as a school of theology. It was reorganized in the 1600s

By 1500 universities had been founded throughout the continent. Of these, the ones that survive today include the universities of Cambridge and Oxford in England; those at Montpellier, Paris, and Toulouse, France; Heidelberg, Germany; Bologna, Florence, Naples, Padua, Rome, and Siena, Italy; and Salamanca, Spain. The methods and techniques developed in these early institutions set standards of academic inquiry that remain part of higher education in the world today.

What was the first **university** in the **Western Hemisphere**?

It was the University of Santo Domingo, founded in 1538 by the Spaniards in the Dominican Republic (which occupies the eastern half of the Caribbean island of Hispaniola).

What was the **first American** university?

It was Harvard, chartered on October 28, 1636, by the Massachusetts general court, which passed a legislative act to found a college. It was not until November of the following year, however, that there was further action; it was then that the general court decreed that the college be built in New Towne, Massachusetts, which in 1638 was renamed Cambridge after England's Cambridge University, where some colonists had studied. In fall of that year, Harvard's first professor, Nathaniel Eaton, began classes, at which time the first building was under construction and a library was being assembled. The university got its name not from a founder, but from a newly arrived British philanthropist and colonial clergyman, John Harvard (1607–1638), who left the library some 400 volumes and donated about 800 pounds sterling to the college. The institution was named in his honor in 1639, the year after he died.

The first state university was the University of North Carolina at Chapel Hill; it was founded in 1789. And in 1795 it became the first public institution of higher education in the United States to begin enrolling students.

What was the **lyceum movement**?

It was a public education movement that began in the 1820s and is credited with promoting the establishment of public schools, libraries, and museums in the United States. The idea was conceived by Yale-educated teacher and lecturer Josiah Holbrook (1788–1854), who in 1826 set up the first "American Lyceum" in Millbury, Massachusetts. He named the program for the place—a grove near the temple of Apollo Lyceus—

where the ancient Greek philosopher Aristotle (384–322 B.C.) taught his students. The lyceums, which were programs of regularly occurring lectures, proved to be the right idea at the right time: They got under way just after the completion of the Erie Canal (1825), which permitted the settlement of the nation's interior, just as the notion that universal, free education was imperative to the preservation of American democracy took hold. The movement spread quickly. At first the lectures were home-grown affairs, featuring local speakers. But as the movement grew, lyceum bureaus were organized, which sent paid lecturers to speak to audiences around the country. The lyceum speakers included such noted Americans as writers Ralph Waldo Emerson (1803–1882), Henry David Thoreau (1817–1862), and Nathaniel Hawthorne (1804–1864), as well as activist Susan B. Anthony (1820–1906). After the Civil War (1861–65), the educational role of the lyceum movement was taken over by the Protestant-led chautauquas.

What was the **chautauqua movement**?

It was a cultural, religious, and political education movement that began in the 1870s and lasted into the 1920s. An estimated 45 million Americans participated in the chautauqua, making it a dominant force in American life during its day. Theodore Roosevelt (1858–1919) hailed it as "the most American thing in America," and, during World War I (1914–18), Woodrow Wilson (1856–1924) claimed that it was "an integral part of the national defense." Some scholars credit the chautauqua movement with sowing the seeds of liberal thought in America.

The movement began in 1874 at a Methodist Episcopal campsite on the shores of Lake Chautauqua, New York. There a young minister named John H. Vincent (1932–1920), of Camden, New Jersey, endeavored to train Sunday-school teachers in a summer camp atmosphere. The program grew in popularity and was expanded beyond Bible study and religious training to include lessons in literacy, history, and sociology. Chautauqua-style summer camps, commonly called Sunday-school assemblies, began popping up across the nation; all of them featured a general meeting hall or pavilion set in a campground. By 1900 there were 200 pavilions in 31 states. Attendees of all ages would attend the summer programs, which featured speakers on a wide variety of subjects, including the arts, travel, and politics. Performances also became part of the movement, with a variety of musicians and entertainers joining the lecturers.

Early in the twentieth century the chautauqua became increasingly secular and went on the road as an organized lecture and entertainment circuit. Speakers and performers traveled from town to town, where tents were set up for weeks at a time to house the summer programs. Many Americans saw their first movies in chautauqua tents. The movement died out in the mid-1920s, with the improvement of communications and transportations. Some consider the chautauqua the first form of American mass culture. The Chautauqua Institute in New York continues to host a summer education program in the spirit of the original.

FOLKTALES

How old are **Aesop's *Fables***?

They date back to the sixth century B.C. However, it was not until the late 1600s that English-language versions appeared: In 1692 a complete translation of the stories, which are believed to have been written by a Greek slave, were published in London by Sir Roger L'Estrange (1616–1704). The short, moralistic tales, which were handed down through the oral tradition, include the well-known story of the tortoise and the hare (which teaches the lesson slow and steady wins the race) and the one about a wolf in sheep's clothing (people are not always what they seem). Since some of the timeless fables have been traced to earlier literature, many believe it is almost certain that Aesop is a legendary figure.

Who were the **Brothers Grimm**?

The German brothers Jacob (1785–1863) and Wilhelm (1786–1859), best known for their fairytales, were actually librarians and professors who studied law, together wrote a dictionary of the German language, and lectured at universities.

In 1805 Jacob traveled to Paris to conduct research on Roman law, and in a library there he found medieval German manuscripts of old stories that were slowly disintegrating; he decided the tales were too valuable to lose, and he vowed to collect them. The brothers' interest in fairytales also led them to search for old traditions, legends, and tales, especially those meant for children. They traveled the German countryside, interviewing villagers in an effort to gather stories—most of which were from the oral tradition and had never been written down. The brothers were diligent in their efforts, recording everything faithfully so that nothing was added and nothing was left out. When the first volume of *Kinderund Hausmärchen* (literally, the Children's Household Tales, but known better as *Grimm's Fairy Tales*) was published in 1812, children loved it. Subsequent volumes were published in German through 1815. The fairytales collected in the multivolume work included such classics as "The History of Tom Thumb," "Little Red Riding-Hood," "Bluebeard," "Puss in Boots," "Snow White and the Seven Dwarfs," "Goldilocks and the Three Bears," "The Princess and the Pea," "The Sleeping Beauty in the Wood," and "Cinderella."

LITERATURE

What is **"the Homeric question"**?

During the eighteenth and nineteenth centuries, scholars became involved in a debate, referred to as "the Homeric question," about whether the *Iliad* and the *Odyssey* were written by the same author, or even if any one author can be credited with the entire composition of either poem, and what kind of an author Homer was. The dispute continues today. Scholars believe the *Iliad* was probably written much earlier than the *Odyssey,* though there is not enough evidence to prove that the Greek poet Homer (c. 850–? B.C.) did not write both epics. Further, it was suggested that Homer was a bard (oral poet) who was unable to read or write and who sang the great stories of the *Iliad* and *Odyssey* to the accompaniment of a lyre. According to this theory, the tales would have been dictated by Homer to a scribe late in the poet's life. However, some have left open the possibility that the human histories told in the *Iliad* and the *Odyssey* were in fact the composite result of the storytelling of numerous bards.

Several other poems, including the *Margites* and the *Batrachomyomachia,* have also been attributed to Homer, but they were most likely written by his successors.

What is known about **Homer**?

It is most likely that Homer was an oral poet and performer. Though little is known about Homer, it's believed that he was an Ionian Greek who lived circa the eighth or ninth century B.C. In the 1920s scholar Milman Parry proved that Homer's poems were "formulaic in nature, relying on generic epithets (such as 'wine-dark sea' and 'rosy-fingered dawn'), repetition of stock lines, and descriptions and themes typical of oral folk poetry." All of this suggested that Homer was most likely a bard or rhapsode—an itinerant professional reciter—who improvised pieces to be sung at Greek festivals.

Why is the ***Iliad*** studied today?

Greek poet Homer's (c. 850–? B.C.) *Iliad* and *Odyssey* (both works credited to him) are considered to be among the greatest works of literature and have had a profound influence on western poetry, serving as the primary models for subsequent works, including the *Aeneid* (Virgil) and the *Divine Comedy* (Dante).

The *Iliad* in particular can be seen as both the beginning of western literature as we know it and the culmination of a long tradition of oral epic poetry that may date as far back as the thirteenth century B.C. The *Iliad* has been a part of western education for nearly 3,000 years. The epic poem, telling the story of a 10-year Trojan War, reveals the author's keen understanding of human nature.

Is Virgil's ***Aeneid*** an unfinished work?

Yes, the *Aeneid* was technically unfinished by its author, Virgil (70–19 B.C.), who is considered the greatest Roman poet. Virgil spent the last ten years of his life working on the *Aeneid,* and he planned to devote three more years making revisions to this epic when, during his travels to gather new material for the poem, he became ill with fever and died. On his deathbed, Virgil requested that his companions burn the *Aeneid.* However, Augustus (63 B.C.–A.D. 14), the emperor of Rome, countermanded the request, asking Virgil's friends to edit the manuscript. Augustus did specify that the writers not add, delete, or alter the text significantly. The *Aeneid,* Virgil's great epic about the role of Rome in world history, was first published in 17 B.C. The work consists of 12 books, each between 700 and 1,000 lines long.

What **innovations** are credited to **Virgil**?

Scholars acclaim Virgil (70–19 B.C.) for transforming the Greek literary traditions, which had long provided Roman writers with material, themes, and styles. Virgil populated his pastoral settings (always idealized by other writers) with contemporary figures; he combined observation with inquiry; employed a more complex syntax than had been in use previously; and developed realistic characters. These technical innovations informed all subsequent literature.

However, writing was not supposed to have been Virgil's occupation: In his youth, he studied rhetoric and philosophy, and he planned to practice law, but proved too shy for public speaking. So he returned to the small family farm his mother and father operated, where he studied and wrote poetry.

In addition to the *Aeneid,* Virgil wrote *Eclogues* (or *Bucolica*), a set of 10 pastoral poems written (from 42–37 B.C.) as a response to the confiscation of his family's lands; and *Georgics,* a four-volume work (written from 36–29 B.C.) glorifying the Italian countryside. Within 50 years of his death in 19 B.C.,Virgil's poems became part of the standard curriculum in Roman schools, ensuring the production of numerous copies. Virgil's works have remained accessible to scholars and students ever since.

Why is ***Beowulf*** considered an important work?

Beowulf, the earliest manuscript of which dates to about A.D. 1000, is the oldest surviving epic poem in English or any other European vernacular. It was written in Old English (the language of the Anglo-Saxons in England; used c. A.D. 500–1100); its author is unknown. Categorized as a folk-epic, *Beowulf* tells the story of a Scandinavian warrior hero who, on behalf of the Danish king, fights and kills the fearsome monster Grendel, then slays the monster's mother, and finally engages a fire-breathing dragon in mortal combat. Because of its combination of Christian and pagan themes, scholars believe the epic may have been written as early as 700 or 750.

Why is ***The Divine Comedy*** widely studied?

Simply put, *The Divine Comedy,* which consists of 100 cantos arranged in three books (*Inferno, Purgatorio,* and *Paradiso*), is studied not only for the beauty of its verse, but for its timeless message.

In a letter to his benefactor, Dante Alighieri (1265–1321) explained that by writing *The Divine Comedy* (*Divina Commedia,* begun c. 1308) he would attempt "to remove those living in this life from the state of misery and lead them to the state of felicity." While the subject of the poem, according to Dante, is "the state of souls after death," allegorically, the poem is about humankind, who can exercise free will to bring "rewarding or punishing justice" upon themselves.

Dante's masterpiece is considered the seminal work of Italian literature: At the time that he wrote *The Divine Comedy,* Latin was the undisputed language of science and literature. Italian, on the other hand, was considered vulgar. By skillfully writing this poem in the vernacular (Tuscan Italian) rather than Latin, Dante parted from tradition, marking a critical development for vernacular writing. In its translations *The Divine Comedy* has become a point of reference for writers in any language. Scholars and students agree that Dante expresses universal truths in this work, which is also a finely crafted piece of literature.

Who was **Dante's Beatrice**?

In Dante's masterpiece *The Divine Comedy,* the central figure is led to redemption by a character named Beatrice (his earlier guide through hell and purgatory was the great Roman poet Virgil). Dante Alighieri (1265–1321) was born in Florence, Italy, where he also spent much of his life. In 1274, at the age of nine, he was introduced to Beatrice Portinari; they met again nine years later, and Dante was profoundly affected by her beauty and grace. When she died in 1290, Dante was inspired to commemorate her in several works, most notably *The Divine Comedy* (c. 1308–1321). Beatrice is also depicted in Dante's *The New Life* (c. 1293), a collection of 31 love poems. He wrote *The Banquet* (c. 1304–1307), another collection of lyrical poems, to commemorate Beatrice's death.

Why is **Chaucer's *Canterbury Tales*** important to literature?

The unfinished work, which was begun about 1486 and written during the last 14 years of its author's life, is considered a masterpiece of Middle English, the language spoken by Anglo-Saxons in England from c. 1200 into the late 1400s. In the *Canterbury Tales,* Geoffrey Chaucer (1340–1400), who was the son of a wealthy wine merchant, weaves together stories told by 28 pilgrims whom the storyteller (the poet himself) met at an inn. The pilgrims, along with the innkeeper who joins them and the poet, represent all facets of English social life—aristocracy, clergy, commoners, and even a middle class, which was not officially recognized by the social structure of the day, but which, in fact, existed.

To connect the tales, Chaucer uses the framing device of a journey to the shrine of Thomas Becket (c.1118–1170), the archbishop of Canterbury who in 1170 was killed by overzealous knights in the service of England's King Henry II. In his prologue, Chaucer indicates that each traveler was to tell two tales out and two tales back from

It is thought that William Shakespeare is widely studied because his words express universal and unchanging human concerns.

Canterbury Cathedral, for a total of 120 stories, which were intended to entertain the pilgrims during their trip. But Chaucer wrote only 24 tales, two of which are incomplete. The tales include bawdy humor, fables, and lessons. While the pilgrims reach the shrine, they do not return—a device some scholars have interpreted as deliberate on the author's part, as it suggests the human journey from earth to heaven. Whatever the intention, *The Canterbury Tales* reveal the author's ability as a storyteller, as the editors of the *Norton Anthology of English Literature* assert, rivaling Shakespeare "in the art of providing entertainment on the most primitive level, and at the same time, of significantly increasing the reader's ability to comprehend reality."

The tales were extremely popular in late medieval England, printed and reprinted numerous times, particularly during the 1400s. Chaucer's rendering of details reveal both story and storyteller (pilgrim) at once, giving the reader a remarkable insight into the characters and revealing basic human paradoxes—which transcend time.

Why is **Shakespeare widely studied**?

English dramatist Ben Jonson (1572–1637) said it best when he proclaimed that Shakespeare "was not of an age, but for all time." Most teachers and students, not to mention critics and theatergoers down through the ages, likely agree with Jonson's remark: Shakespeare's canon (consisting of 37 plays, divided into comedies, tragedies, or histories, plus poems and sonnets) expresses universal and unchanging human concerns as no other works have. Shakespeare's words are familiar even to those who have not studied them, not simply because of the many contemporary adaptations of his works, but because Shakespearean phrases and variations thereof have, through the years, fallen into common usage. Consider these few examples from *Hamlet* alone: "Neither a borrower nor a lender be"; "To thine own self be true"; and "The play's the thing." No other writer's plays have been produced so often or read so widely in so many countries.

What did **Shakespeare study**?

It is thought that William Shakespeare (1564–1616) attended the King's New School, the local grammar school in Stratford-upon-Avon, England, where the main course of

Was Shakespeare famous in his own time?

Yes, by 1592 he was well known as a dramatist. William Shakespeare (1564–1616) was the son of John Shakespeare, who belonged to the merchant class, and Mary Arden, who came from a family of slightly higher social standing. His first plays, the three parts of the Henry VI history cycle, were presented in London in 1590 to 1592. The first reference to Shakespeare in the London literary world dates from 1592, when dramatist Robert Greene (c. 1558–92) referred to him as "an upstart crow."

The critical remark notwithstanding, Shakespeare's literary reputation and his acclaim grew over the next few years. He experimented with classical dramatic forms in the early tragedy *Titus Andronicus* (1593–1594) and issued a pair of narrative poems, *Venus and Adonis* (1593) and *The Rape of Lucrece* (1594). These works, which played to the fashion for poems on mythological themes, were immensely successful, establishing "honey-tongued Shakespeare"—as his contemporary Francis Meres (1565–1647) called him—as a prominent writer.

Shakespeare further established himself as a professional actor and playwright when he joined the Lord Chamberlain's Men, an acting company formed in 1594 when they began performing at theaters in London (in 1603 the group was renamed the King's Men). They became the foremost London company, largely attributable to the fact that, after joining the group in 1594, Shakespeare wrote for no other company.

instruction was in Latin. There, students were taught rhetoric, logic, and ethics, and studied works by classical authors Terence, Plautus, Cicero, Virgil, Plutarch, Horace, and Ovid. It is believed that this was the extent of Shakespeare's education; there is no evidence that he attended a university.

What is a **poet laureate**?

A poet laureate is someone who is recognized by his or her country or state as its most eminent and representative poet. Officially, a poet laureate is appointed or named by the government. England's first, if unofficially titled, poet laureate was Ben Jonson (1572–1637), a contemporary of Shakespeare. (Shakespeare acted a leading role in the first of Jonson's great plays, *Every Man in His Humour,* 1598.) In 1605 Jonson began writing a series of masques (short, allegorical dramas that were performed by actors wearing masks) for the court. Years later, in 1616, he was appointed poet laureate and in that capacity received a "substantial pension." Among Jonson's works are *Volpone* (1605), *Works* (a collection of poetry published in 1616, and which includes the oft-quoted line, "Drink to me only with thine eyes"), and *Pleasure Reconciled to Virtue* (1618).

Some sources trace the first British poet laureate back to Edmund Spenser (1552 or 1553–1599), who is called the "Poet's Poet." However, the title of poet laureate was not officially conferred on an English writer until 1638, when poet and dramatist William Davenant (1606–1668), who was reputed to be the godson or even the illegitimate son of Shakespeare, was given the honor. Other poet laureates of England include John Dryden (1631–1700), William Wordsworth (1770–1850), and Lord Alfred Tennyson (1809–1892).

Is there an **American poet laureate**?

Yes, in 1985 the U.S. Congress authorized the naming of a national poet laureate. In 1986 Kentucky-born man of letters Robert Penn Warren (1909–1989) became the country's first poet laureate. Among his works are the novels *All the King's Men* (1946, Pulitzer prize) and *A Place to Come to* (1977); several volumes of poetry; and essays published in the anthology *I'll Take My Stand* (1930). He was also the editor (1935–42) of the literary journal *The Southern Review.* Warren's successors have included Joseph Brodsky; Mona Van Duyn, the first woman to receive the honor; and Rita Dove, the first African American to receive the honor. The complete list of poet laureates is available on the Library of Congress's Web site (http://www.loc.gov/poetry/laureate.html).

How did the word **"Machiavellian"** get its meaning?

Machiavellian is defined as "characterized by cunning, duplicity, or bad faith." It's based on the theory of Italian diplomat Niccolò Machiavelli (1469–1527), who developed a code of political conduct that operates independent of ethics, thus disregarding moral authorities such as classical philosophy and Christian theology.

In 1513, after having been exiled from Florence, Italy, by the powerful Medici family, Machiavelli abruptly turned his attention to writing *The Prince,* which puts forth a calm and uncompromising analysis of techniques and methods that the successful ruler must use in order to gain—and keep—power. Written in the form of advice to the ruler, Machiavelli advises the Prince that only one consideration should govern his decisions: the effectiveness of a particular course of action, regardless of its ethical character. The book had little immediate impact in Italy, although it soon became legendary throughout Europe, and its major ideas—the power of politics—are familiar today even to people who have never read the book.

Why is **Milton** important to **English literature**?

Except for Shakespeare, the works of John Milton (1608–1674) have been the subject of more commentary than those of any other English writer. Milton is considered one of only a few writers to take their place in "the small circle of great epic writers." According to *Norton Anthology of English Literature,* in Milton's writings "two tremendous intellectual and social movements come to a head." The movements referred to are the Renaissance and the Reformation. Scholars point to Milton's use of classical references and the rich tapestry of his works as being Renaissance in nature, while his "earnest and individually minded Christianity" are resonant of the Reformation. For example, in his masterpiece *Paradise Lost* (1667), Milton, like poets Homer and Virgil before him, takes on humankind's entire experience: war, love, religion, hell, heaven, and the cosmos. But rather than having Adam triumph over evil through an act of heroism, he "accepts the burden of worldly existence, and triumphs over his guilt by admitting it and repenting it."

In addition to his famous epics, Milton wrote sonnets and other short poems, including "On Shakespeare," "L'Allegro," "Il Penseroso," and "Lycidas." His writings also include political discourse, chief of which is the essay *Areopagitica* (1644). Among the ideas that Milton championed were the limitation of the monarchy, dethroning of bishops, freedom of speech, and the institution of divorce. One commentator mused that "the guarantees of freedom in the United States Constitution owe more to Milton's *Areopagitica* than to John Locke."

What were **Voltaire's beliefs**?

The prolific French writer's corpus of 52 works were produced as part of his lifelong effort to expose injustices. Voltaire's famous words, *"Ecrasez l'infame"* (squash that which is evil), encapsulate his tenets: He believed in God, but abhorred priestly (high church) traditions; he spread the doctrines of rational skepticism to the world; he strongly advocated religious and political tolerance; and he held great faith in humankind's ability to strive for perfection. To the European literary world, he embodied the highest ideal of the Age of Reason (also called the Enlightenment). But victims of his wit feared and denigrated him. Celebrated by some during his lifetime, he has certainly been celebrated since. His masterpiece, *Candide* (1759), a satirical tale exploring the nature of good and evil, has been translated into more than 100 languages.

Why was **Voltaire** exiled from France?

The French writer Voltaire (1694–1778), born Francois-Marie Arouet (Voltaire was an assumed name), was imprisoned twice during his lifetime; he was released the second time on the condition that he leave the country. The prison terms and expulsion were the result of Voltaire's "expert satire," which first got him into trouble when he was a young man. After finishing a course of study at the Jesuit school College Louis-le-Grand (1704–11), Voltaire joined a group of aristocrats in Paris who valued the young writer's wit. He wrote and circulated verse criticizing the regent, the Duke d'Orleans. As a result of these offensive works, Voltaire was put into the Bastille (in 1717), where he began writing an epic (the *Henriade*) about France's King Henry IV (1553–1610). Full of indictments of religious fanaticism and praise for toleration, the work proved highly controversial in its day. Such antiestablishment protests eventually led the writer to have an argument with the chevalier de Rohan, a member of one of France's most powerful families. This conflict resulted in Voltaire's arrest, imprisonment (again in the Bastille), and exile to England in 1726.

He stayed in London until 1729. Returning to France, the writer penned his observations on English social and political beliefs (*Letters Concerning the English Nation,* 1734), again stirring a controversy—his exaltation of English liberalism was viewed by the authorities as a criticism of French conservatism. He fled the trouble by going into seclusion in Lorraine, where he stayed through 1749. The biting criticism of his works won the writer fame as well as controversy, both of which followed him throughout his life. In 1750 he was invited to visit Prussian King Frederick the Great at court; accepting, he stayed there only two years—he was forced to leave in 1753 after quarreling with the man he called the "Philosopher King." He spent the last 20 years of his life in Switzerland, returning to Paris to see a performance of one of his plays (*Irene*) just before his death.

What was **Goethe's** contribution to world literature?

Johann Wolfgang Goethe (1749–1832) is considered Germany's greatest writer. He also was a scientist, artist, musician, and philosopher. As a writer, Goethe experimented with many genres and literary styles, and his works became a shaping force of the major German literary movements of the late eighteenth and early nineteenth centuries. His masterwork, the poetic drama *Faust* (1808; rewritten 1832), embodies the author's humanistic ideal of a world literature—one that transcends the boundaries of nations and historical periods. Indeed, the story of Faust, a German astrologer, magician, and soothsayer (c. 1480–1540), remains one of universal interest, and has been treated often in both literature and music: the legendary figure was believed to have sold his soul to the devil in exchange for the opportunity to experience all of life's pleasures.

Who was Alexis **de Tocqueville**?

Aristocrat Alexis de Tocqueville (1805–1859) was only 26 years old when he traveled to New York with his colleague and friend, Gustave de Beaumont (1802–1866), to study and observe American democracy.

Though Tocqueville set out with the pretext of studying the American penal system on behalf of the French government (both he and Beaumont were magistrates at the time), he had the deliberate and personal goal of conducting an onsite investigation of the world's first and then only completely democratic society: the United States. Tocqueville and Beaumont traveled for nine months through New England,

Which came first—the word "scrooge" or Dickens's character Scrooge?

The character Ebenezer Scrooge came first, brought to life in Charles Dickens's extremely popular story A Christmas Carol, published in 1843. By 1899 the term "scrooge," meaning a miserly person, had entered into usage.

Dickens (1812–1870) created many memorable characters: Oliver Twist, Tiny Tim, and Little Nell, to name a few. Among the English writer's most notable works are Oliver Twist (1837–39), The Old Curiosity Shop (1840–41), Bleak House (1852–53), A Tale of Two Cities (1859), and Great Expectations (1860–61). Dickens was popular during his own time and is still popular today—attributable not only to the vivid characters he created, but for his expression of social concerns. Though he grew more pessimistic in his later works, Dickens continued to demonstrate his profound sympathy for the oppressed and his belief in the dignity of man.

eastern Canada, and numerous American cities, including New York; Philadelphia, Pennsylvania; Baltimore, Maryland; Washington, D.C.; Cincinnati, Ohio; and New Orleans, Louisiana.

The pair returned to France in 1832 and the following year published their study, *On the Penitentiary System in the United States and Its Application in France.* Once this official obligation was behind him, Tocqueville left his post as magistrate and moved into a modest Paris apartment. There he devoted two years to writing *Democracy in America* (1835, 1840). The work was soon proclaimed the classic treatment of its subject throughout the Western world and secured Tocqueville's fame as political observer, philosopher, and, later, sociologist.

Tocqueville proclaimed that during his travels, "Nothing struck me more forcibly than the general equality of conditions.... All classes meet continually and no haughtiness at all results from the differences in social position. Everyone shakes hands...." But he also foresaw the possibility that the principles of economic equality could be undermined by the American passion for equality, which not only "tends to elevate the humble to the rank of the great," but also "impels the weak to attempt to lower the powerful to their own level." While he warned against the possible "tyranny of the majority" as a hazard of democracy, he also added that law, religion, and the press provide safeguards against democratic despotism.

Who was the first to write a **modern novel**?

While there are differing opinions on the answer to this question, it is generally accepted that the credit for the novel as we know it belongs to Spanish writer Miguel de Cervantes (1547–1616). Cervantes wrote *Don Quixote* (in two parts, 1605 and 1615): It was the first extended prose narrative in European literature in which characters and events are depicted in what came to be called the modern realistic tradition. Considered an epic masterpiece, *Don Quixote* had an undeniable influence on early novelists, including English novelist and playwright Henry Fielding (who wrote the realistic novel *Tom Jones,* 1749). *Don Quixote* is also said to have anticipated later fictional masterpieces, including French novelist Gustave Flaubert's *Madame Bovary* (1857), Russian novelist Fyodor Dostoevsky's *The Idiot* (1868–69), and American writer Mark Twain's *The Adventures of Tom Sawyer* (1876) and *The Adventures of Huckleberry Finn* (1884).

How did the **novel develop**?

Critics and scholars agree that it is French writer Gustave Flaubert (1821–1880) who developed the modern novel into a "conscious art form." Flaubert's *Madame Bovary* is recognized for its objective characterization, irony, narrative technique, and use of imagery and symbolism. American writer (and naturalized British citizen) Henry James (1843–1916) is acknowledged for having enlarged the scope of the novel, introducing dramatic elements to the narrative, developing point of view technique, and advocating realism in literature. James's works include *The American* (1877), *Daisy Miller* (1879), *The Portrait of a Lady* (1881), and *The Ambassadors* (1903).

Irish writer James Joyce (1882–1941), considered the most prominent English-speaking literary figure of the first half of the twentieth century, is often credited with redefining the modern novel. Joyce experimented with the form—and revolutionized it—through his first novel, *A Portrait of the Artist as a Young Man* (1916), and with his masterpiece, *Ulysses* (1922), in which he developed the techniques of interior monologue and stream-of-conscious narrative.

Writer William Faulkner (1897–1962) was the American counterpart to Joyce's experimentation with the form of the novel. The author of *The Sound and the Fury* (1929), *Light in August* (1932), and *Absalom, Absalom!* (1936), among others, Faulkner, in his acceptance speech for the Nobel prize in literature in 1949, stated that the fundamental theme of his fiction is "the human heart in conflict with itself." This he explored by employing a variety of narrative techniques, which, like Joyce's, departed radically from traditional methods.

Why is **Jane Austen widely read** today?

Austen is considered one of the greatest novelists in English. She wrote just six books during her lifetime, including her best-known works *Sense and Sensibility* (published 1811), *Pride and Prejudice* (1813), and *Emma* (1816), but in so doing she created the novel of manners, which continues to delight readers today. The daughter of a clergyman, Jane Austen (1775–1817) rejected the literary movement of the day, romanticism, opting instead to portray life as she knew it. As such, she was the first realist in the English novel. Austen's works are ripe with shrewd observation, wit, and an appreciation for the charms of everyday life, making her an engaging storyteller for all time.

Why is *Moby Dick* considered the greatest American novel?

The 1851 novel by Herman Melville (1819–1891), which opens with the familiar line "Call me Ishmael," has been acclaimed as one of the greatest novels of all time; many regard it as the best American novel. Of course, determining the best is a purely subjective matter, and Melville's work has many worthy rivals for the distinction, but *Moby Dick* remains a compelling and finely wrought work—in spite of the fact that it was not appreciated in its day. The story of a whaling captain's obsessive search for the whale that ripped off his leg, *Moby Dick* is both an exciting tale of the high seas and an interesting allegory, interpreted as the human quest to understand the ultimately unknowable ways of God. The work first received notoriety some 30 years after Melville's death.

Why was James Joyce's *Ulysses* banned in the United States?

Irish writer James Joyce's (1882–1941) masterpiece was originally published in 1922 (it had been serialized prior to then) by the Paris bookstore Shakespeare and Company. By 1928 it was officially listed as obscene by the U.S. Customs Court. The reason was twofold: the use of four-letter words and the stream-of-consciousness narrative of one of the characters, revealing her innermost thoughts. When the official stance on the book was challenged in U.S. court in 1933, the judge (John Woolsey) called it a "sincere and honest book," and after long reflection he ruled that it be openly admitted into the United States. Random House, the American publisher who had advocated the obscenity charge be challenged in court, promptly began typesetting the work in order to release a U.S. edition. But the court decision had important and lasting legal impact as well: it was a turning point in reducing government censorship. Prior to the case, laws that prohibited obscenity were not seen to be in conflict with the First Amendment of the U.S. Constitution (which is most often interpreted as a guarantee of freedom of speech), and the U.S. Post Office and the Customs Service alike both had the power to determine obscenity. The government appealed the decision to the U.S. Circuit Court of Appeals, but Judge Woolsey's decision held.

What is **Proust**'s claim to literary fame?

Marcel Proust (1871–1922) is generally considered the greatest French novelist of the twentieth century and is credited with introducing to fiction the elements of psychological analysis, innovative treatment of time, and multiple themes. Proust is primarily known for his multivolume work *A la recherche du temps perdu* (1954), which was published in English as *Remembrance of Things Past.* Proust was an creative stylist as well as shrewd social observer.

In the mid-1890s Proust joined other prominent artists, including the great French novelist of the nineteenth century, Emile Zola (1840–1902), to form the protest group known as the Revisionists or Dreyfusards. The artists were staunch supporters of Alfred Dreyfus (1859–1935), and therefore vocal critics of the French military, who they accused of anti-Semitism for keeping the French army officer, wrongly accused of treason, imprisoned on Devil's Island.

When did **American poetry** begin?

As the self-described poet of democracy, Walt Whitman (1819–1892) was the first to compose a truly American verse—one that showed no references to European antecedents (throwing off both the narrative and ode forms of verse) and that clearly articulated the American experience.

His first published poetry was the self-published collection *Leaves of Grass* (1855). In an effort to gain recognition, Whitman promptly sent a copy to the preeminent man of American letters, Ralph Waldo Emerson (1803–82), who could count as his acquaintances and friends the great British poets William Wordsworth (1770–1850) and Samuel Taylor Coleridge (1772–1834), the renowned Scottish essayist Thomas Carlyle (1795–1881), and prominent American writers Henry David Thoreau (1817–1862) and Nathaniel Hawthorne (1804–1864). It was a bold move on Whitman's part, but it paid off: While *Leaves of Grass* had been unfavorably received by reviewers, Emerson composed a five-page tribute, expressing his enthusiasm for the poetry and remarking that Whitman was "at the beginning of a great career." Thoreau, too, praised the work. More than a century later, biographer Justin Kaplan acclaimed that in its time *Leaves of Grass* was "the most brilliant and original poetry yet written in the New World, at once the fulfillment of American literary romanticism and the beginnings of American literary modernism." Whitman's well-known and frequently studied poems include "Song of Myself," "O Captain! My Captain!," "Song of the Open Road," and "I Sing the Body Electric."

While she was virtually unknown for her poetry during her lifetime, Emily Dickinson (1830–1886) was writing at about the same time as Whitman (the 1850s), publishing only a handful of poems before her death. Collections of Dickinson's works were published posthumously, and today she, too, is regarded as one of the great early poets of the United States. Had more of her work been brought out in print, perhaps she would have been recognized as the first truly American poet.

What were the lasting effects of the **Harlem Renaissance**?

The Harlem Renaissance (1925–35) marked the first time that white Americans (principally intellects and artists) gave serious attention to the culture of African Americans.

The movement, which had by some accounts begun as early as 1917, was noted in a 1925 *New York Herald Tribune* article that announced, "We are on the edge, if not in the midst, of what might not improperly be called a Negro Renaissance." The first African American Rhodes scholar, Alain Locke (1886–1954), who was a professor of philosophy at Howard University, led and shaped the movement during which Upper Manhattan became a hotbed of creativity in the post–World War I (1914–18) era.

Zora Neale Hurston (pictured c. 1940) was the first black woman to be honored for creative writing with a prestigious Guggenheim Fellowship; she was one of the key figures of the Harlem Renaissance.

Not only was there a flurry activity, but there was a heightened sense of pride as well. The movement left the country with a legacy of literary works including those by Jean Toomer (his 1923 work *Cane* is generally considered the first work of the Harlem Renaissance), Langston Hughes ("The Negro Speaks of Rivers," 1921; *The Weary Blues,* 1926), Countee Cullen (*Color,* 1925; *Copper Sun,* 1927), Jessie R. Fauset (novelist and editor of *The Crisis,* the journal of the National Association for the Advancement of Colored People, or NAACP), Claude McKay (whose 1928 novel *Home to Harlem* evoked strong criticism from W. E. B. Du Bois and Alain Locke for its portrayal of black life), and Zora Neale Hurston (the author of the highly acclaimed 1937 novel *Their Eyes Were Watching God,* who was the first black woman to be honored for her creative writing with a prestigious Guggenheim Fellowship).

The Harlem Renaissance was not only about literature: jazz and blues music also flourished during the prosperous times of the postwar era. During the 1920s and 1930s Louis Armstrong, "Jelly Roll" Morton, Duke Ellington, Bessie Smith, and Josephine Baker rose to prominence. Their contributions to music performance are still felt by artists and audiences, regardless of color, today.

How big is the **Harry Potter** sensation?

British author J. K. Rowling's (1965–) Harry Potter series, following the adventures of a young wizard, debuted in 1997 and has been so popular with readers that it set new records in the publishing industry. It also made Rowling one of the wealthiest people in the world.

On December 21, 2004, it was announced that the sixth of seven planned books in the series would be released on July 16, 2005. That volume, titled *Harry Potter and the Half-Blood Prince,* immediately shot to the top of bestseller lists in the United States and Great Britain, based only on advance orders. The announcement also gave a boost to the stock prices of Rowling's U.K. and U.S. publishers as well as major book retailers. At that time about 260 million copies of the first five books had been sold worldwide, and it was anticipated that the sixth would be the largest-selling trade book of 2005, selling at least 11 million copies in the United States alone. In sizing up the runaway publishing success that is Harry Potter, Steve Riggio, chief executive officer of Barnes & Noble book retailer, said, "Sales from the fifth book grossed as much as a major Hollywood movie in its first week of release." Indeed that book, *Harry Potter and the Order of the Phoenix,* was released on June 21, 2003, and became the fastest-selling title in history on the first weekend of its publication. And all Harry Potter books have been number-one bestsellers.

These numbers translated to great personal wealth for the British author who, by her own account, had been on the dole when she began planning the series in the mid-1990s. In spring 2004 Rowling made her debut on *Forbes* magazine's annual list of the world's richest people; her $1-billion fortune ranked her 552nd on the list of 587 billionaires.

FINE ART

What are the characteristics of **Botticelli's paintings**?

The works of Sandro Botticelli (1445–1510), one of the early painters of the Italian Renaissance, are known for their serene compositions, refined elegance, and spirituality. A student of Florentine painter Fra Filippo Lippi (1406–1469), Botticelli refined Lippi's method of drawing such that he is considered one of the great "masters of the line."

Botticelli's work was soon eclipsed by that of Leonardo da Vinci (1452–1519), who was just a few years younger than he, but whose range of talents made Botticelli's work seem dated. Nevertheless, late in the nineteenth century, Botticelli began to be revered again by artists and critics alike, who hailed his works for their simplicity and sincerity. English art critic John Ruskin (1819–1900) held Botticelli up as an example of an artist who presented nature as an expression of a divinely created world.

Why is Botticelli's ***The Birth of Venus*** famous?

This immediately recognizable painting (c. 1482) is most likely known for its elegant figures, use of pictorial space, and decorative detail, which give the painting a tapestry effect. At the time it was painted, the presentation of a nude Venus was an innovation since the use of unclothed figures in art had been prohibited during the Middle Ages

Botticelli's well-known painting *Birth of Venus* was commissioned by Florence's powerful Medici family.

(500–1350). Botticelli, however, felt free to render Venus in this way since the work was commissioned by Florence's powerful Medici family, who were his patrons. Under their protection, Botticelli could pursue the world of his imagination without fearing charges of paganism and infidelity.

Though *The Birth of Venus* is extremely well known, *The Magnificat* (1483), Botticelli's round picture of the Madonna with singing angels, is his most copied work.

Why was the **Medici family** important to **Renaissance art**?

The Medici family was powerful in Florence, Italy, between the fourteenth and sixteenth centuries. The founder of the family was Giovanni di Bicci de Medici (1360–1429), who amassed a large fortune through his skill in trade and who virtually ruled Florence between 1421 and 1429.

Later, Lorenzo de Medici (1449–1492) ruled Florence between 1478 and 1492. Though he was tyrannical, he was a great patron of the arts and letters. Lorenzo (also called "the Magnificent") maintained Fiesole, a villa outside Florence, where he surrounded himself with the great talents and thinkers of Florence, including a young artist named Sandro Botticelli (1445–1510). Lorenzo was also a patron of Michelangelo.

Why is da Vinci called the "universal man"?

Leonardo da Vinci (1452–1519) possessed an intensely curious mind and an inventive imagination. He is known by students both for his famous works of art including *The Last Supper* (1495–98) and *Mona Lisa* (1503–05), as well as for his scientific notes and drawings dealing with matters of botany, anatomy, zoology, hydraulics, and physiology. By his own claim, he pursued scientific investigations only to make himself a better painter. Nevertheless, he clearly endeavored to understand the laws of nature. Consequently, he made a study of man, contributing to the understanding of physiology and psychology.

Leonardo da Vinci's body of work provided the foundation of High Renaissance sculpture, painting, drawing, and architecture. As an artist-genius, da Vinci earned the epithet "universal man," and has become a wonder of the modern world, for, as *Gardner's Art through the Ages* put it, having stood at the beginning of "a new epoch like a prophet and a sage."

Did **Michelangelo** study anatomy?

Yes: In 1492 Michelangelo Buonarroti (1475–1564), a master sculptor of the human form, undertook the study of anatomy based on the dissection of corpses from the Hospital of Santo Spirito.

Perhaps most well known for his sculptures of *David* (1501–04) and *Moses* (1515–15), as well as his frescoes on the ceiling and walls of the Sistine Chapel, Michelangelo was also an architect who believed that buildings should follow the form of the human body "to the extent of disposing units symmetrically around a central and unique axis, in a relationship like that of the arms to the body." He also wrote poetry; he was a true Renaissance man.

Michelangelo was totally absorbed in his work and was known to be impatient with himself and with others. He has been likened to German composer Ludwig van Beethoven (1770–1827) since the personal letters of both men reveal a "deep sympathy and concern for those close to them, and profound understanding of humanity informs their works" (*Gardner's Art through the Ages*).

Of the great trio of **High Renaissance artists**—Leonardo, Michelangelo, and Raphael—who is considered *the* master?

Most historians and critics agree that it was Raphael Sanzio (1483–1520) who most clearly stated the ideals of the High Renaissance. Though arch rivals Leonardo da Vinci (1452–1519) and Michelangelo Buonarroti (1475–1564) influenced the younger Raphael, he developed his own style. A prolific painter, he was also a great technician whose work is characterized by a seemingly effortless grace. His most well-known work is *The School of Athens* (1509–11), which has been called "a complete statement of the High Renaissance in its artistic form and spiritual meaning." The painting, which projects a stagelike space onto a two-dimensional surface, reconvenes the great minds of the ancient world—Plato, Aristotle, Pythagoras, Herakleitos, Diogenes, Euclid—for an exchange of ideas. Raphael even included himself in this gathering of greatness. But it seems only appropriate for the master to be in such company: In this work, Raphael has achieved the art of perspective, bringing the discipline of mathematics to pictorial space where human figures appear to move naturally.

Michelangelo's statue of *David* is housed at the dome of Florence's Accademia Gallery, Italy.

Why is **Titian** thought of as the "father of modern painting"?

During Titian's time (1488 or 1490–1576), artists began painting on canvas rather than on wood panels. A master of color, the Venetian painter was both popular and prolific. His work was so sought after that even with the help of numerous assistants, he could not keep up with demand.

His body of works established oil color on canvas as the typical medium of western pictorial tradition. Among his most well-known paintings are *Sacred and Profane Love* (c. 1515) and *Venus of Urbino* (1538).

Which **van Eyck**—Hubert or Jan—painted *The Ghent Altarpiece*?

The large, multipaneled altarpiece is as controversial as it is admired. The controversy stems from an 1832 discovery (under a coat of paint on one of its outside panels) of a Latin poem that indicated that Hubert (1395–1441) had begun the work and Jan (c. 1370–1426) had completed it. So it was believed that *The Ghent Altarpiece* (1432) was a collaboration between the Flemish brothers. But the question of attribution continued to puzzle art historians for a century and a half as attempts to assign different parts of the polyptych (multipaneled work) to either of the brothers failed to gain acceptance. One art historian suggested that Hubert may not have been a painter at all, but rather a sculptor. This theory posited that Hubert's contribution was only in crafting the frames—from which the paintings had been removed in 1566 and which were subsequently lost. However, scholars seem to have now reached the consensus that Hubert was largely responsible for the design of the altarpiece and for much of its execution, while Jan was the designer and painter of most of the figures. This elaborate altarpiece, which is composed of 20 folding panels, was typical of northern European art during the Middle Ages (500–1350). However, both van Eycks contributed to the flowering of Renaissance art in northern Europe as well.

In Jan's works, which are finely detailed and ornamental (he was originally a miniaturist and illuminator), the progression from medieval to Renaissance art can be seen. In particular, his painting *Man in a Red Turban* (1433), which may be a self-portrait, marks an important step in the humanization of art. Prior to this, the artist's subjects had been religious in nature; here the painting is simply a record of a living individual. This kind of portraiture began to multiply as artists and patrons alike became increasingly interested in the reality revealed by them. Through such portraits, man began to confront himself—rather than the "otherworldly anonymity of the Middle Ages." Renaissance art—in Italy as well as in northern Europe—marks the "climax of the slow but mighty process that brings man's eyes down from the supernatural to the natural world" (*Gardner's Art through the Ages*).

Why is **Rembrandt** considered the archetype of the modern artist?

To understand the similarities between Rembrandt van Rijn (1606–1669) and the modern artist, it's important to note that this master portrait-painter, who broke ground in his use of light and shadow, was in his own time criticized for his work: Some thought it too personal or too eccentric. An Italian biographer asserted that Rembrandt's works were concerned with the ugly, and he described the artist as a tasteless painter. Rembrandt's subjects included lower-class people, the events of everyday life and everyday business, as well as the humanity and humility of Christ (rather than the choirs, trumpets, and celestial triumph that were the subjects of other religious paintings at the time). His portraits reveal his interest in the effects of time on human features—including his own. In summary, the Dutch artist approached his work with "psychological insight and...profound sympathy for the human affliction." He was also known to use the butt end of his brush to apply paint. Thus, he strayed outside the accepted limits of great art at the time.

Art critics today recognize Rembrandt as not only one of the great portrait painters, but a master of realism. The Dutch painter, who also etched, drew, and made prints, is regarded as an example for the working artist; he showed that the subject is less important than what the artist does with his materials.

Among his most acclaimed works are *The Syndics of the Cloth Guild* (1662) and *The Return of the Prodigal Son* (c. 1665). The first painting shows a board of directors going over the books, and Rembrandt astutely captures the moment when the six businessmen are interrupted, thus showing a remarkably real everyday scene. *The Return of the Prodigal Son* is one of the most moving religious paintings of all time. Here Rembrandt has with great compassion rendered the reunion of father and son, capturing that moment of mercy when the contrite son kneels before his forgiving father. Through his series of self-portraits, Rembrandt documented his own history—from the confidence and optimism of his youth to the "worn resignation of his declining years."

So much art is called impressionistic today—what exactly is **impressionism**?

The term "impressionism" was derived by a rather mean-spirited art critic from the title of one of Claude Monet's (1840–1926) early paintings, *Impression, Fog* (*La Havre,* 1872). The French impressionist painters were interested in the experience of the natural world and in rendering it exactly as it is seen—not fixed and frozen with an absolute perspective, but rather as constantly changing and as it is glimpsed by a moving eye.

Georges Seurat (1859–1891) and Paul Signac (1863–1935) are also typically thought of as impressionists; however, they are more appropriately dubbed neoimpressionists since they, along with Camille Pisarro (1830–1903), advanced the work of the original group through more scientific theories of light and color, introducing deliberate optical effects to their works. Seurat and Signac are commonly referred to as pointillists for the technique, pioneered by Seurat, of using small brush strokes to create an intricate mosaic effect. The postimpressionists, artists representing a range of explorations but all having come out of the impressionist movement, included both Seurat and Signac, as well as Henri de Toulouse-Lautrec (1864–1901), Paul Gaugin (1848–1903), Vincent van Gogh (1853–1890), and Paul Cézanne (1839–1906, who was also associated with the original impressionists).

Together the impressionists paved the way for the art of the twentieth century, since as a group they "asserted the identity of a painting as a thing, a created object in its own right, with its own structure and its own laws beyond and different from...the world of man and nature" (*History of Modern Art*).

Was **Monet** the "father of French impressionism"?

Though the movement was named for one of Claude Monet's (1840–1926) paintings and his *Water Lilies* (1905) are arguably the most well-known and highly acclaimed impressionist works, impressionism is actually rooted in the works of the group's spiritual leader, Édouard Manet (1832–1883), who first began experimenting with color and light to bring a more naturalistic quality to painting.

In 1863 Manet exhibited two highly controversial and groundbreaking works: *Déjeuner sur l'herbe* and *Olympia.* Both paintings were based on classic subjects, but Manet rendered these pastoral scenes according to his own experience, giving them a decidedly more earthy and blatantly erotic quality than the Parisian critics and academicians of the day could accept. He was roundly criticized for his scandalous exhibition. Nevertheless, Manet persevered, and in 1868, with his portrait of the French writer Emile Zola, he again challenged the art world and its values. A critic for *Le National* denounced the portrait and cited among his complaints that Zola's trousers were not made of cloth. This, the artists observed, was both truth and revelation: the pants were made of paint. A few years later, in 1870, Manet began experimenting with painting outside, in the brilliance of natural sunlight. Manet pioneered many of the ideas and techniques taken up by the impressionists.

How did American **Mary Cassatt** join the Paris art world of the impressionists?

Mary Cassatt (1844–1926), the daughter of a wealthy investment banker from Pittsburgh, Pennsylvania, traveled to Paris in 1866 in the company of her mother and some women friends; the young Cassatt was determined to join the city's community of artists. Since women were not allowed to enroll in classes at Paris's Institute of Beaux Arts (the policy was changed in 1897), Cassatt privately studied painting and traveled in Europe, pursuing her artistic interests. Returning to Paris in 1874, she became acquainted with Edgar Degas (1834–1917), who remarked that the American artist possessed an "infinite talent" and that she was "a person who feels as I do." He made these observations after viewing one of her paintings at the Salon d'Automne in Paris. Cassatt went on to exhibit with the impressionists in 1879, 1880, 1881, and 1886, gaining her first solo exhibit in 1891.

Judith Barter, curator of American arts at the Art Institute of Chicago and organizer of the traveling exhibit "Mary Cassatt: Modern Woman," describes Cassatt as "a very good businesswoman…who knew how to market her career." During three and a half years of research, which she conducted to launch the exhibit, Barter explored the prevailing social climate of the day: The late nineteenth century was a time when feminists, who organized to campaign for political and social reforms (eventually winning women the vote in 1920), focused on maternity, encouraging women to be involved in caring for their children. To Cassatt, observed Barter, maternity was "the highest expression of womanhood." Women and children were the subjects of Cassatt's body of works, which includes oil paintings, pastels, prints, and etchings.

Cassatt's place among the impressionists has often been overshadowed by her male colleagues, and her contributions to the art world are mentioned only in passing in many art books, but her talent, insights, and sheer determination combined to create an impressive legacy. As Gaugin quipped, "Mary Cassatt has charm but she also has force."

Why were **Matisse's** paintings considered so **shocking** when they were debuted?

Even if they seem commonplace to art today, the color and style of the paintings of French expressionist Henri Matisse (1869–1954) were revolutionary in their day.

In 1905 Matisse, along with several other artists, exhibited works at Paris's Salon d'Automne. The wildly colorful paintings on display there are said to have prompted an art critic to exclaim that they were *fauves,* or "wild beasts." The name stuck: Matisse and his contemporaries who were using brilliant colors in an arbitrary fashion became known as the fauves. His famous work *Madame Matisse,* or *Green Stripe* (1905), showed his wife with blue hair and a green stripe running down the middle of her face, which was colored pink on one side of her nose and yellow on the other. Matisse was at the forefront of a movement that was building new artistic values. The fauves were not using color in a scientific manner (as Georges Seurat had done), nor were they using it in the nondescriptive manner of Paul Gaugin (1848–1903) and Vincent Van Gogh (1853–1890). The fauves were developing the concept of abstraction.

Throughout his career, Matisse continued to experiment with various art forms—painting, paper cutouts, and sculptures. All of his works indicate a progressive elimination of detail and simplification of line and color. So influential was his style on modern art that some 70 years later one art critic commented that it was as if Matisse belonged to a later generation—and a different world.

How is **Picasso's** work characterized?

It's impossible to characterize or classify the work of Spaniard Pablo Picasso (1881–1973) since his career as an artist spanned his entire life and he experimented with many disciplines. Picasso often claimed that he could draw before he could speak, and by all accounts he spent much of his childhood engaged in drawing. He was only 15 years old when he submitted his first works for exhibition. And by the turn of the century, when he was still a young man, he began exploring the blossoming modern art movement. The rest of his career breaks into several periods. His Blue period (1901–04) was named for the monochromatic use of the color for its subjects, and was likely the result of a despair brought on by the suicide of a friend. Next came his Rose period (beginning 1905), when images of harlequins and jesters appear in his works—

Many of Picasso's works fall into the category of cubism, exemplified by this 1948 painting, *Woman in an Armchair.*

all to a somewhat melancholic effect. He soon began to incorporate aspects of primitive art, and later experimented with geometric line and form in his works, which were constructions—or deconstructions—sometimes only identifiable by their title.

In the spring of 1912 cubism exploded, and Picasso was on its forefront. In 1923 he broke new ground with surrealism. The key masterpiece in his body of works came in 1937 when he painted *Guernica,* his rendering of the horror of the German attack (supported by Spanish fascists) on the small Basque town (of Guernica) in Spain. His career reached its height during the 1940s, during which he lived in Nazi-occupied Paris.

Biographer Pierre Cabbane summed up the last period (1944–73) of Picasso's work: "He invented a second classicism: autobiographical classicism.... His final 30 years were to be a dizzying, breakneck race toward creation." During this time, Picasso did not chart any new artistic territory, but simply created art at an amazing rate. After his death in 1973, his estate yielded an inventory of 35,000 remaining works—paintings, drawings, sculptures, ceramics, prints, and woodcuts.

He left an enormous—even mind-boggling—legacy to the art world. In a 1991 article in *Vanity Fair,* Picasso's friend and biographer John Richardson observed, "Almost every artist of any interest who's worked in the last 50 years is indebted to Picasso...whether he's reacting against him knowingly or is unwittingly influenced by him. Picasso sowed the seeds whose fruits we are continuing to reap."

PHOTOGRAPHY

When was **photography** invented?

The concept of still photography dates back to the tenth century when Islamic scientists developed the camera obscura (Latin for "dark chamber"), a darkened enclosure with a small aperture (opening) to admit light. The light rays would cast an inverted image of external objects onto a flat surface opposite the aperture. This image could be studied and traced by someone working inside the camera obscura, or the image could be viewed from the outside of the camera, through a peephole.

Thomas Edison and George Eastman stand with a motion picture camera, c. 1925. Both inventors contributed to the innovation of movies.

In the sixteenth century, the Italian scientist Giambattista della Porta (c. 1535–1615) published his studies on fitting the aperture of the camera obscura with a lens to strengthen or enlarge the image projected. Made increasingly versatile through additional improvements, the camera obscura become popular among seventeenth- and eighteenth-century European artists.

But the camera obscura could only project (rather than reproduce) images onto a screen or a piece of paper. During the 1800s scientists experimented with ways of making the images permanent. Among those who made advances in the photographic process were French physicist Joseph-Nicéphore Niepce (1765–1833), who produced the first negative image in 1826; French painter Louis-Jacques Daguerre (1759–1851), who in 1839 succeeded in making a direct positive image on a silver plate, known as the daguerreotype; English scientist William Henry Fox Talbot (1800–1877), who developed a paper negative (c. 1841) that could be used to print any number of paper positives; and English astronomer Sir John Herschel (1792–1891), who was the first to produce a practical photographic fixing agent and the first to apply the terms "positive" and "negative" to photographic images. All of these milestones made photography a practical way of permanently recording real-life images.

The breakthrough in still photography was the Kodak, introduced in 1888 by American inventor George Eastman (1854–1932). The Kodak camera used film that was wound on rollers, eliminating the glass photographic plates that had been in use. The box-shaped camera made photography accessible to everyone—including amateurs. By the early 1900s the Eastman Kodak Company had become the largest photographic film and camera producer in the world. George Eastman has been credited with mass-producing the moment: Before the Kodak (a word he made up because he was fond of the letter *k*), photography had largely been the domain of professionals who were commissioned to take portraits of the well-to-do prominent members of society. Once the Kodak became widely available, photographs preserved the faces of ordinary people and the events of everyday life.

When was **photography** established as an **art form**?

In the early 1900s. Alfred Stieglitz (1864–1946) is the acknowledged "father of modern photography." His interest in the medium began when he was just a toddler: at the age of two, he became obsessed with a photo of his cousin, carrying it with him at all times. When he was nine years old, he took exception to a professional photographer's practice of using pigment to color a black-and-white photo, complaining that this spoiled the quality of the print.

Between 1887 and 1911 Stieglitz worked to establish photography as a valid form of artistic expression, a pursuit for which he was sometimes publicly derided. He believed that photography should be separate from painting, but on an equal footing as an art form. He also strove to differentiate photography by instilling it with an American essence; the streets of New York City became his subject. By the time Stieglitz founded the Photo-Secession Group in 1902, he had developed a uniquely American art form. Stieglitz also published and edited photography magazines, most notably *Camera Work* (1903–17). After an unhappy first marriage, in 1924, Stieglitz married American artist Georgia O'Keeffe (1887–1986), who became the subject of one of his best-known series of works.

ARCHITECTURE

How old is the **Great Wall of China**?

The immense structure, built as a barricade of protection against invasion, was begun during the third century B.C. by Emperor Shih Huang Ti (Cheng; c. 259–210 B.C.) of the Ch'in dynasty, and was expanded over the course of succeeding centuries. The wall stretches 1,500 miles, ranges in height between 20 to 50 feet, and is between 15 and 25 feet thick. In the thirteenth century, the wall was penetrated when Mongols conquered China, expanding their empire across all of Asia.

How old is the **Parthenon**?

The ancient temple, originally built on the Acropolis, a hill overlooking the city of Athens, was constructed between 447 and 432 B.C. by Greeks. The white marble edifice,

The Great Wall of China, originally built as a barricade against invasion, stretches 1,500 miles and is visible from space.

considered a prime example of Greek architecture, has an interesting history: About A.D. 500, it became a Christian church; in the mid-1400s, when the region was captured by Turkish Muslims, it was turned into a mosque; and in 1687, when Venetians tried to take the city, the Parthenon was severely damaged; only ruins remain today.

How was the **Colosseum** ruined?

The Roman structure, begun during the reign of Vespasian (ruled A.D. 69–79), was disassembled during the Middle Ages (500–1350) when its stones and brocks were removed and used to construct other buildings. The Colosseum, situated in the center of the city of Rome, was a giant, outdoor theater. Between 80 and 404, it was an entertainment center where battles were staged, gladiators competed, and men fought wild animals. It could seat 50,000 spectators who were separated from the arena by a 15-foot wall.

When was London's **Westminster Abbey** built?

The famed national church of England was begun between 1042 and 1065 when Edward the Confessor (c. 1003–1066) built a church on the site of the Abbey. King Henry III (1207–1272) began work on the main part of Westminster in 1245. Since the time of William the Conqueror (1066), all of England's rulers, except Edward V and Edward VIII, have been crowned at the church. The Abbey is also a burial place of great English statesmen and literary giants (the latter are buried in the Poet's Corner).

Why does the **Leaning Tower of Pisa** lean?

The famous bell tower in Pisa (in northwestern Italy) leans because of the unstable soil on which it was built. Construction began in 1173 on the approximately 180-foot campanile; it began to lean as soon as the first three floors were completed. Nevertheless, building continued, and the seven-story structure was finished between 1360 and 1370. Leaning a bit more each year, by the time it was closed for repairs in 1990, the tower tilted 14.5 feet out of line when measured from the top story. Engineers on the project worked to stabilize the foundation and straighten it slightly (to prevent damage). The tower, which was built alongside a church and a baptistery, would probably not be remarkable if it were not for its slant. But with its characteristic angle, it continues to attract tourists to the small town on the Arno River.

Why is the **Cathedral of Notre-Dame** famous?

The Paris cathedral was built using the first true flying buttresses (masonry bridges that transmit the thrust of a vault or a roof to an outer support). The device allowed the structure to achieve a great height—one of the first Gothic churches to do so. Gothic was a medieval architectural style that predominated in northern Europe from the early twelfth century until the sixteenth century; it was epitomized in elaborate churches with stained-glass windows—ornamentation meant to instill the building itself with transcendental qualities. One of the leading examples of Gothic architecture is the Amiens Cathedral (in Amiens, northern France), which was begun in 1220. Its soaring nave (the central area of a church) epitomizes the era's drive for height. The Amiens cathedral is France's largest.

Why is Spain's **Alhambra** historically important?

The elaborate palace, built east of the city of Granada, in southern Spain, was built by Moors, Muslim North Africans who occupied the Iberian Peninsula (Spain and Portugal) for hundreds of years during the Middle Ages (500–1350). The fortified structure, built between 1238 and 1354, is a monument of Islamic architecture in the Western world. Its name is derived from an Arabic word meaning "red"; the highly ornamental palace, with its decorative columns, walls, and ceilings, was constructed of red brick. Perched on a hilltop, the Alhambra was the last stronghold of the Moors in Spain. In 1492 the palace was captured by forces of Spain's King Ferdinand (1452–1516) and Queen Isabella (1451–1504).

Upon completion in 1883 the Brooklyn Bridge was considered a feat of modern engineering. (Photo, c. 1900.)

When was the **Brooklyn Bridge** completed?

The bridge, which spans New York's East River to connect Manhattan and Brooklyn, was completed in 1883. Upon opening, it was celebrated as a feat of modern engineering and, with its twin gothic towers, as an architectural landmark of considerable grace and beauty. It is a high statement of the era—an expression of the optimism of the Industrial Revolution. It was designed by German American engineer John Augustus Roebling (1806–1869), who, upon his death, was succeeded on the project by his son Washington Augustus Roebling (1837–1926). When the Brooklyn Bridge was finished, it was the longest suspension bridge in the world: it measures 1,595 feet. The bridge hangs from steel cables that are almost 16 inches thick. The cables are suspended from stone and masonry towers that are 275 feet tall. Specially designed watertight chambers allowed for the construction of the two towers—whose bases are built on the floor of the East River. The project proved to be an enormous and dangerous undertaking. Underwater workers suffered from the bends, a serious and potentially fatal blood condition caused by the decrease in pressure that results from rising from the water's depth too quickly. But man prevailed against the elements and, following 14 laborious years, on May 24, 1883, the Brooklyn Bridge was inaugurated. Five years later, Brooklyn became a borough of New York City, and in 1964 the bridge was designated a national historic landmark.

When did **modern architecture** begin?

The term "modern architecture" is used to refer to the architecture that turned away from past historical designs in favor of designs that are expressive of their own time. As such, it had its beginnings in the late nineteenth century when architects began reacting to the eclecticism that was prevalent at the time. Two "schools" emerged: art nouveau and the Chicago school.

Art nouveau, which had begun about 1890, held sway in Europe for some 20 years and was evident not only in architecture and interiors, but in furniture, jewelry, typography, sculpture, painting, and other fine and applied arts. Its proponents included Belgian architects Victor Horta (1861–1947) and Henry Van de Velde (1863–1957), and Spaniard Antonio Gaudi (1852–1926).

But it was the Chicago school that, in the rebuilding days after the Great Chicago Fire (1871), created an entirely new form. American engineer and architect William Le Baron Jenney (1832–1907) led the way. Four of the five younger architects who followed him had at one time worked in Jenney's office: Louis Henry Sullivan (1856–1924), Martin Roche (1855–1927), William Holabird (1854–1923), and Daniel Hudson Burnham (1846–1912). Burnham was joined by another architect, John Wellborn Root (1850–1891). Together these men established solid principles for the design of modern buildings and skyscrapers where "form followed function." Ornament was used sparingly, and the architects fully utilized iron, steel, and glass.

By the 1920s modern architecture had taken firm hold, and in the mid-twentieth century it was furthered by the works of Walter Adolf Gropius (1883–1969), Le Corbusier (Charles-Édouard Jeanneret; 1887–1965), Ludwig Mies van der Rohe (1886–1969), and Frank Lloyd Wright (1867–1959). For practical purposes, modern architecture ended in the 1960s with the deaths of the aforementioned masters.

Examples of modern architecture include Chicago's Monadnock Building (1891), Reliance Building (1895), Carson Pirie Scott store (1904), and Robie House (1909); New York City's Rockefeller Center (1940), Lever House (1952), and Seagram Building (1958); as well as Taliesin West (1938–59) in Arizona, Johnson Wax Company's Research Tower (1949) in Wisconsin, and the Lovell House (1929) in Los Angeles.

Who invented the **skyscraper**?

The credit is usually given to American architect William Jenney (1832–1907), who designed the 10-story Home Insurance Building, erected on the corner of LaSalle and Monroe Streets in Chicago in 1885. The building was the first in which the entire structure was of skeleton construction—of cast iron, wrought iron, and Bessemer steel. However, some experts believe the first skyscraper to have been designed was one by the American firm Holabird and Roche, also in Chicago. The firm, founded by two former students of Jenney, designed the skeleton-framed Tacoma Building, which was actually not completed until 1889. Both the Home Insurance Building and the Tacoma Building were demolished in 1931 and 1929, respectively.

It was the use of steel, the innovation of a safe elevator, and the use of central heating that combined to make possible the construction of tall buildings toward the end of the nineteenth century. Once the trend had started, it quickly took off: Another Chicago firm, Burnham and Root (Burnham, too, had been a student of Jenney), completed the 14-story Reliance Building in 1895; it had a steel skeleton frame. The further development of the skyscraper is visible in the Gage Buildings in Chicago—two of which were designed by Holabird and Roche, and one by Louis Henry Sullivan (1856–1924), the Chicago architect often credited for mastering the skyscraper. Other Chicago skyscrapers built by Holabird and Roche during the early days of modern architecture include the Marquette Building (1894) and the Tribune Building (1901).

Standing at 1,671 feet, Taiwan's Taipei 101 became the world's tallest building when its pinnacle was completed in 2003.

What is the **world's tallest building**?

The honor belongs, at least for a time, to Taipei 101 (also known as Taipei Financial Center), in Taiwan. The soaring spire, which rises to a height of 1,670 feet and includes 101 stories, was completed in 2004, besting Kuala Lumpur, Malaysia's twin Petronas Towers (completed in 1998), which measure 1,483 feet and 88 stories; for six years, the Petronas laid claim to the title of world's tallest building. The next tallest is the Sears Tower (completed in 1974), which rises 1,450 feet and 110 stories above Chicago's sidewalks; for a time, it, too, was the world's tallest.

The skyscraper is a decidedly American contribution to world architecture. When Chicago rebuilt following the Great Fire of October 1871, a new brand of architecture emerged, which focused on commercial buildings. Architects of the so-called Chicago school used new building materials and the innovative elevator (first patented by Elisha Otis in 1861) to construct vertical office buildings, making the most of city real estate. Though many designers worked on the form, engineer/architect William Jenney (1832–1907) is called the "father of the skyscraper." His Home Insurance Building, completed in 1885 (it was demolished in 1931), rose to 10 stories and used mass-produced steel beams, cast iron, and wrought iron with masonry walls. Soon Chicago's and New York's skylines were punctuated by rising towers. By the end of the nineteenth century, the race for tallest was on: New York's Park Row Building (which still stands today in lower Manhattan) was completed in 1899 and rises to a height of 386 feet and 30 stories.

The twentieth century saw the skyscraper reach ever upward. For a short time, the title of world's tallest building was held by Manhattan's celebrated Chrysler Building (completed in 1930), considered the height of Art Deco design. The Chrysler's 1,046 feet and 77 stories of glory were soon bettered by the Empire State Building (completed in 1931), which boasts a total height of 1,224 feet and 102 stories. That New York City landmark held onto its position for four decades, becoming second-tallest with the 1973 completion of the twin towers of the World Trade Center, which soared to 1,368 and 1,362 feet, and 110 stories each, before they fell in the terrorist attacks of September 11, 2001. Despite the fear that skyscrapers, bold symbols of capitalism, could become targets for other terrorist strikes, architects and developers continued to reach skyward, assuring that Taipei 101 would be surpassed.

THEATER

How old is the **dramatic** form of **tragedy**?

Tragedy, a form of drama central to western literature, dates to ancient times—the fifth century B.C., when Greeks held a religious festival to honor the god Dionysus (god of fertility, wine, and, later, drama). Famous ancient tragedies include *Oresteia* by Aeschylus (who is credited with inventing tragedy), *Oedipus Rex* by Sophocles, and *Medea* and *Trojan Women* by Euripides. The philosopher Aristotle observed that tragedy's function is a cathartic one—by participating in the drama, the spectators are purged of their emotions of pity and fear. The well-known Renaissance tragedies of William Shakespeare (1564–1616) harken back to the works of Roman statesman and playwright Seneca (c. 4 B.C.–A.D. 65), who wrote during the first century. He is credited with creating dramatic conventions including unity of time and place, violence, bombastic language, revenge, and ghostly appearances.

How old is **comedy**?

Like tragedy, comedy as a form of drama dates back to ancient Greece. While tragedy was meant to engage human emotions, thereby cleansing spectators of their fears (according to Aristotle), comedy's intent was simply to entertain and amuse audiences. Athenian poet Aristophanes (who flourished circa the fifth century B.C.) is considered the greatest ancient writer of comedy. His plays, written for the festival of Dionysus (the god of fertility, wine, and, later, drama), were a mix of social, political, and literary satire. Performance vehicles included farce, parody, and fantasy. During the fourth century B.C., this old comedy evolved into a new comedy, which was less biting and more romantic and realistic in nature. New comedy, which was marked by strong character development and often subtle humor, includes the works of Greek playwright Menander (flourished during the fourth century B.C.) and those of Roman comic writers Plautus (flourished third century B.C.) and Terence (flourished second century B.C.), all of whom were influences on Ben Jonson, William Shakespeare, Jean Molière, and other writers of the sixteenth and seventeenth centuries.

What is ***No*** drama?

It is the oldest form of traditional Japanese drama, dating to A.D. 1383. It is rooted in the principles of Zen Buddhism, a religion emphasizing meditation, discipline, and the transition of truth from master to disciple. History and legend are the subjects of *No* plays, which are traditionally performed on a bare, wooden stage by masked male actors who performed the story using highly controlled movements. The drama is accompanied by a chorus, which chants lines from the play. The art form was pioneered by actor-dramatist Motokiyo Zeami (1363–1443) when he was 20 years old. Zeami had begun acting at age seven and went on to write more than half of the roughly 250 *No* dramas that are still performed today.

What are the elements of **Japanese kabuki**?

The most popular traditional form of Japanese drama, kabuki features dance, song, mime, colorful costumes, heavy makeup, and lively, exaggerated movements to tell stories about historical events. The drama had its beginnings in 1575 when Okuni, a woman, founded a kabuki company. In 1603 at Kyoto women danced at the Kitani shrine, playing men's roles as well as women's. In October 1629 kabuki became an all-male affair by order of the shogun Iemitsu, who decided that it was immoral for women to dance in public. Just as in Elizabethan England, women's roles were then performed by men. The performing art became increasingly popular during the 1600s, eclipsing *bunraku* (puppet theater), in which a narrator recites a story, which is acted by large, lifelike puppets. Today kabuki remains a viable art form, borrowing from other forms of drama to adapt to changing times.

Japanese kabuki tells stories through dance, song, mime, colorful costumes, heavy makeup, and lively, exaggerated movements.

What is a **passion play**?

A passion play is a dramatization of the scenes connected with the passion and crucifixion of Jesus Christ. The roots of the passion play can be traced to ancient times: early Egyptians performed plays dedicated to the god Osiris (god of the underworld and judge of the dead), and the Greeks also acted out plays to honor their god Dionysus (the god of fertility, wine, and, later, drama). During the Middle Ages (500–1350) liturgical (religious ceremonial) dramas were performed. Toward the end of the tenth century, the Western church began to dramatize parts of the Latin mass, especially for holidays such as Easter. These plays were performed in Latin by the clergy, inside the church building. Eventually the performances became more secular, with laymen acting out the parts on the steps of the church or even in marketplaces.

The liturgical dramas developed into so-called miracle plays or mystery plays. As a symbol of gratitude or as a request for a favor, villagers would stage the life story of the Virgin Mary or of a patron saint. When the plague (also called the Black Death) ravaged Europe, the villagers at Oberammergau, Germany, in the Bavarian Alps, vowed to enact a passion play at regular intervals in the hope that by so doing they would be spared the Black Death. They first performed this folk drama in 1634 and have continued to stage it every 10 years, attracting numerous tourists to the small town in southern Germany.

Why is the **Globe Theatre** famous?

The Globe is known because of William Shakespeare's (1564–1616) involvement in it. In the 1590s an outbreak of the plague prompted authorities to close London theaters. At the time Shakespeare was a member of the Lord Chamberlain's Men, an acting company. With other members of the troupe, he helped finance the building of the Globe (on the banks of the Thames River), which opened in 1599 as a summer playhouse. Plays at the Globe, then outside of London proper, drew good crowds, and the Lord Chamberlain's Men also gave numerous command performances at court for King James. By the turn of the century, Shakespeare was considered London's most popular playwright, and by 1603 the acting group, whose summer home was the Globe Theatre, was known as the King's Men.

What was **vaudeville**?

Light, comical theatrical entertainment, vaudeville flourished at the end of the nineteenth century and beginning of the twentieth century. Programs combined a variety of music, theater, and comedy to appeal to a wide audience. Script writers attracted immigrant audiences by using ethnic humor, exaggerating dialects, and joking about the difficulties of daily immigrant life in America. (The word *vaudeville* is derived from an old French term for a satirical song, *vaudevire,* which is a reference to the Vire valley of France, where the songs originated.)

Vaudeville made its way to the American stage by the 1870s, when acts performed in theaters in New York, Chicago, and other cities. Troupes traveled a circuit of nearly 1,000 theaters around the country. As many as 2 million Americans a day flocked to the shows to see headliners such as comedians Eddie Cantor (1892–1964) and W. C. Fields (1880–1946), singer Eva Tanguay (1878–1947), and French actress Sarah Bernhardt (1844–1923).

During the first two decades of the twentieth century, vaudeville was the most popular form of entertainment in the country. In the 1930s, just as New York opened the doors of its famous Radio City Music Hall, which was intended to be a theater for vaudeville, the entertainment form began a quick decline. Motion pictures, radio, and, later, television took its place, with numerous vaudeville performers parlaying their success into these new media. Among those entertainers who had their origins in vaudeville acts were actors Rudolph Valentino, Cary Grant, Mae West, Jack Benny, George Burns, Gracie Allen, Ginger Rogers, Fred Astaire, Will Rogers, and Al Jolson.

MUSIC

When was our system for **notating music** developed?

The innovation came in the early eleventh century, when Guido of Arezzo (c. 991–1050), an Italian monk, devised a precise system for defining pitch. Guido was a leading music teacher and theorist in his day. As such, he was invited in about 1028 to Rome, where he presented a collection of religious anthems to Pope John XIX. Guido used a system of four horizontal lines (a staff) on which to chart pitch, and he used the syllables *ut* (later replaced with *do*), *re, mi, fa, sol,* and *la* to name the first six tones of the major scale. Before Guido developed his precise method for teaching music, singers had to learn melodies by memorizing them, a process that took many years. Using his notating system, singers were able to sight-read melodies. Guido's famous treatise, *Micrologus,* was one of the most widely used instruction books of the Middle Ages (500–1350).

How many **musical Bachs** were there?

Johann Sebastian Bach (1685–1750) was only one of a long and extended line of competent musicians—some 14 of them. The Bach family was a musical dynasty. J. S.'s father, Ambrosius Bach (1645–1695), was a court musician for the Duke of Eisenach, and several of J. S. Bach's close relatives were organists in churches. His eldest brother, Johann Christoph Bach (1671–1721), was apprenticed to the famous German composer Johann Pachelbel (1653–1706).

J. S. Bach left a musical legacy even beyond the vast body of church, vocal, and instrumental music that he composed: Four of his sons and one grandson were also accomplished musicians. "The English Bach" refers to J. S.'s son Johann Christian Bach (1735–1782), who composed operas, oratorios, arias, cantatas, symphonies, concertos, and chamber music. A proponent of Rococo style music, J. C. Bach influenced Wolfgang Amadeus Mozart (1756–1791).

Why do music historians talk about **"before Bach"** and **"after Bach"**?

Some scholars use these terms to classify music history since the life work of Johann Sebastian Bach (1685–1750) was so substantial, consisting of some 1,100 works, and has had lasting and profound influence on music composition. While he was not famous during his lifetime and had disagreements with employers throughout his career, J. S.

Bach's works and innovations in many ways defined music as people now know it. The tempered scale is among his inventions, and he initiated a keyboard technique that is considered standard today. Chronologically, J. S. Bach marks the end of the "prolific and variegated" baroque era, which began about 1600 and ended the year of his death, 1750.

A devout Christian, J. S. Bach believed that all music was to "the glory of God and the re-creation of the human spirit." As a spiritual person and true believer in eternal life, he left behind an impressive body of church music, including 300 cantatas (or musical sermons) as well as passions and oratorios. As a devoted family man who believed all his children were born musicians (and therefore, the Bachs could stage drawing-room music at any time), J. S. Bach also wrote chamber music, including instrumental concertos, suites, and overtures. Among his most well-known and beloved works are *The Saint Matthew Passion; Jesu, Joy of Man's Desiring; Sheep May Safely Graze;* and his *Christmas Oratorio.*

How old was **Mozart** when he composed his first work?

A child prodigy, Wolfgang Amadeus Mozart (1756–1791) was composing at the age of five. He had been playing the harpsichord since the age of three. His father, Leopold (1719–1787), was a composer and violinist who recognized his son's unusual music ability and encouraged and taught young Mozart. In 1762 Leopold took his son and daughter, Maria Anna (nicknamed "Nannerl"; 1751–1829), on tour to Paris. While there, young Wolfgang Mozart composed his first published violin sonatas and improvisations.

However, the image of the effortless and "artless child of nature" is not altogether true. Contrary to the reports that the gifted composer never revised first-and-only drafts, he did work at his craft. In a letter to his father, he wrote, "It is a mistake to think that the practice of my art has become easy to me—no one has given so much care to the study of composition as I have. There is scarcely a famous master in music whose works I have not frequently and diligently studied." The fact is that he did make revisions to his works, though it is also true that he composed at a rapid pace. The result is an impressive body of works, unequaled in beauty and diversity. The complete output—some 600 works in every form (symphonies, sonatas, operas, operettas, cantatas, arias, duets, and others)—would be enough to fill almost 200 CDs. Among his most cherished works are *The Marriage of Figaro* (1786), *Don Giovanni* (1787), *Così fan tutte* (1790), and *The Magic Flute* (1791).

Was **Beethoven** really deaf for much of his life?

Yes, Ludwig van Beethoven (1770–1827) suffered a gradual hearing loss during his twenties, and eventually lost his hearing altogether (in his early thirties). The loss was devastating to the German composer. In a letter to his brother he wrote, "But how humbled I feel when someone near me hears the distant sound of a flute, and I hear

Ludwig van Beethoven, considered a genius of musical composition, was deaf for much of his life.

nothing; when someone hears a shepherd singing, and I hear nothing!" At one point he even contemplated suicide but instead continued his work.

He had studied briefly with Mozart (in 1787) and Joseph Haydn (in 1792), and appeared for the first time in his own concert in 1800. While the loss of his hearing later prevented him from playing the piano properly, it did nothing to hold back his creativity. Between 1800 and 1824, Beethoven wrote nine symphonies, and many believe that he developed the form to perfection. His other works include five piano concertos and 32 piano sonatas, as well as string quartets, sonatas for piano and violin, opera, and vocal music, including oratorios. It was about the time that he completed his work on his third symphony, the *Eroica* (1804), that he went completely deaf. Though he was himself a classicist, music critics often refer to a turning point marked by the *Eroica,* which shows the complexity of the romantic age of music.

A true genius, Beethoven's innovations include expanding the length of both the symphony and the piano concerto, increasing the number of movements in the string quartet (from four to seven), and adding instruments—including the trombone, contrabassoon, and the piccolo—to the orchestra, giving it a broader range. Through his adventurous piano compositions, Beethoven also heightened the status of the instrument, which was a relatively new invention (1710). Among his most well-known and most-often-performed works are his third (*Eroica*), fifth, sixth (*Pastoral*), and ninth (*Choral*) symphonies, as well as the fourth and fifth piano concertos.

It is remarkable—even unfathomable—that these works, so familiar to so many, were never heard by their composer. A poignant anecdote tells of Beethoven sitting on stage to give tempo cues to the conductor during the first public performance of his ninth symphony. When the performance had ended, Beethoven—his back to the audience—was unaware of the standing ovation his work had received until a member of the choir turned Beethoven's chair around so he could see the tremendous response.

Why is **Brahms's** first symphony sometimes called **"Beethoven's tenth"**?

In many ways, Johannes Brahms (1833–1897) was the inheritor of Beethoven's genius, prompting some music historians to refer to Brahms's first symphony as "Beethoven's tenth." This is not to diminish the work of the great nineteenth-century composer, who left an enduring corpus of works. Brahms demonstrated that classicism continued to have artistic validity—and was not incompatible with—the romanticism of the late nineteenth century.

What does **Wagnerian** mean?

It is a reference to anything that is in the style of German composer Richard Wagner (1813–1883). Wagner was an enormously creative composer, conductor, and artistic manager who is credited with no less than originating the music drama. His interest in theater began in his boyhood, and by his teens he was writing plays. So that he could put music to these works, he sought out composition teachers. It is no surprise then that Wagner later conceived of the idea of the "total work of art," where music, poetry, and the visual arts are brought together in one stunning performance piece.

As an adult, Wagner led a scandalous life—even today challenging the music-listening public to separate his life from his art. He was someone modern audiences would recognize as a truly gifted and charismatic—if amoral—artist, working on a grand scale. Were he alive now, Wagner might well be creating blockbusters. In fact his musical compositions are heard in movies (including Francis Ford Coppola's *Apocalypse Now*) and are familiar to even the youngest audience today—or at least those who watch Bugs Bunny cartoons.

But this is not to take away from Wagner's serious accomplishments: His most widely recognized operatic works include *Lohengrin* (1848), the *Ring* cycle (1848–74), and *Tristan und Isolde* (1859). In the decades after his death, Wagner's reputation grew to the point that through the end of the nineteenth century his influence was felt by most every composer, who often referred to Wagner's works in measuring the value of their own.

Why did **Schoenberg** face sharp criticism in his day?

The Vienna-born American composer Arnold Schoenberg (1874–1951), now considered one of the great masters of the twentieth century, was derided for having thrown out the rules of composition—for working outside the confines of traditional harmony.

In his youth, he was a fan of Wagner's compositions, seeing each of his major operas repeatedly. A series of Schoenberg's early works reflect the Wagnerian influence. But just after the turn of the century, Schoenberg set out on his own path. The result was the 1909 composition *Three Pieces* for piano, which some music historians argue is the single most important composition of the twentieth century. The work is atonal, which is to say it is organized without reference to key. Schoenberg abandoned the techniques of musical expression as they had been understood for hundreds of years. This was no small moment for the music world, and many reacted with vocal and vehement criticism. (Of the outcry Schoenberg remarked in 1947 that it was as if "I had fallen into an ocean of boiling water.")

But he had his followers, too, among them his students. Though he was essentially self-taught as a composer, he became one of the most influential teachers of his time. It's interesting to note, however, that his teaching approach was grounded in the traditional practices of tonal harmony. He later brought order to the chaos of atonalism by developing a 12-tone serialism, showing how entire compositions could be organized around an ordained sequence of 12 notes. However, he never taught the method and rarely lectured or wrote about it.

Why does the music of **Bartók** figure prominently in concert programs today?

Béla Bartók (1881–1945) is revered today not only for his ability as a pianist (his teacher compared him to Franz Liszt [1811–1886], who was perhaps the greatest pianist of the nineteenth century), but for his compositions, which are steeped in the tradition of Hungarian folk music. Bartók studied and analyzed Hungarian, Romanian, and Arabian folk tunes, publishing thousands of collections of them in his lifetime. While ethnic music had influenced the works of other composers, Bartók was the first to make it an integral part of art music composition. His works were unique in that the folk music provided the sheer essence and substance of the music, lending the compositions a primitive quality. Among his masterpieces are his three stage works: the ballets *The Wooden Prince* and *The Miraculous Mandarin,* and the one-act opera *Duke Bluebeard's Castle.*

The introduction of folk music as the core of a musical composition has had far-reaching influence, which must have been felt by American composer Aaron Copland (1900–1990), whose *Appalachian Spring* (1944) features a simple Shaker tune, front and center.

Is **Stravinsky** the twentieth century's foremost composer?

The Russian-born American composer Igor Stravinsky (1882–1971) is certainly one of the greatest composers of the twentieth century. Stravinsky wrote concerts, chamber music, piano pieces, and operas, as well as ballets, for which he may be most well known.

Between 1903 and 1906 Stravinsky studied under the great Russian composer Nikolay Rimsky-Korsakov (1844–1908). In 1908 Stravinsky wrote his first work of note, the orchestral fantasy *Fireworks,* which was in honor of the marriage of Rimsky-Korsakov's daughter. The piece caught the attention of Sergey Diaghilev (1872–1929) of the Ballets Russes, who invited the young composer to participate in the ballet company's 1910 season (Ballets Russes had dazzled audiences the year before, bringing new energy to the art form). In collaboration with Diaghilev, Stravinsky went on to create masterpieces—*The Firebird* (1910), *Petrushka* (1911), and *Rite of Spring* (1913) among them. The partnership served to elevate the role of the ballet composer in the art world.

Rite of Spring is either Stravinsky's most famous or most infamous work. It was first performed by the Ballets Russes in the third week of its 1913 season. The choreography was arranged by the famous dancer Vaslav Nijinsky (1890–1950). But the performance stunned both the music and dance worlds. So extreme was the audience's reaction to this premier work that a riot nearly broke out inside the theater. Stravinsky had composed his music not to express spring's idyllic qualities but rather its turmoil and dissonance—similar to childbirth. Nijinsky paired Stravinsky's composition with complicated and visually frenzied dance movements, later characterized by the composer as a jumping competition. Though many thought it a disastrous performance, when the Ballets Russes continued to London, *Rite of Spring* was more widely accepted there—largely because the audience had been duly prepared for it.

The following year, *Rite of Spring* was performed in concert in Russia, but the reaction was mixed. The young composer Sergei Prokofiev (1891–1953) was in the audience and later wrote that he had been so moved by the work that he could not recover from the effects. Listeners today are still moved by the elevated rhythm of *Rite of Spring,* which makes an entire orchestra into a kind of sustained percussion instrument. Ultimately, most musicians and critics came to regard the watershed work as one of the finest compositions of the twentieth century.

Who invented **jazz**?

Ferdinand "Jelly Roll" Morton (1885–1941), a New Orleans pianist, claimed credit for having invented jazz. And to some degree, it was fair of him to think so—after all, his recordings with the group the Red Hot Peppers (1926–30) are among the earliest examples of disciplined jazz ensemble work. But in truth, the evolution of jazz from ragtime and blues was something that many musicians, in several cities, took part in. Most regard Morton as *one* of the founders of jazz, the other founders include Bennie Moten (1894–1935), Eubie Blake (1883–1983), Duke Ellington (1899–1974), and Thomas "Fats" Waller (1904–1943).

Some would go back even farther to trace the roots of jazz: From 1899 to 1914 Scott Joplin (1868–1917) popularized ragtime, which was based on African folk music. Even astute music critics may not be able to draw a clear-cut distinction between ragtime and early jazz. Both musical forms rely on syncopation (the stressing of the weak beats), and either style can be applied to an existing melody and transform it. The definitions and boundaries of the two terms have always been subject to debate, which is further complicated by the fact that some musicians of the time considered ragtime to be more or less a synonym for early jazz.

Jazz greats Duke Ellington (piano) and Louis Armstrong (trumpet) rehearse in New York City in January 1946.

But there are important, albeit not strict, differences between the two genres as well: Rags were composed and written down in the European style of notation, while early jazz was learned by ear (players would simply show one another how a song went by playing it); jazz encourages and expects improvisation, whereas ragtime, for the most part, did not; and the basic rhythms are also markedly different, with jazz having a swing or "hot" rhythm that ragtime does not.

Whatever its origins, jazz became part of the musical mainstream by the 1930s and influenced other musical genres as well—including classical. American composer George Gershwin (1898–1937) was both a songwriter and composer of rags as well as a composer of symphonic works. Many of his works, including *Rhapsody in Blue* (1924) and his piano preludes, contain ragtime and jazz elements.

Perhaps more than any other composer and musician, Miles Davis (1926–1991) expanded the genre: Through decades of prolific work, Davis constantly pushed the boundaries of what defines jazz and in so doing set standards for other musicians.

Is **blues music** older than jazz?

Only slightly (and only if your definition of jazz doesn't include ragtime). Really, the two musical traditions developed side by side, with blues emerging about the first decade of the 1900s and hitting the height of its early popularity in 1920s Harlem, where the songs were seen as an expression of African American life. Great blues singers like Ma Rainey (1886–1939) and Bessie Smith (1894 or 1898–1937) sang of the black reality—determined but weary. During the Harlem Renaissance the music was a symbol for African American people who were struggling to be accepted for who they were. Poet Langston Hughes (1902–1967) saw the blues as a distinctly black musical genre, and as helping to free blacks from American standardization.

As the first person to codify and publish blues songs, American musician and composer W. C. Handy (1873–1958) is considered the "father of the blues." The Florence, Alabama, native produced a number of well-known works, including "Memphis

Blues," "St. Louis Blues" (which is one of the most frequently recorded songs in popular music), "Beale Street Blues," and "Careless Love."

When did the **Big Band** era begin?

On December 1, 1934, Benny Goodman's *Let's Dance* was broadcast on network radio, which effectively launched the swing era, in which Big Band music achieved huge popularity. Goodman (1909–1986) was a virtuoso clarinetist and bandleader. His jazz-influenced dance band took the lead in making swing the most popular style of the time.

How old is **country music**?

Old-time music or "hillbilly music," both early names for country music, emerged in the early decades of the 1900s. By 1920 the first country music radio stations had opened, and healthy record sales in rural areas caused music industry executives to take notice. But it was an event in 1925, in the middle of the American Jazz Age, that put country music on the map: On November 28, WSM Radio broadcast *The WSM Barn Dance,* which soon became known as *The Grand Ole Opry* when the master of ceremonies, George D. Hay, took to introducing the program that way—since it was aired immediately after an opera program. The show's first performer was Uncle Jimmy Thompson (1848–1931). Early favorites included Uncle Dave Macon (1870–1952), who played the banjo and sang, and Roy Acuff (1903–1992), who was the Opry's first singing star. Millions tuned in and soon the Nashville-based show had turned Tennessee's capital city into Music City U.S.A. In the 1960s and again in the late 1980s and 1990s, country music reached the height of popularity, while holding on to its small-town, rural-based audience who were the show's first fans.

Is **bluegrass** music a distinctly American genre?

Yes, the style of music developed out of country music during the late 1930s and throughout the 1940s. Bill Monroe (1911–1996), a country and bluegrass singer-songwriter, altered the tempo, key, pitch, and instrumentation of traditional country music to create a new style—named for the band that originated it, Bill Monroe and the Blue Grass Boys (Monroe's home state was Kentucky). Bluegrass was first heard by a wide audience when in October 1939, Monroe and his band appeared on the popular country music radio program *The Grand Ole Opry.*

Although bluegrass evolved through several stages and involved a host of contributors, through it all Bill Monroe remained the guiding and inspirational force, and therefore merits the distinction of being the "father of bluegrass."

Elvis Presley (shown in 1956) brought to music an exciting and fresh combination of country, gospel, blues, and rhythm and blues.

Who was more important to **rock and roll**—Elvis Presley or the Beatles?

While music historians—and fans of either or both—may be willing to offer an opinion, the question cannot be definitively answered. The fact is that popular music today would not be what it is had it not been for both Elvis Presley and the Beatles. And the influences of both are still felt.

Elvis Presley (1935–1977) brought to music an exciting and fresh combination of country, gospel, blues, and rhythm and blues music, and topped it all off with a style and sense of showmanship that dazzled young audiences. His first commercial recording was "That's All Right, Mama" in 1954, which was followed in 1956 by the success of "Heartbreak Hotel." Between 1956 and 1969 he had 17 number-one records. Presley defined a new musical style—and an era.

Among those the American Presley had influenced were four English musicians who called themselves the Beatles. Originally founded as the Quarrymen by John Lennon (1940–1980) in 1956, the group became the most popular rock-and-roll band of the 1960s. Their first single was "Love Me Do," released on October 5, 1962, and producer George Martin was encouraged that the Beatles could produce a number-one record. In 1963 they did: "Please Please Me" was released in Britain on January 12 and was an immediate hit. Other hits off their first album included "She Loves You" and "I Want to Hold Your Hand." The follow-up album, *With the Beatles,* was released in 1964 and established them as Britain's favorite group.

Already popular in their homeland, "Beatlemania" began in the United States on February 7, 1964, when the mop-topped "Fab Four" (Lennon along with Paul McCartney, b. 1942, George Harrison, 1943–2001, and Ringo Starr, b. 1940) arrived at New York's Kennedy International Airport and were met by a mob of more than 10,000 screaming fans and 110 police officers. Two days later, on February 9, the Beatles made their legendary appearance on *The Ed Sullivan Show.* By April the group held onto the top five positions on the U.S. singles charts. The British Invasion had begun.

In their early years, the Beatles brought a new energy to rock and roll and picked up where Presley, Buddy Holly, and Little Richard had left off. The instrumentation and orchestration of Beatles songs (for which their producer George Martin deserves at least some of the credit) were innovative at the time, and are common for rock music today. Their rock movies, *A Hard Day's Night* (1964) and *Help!* (1965), were a precursor to the music videos of today. When the band decided to break up, the April 10, 1970, announcement proved to be the end of an era.

DANCE

Why is the **Ballets Russes** famous?

The notoriety of the Ballets Russes began on a May night in 1909. It was then that the company, created by Russian impresario Sergey Diaghilev (1872–1929), performed innovative ballet choreographed by Michel Fokine (1880–1942). The Parisian audience, made up of the city's elite, was wowed by the choreography, set design, and musical scores, as well as the performances of the lead dancers—the athletic vigor of Vaslav Nijinsky, the delicate beauty of Tamara Karsavina, the expressiveness of Anna Pavlova, and the exotic quality of Ida Rubinstein. Ballet had been freed of the constraints and conventions that had held it captive. The art form was reawakened.

The reforms were on every level: choreography, performance, costuming, and design. The company's chief set designer was Léon Bakst (1866–1924), whose sense of color had influenced not only stage designs but even women's fashions. Soon Diaghilev and the Ballets Russes were at the center of the art world: Major twentieth-century painters, including Robert Edmond Jones, Pablo Picasso, Andre Derain, Henri Matisse, and Joan Miró, created set and costume designs for the dance company. And Diaghilev commissioned music that could match the spectacular dancing, choreography, and decor of his ballets. History's most celebrated composers, including Maurice Ravel, Claude Debussy, Richard Strauss, Sergei Prokofiev, and Igor Stravinsky, provided the scores for the dances performed by Ballets Russes. The company, under Diaghilev's direction, had created a completely different kind of dance drama, bringing ballet out of the shadows of opera and asserting it as an art form unto itself.

The ballet companies of today are the lasting legacy of the Ballets Russes. Diaghilev illustrated that through a collaborative process, excellent art could be created outside the traditional academy. The Ballets Russes provided twentieth-century dance with the model of the touring ballet company and seasonal repertory.

Who was **Balanchine**?

The name of the Russian-born choreographer is synonymous with modern American ballet: George Balanchine (1904–1983) was one of the most influential choreographers of the twentieth century, creating more than 200 ballets in his lifetime and choreographing 19 Broadway musicals as well as four Hollywood films. He co-founded three of the country's foremost dance institutions: the School of American Ballet (in 1934); the American Ballet Company (1935); and the New York City Ballet (1948), the first American ballet company to become a public institution.

His entrance into the world of dance was entirely accidental: In August 1914 Balanchine accompanied his sister to an audition at the Imperial School of Ballet and was invited to audition as well. Though his sister failed, he passed and, against his own wishes, was promptly enrolled. However, Balanchine remained uninterested in the art form, even running away from school shortly after starting. The turning point for the young dancer came with a performance of Tchaikovsky's ballet *The Sleeping Beauty* (1890). He was dazzled by the experience and chose to stay with the school's rigorous training program.

Serenade (1935; music by Tchaikovsky) is considered by many to be Balanchine's signature work. His other well-known works include *Apollo* (1928), *The Prodigal Son* (1929), *The Nutcracker* (1954), and *Don Quixote* (1965), as well as *Jewels,* the first full-length ballet without a plot.

Remembering the opportunity he had been given as a child, Balanchine was known for choreographing children's roles into many of his ballets. His outreach did not end there: He organized lecture-demonstration tours for schools, gave free ballet performances for underprivileged children, conducted free annual seminars for dance teachers, and gave free advice and use of his ballets to other ballet companies. Balanchine's unparalleled body of work was instrumental in establishing the vibrant style and content of contemporary ballet in America, where he brought ballet to the forefront of the performing arts.

Who was **Dame Margot Fonteyn**?

Fonteyn (1919–1991) has been called an "international ambassador of dance." The British-trained ballerina achieved worldwide fame and recognition during more than 34 years with the Royal Ballet, expanding the company's female repertoire and becoming the model for the modern ballerina. In 1962, at the age of 43, Fonteyn formed a dance partnership with Soviet defector Rudolf Nureyev (1938–1993), challenging traditional assumptions about the ability of mature dancers to continue vigorous performance careers. In her later years, she continued to be active in the world of dance, helping set up dance scholarships, fostering international artistic relations, and encouraging the growth of dance institutions around the world.

How did **modern dance** begin?

American dancer and choreographer Martha Graham (1894–1991) is the acknowledged creator of modern dance. She was 35 years old when the Martha Graham Dance

Group made its debut on April 14, 1929, ushering in a new era in dance performance. The new form of dance dissolved the separation between mind and body and relied on technique that was built from within.

Graham's interest in dance had begun in her youth, and as an astute observer and manipulator of light and space, she came to be regarded later in life as one of the masters of the modernist movement—on a par with artist Pablo Picasso (1881–1973). She is credited with revolutionizing dance as an art form; in her hands it had become nonlinear and nonrepresentational theater. Choreographing some 180 works in her lifetime, she also taught many students who rose to prominence as accomplished and masterful dancers, including Merce Cunningham and Paul Taylor.

Who founded the **Dance Theater of Harlem**?

The Dance Theater of Harlem, the first world-renowned African American ballet company, was founded by Arthur Mitchell (1934–), a principal dancer with the New York City Ballet, along with Karel Shook (1920–1985), a dance teacher and former director of the Netherlands Ballet. The impetus for the creation of the company came on April 4, 1968, while Mitchell was waiting to board a plane from New York City to Brazil (where he was establishing that country's first national ballet company) and he heard that Martin Luther King Jr. (1929–1968) had been assassinated. Mitchell later said that as he sat thinking about the tragic news, he wondered to himself, "Here I am running around the world doing all these things, why not do them at home?" Mitchell had spent his youth in Harlem, and he felt he should return there to establish a school to pass on his knowledge to others and to give black dancers the opportunity to perform. The primary purpose of the school was "to promote interest in and teach young black people the art of classical ballet, modern and ethnic dance, thereby creating a much-needed self-awareness and better self-image of the students themselves."

The idea was a success: During the 1970s and 1980s the company toured nationally and internationally, often performing to sell-out crowds and participating in prestigious events including international art festivals, a state dinner at the White House, and the closing ceremonies of the 1984 Olympic Games.

Today, the Dance Theater of Harlem is acknowledged as one of the world's finest ballet companies. Not only did Mitchell succeed in giving black dancers the opportunity to learn and to perform, he effectively erased color barriers in the world of dance, testimony to the universality of classical ballet.

How long has the **waltz** been danced?

Considered the quintessential ballroom dance, the waltz first became popular in Europe in 1813. But it dates as far back as the mid-1700s (the first written occurrence of the word *waltz* was in 1781). In the 1850s the dance captivated Vienna, and the pro-

The jitterbug is a variation of the two-step: couples swing and twirl in standardized patterns, sometimes incorporating acrobatics.

lific Johann Strauss (1825–1899), also known as the "Waltz King," produced scores of new waltzes to meet the increasing demand. Many of the compositions were named for professional associations and societies.

One of the most well-known waltzes is "The Blue Danube," first performed by Strauss on February 15, 1867, in Vienna. The lyrics, from a poem by Karl Beck, were sung by the Viennese Male Singing Society. The new waltz created an immediate sensation. It is an Austrian tradition whenever "The Blue Danube" is played that the opening strain is played first, followed by a pause before the work is played by the full orchestra. The pause is so that the audience may applause.

How did the **Charleston** get started?

The Charleston, a lively ballroom dance, emerged in 1923 as one of the flashy elements of the period in American history that writer F. Scott Fitzgerald (1896–1940) dubbed the Jazz Age. Other hallmarks of the Jazz Age, also called the Roaring Twenties, were speakeasies (since the country was in the midst of Prohibition), flappers, roadsters, raccoon coats, hedonism, and iconoclasm. The optimism of the Jazz Age came to a screeching halt on October 29, 1929, when the U.S. stock market crashed.

How did the **jitterbug** get started?

During the height of swing music's popularity in the late 1930s and early 1940s, there were at least 50 dance bands with national reputations and significant followings. Dance styles such as the jitterbug were based on Big Band music, the dominant form of American musical entertainment during those decades. The dance itself is a variation of the two-step; couples swing and twirl in standardized patterns, which sometimes include acrobatics.

MOVIES

When was the **first movie** shown?

On March 22, 1895, the first in-theater showing of a motion picture took place in Paris, when the members of the Société d'Encouragement à l'Industrie Nationale (National Society for the Promotion of Industry) gathered to see a film of workers leaving the Lumière factory at Lyons for their dinner hour. The cinematography of inventors Louis (1864–1948) and Auguste (1862–1954) Lumière, ages 31 and 33 respectively, was a vast improvement over the kinetoscope, introduced in 1894 by Thomas Edison, whose film could only be viewed by one person at a time. The 16-frame-per-second mechanism developed by the Lumière brothers became the standard for films for decades. The following year, on April 20, 1896, the first motion picture showing in the United States took place in New York; the film was shown using Thomas Edison's vitascope, which was an improvement on his kinetoscope, and a projector made by Thomas Armat.

What are *the* **milestones in the motion picture industry**?

Motion pictures continue to develop as new, sophisticated technologies are introduced to improve the moviegoing experience for audiences. In the decades following their rudimentary beginnings, there were many early milestones, including not only advancements in technology but improvements in conditions for those working in the then-fledgling industry:

1903: Edwin S. Porter's *The Great Train Robbery* was the first motion picture to tell a complete story. Produced by Edison Studios, the 12-minute epic established a pattern of suspense drama that was followed by subsequent movie makers.

1907: Bell & Howell Co. was founded by Chicago movie projectionist Donald H. Bell and camera repairman Albert S. Howell with $5,000 in capital. The firm went on to improve motion picture photography and projection equipment.

1910: *Brooklyn Eagle* newspaper cartoonist John Randolph Bray pioneered animated motion picture cartoons, using a "cel" system he invented and which was subsequently used by all animators.

1912: *Queen Elizabeth,* starring Sarah Bernhardt, was shown July 12 at New York's Lyceum Theater and was the first feature-length motion picture seen in America.

1915: D. W. Griffith's *The Birth of a Nation* provided the blueprint for narrative films.

1925: The new editing technique used in *Potemkin* revolutionized the making of motion pictures around the world. Soviet film director Sergei Eisenstein created his masterpiece by splicing film shot at many locations, an approach subsequently adopted by most film directors.

1926: The first motion picture with sound ("talkie") was demonstrated.

1927: The Academy of Motion Picture Arts and Sciences was founded by Louis B. Mayer of MGM Studios. The first president of the academy was Douglas Fairbanks.

1927: The first full-length talking picture, *The Jazz Singer,* starring vaudevillian Al Jolson, was released. By 1932 all movies talked.

1929: The first Academy Awards (for 1928 films) were held: Winners were William Wellman for *Wings,* Emil Jannings for best actor (in *Last Command*), and Janet Gaynor for best actress (in *Sunrise*). Movie columnist Sidney Skolsky dubbed the awards "the Oscars."

1928: Hollywood's major film studios signed an agreement with the American Telephone & Telegraph Corporation (AT&T) to use their technology to produce films with sound, leading to an explosion in the popularity of motion pictures.

1929: Eastman Kodak introduced 16-millimeter film for motion picture cameras.

1933: The Screen Actors Guild (SAG) was formed when six actors met in Hollywood to establish a self-governing organization of actors. The first organizing meeting yielded 18 founding members.

1935: The first full-length Technicolor movie was released, *Becky Sharpe.* The technology, however, was still in development, and the colors appeared garish.

1939: *Gone with the Wind* was released in Technicolor, which had come along way since its 1935 debut.

What was the **Hollywood blacklist**?

In 1947 studio executives assembled at the Waldorf-Astoria Hotel in New York put together a list of alleged Communist sympathizers, naming some 300 writers, directors, actors, and others known or suspected to have Communist Party affiliations or of having invoked the Fifth Amendment against self-incrimination when questioned by the House Committee to Investigate Un-American Activities. The "Hollywood Ten" who refused to tell the committee whether or not they had been Communists were Alvah Bessie, Herbert Biberman, Lester Cole, Edward Dmytryk, Ring Lardner Jr., John Howard Lawson, Albert Maltz, Samuel Ornitz, Adrian Scott, and Dalton Trumbo. The film industry blacklisted the Hollywood Ten on November 25, and all of them drew short prison sentences for refusing to testify.

When was Hollywood's golden age?

Hollywood had its heyday in the 1930s: In the same decade that the Great Depression crippled the world economy, the American film industry enjoyed its golden age. The era was marked by technical innovations: "talking movies" had made their debut in 1927 with the first full-length film with sound, *The Jazz Singer,* and by 1932 all films were "talkies"; the first Technicolor film, *Becky Sharpe,* debuted in 1935, and by 1939 was perfected when *Gone with the Wind* was released; and special effects were brought to the screen in 1933 with *King Kong,* which was the result of painstaking stop-motion and rear-projection photography.

In the meantime, movie stars such as Clark Gable, Claudette Colbert, Greta Garbo, and the Marx Brothers achieved public followings that were "the envy of political and business leaders." The MGM, Warner Bros., and RKO studios led Hollywood production, but other studios, including Fox, Paramount, Universal, Columbia, and United Artists, also fared well during these difficult times. In 1939 Hollywood had what has often been called its greatest year: Among the top releases that year were the classics *Gone with the Wind, The Wizard of Oz, Stagecoach, Ninotchka, Mr. Smith Goes to Washington,* and *Gunga Din.* By the end of the decade Hollywood had become a "major contributor to popular culture, an occasional contributor to high culture, and a dynamic, if unsteady, force in the nation's economy."

What were **newsreels**?

Newsreels got their start in 1910 when the pioneer film newsreel *Pathé Gazette* was shown in Britain and the United States. French cinematographer Charles Pathé (1863–1957) and his brother Emil (1860–1937) were Paris agents for the Edison phonograph. They visited London to acquire filmmaking equipment and secured financial support in order to set up production units in Britain, the United States, Italy, Germany, Russia, and Japan. These short movies, covering current events, were predominately used during wartime and were shown in theaters before motion pictures were shown. Superseded by television newscasts, the last newsreels were screened in 1967.

RADIO AND TELEVISION

What was **radio's** immediate cultural impact?

During the 1930s, radio, pioneered in the late 1800s, was woven into the fabric of everyday American life. People across the country—in cities, suburbs, and on farms—

Between the 1920s and 1950s, gathering around the radio in the evenings was as common to Americans as watching television is today.

tuned in for news and entertainment, including broadcasts of baseball games and other sporting events as well as comedy and variety shows, dramas, and live music programs. President Franklin D. Roosevelt (1882–1945) used the new medium to speak directly to the American public during the trying times of the Great Depression, broadcasting his "fireside chats" from the White House. Between the 1920s and the 1950s, gathering around the radio in the evenings was as common to Americans as watching television is today. Networks offered advertisers national audiences, and corporate America eagerly seized the opportunity to speak directly to people in their own homes. The advent of television in the 1950s and its growing popularity over the next two decades changed the role of radio in American life. Having lost their audience to TV, radio programmers seized rock music as a way to reach a wide, albeit a very young, audience. Many argue that the rise of the musical genre kept radio alive. In the decades since, radio programming has become increasingly music-oriented; talk and news programming are also popular.

What was the immediate impact of **television**?

The publicity surrounding the World's Fair television broadcast (April 30, 1939) inspired a flurry of broadcasting activity but reached limited audiences. On May 17 the

National Broadcasting Company (NBC) televised a baseball game between Princeton University and Columbia University, which NBC billed as the world's first televised sporting event. On August 26 NBC telecast a professional baseball game between the Brooklyn Dodgers and the Cincinnati Reds. More NBC broadcasts from Radio City featured live opera, comedy, and cooking demonstrations. Crowds waiting outside the 1939 New York premiere of *Gone with the Wind* were also televised. Feature films were aired, including a dramatization of *Treasure Island, Young and Beautiful,* and the classic silent film *The Great Train Robbery* (1903). Soon television stations had proliferated: by May 1940, 23 stations were broadcasting.

In 1941, after considerable deliberation on its part and that of the industry itself, the Federal Communications Commission (FCC) adopted transmission standards. Commercial operations were approved effective July 1, and two New York stations, National Broadcasting Company (NBC) and Columbia Broadcasting System (CBS) affiliates, went on the air. By the end of that year, the first commercial on television, financed by watch-manufacturer Bulova, was aired. In December, with the bombing of Pearl Harbor and the entrance of the United States into World War II (1939–45), commercial development of television was put on hold while American industry devoted its resources to the war effort.

But the television industry was eager, and as soon as allied victory looked like a sure thing, Radio Corporation of American (RCA) reopened its NBC television studio on April 10, 1944. CBS followed suit, reopening its operations on May 5. At the war's end in 1945, nine part-time and partly commercial television stations were on the air, reaching about 7,500 set owners in the New York, Philadelphia, Pennsylvania, and Schenectady, New York.

By 1947 the four networks that then existed—ABC, CBS, NBC, and DuMont (a short-lived competitor)—could still provide only about 10 hours of prime-time programming a week, much of it sporting events. In late 1948 it was estimated that only 10 percent of the population had even seen a television show. However, as the networks stepped up their programming with live-drama programs, children's shows, and variety shows (a format that was familiar and popular to the American radio-listening audience), interest in television grew rapidly. By the spring of 1948 industry experts estimated that 150,000 sets were in public places such as bars and pubs, accounting for about half of the total number of sets in operation. Just a year later, 940,000 homes had televisions. And by 1949 production of sets had jumped to 3 million.

Which was the **first TV network**?

It was the National Broadcasting Company (NBC), founded November 11, 1926, by David Sarnoff (1891–1971), who was then president of Radio Corporation of America (RCA). Sarnoff, considered one of the pioneers of radio and television broadcasting, created NBC to provide a program service to stimulate the sale of radios. In the 1940s he reorganized the network to provide TV programming, again to stimulate sales of RCA products—this time, televisions. It was Sarnoff who demonstrated television at the World's Fair in New York in 1939.

Next came Columbia Broadcasting System (CBS), on September 26, 1928, which was established by William S. Paley (1901–1990), an advertising manager for Congress Cigar Company. Paley sold some of his stock in the cigar company in order to raise $275,000 to buy into the beleaguered United Independent Broadcasters (which controlled Columbia Phonograph, hence the name). He built the floundering radio network into a powerful and profitable broadcasting organization.

The American Broadcasting Corporation (ABC) television network was last, in 1943. It was only by government order that the third network, ABC, was created at all. In 1943, when RCA was ordered to give up one of its two radio networks, it surrendered the weaker of the two (NBC Blue), which was bought by Edward J. Noble, the father of Life Savers candy. In 1945 Noble formally changed the name to the American Broadcasting Company, which three years later began broadcasting television from its New York flagship station.

Is there a **golden age** of **television**?

Yes, people commonly refer to the 1950s as TV's golden age, which is the decade when Americans embraced television and the networks responded with a rapid expansion of programming. Critics still hail the programs of the golden age to be the most innovative programming in television history. It was during this decade that anthology programs such as *Kraft Television Theatre, Playhouse 90,* and *Studio One* made live drama part of the nightly fare on prime-time television. Americans could tune in to watch original screenplays such as *Twelve Angry Men* (1954), *Visit to a Small Planet* (1955), and *The Miracle Worker* (1957). And tune in they did, prompting the production of more than 30 anthology programs sponsored by the likes of Goodyear, Philco, U.S. Steel, Breck, and Schlitz. Since the production work was based in New York, the anthologies drew young playwrights including Gore Vidal, Rod Serling, Arthur Miller, and A. E. Hotchner. A new group of prominent television directors and producers emerged. And the studio dramas attracted the talents of actors George C. Scott, James Dean, Paul Newman, Grace Kelly, Eva Marie Saint, Sidney Poitier, Lee Remick, and Jack Lemmon.

The other cornerstone of 1950 television programming was the variety show—also done live. Comedians Jack Benny, Red Skelton, Jackie Gleason, George Burns, Sid Caesar, and "Mr. Television" Milton Berle thrived in the format.

But the cost of producing live programs and the growing popularity of television, which created a new mass market that demanded even more programming, combined to spell the end of television's golden age. Soon, live dramas and variety shows were replaced by situation comedies, westerns, and other set-staged programs that could be taped in advance, and could be produced in quantity.

What was the impact of **network television**?

"Mr. Television" Milton Berle (pictured in 1952) hosted a popular weekly variety show.

The Big Three networks rose to power in the 1950s and dominated television for the next two decades. During much of this period, they captured more than 90 percent of the total viewing audience. Americans had turned away from the radio programs and films that had diverted them in the postwar 1940s and were tuning in to television—"the tube"—at an average rate of 25 hours per week, leaving little time for any other recreational pursuits. In short, television had become not only an American pastime, but an American obsession.

Radio, which had given birth to television (both NBC and CBS were radio networks before they began developing TV programming), saw its revenue cut in half almost overnight. It not only lost audiences and advertisers to TV, but lost popular programs and stars as well. Radio turned its attention to an emerging new art form: rock and roll. The move was a success, as young listeners tuned in to hear the music that was considered too raw to be included in the evening TV line-up.

Film felt the effects of television as well, as audiences stayed home to be entertained. Moviemakers attempted to lure audiences back into the theaters with gimmicks including 3-D movies, Panavision, Cinemascope, and Circle-Vision. Hollywood abandoned the western and other B-movies in favor of big-budget blockbusters, many of which were filmed on location rather than at studios or on back lots. Studios even forbade their stars from appearing on TV, but they soon relented. Cooperation between the two industries is what saved film: Studios sold old movies to the networks for broadcast and provided production talent and facilities to television.

Newspapers were least affected by the tremendous popularity of television, since programming time was at first limited to 8:00 P.M. to 11:00 P.M., which still left time to read the paper. As soon as television expanded programming beyond that three-hour window, however, newspapers felt the pinch—more sharply in 1963, when both CBS and NBC began airing news shows. Daytime programming spelled the demise of the evening paper.

While television had no impact on the number of books published, it did prompt a decrease in the number of fiction titles that were published (and a corresponding increase in the number of nonfiction titles). The upward trend of nonfiction titles is a lasting effect of television on the publishing industry.

Today experts disagree over the impact of television on our lives. Some argue that increased crime is a direct outcome of television since programs show crime as an everyday event and advertisements make people more aware of what they don't have. Critics also maintain that television stimulates aggressive behavior, reinforces ethnic stereotyping, and leads to a decrease in activity and creativity. Proponents of television counter, citing increased awareness in world events, improved verbal abilities, and greater curiosity as benefits of television viewing. By the end of the 1950s, more the 50 million American families owned a television.

When was **public broadcasting** started?

In the United States it was started in 1967, when the Public Broadcasting Act was signed into law by President Lyndon Johnson (1908–1973) on November 7, creating a Corporation for Public Broadcasting to broaden the scope of noncommercial radio and TV beyond its educational role. Within three years, and as a result of federal grants, plus funds from foundations, business, and private contributions, Public Broadcasting Service (PBS) rivaled the Big Three networks, NBC, CBS, and ABC, for viewers.

In England the British Broadcasting Corporation (BBC) took control of the development of television in 1932, launching BBC-TV. The BBC had been founded as a radio broadcaster in 1922 under the leadership of English engineer John Charles Reith (1889–1971). Reith remained at the helm of the BBC for 16 years after its founding and under his guidance, it became one of Britain's most revered institutions, supported by the public with license fees.

How did **cable TV** develop?

The industry had its beginnings in the 1970s, when Home Box Office (HBO) started to beam its signal to customers on a subscription basis. It was a radical concept: Let audiences pay direct for the programming (in this case, movies) rather than having it wholly supported through advertising and, in some cases, syndication fees. Another development during that decade proved to have lasting effect: Southern businessman Ted Turner (1938–) bought an independent TV station based in Atlanta and dubbed it a "superstation." WTBS was soon made available to some 10 million subscribing households and was the beginning of Turner's cable TV empire, which would later include Cable News Network (CNN), CNN's Headline News, Turner Network Television (TNT), and, briefly, the music network VH-1.

Soon there were a number of niche-based cable networks airing programming around the clock: Music Television (MTV) was reaching teens and twenty-somethings;

Lifetime targeted women; ESPN focused on sports; Black Entertainment Television (BET) catered to the African American audience; and American Movie Classics (AMC) appealed to a cross-section of American viewers who share a fondness for old, mostly black and white, movies.

By the 1980s the three major television networks, which had risen to power in the 1950s and had remained on the top of their game for the next two decades, found that their audience share was dropping. This was aided by the introduction of the upstart Fox television network, backed by media mogul Rupert Murdoch (1931–). Fox's programming, geared toward a young, hip, and increasingly multicultural audience, found its own following.

Soon Americans had more choices than every before—and for a relatively low subscription cost. By the 1990s the Big Three networks (NBC, ABC, and CBS) were reaching only 61 percent of the television audience. Not to be beaten at their own game, the networks responded to the new competition by producing more cutting-edge shows—programs that likely would not have made it on the air before the advent of cable. Further, they began charging their affiliate stations for programming—a trend begun by CBS in 1992.

How did **CNN** change television news?

When Ted Turner's Cable News Network (CNN) went on the air June 1, 1980, it was amid a fair amount of skepticism. Some thought the maverick businessman was ill advised to air news around the clock to cable television subscribers. History would soon prove Turner's detractors wrong.

Twenty-four hours of air time brought CNN something other news entities didn't have—the time to do more stories and more in-depth news stories. The American public embraced the concept and soon began to rely on CNN not only to provide more information than other TV news sources, but for breaking news and up-to-the minute updates on top stories. In 1991, during the CNN coverage of the Persian Gulf War (which CNN had more or less aired live), newspapers reported a phenomenon—Americans couldn't turn the news station off.

Gone were the days of planning dinner around the evening network news or waiting until 11:00 P.M. to learn the latest. CNN and its sister station, Headline News Network were news at-the-ready. At a time when the term "global marketplace" was quickly becoming part of the vocabulary of every working American, CNN was uniquely able to capitalize on the growing sensibility of a world community. In 1985 CNN International was launched as a 24-hour global news service. At first reaching only to England, by 1989 the signal was beamed via satellite to Africa, Asia, and the Middle East.

CNN has continued to evolve its programming to cover news in every area of endeavor with programs such as *Business Day, Larry King Live, World Today,* and

Science and Technology Week, proving that the concept has staying power. The network, which began turning a profit after five years, has picked up multiple journalism awards, including the coveted Peabody Award. One of the early harbingers of CNN's success came in April 1982 when CNN won the right to be on equal footing with the major network news organizations in the White House press pool.

When did **MTV first air**?

Music Television (MTV) made its debut August 1, 1981, when it was made available to 2.1 million cable-subscribing households in the United States. The format was all music, 24 hours a day. Audiences could tune in any time to watch popular rock artists performing hit songs.

The notion of pairing music with video was not without precedent, the most obvious of which is the Beatle's 1964 critically acclaimed pseudo-documentary, *A Hard Day's Night.* What was new was the idea of airing music videos around the clock. MTV was the brainchild of John Lack, vice president of Warner-Amex-Satellite Entertainment, which owned the cable station Nickelodeon. He'd taken interest in the Nickelodeon program *Popclips,* a music and video show developed by Michael Nesmith, a former member of the pop group the Monkees. Lack thought the format had potential, and soon a young executive, Robert Pittman, just 27 years old, was given charge of the project. MTV was launched with 13 advertisers and a meager library of only 125 videos, all provided by the record labels. But MTV caught on, and by 1984 it had captured an audience of more than 24 million viewers, was showing a profit, and was soon spun off into a separate company by parent company Warner.

The video colossus thrived during the 1980s, helping launch more than a few music careers. However, in the following decade "veejays" (video jockies) had to make way for alternate programming on the music channel in order to keep audiences interested. In addition to the standard video programs (including theme shows such as *Yo! MTV Raps*), the cable channel broadened its offerings to include specials (*MTV Spring Break* and *MTV Video Music Awards*) as well as series such as *Real World, Road Rules,* and *The Osbournes*). All of the new programming was aimed at the Generation-X audience—the very group of teens and twenty-somethings who, in the 1990s, couldn't remember *not* watching MTV. Even though its audience was estimated to have declined by more than half since its peak, MTV remained a going—and profitable—concern into the 2000s, as advertisers continued to rely on the medium to reach the youth market.

How has **MTV affected the music industry**?

MTV's almost immediate impact was to launch the music careers of fledgling artists. Some critics believe that superstar Madonna, who showed up on the music scene at about the same time as MTV, would not have risen to the heights of fame that she has were it not for Music Television. Or at least her star might not have risen so quickly. But she and other media-savvy artists exploited the new format to reach the music-buying public—the world over, for soon MTV had a global presence. The video channel also gave established artists a boost by airing more than one single off a given album—resulting in several hits from any one recording. Such was the case for artists like Billy Joel, Bruce Springsteen, U2, and Peter Gabriel. Increasingly creative videos gave the works of album-oriented musicians longer lives and steady sales.

MTV quickly established itself—and the format—as an integral part of the music industry. Once the format was proven viable, other music television channels emerged—including VH-1 (which was begun by media mogul Ted Turner and was later bought by MTV and molded into an adult-oriented music station), The Nashville Network (TNN, which aired country music videos and programming in the 1980s and 1990s), and Country Music Television (CMT). Today, the music video remains an important tool for new and established artists alike.

Media analysts also believe MTV has had an impact on modern culture. Since the channel relies on interesting visuals to capture viewing audiences, MTV is constantly upping the creative ante for artists who freely experiment with bright colors, images, rapid-fire editing, motifs, dreamlike imagery, and other visual techniques, which began showing up on other television shows, in movies, and in advertising. Some observers believe the phenom has ushered in a new visual order.

Of course, MTV has its detractors: Critics argue that the MTV aesthetic is superficial and that it is accelerating the movement away from traditional forms of literacy. While MTV is praised and panned in the worldwide media, there's no arguing that the music channel continues to be a window on what's hip and hot to the American youth.

What impact have **VCRs and digital recorders** had on **television viewing**?

Since VCRs (invented 1975) and the more recent digital recorders such as TiVo give viewers the ability to record television programs and watch them when they choose, a phenomenon known as "time-shifting" has emerged. As Americans watch previously recorded programming, they miss prime-time broadcasts. Commentators mused that Americans were losing a sense of community—gone were the days when office workers would stand around the water cooler to chat about last night's episode of a popular program. Since these recording devices also gave viewers the ability to fast-forward through commercials (a practice dubbed "zapping"), they posed challenges to the advertising industry.

GAMES

How old are **card games**?

Card games are believed to have originated in China in about A.D. 800. By the late 1200s cards had made their way to Italy (perhaps carried aboard merchant ships), and from there spread to other parts of Europe. The present-day variety of four suits—hearts, diamonds, clubs, and spades—was adopted in France during the 1500s.

When was **poker** invented?

The card game, in which a player bets that the value of his or her hand is higher than those held by the other players, was invented in the 1820s by sailors in New Orleans. The sailors combined the ancient Persian game *as nas* with the French game *poque,* which itself was a derivation of the Italian game *primiera,* and a cousin of the English game *brag.*

Poker was originally played with 3 cards from a deck of only 32 and included combinations such as pairs and three of a kind. The game was later played with five cards from a deck of 52 cards, and the draw was also added. Stud poker, where each player is dealt their first card face-down and the next four cards face-up, didn't come into existence for many years (until about 1864). Other later additions to the game were the straight (a hand consisting of five cards in sequence but not of the same suit) and the flush (a hand consisting of five cards of the same suit but not in sequence).

What is **contract bridge**?

It's the game most people play when they play bridge, a card game for four players in two partnerships who bid to name the trump suit. The other kind of bridge is called auction bridge, which was invented in 1904 as a variation on the card game whist (whist had been played since the early sixteenth century, if not longer). Auction bridge differs from contract bridge in that tricks made in excess of the contract are scored toward game; in contract bridge they are not. It's believed that contract bridge originated in 1926 when railroad heir and yachtsman Harold S. Vanderbilt (1884–1970) invented the variation while on a Caribbean cruise. The game did not catch on until 1930, when Romanian-American contract bridge expert Eli Culbertson defeated Lieut. Colonial W. T. M. Butler in a challenge match at London's Almack's Club; the match was highly publicized.

Who was **Hoyle**?

Edmond Hoyle (1671 or 1672–1769) was an English cardplayer who, in 1742 at the age of 70, published *Short Treatise on Whist,* which provided the rules for a game that developed into auction bridge and later into contract bridge. Hoyle's name has survived among bridge players in the phrase "according to Hoyle."

How old is **chess**?

It dates to the Middle Ages: In 1283 Alfonso X (1221–1284), king of Castile and León (Spain), commissioned the *Libro de ajedrez, dados y tablas* (Book of Chess, Dice, and Backgammon), based on an Arabic text. This book is still considered an important source on leisure activities in the Middle Ages (500–1350).

How old is the game of **billiards**?

The Italians were the first to play this table game, in the 1550s. It's different than pool, since billiards requires the players to cause the cue ball to hit two object balls in succession. There are no pockets, as there are in pool.

SPORTS

When were the **first Olympic Games**?

The Olympics date to about 900 B.C., when, in Ancient Greece, tens of thousands of sandal-wearing spectators descended on Olympia to cheer the runners, wrestlers, and bare-skinned boxers competing there. The games at Olympia were one of four athletic festivals in Greece, the others being the Isthmian games at Corinth, the Nemean games, and the Pythian games at Delphi, all of which alternated to form the *periodos*, or circuits, which guaranteed sports fans the opportunity to attend an athletic festival every year.

Winning was everything then: Athletes were required to register in order to compete, and rumors of Herculean opponents sometimes prompted competitors to withdraw. Victors were awarded crowns of olive leaves, and the second- and third-place finishers returned home undecorated.

The modern Olympic Games, begun by diminutive Frenchman Baron Pierre de Coubertin (1862–1937) possess a decidedly different spirit than did their ancient counterpart where the only rules were that participants were not allowed to gouge, bite, put a knee to the groin, strangle, or throw sand at their opponent. The modern Olympic Games, publicly proposed by Coubertin on November 25, 1892, in Paris, and first held in Athens, Greece, in 1896, are based on their initiator's vision of the Olympic competition as an occasion to promote peace, harmony, and internationalism.

In April 1896 some 40,000 spectators pressed into the Panathenean Stadium, which had been constructed on the site of an ancient stadium in Athens, to witness the athletic feats of the first modern Olympic heroes. Thirteen nations participated; only male athletes (just more than 300 of them) competed; and Greece received the most medals (47). The second Olympic Games were held in 1900 in Paris.

When were the **first Winter Games** held?

The Winter Olympic Games had a slow birth, making their first official appearance almost three decades after the first modern Games were held in Athens (1896). In 1901 Nordic Games were held in Sweden. However, only Scandinavian countries participated in the events, which organizers intended to hold every four years. The Nordic Games constituted the first organized international competition involving winter sports. Then, as part of the Summer Olympic Games in 1908, host-city London held a figure skating competition that October. Three years later, an Italian member of the International Olympic Committee (IOC) encouraged Sweden, the next host of the Summer Games, to include winter sports in 1912 or hold a separate event for them. Since Sweden already played host to the Nordic Games, they declined to pursue the IOC suggestion. The sixth Olympiad was slated to be held in 1916 in Berlin, and Germany vowed to stage winter sports competition as part of the event. But in 1914 World War I began, and the Berlin Games were canceled altogether.

After an eight-year hiatus, the Olympics resumed in 1920: Antwerp, Belgium, played host to athletes, which included figure skaters and ice hockey players along with the usual contingency of gymnasts, runners, fencers, and other summer sports competitors. The first IOC-sanctioned competition of winter sports was held in Chamonix, France, from January 25 to February 4, 1924. When the Games were staged next, in St. Moritz, Switzerland, in 1928, they were formally designated the second Winter Olympics.

From that year, the Winter Games were held every four years in the same calendar year as the Summer Games—until 1994. In 1986 IOC officials voted to change the schedule. The result was that the 1992 Winter Olympics in Albertville, France, were followed only two years later by the Games in Lillehammer, Norway. The Winter and Summer Games are now each held every four years, alternating in even-numbered years.

Have the **Olympic Games been held regularly** since 1896?

No. In spite of the fact that international harmony ("Truce of God") is one of the hallmarks of the modern Olympic movement, the Games have been canceled by the governing body, the International Olympic Committee (IOC), due to world events: In

1916 the Games were canceled because of World War I (1914–18); the 1940 and 1944 Games were called off due to World War II (1939–45).

The Games have been affected by international politics, occasional boycotts, and demonstrations as well. Though the 1980 Summer Games continued as planned, the United States and as many as 62 other noncommunist countries (including Japan and the Federal Republic of Germany) boycotted them in protest of the 1979 Soviet invasion of neighboring Afghanistan. The following Summer Games, held in Los Angeles in 1984, were boycotted by the Soviets. The official reason was cited as "fear," though some skeptics believed the reason to be more specific: a fear of drug testing. In 1968, in Mexico City, two African American track medalists rose gloved-and-clenched fists of support for black power, which earned them suspension and expulsion from the Olympic Village. Olympic history turned dark when the 1972 Summer Games in Munich, 11 Israeli athletes were killed in the Olympic Village by the Arab terrorist group Black September. A 1996 bombing at Olympic Park in Atlanta, Georgia, also cast a shadow over the Games.

How old is **baseball**?

Baseball, America's pastime, is more than 200 years old. According to legend, the sport's originator was U.S. Army officer Abner Doubleday (1819–1893), who was credited with inventing and naming the game in 1839, while he was attending school in Cooperstown, New York (the site of the Baseball Hall of Fame and Museum). But in 2004 a document was uncovered in Pittsfield, Massachusetts, citing a 1791 bylaw prohibiting the playing of baseball too close to (within 80 yards of) the town's meeting hall. Historians verified the authenticity of the document and its date. This is believed to be the earliest written record of the game—and it establishes that the stick-and-ball sport was being played 42 years before Doubleday's involvement. Baseball historians have long acknowledged that the sport, which is similar to the English games of cricket and rounders, had not one father, but thousands. Although the 2004 discovery indicates that the game was already in existence in 1791, and popular enough to be the subject of a town ordinance, it was in the 1800s that baseball developed into the game Americans still love today.

The first baseball club, the Knickerbocker Base Ball Club, was organized by American sportsman Alexander Cartwright (1820–1892) in 1842 in New York City. By 1845 the team had developed a set of 20 rules, which included specifications for where the bases are positioned and how runners can be tagged as out. The rules also defined a field of play, outside of which balls are foul. The so-called New York game spread in popularity after a famous 1846 match in Hoboken, New Jersey. By 1860 there were at least 50 organized ball clubs in the country. Union soldiers helped spread the game during the American Civil War (1861–65), and the popularity of the sport greatly increased during the last three and a half decades of the nineteenth century. The first professional baseball team was the Cincinnati Red Stockings, which began play in

America's sport: The Chicago Cubs play the St. Louis Cardinals, 1930. Historians recently discovered that baseball dates back to the 1700s.

1869. In 1876 the National League (NL) was founded; it included teams in Boston; Chicago; Cincinnati, Ohio; Hartford, Connecticut; Louisville, Kentucky; New York; Philadelphia; and St. Louis, Missouri. By the 1880s the sport had evolved into big business: An 1887 championship series between St. Louis and Detroit drew 51,000 paying spectators. The American League (AL) was formed in 1901, and two years later the two leagues staged a championship between their teams: In 1903, the Boston Red Socks beat the Pittsburgh Pirates in the first World Series.

An overall increase in American leisure time, created by the innovation of labor-saving household devices as well as a reduction in the average laborer's workweek helped baseball become the national sport and its favorite pastime. Played on an open field, the game harkened back to the nation's agrarian roots; but with its standardized rules and reliance on statistics, it looked forward to a modern, industrialized future.

Who invented **basketball**?

The ball-and-hoop game was invented by Canadian American James Naismith (1861–1939) in December 1891. An instructor at the YMCA College, in Springfield, Massachusetts, Naismith was asked by the head of the physical education department to come up with a game to keep students active indoors during the winter months. It had to fit inside the confines of a gym, have no physical contact, use a soft ball, and give everyone who participated a chance to handle the ball. Naismith nailed two peach baskets, which he found in the storeroom, to balcony railings at each end of the school's gym, found a soccer ball, divided his class of 18 men into two teams, and introduced them—and as it would turn out, the rest of the world—to the game, which was later dubbed basket ball (two words). Improvements to the game came over the next two decades as it spread in popularity. In 1910 the important change of allowing ball handlers to move by dribbling was made. In 1916 the rules were changed to allow dribblers to shoot the ball.

When did **football** begin?

In ancient Greece and Rome, a game was played in which the object was to move a ball across a goal line by throwing, kicking, or running with it. Several modern games were derived from this, including rugby and soccer, from which American football directly evolved (in much of the world football refers to soccer, in which players are allowed to advance the ball only with their feet or heads). Historians generally agree that the first game of American football was played on November 6, 1869, in New Brunswick, New Jersey, when Rutgers defeated the College of New Jersey (present-day Princeton University), 6-4. They played on a field 120 yards long and 75 yards wide and used a round, soccerlike ball. Other eastern colleges, including Columbia, Harvard, and Yale, soon added the sport to their athletic programs. In 1876 a set of official rules were compiled. In the 1880s Yale coach Walter Camp (1859–1925) revised the rules, giving the world the game played today. He limited teams to 11 players, established the scrimmage system for putting the ball into play, introduced the concept of requiring a team to advance the ball a certain number of yards within in given number of downs, and came up with the idea of marking the field with yard lines.

How old is **golf**?

Some historians trace golf back to a Roman game called *paganica*. When they occupied Great Britain between roughly A.D. 43 until 410, Romans played the game in the streets, using a stick and a leather ball. But there are other possible predecessors as well, including an English game (called *cambuca*), a Dutch game (*kolf*), a French and Belgian game called (*chole*), and a French game (*jeu de mail*). But the game as we know it, the rules, equipment, and 18-hole course, certainly developed in Scotland, where it was played as early as the early 1400s. The rules of the game were also codified there: *The Rules of Golf* was published in 1754 by the St. Andrews Golfers, later called the Royal & Ancient Golf Club. The first golf club (formed 1744) was the Honourable Company of Edinburgh Golfers in Edinburgh, Scotland. And it was none other than Mary, Queen of Scots (1542–1587), who is credited with being both the first woman golfer and the originator of the term "caddie." (The term is derived from the French term for the royal pages, *cadets,* who carried the queen's clubs.)

Where does golf sensation **Tiger Woods** rank among the greats?

Woods (1975–) ranks among the best players of all time. In 1999 he won 8 tournaments, including the Professional Golfers' Association (PGA) and Tour championships, and he was the first player to win 4 consecutive starts since Ben Hogan (1912–1997) in 1953 (Hogan won a total of 5 tournaments that year). In 2000 Woods went on to win 9 tournaments, and in 2001 and 2002 he had 5 wins each season, giving him a total of 27 wins in four years. (In July 2005, with his victory at the British Open, Woods became only the fifth player in golf history to complete a career grand slam, winning golf's four major tournaments: the Masters, the U.S. Open, the British Open, and the PGA Championship. In 2001 he became the first player to consecutively win the grand slam.) This impressive run alone assures Woods a place in the annals of golf history.

As of mid-2005 Arnold Palmer (1929–) remained the PGA leader for the number of wins in any four-year period; between 1960 and 1964 he racked up 29 first-place finishes. He was also the first player to win the Masters four times (1958, 1960, 1962, and 1964). A couple of other legendary names in golf tallied record numbers of wins in short periods to dominate the sport: Ben Hogan won 30 tournaments in three seasons, 1946–48; and Byron Nelson (1912–) won 26 tournaments in just two years, 1944 and 1945. Sam Snead (1912–2002) and Jack Nicklaus (1940–) dominated the sport over the course of decades: Snead holds the record for the most wins in a career (81); and Nicklaus follows with 71.

At the age of 24 Woods was already the PGA's career earnings leader. To put this in perspective, in 1999 he won a million dollars more for the year than golf legend Jack Nicklaus made in his entire PGA Tour career. Woods's ability and appeal combined to help raise earnings overall for the sport. He was credited with expanding golf's horizons by bringing new fans to the game. According to *ESPN The Magazine,* ratings for the final round of each of the four majors collectively jumped 56 percent in 1997 when Woods broke the Masters scoring record (he shot 270 over 72 holes to finish 12 strokes ahead of the second-place winner). At the end of the 2004 season and despite a slump, Woods remained on top for career earnings, but he was followed very closely by Vijay Singh (1963–).

How old is **tennis**?

The game is believed to have originated in France, where it was called *jeu de paume* (meaning "game of the palm") in the 1100s or 1200s: Players hit a small ball back and forth over a net using the palms of their hands. The "father of modern tennis" is English soldier and sportsman Walter Clopton Wingfield (known as Major Wingfield [1833–1912]), who in 1873 published a rule book for lawn tennis and in 1874 patented equipment for the game.

Further Reading

Your local library and neighborhood bookstore shelves are filled with history books. Some authors have taken a survey approach; others have narrowed their focus to a person or event; still others have written novels that so faithfully recount history that they, too, have been acclaimed as serious works of record. The following list is a sampling of what is available. The pages of this book also include numerous mentions of books by and about people who have made history.

Alexander, Caroline, and Frank Hurley. *Endurance: Shackleton's Legendary Antarctic Expedition.* New York: Alfred A. Knopf, 1998. Accompanies a 1999 American Museum of Natural History exhibit of more than 150 photographs of the heroic Antarctica expedition of 1915–16.

Ambrose, Stephen E. *Citizen Soldiers: The U.S. Soldiers from the Normandy Beaches to the Bulge to the Surrender of Germany, June 7, 1944–May 7, 1945.* New York: Simon & Schuster, 1997.

——— *D-Day: June 6, 1944: The Climactic Battle of World War II.* New York: Simon & Schuster, 1994.

——— *Eisenhower: Soldier, General of the Army, President-Elect, 1890–1952.* New York: Simon & Schuster, 1983.

——— *Nixon: The Education of a Politician, 1913–62.* New York: Simon & Schuster, 1987.

——— *Nixon: Ruin and Recovery, 1973–90.* New York: Simon & Schuster, 1991.

——— *Nixon: The Triumph of a Politician, 1962–72.* New York: Simon & Schuster, 1989.

——— *Undaunted Courage: Meriwether Lewis, Thomas Jefferson, and the Opening of the American West.* New York: Simon & Schuster, 1996.

American Life: A Social History. A Macmillan Information Now Encyclopedia. New York: Macmillan Library Reference USA, 1993. Selections from the *Encyclopedia of American Social History*, a home edition of the library reference.

Anderson, John Lee. *Che Guevara: A Revolutionary Life.* New York: Grove Press, 1997.

Aranson, H. H. *History of Modern Art.* New York: Harry N. Abrams, 1977.

Armitage, Michael, et al. *World War II Day by Day.* New York: DK Publishing, 2004 revised edition.

Axelrod, Alan, and Charles Phillips. *What Every American Should Know About American History.* Holbrook, Mass.: Bob Adams Publishers, 1992.

Bailyn, Bernard. *Voyagers to the West: A Passage in the Peopling of America on the Eve of Revolution.* New York: Alfred A. Knopf, 1986.

Ball, Edward. *Slaves in the Family.* New York: Ballantine Books, 1998. A National Book Award winner.

Berg, A. Scott. *Lindbergh.* New York: G. P. Putnam & Sons, 1998. A Pulitzer Prize–winning biography of American aviator Charles Lindbergh.

Bernstein, Sara Tuvel, with Louise Loots Thornton and Marlene Bernstein Samuels. *The Seamstress: A Memoir of Survival.* New York: Berkley Books, 1999 reprint. The memoirs of a holocaust survivor.

Boorstin, Daniel J. *The Americans: The Colonial Experience.* New York: Random House, 1958.

——— *The Americans: The Democratic Experience.* New York: Random House, 1973. Pulitzer prize winner.

——— *The Americans: The National Experience.* New York: Random House, 1965.

——— *The Creators: A History of Heroes of the Imagination.* New York: Random House, 1992.

——— *The Discoverers: A History of Man's Search to Know His World and Himself.* New York: Random House, 1983.

——— *The Image: What Happened to the American Dream?* New York: Atheneum, 1961.

———, editor. *An American Primer.* Chicago: University of Chicago Press, 1966.

———, et al. *Hidden History.* New York: Harper & Row, 1987.

Boyden, Matthew, et al. *The Rough Guide to Classical Music on CD,* 3rd ed. London: Rough Guides, 2001.

Branch, Taylor. *Parting the Waters: America in the King Years 1954–63.* New York: Simon & Schuster, 1988.

——— *Pillar of Fire: America in the King Years 1963–65.* New York: Simon & Schuster, 1998.

Brokaw, Tom. *The Greatest Generation.* New York: Random House, 1998. Acclaimed profile of the generation of Americans who came of age during the Great Depression and unquestioningly shouldered the burden of World War II.

Brown, Les. *Les Brown's Encyclopedia of Television,* 3rd ed. Detroit: Visible Ink Press, 1992.

Brownstone, David, and Irene Franck. *Dictionary of 20th Century History.* New York: Prentice Hall, 1990.

Bruce, Robert C. *The Launching of Modern American Science, 1846–1876.* New York: Alfred A. Knopf, 1987.

——— *Bell: Alexander Graham Bell and the Conquest of Solitude.* Boston: Little, Brown, 1973.

Burrell, Brian. *The Words We Live By: The Creeds, Mottoes, and Pledges That Have Shaped America.* New York: The Free Press, 1997.

Burrows, Edwin G., and Mike Wallace. *Gotham: A History of New York City to 1898.* New York: Oxford University Press, 1999.

Cahill, Thomas. *How the Irish Saved Civilization: The Untold Story of Ireland's Heroic Role from the Fall of Rome to the Rise of Medieval Europe.* New York: Nan A. Talese/Doubleday, 1995.

Cantor, Norman F. *The American Century.* New York: HarperCollins, 1997. A retrospective on the cultural movements of the twentieth century; called "an intellectual roadmap" by *Publisher's Weekly.*

Carey, John, ed. *Eyewitness to History.* Cambridge, Mass.: Harvard University Press, 1987. Try your local library for this book, described as "history with the varnish removed": Three hundred first-hand accounts of events—culled from hundreds of memoirs, letters, travel books, and newspapers.

Carnes, Mark C., ed. *U.S. History.* A Macmillan Information Now Encyclopedia. New York: Macmillan Library Reference USA, 1996. Selections from the eight-volume *Dictionary of American History,* a home edition of the library reference.

Carruth, Gorton, ed. *The Encyclopedia of American Facts and Dates,* 10th ed. New York: HarperCollins, 1997. Covers more than 100 years of American history and popular culture. Arranged chronologically; thoroughly indexed.

Carson, Rachel. *Silent Spring.* Boston: Houghton Mifflin, 1962, 1994 reprint. Seminal work credited with heightening environmental awareness in the second half of the twentieth century.

Chang, Iris. *The Rape of Nanking: The Forgotten Holocaust of World War II.* New York: Basic Books, 1997.

Chernow, Ron. *The House of Morgan: An American Banking Dynasty and the Rise of Modern Finance.* New York: Atlantic Monthly Press, 1990.

——— *Titan: The Life of John D. Rockefeller, Sr.* New York: Random House, 1998.

Churchill, Winston. *The Second World War.* 6 vols. New York: Time, Inc., 1959.

Coles, Harry L. *War of 1812.* Chicago: University of Chicago Press, 1965.

Davis, Kenneth C. *Don't Know Much about History: Everything You Need to Know about American History but Never Learned.* New York: Crown, 1990.

de la Croix, Horst, and Richard G. Tansey. *Gardner's Art through the Ages,* 7th ed. New York: Harcourt Brace Jovanovich, 1980.

The Debate on the Constitution. 2 vols. Part One (September 1787–February 1788); Part Two (January–August 1788). Selected and annotated by Bernard Bailyn. New York: Library of America, 1993. Federalist and antifederalist speeches, articles, and letters during the struggle over ratification.

Diamond, Jared. *Guns, Germs, and Steel: The Fates of Human Societies.* New York: W. W. Norton, 1997. A review of human history in an attempt to explain why the West rose to power.

Donald, David Herbert. *Lincoln.* New York: Touchstone/Simon & Schuster, 1995. The *New York Times Book Review* said it is "hard to imagine a more satisfying life of our most admired and least understood president." Written by a two-time Pulitzer Prize winner.

Dorson, Richard Mercer. *American Folklore.* Edited by Daniel J. Boorstin. New York: Anchor Books, 1972.

Douglass, Frederick. *Narrative of the Life of Frederick Douglass, an American Slave.* Garden City, N.Y.: Doubleday, 1972.

DuBois, Ellen Carol. *Feminism and Suffrage: The Emergence of an Independent Women's Movement in America, 1848–1869.* Ithaca, N.Y.: Cornell University Press, 1999 reprint (with a new preface).

——— *Woman Suffrage and Women's Rights.* New York: New York University Press, 1998.

Einstein, Albert. *Ideas and Opinions.* New York: Crown Publishers, 1954.

Ellis, John Tracy. *American Catholicism,* 2nd ed. Chicago: University of Chicago Press, 1969.

Evans, Harold. *The American Century.* New York: Alfred A. Knopf, 1998.

Ferguson, Niall. *Pity of War: Explaining World War I.* New York: Basic Books, 1999.

Fletcher, Richard. *The Barbarian Conversion: From Paganism to Christianity.* New York: Henry Holt, 1997. Hailed by Robert Runcie, the Archbishop of Canterbury (1980–91), as an "enthralling book" in which "epic moments and tiny vignettes mingle to convey the flavour of Christian life as the people of Europe were experiencing conversion."

Foote, Shelby. *The Civil War.* 3 vols. Fort Sumter to Perryville; Fredericksburg to Meridian; Red River to Appomattox. New York: Random House, 1958.

——— *Stars in Their Courses: The Gettysburg Campaign, June–July 1863.* New York: Modern Library, 1994 reprint.

Fossier, Robert, ed. *The Cambridge Illustrated History of the Middle Ages.* 3 vols. 350–950, 950–1250, 1250–1520. New York: Cambridge University Press, 1997.

Frank, Anne. *The Diary of a Young Girl.* Translated by B. M. Mooyaart. Garden City, N.Y.: Doubleday, 1967.

Fromkin, David. *The Way of the World: From the Dawn of Civilizations to the Eve of the Twenty-First Century.* New York: Alfred A. Knopf, 1998.

Fussell, Paul. *The Great War and Modern Memory.* New York: Oxford University Press, 1975.

Gandhi, Mohandas Karamchand. *An Autobiography: The Story of My Experiments with Truth.* Translated by Mahadev Desai. Beacon Press, 1993.

Garner, Joe. *We Interrupt This Broadcast: Relive the Events that Stopped Our Lives…from the Hindenburg to the Death of Princess Diana.* Naperville, Ill.: Sourcebooks, 1998. Includes two CDs.

Gibbon, Edward, et al. *The Decline and Fall of the Roman Empire.* 3 vols. New York: Modern Library, 1995 reprint.

Gilbert, Martin. *A History of the Twentieth Century.* 3 vols. New York: William Morrow, 1998.

Glazer, Nathan. *American Judaism.* Chicago: University of Chicago Press, 1972.

——— *We Are All Multiculturalists Now.* Cambridge, Mass.: Harvard University Press, 1997.

Goodwin, Doris Kearns. *The Fitzgeralds and the Kennedys: An American Saga.* New York: Simon & Schuster, 1987.

——— *Lyndon Johnson and the American Dream.* New York: Harper & Row, 1976.

——— *No Ordinary Time: Franklin and Eleanor Roosevelt, the Home Front in World War II.* New York: Simon & Schuster, 1994. Pulitzer Prize winner.

——— *Wait Till Next Year.* New York: Touchstone/Simon & Schuster, 1997. Historian's memoir evokes postwar America.

Goodwin, Jason. *Lords of the Horizons: A History of the Ottoman Empire.* New York: Henry Holt, 1998.

Grun, Bernard. *The Timetables of History: A Horizontal Linkage of People and Events.* New York: Simon & Schuster, 1991.

Gurko, Miriam. *The Ladies of Seneca Falls: The Birth of the Woman's Rights Movement.* New York: Macmillan, 1974. (Paperback issued by Pantheon, 1987.)

Halberstam, David. *The Best and the Brightest.* New York: Random House, 1972. Chronicle of how America became involved in the conflict in Vietnam.

——— *The Children.* New York: Random House, 1998. Traces the civil rights movement and profiles its leaders.

——— *The Fifties.* New York: Villard Books, 1993. Charts the course of the decade in American history.

——— *Next Century.* New York: Morrow, 1991. Pulitzer Prize–winning journalist considers the recent past and suggests changes for the twenty-first century.

Harbage, Alfred, ed. *William Shakespeare, the Complete Works.* New York: The Viking Press, 1969.

Harris, John F. *The Survivor: Bill Clinton in the White House.* New York: Random House, 2005.

Hibbert, Christopher. *The Virgin Queen: Elizabeth I, Genius of the Golden Age.* Reading, Mass.: Perseus Books, 1991. The *Philadelphia Inquirer* called this a "highly readable account"; the *Chicago Tribune* hailed Hibbert as a "master at bringing together details in a way that makes his subjects recognizable as people."

Howard, Michael, and William Roger Louis, eds. *The Oxford History of the Twentieth Century.* New York: Oxford University Press, 1998.

Hunt, William Dudley. *Encyclopedia of American Architecture.* New York: McGraw-Hill, 1980.

Jennings, Peter, and Todd Brewster. *The Century.* New York: Doubleday, 1998. Similar in scope to Evans's *American Century,* but includes the perspectives of ordinary men and women.

Johnson, Paul. *Modern Times: The World from the Twenties to the Eighties.* New York: Harper & Row, 1983.

Kennedy, David M. *Freedom from Fear: The American People in Depression and War, 1929–45.* New York: Oxford University Press, 1999.

Kennedy, John F. *Profiles in Courage.* New York: Harper, 1956. Pulitzer Prize–winning portraits of eight American politicians.

Kennedy, Paul M. *Rise and Fall of the Great Powers: Economic Change and Military Conflict from 1500 to 2000.* New York: Random House, 1988.

Ketchum, Robert M. *The Battle for Bunker Hill.* Garden City, N.Y.: Doubleday, 1962.

——— *The Borrowed Years, 1938–41: America on the Way to War.* New York: Random House, 1989.

——— *Saratoga: Turning Point of the American Revolution.* New York: Henry Holt, 1997.

——— *The Winter Soldiers: The Battles for Trenton and Princeton.* Garden City, N.Y.: Doubleday, 1973.

Keynes, John Maynard. *The General Theory of Employment, Interest, and Money.* San Diego: Harcourt Brace Jovanovich, 1964, 1991 reprint.

King, Martin Luther Jr. *Why We Can't Wait.* New York: Harper & Row, 1964.

Kluger, Richard. *Simple Justice: The History of Brown v. the Board of Education and Black America's Struggle for Equality.* New York: Alfred A. Knopf, 1975.

Knappman, Edward W., ed. *Great American Trials.* Detroit: Visible Ink Press, 1994.

——— *Great World Trials.* Detroit: Visible Ink Press, 1997.

Lacey, Robert. *Ford: The Men and the Machine.* Boston: Little, Brown, 1986. Biography of Henry Ford and his ancestors.

——— and Danny Danziger. *Year 1000: What It Was Like at the Turn of the First Millennium, An Englishman's World.* Boston: Little, Brown, 1999.

Lamar, Howard, R. *The New Encyclopedia of the American West.* New Haven, Conn.: Yale University Press, 1998.

Landes, David S. *The Wealth and Poverty of Nations: Why Some Are So Rich and Some So Poor.* New York: W. W. Norton, 1998. Highly acclaimed examination of one of the most hotly debated questions of our time.

Larson, Edward J. *Summer for the Gods: The Scopes Trial and America's Continuing Debate over Science and Religion.* New York: Basic Books, 1997. Pulitzer Prize winner.

Larson, Eric V. *Casualties and Consensus: The Historical Role of Casualties in Domestic Support for U.S. Military Operations.* Santa Monica, Calif.: Rand Corporation, 1996.

Levi, Primo. *Survival in Auschwitz.* Translated by Stuart Woolf. New York: Summit Books, 1986.

Lewis, David L., ed. *W. E. B. Du Bois: A Reader.* New York: Henry Holt, 1995.

Malone, Dumas. *Jefferson and His Time.* Boston: Little, Brown, 1948. Pulitzer Prize winner.

Manchester, William. *American Caesar: Douglas MacArthur, 1880–1964.* Boston: Little, Brown, 1978.

——— *The Last Lion.* Vol. 1. Winston Spencer Churchill: Visions of Glory, 1874–1932. Boston: Little, Brown, 1983.

——— *The Last Lion.* Vol. 2. Winston Spencer Churchill: Alone, 1932–40. Boston: Little, Brown, 1988.

——— *A Rockefeller Family Portrait: From John D. to Nelson.* Boston: Little, Brown, 1959.

——— *World Lit Only by Fire: The Medieval Mind and the Renaissance—Portrait of an Age.* Boston: Little, Brown, 1992.

Mandela, Nelson. *Long Walk to Freedom.* Boston: Little, Brown, 1994.

Matthews, Glenna. *American Women's History: A Student Companion.* New York: Oxford University Press, 2000.

McCullough, David. *Truman.* New York: Simon & Schuster, 1992.

McDougall, Walter A. *The Heavens and the Earth: A Political History of the Space Age.* New York: Basic Books, 1985. Pulitzer Prize winner.

McNamara, Robert S. *In Retrospect: The Tragedy and Lessons of Vietnam.* New York: Times Books, 1995.

McNeill, William H. *The Rise of the West: The History of the Human Community.* Chicago: University of Chicago Press, 1963.

McPherson, James M. *Battle Cry of Freedom: The Civil War Era.* New York: Oxford University Press, 1988.

Morgan, Edmund Sears. *The Birth of the Republic, 1763–89.* Chicago: University of Chicago Press, 1977.

——— *The Challenge of the American Revolution.* New York: W. W. Norton, 1976.

The 9/11 Commission Report: Final Report of the National Commission on Terrorist Attacks upon the United States. New York: W. W. Norton, 2004.

Norwich, John Julius. *A Short History of Byzantium.* New York: Vintage Books, 1999 reprint. The *Wall Street Journal* hailed this as a "grand and exciting story."

Olmert, Michael. *Milton's Teeth and Ovid's Umbrella: Curiouser and Curiouser Adventures in History.* New York: Simon & Schuster, 1996. Leads readers "through the back door and into the kitchen of history—where people really lived." Examines the histories of articles of everyday life, leisure activities, and celebrations.

O'Neill, Jaime. *We're History! The 20th Century Survivor's Final Exam.* New York: Fireside, 1998. Quizzes and answers on history and pop culture—from Geronimo to *Larry King Live*.

Peckham, Howard Henry. *The Colonial Wars: 1689–1762.* Chicago: University of Chicago Press, 1964.

Rachlin, Harvey. *Lucy's Bones, Sacred Stones, and Einstein's Brain: The Remarkable Stories Behind the Great Objects and Artifacts of History, from Antiquity to the Modern*

Era. New York: Henry Holt, 1996. *Publishers Weekly* called it a "pageant of human aspiration, achievement, obsession, and belief."

Rakove, Jack N. *Original Meanings: Politics and Ideas in the Making of the Constitution.* New York: Alfred A. Knopf, 1996. Pulitzer Prize winner.

Remnick, David. *Lenin's Tomb: The Last Days of the Soviet Empire.* New York: Random House, 1993.

Rhodes, Richard. *Dark Sun: The Making of the Hydrogen Bomb.* New York: Simon & Schuster, 1995.

——— *The Making of the Atomic Bomb.* New York: Simon & Schuster, 1986.

Riley-Smith, Jonathan. *The Oxford Illustrated History of the Crusades.* New York: Oxford University Press, 1997.

Rosenberg, Tina. *The Haunted Land: Facing Europe's Ghosts after Communism.* New York: Random House, 1995.

Russ, Martin. *The Last Parallel: A Marine's War Journal.* New York: Rhinehart, 1957. Experience of combat soldiers during the Korean War.

Schlesinger, Arthur M. Jr. *The Age of Jackson.* Boston: Little, Brown, 1945.

——— *The Age of Roosevelt.* Boston: Houghton-Mifflin, 1957.

——— *The Cycles of American History.* Boston: Houghton-Mifflin, 1986.

——— *A Thousand Days: John F. Kennedy in the White House.* Boston: Houghton-Mifflin, 1965.

——— and Dixon Ryan Fox, eds. *A History of American Life.* Abridged and revised by Mark C. Carnes. New York: Scribner, 1996. A comprehensive social history of the American people.

Schlesinger, Arthur M. Sr. *The Birth of the Nation: A Portrait of the American People on the Eve of Independence.* New York: Alfred A. Knopf, 1968.

——— *The Immigrant in American History.* New York: Harper & Row, 1964.

Schom, Alan. *Napoleon Bonaparte.* New York: HarperCollins, 1997.

Shaara, Jeff M. *Gods and Generals.* New York: Ballantine, 1996. A prequel to his father's epic novel, *Killer Angels*; chronicles the events leading up to the Battle of Gettysburg.

——— *The Last Full Measure.* New York: Ballantine, 1998. The last in the Shaara family's Civil War trilogy; begins with Robert E. Lee's retreat from Pennsylvania after Confederate forces were defeated at Gettysburg.

Shaara, Michael. *The Killer Angels.* Demco Media, 1993 reprint. Pulitzer Prize–winning, best-selling novel chronicling the Battle at Gettysburg; highly acclaimed for its historical accuracy.

Sheehan, Neil. *A Bright Shining Lie: John Paul Vann and America in Vietnam.* New York: Random House, 1988. Epic biographical account of the war; Pulitzer Prize winner, also National Book Award.

Shipler, David K. *Arab and Jew: Wounded Spirits in a Promised Land.* New York: Times Books, 1986.

Shirer, William L. *Gandhi: A Memoir.* New York: Simon & Schuster, 1979.

——— *The Rise and Fall of the Third Reich.* New York: Simon & Schuster, 1960.

——— *20th Century: A Memoir of a Life and the Times.* New York: Simon & Schuster, 1976.

Smith, Jessie Carney, ed. *Black Heroes of the Twentieth Century.* Detroit: Visible Ink Press, 1998.

Sontag, Sherry, and Christopher Drew, with Annette Lawrence Drew. *Blind Man's Bluff: The Untold Story of American Submarine Espionage.* New York: Public Affairs, 1998.

Strouse, Jean. *Morgan: American Financier.* New York: Random House, 1999.

Suarez, Ray. *Old Neighborhood: What We Lost in the Great Suburban Migration, 1966–99.* New York: The Free Press, 1999.

Swartz, Mimi. *Power Failure, the Inside Story of the Collapse of Enron.* New York: Doubleday, 2003.

Taylor, Alan. *William Cooper's Town: Power and Persuasion on the Frontier of the Early American Republic.* New York: Alfred A. Knopf, 1995. Pulitzer Prize winner.

Thompson, E. P. *The Making of the English Working Class.* New York: Vintage Books, 1963.

Tindall, George Brown, and David E. Shi. *America: A Narrative History,* 4th ed. New York: W. W. Norton, 1996. From collision of cultures to modern America.

Toland, John. *Adolf Hitler.* Garden City, N.Y.: Doubleday, 1976.

——— *Captured by History: One Man's Vision of Our Tumultuous Century.* New York: St. Martin's Press, 1997.

——— *The Last 100 Days.* New York: Random House, 1965. Traces the final three months of World War II.

——— *The Rising Sun: The Decline and Fall of the Japanese Empire, 1936–1945.* New York: Random House, 1970.

Turner, Frederick Jackson. *The Frontier in American History.* Huntington, N.Y.: R. E. Krieger Publishers, 1920, 1976 reprint.

von Clausewitz, Carl. *On War.* Edited and translated by Michael Howard and Peter Paret. New York: Alfred A. Knopf, 1993 reprint.

Wallis, Michael. *The Real Wild West: The 101 Ranch and the Creation of the American West.* New York: St. Martin's Press, 1999.

Washington, Booker T. *Up from Slavery.* New York: Oxford University Press, 1995.

Weinberg, Gerhard L. *A World at Arms: A Global History of World War II,* 2nd ed. New York: Cambridge University Press, 2005.

Williams, Jonathan, ed. *Money: A History.* New York: St. Martin's Press, 1997. Illustrated companion volume to exhibit at the British Museum.

Wills, Garry. *Lincoln at Gettysburg: The Words That Remade America.* New York: Simon & Schuster, 1992.

Wilson, James. *The Earth Shall Weep: A History of Native America.* New York: Atlantic Monthly Press, 1999.

Wolfe, Tom. *The Right Stuff.* New York: Farrar, Straus and Giroux, 1979. Described as a "nonfiction novel," this best-seller profiles the lives of the first Americans in space.

Wood, Gordon S. *The Radicalism of the American Revolution.* New York: Alfred A. Knopf, 1992. Pulitzer Prize winner.

Woodham-Smith, Cecil. *Florence Nightingale.* New York: McGraw-Hill, 1951.

Worsley, Frank Arthur. *Endurance: An Epic of Polar Adventure.* New York: W. W. Norton, 1999.

Yergin, Daniel. *The Prize: The Epic Quest for Oil, Money & Power.* New York: Touchstone/Simon & Schuster, 1991.

Zhisui Li. *The Private Life of Chairman Mao: The Memoirs of Mao's Personal Physician.* New York: Random House, 1994.

Zinn, Howard. *People's History of the United States, 1492–Present.* New York: HarperPerennial, 1995. Credited with "turning traditional textbook history on its head."

Index

(ill.) refers to illustrations.

B

D

E

F

G

H

I

J

K

M

N

O

P

Q

S

U

V

X

Y

Z